XIV Russian Conference on Petroleum and Mining Engineering 2021

IOP Conference Series: Earth and Environmental Science
Volume 1021

Online
9-12 November 2021

Editors:

Vladimir N. Korotaev
Sergey V. Galkin
Sergey N. Krivoshchekov

ISBN: 978-1-7138-5463-0
ISSN: 1755-1307

Printed from e-media with permission by:

Curran Associates, Inc.
57 Morehouse Lane
Red Hook, NY 12571

Some format issues inherent in the e-media version may also appear in this print version.

Institute of Physics
Dirac House, Temple Back
Bristol BS1 6BE UK

Phone: 44 1 17 929 7481
Fax: 44 1 17 920 0979

techtracking@iop.org

Additional copies of this publication are available from:

Curran Associates, Inc.
57 Morehouse Lane
Red Hook, NY 12571 USA
Phone: 845-758-0400
Fax: 845-758-2633
Email: curran@proceedings.com
Web: www.proceedings.com

TABLE OF CONTENTS

IOP Conf. Series: Earth and Environmental Science 1021 (2022) 011001 doi:10.1088/1755-1315/1021/1/011001

PREFACE

The committee and organizers are pleased to introduce the proceedings of the XIV Russian Conference on Petroleum and Mining Engineering (PETMET). The annual event hosted by Perm National Research Polytechnic University took place on 09-12, November in Perm, Russia.

The goal of the conference was to develop the scientific cooperation between Russian and foreign researchers in the field of information technologies and technical sciences.

The Conference also provides an interdisciplinary forum for researchers, practitioners and scholars to present and discuss the most recent innovations and developments in the field of new oil technologies and modeling methods, techniques and technologies of the petroleum engineering.

Main topics of the conference:

1 - Mineralogy and petrology
2 - Surface and subsurface geological methods of mineral exploration
3 - Seismic methods. well logging and subaerial geophysical methods
4 - Remote land sensing in geophysical surveys
5 - Geodetic techniques
6 - Petrochemistry and petrophysics
7 - Petroleum fluid mechanics and reservoirs fracturing
8 - Solid minerals and deposits

Scientists and practitioners in Russia and abroad submitted over one hundred manuscripts. All the papers were peer-reviewed by two reviewers. As a result, 82 of them were selected for publication.

Finally, we would like to express our deep appreciation and gratitude to all the members of the PETMET 2021 Scientific Committee for their strenuous efforts. Without their assistance and dedication, it would be impossible to hold the discussion at such a high level and to collect quality papers for compiling the conference proceedings. We also appreciate the assistance of IOP Publishing in producing the volume of the collected papers. In the end, we wish to extend appreciation to all the authors for their outstanding contributions to this valuable volume.

With the best regards on behalf of the organizers,

Editors of the PETMET 2021
Vladimir N. Korotaev
Sergey V. Galkin
Sergey. N. Krivoshchekov

PROGRAMME COMMITTEE

The chairman:

Korotaev V. N. - Dr., Vice-Rector, Perm National Research Polytechnic University

Members of the Program committee:

Andreiko S. S. - Dr. Techn. Sciences, Head of the Department of Development of Mineral Deposits of Perm National Research Polytechnic University

Ashikhmin S. G. - Dr. Techn. Sciences, Department of Markscheider, Geodesy and Geoinformation Systems, Perm National Research Polytechnic University

Bashkov A. N. - candidate geol. min. Sciences, chief geologist of AVK Invest LLC

Vukovich D. - Dr. Econ. Sciences, Department of Economics and Industrial Production Management, Perm National Research Polytechnic University

Galkin V. I. - Dr., Head of the Department of Geology of Oil and Gas, Perm National Research Polytechnic University

Galkin S. V. - Dr., Dean of the Mining and Petroleum Faculty of Perm National Research Polytechnic University

Zhulanov E. E. - Dr. econ. Sciences, Head of the Department of Economics and Industrial Production Management of Perm National Research Polytechnic University

Kashnikov Y. A. - Dr. Techn. Sciences, Head of the Department of Markscheider Affairs, Geodesy and Geoinformation Systems of Perm National Research Polytechnic University

Korekov A. V. - Chief Marksman, Head of the Markschader Department of LUKOIL-PERM LLC

Lagunova J. A. - Dr. Techn. Sciences, Professor of the Department of MMC Ural State Mining University (UGSU), Yekaterinburg.

Levin L.Y. - Dr. Techn. Sciences, Department of Mineral Deposits Development, Perm National Research Polytechnic University

Maltsev A. S. - Deputy Head of a Separate Division, EuroChem - Project LLC, Perm

Macheret A. M. - chief marksman of PJSC Uralkaliy.

Pankov I. L. - Candidate. Techn. Sciences, Senior Researcher at the Laboratory of Physical Processes of Development of Georesources, GI Ural Branch of the Russian Academy of Sciences

Plotnikov V. M. - Dr. Techn. Sciences, Department of Oil and Gas Technologies, Perm National Research Polytechnic University

Ponomareva I. N. - Dr. Techn. Sciences, Department of Oil and Gas Technologies, Perm National Research Polytechnic University

Putilov I. S. - Dr. Techn. Sciences, Deputy Director for Scientific Work in the Field of Geology of the Branch of LLC LUKOIL-Engineering "PermNIPIneft" in Perm

Rybin A. A. - Dr. Techn. Sciences, Professor, Department of Mining Electromechanics, Perm National Research Polytechnic University

Seredin V. V. - Dr., Department of Oil and Gas Geology, Perm National Research Polytechnic University

Soloviev V. A. - Dr. Techn. Sciences, Head of Geodynamic Safety Laboratory of Galurgia OJSC

Trifanov G. D. - Dr. Techn. Sciences, Head of the Department of Mining Electromechanics of Perm National Research Polytechnic University

Feinburg G. Z. - Dr. Techn. Sciences, Department of Mineral Deposits Development, Perm National Research Polytechnic University

Khizhnyak G. P. - Dr. Techn. Sciences, Head of the Department of Oil and Gas Technologies, Perm National Research Polytechnic University

IOP Conf. Series: Earth and Environmental Science 1021 (2022) 011001 doi:10.1088/1755-1315/1021/1/011001

Tsvetkov G. A. - Dr. Techn. Sciences, Department of Life Safety, Perm National Research Polytechnic University

Cherniy K. A. - Dr. Techn. of Sciences, Head of the Department of Life Safety of Perm National Research Polytechnic University

ORGANIZING COMMITTEE

The chairmans:

Korotaev V.N. – Vice Rector of Perm National Research Polytechnic University

Galkin S.V. – Dean of MOF Perm National Research Polytechnic University

Deputy chairman:

Melekhin A.A. – Associate Professor of Perm National Research Polytechnic University

Members of the Organizing committee:

Kozlova I.A.– Associate Professor of Perm National Research Polytechnic University

Krivoshchekov S.N. – Associate Professor of Perm National Research Polytechnic University

Melkishev O.A.– Associate Professor of Perm National Research Polytechnic University

Turbakov M.S.– Associate Professor of Perm National Research Polytechnic University

Shishlyannikov D.I. – Associate Professor of Perm National Research Polytechnic University

Shcherbakov A.A. – Senior Lecturer of Perm National Research Polytechnic University

Lebedeva O.O.– Assistant of Perm National Research Polytechnic University

Papulov A.S.– Assistant of Perm National Research Polytechnic University

Peer Review Statement

All papers published in this volume have been reviewed through processes administered by the Editors. Reviews were conducted by expert referees to the professional and scientific standards expected of a proceedings journal published by IOP Publishing.

- **Type of peer review:** Single Anonymous
- **Conference submission management system:** Morressier
- **Number of submissions received:** 142
- **Number of submissions sent for review:** 101
- **Number of submissions accepted:** 88
- **Acceptance Rate (Submissions Accepted / Submissions Received × 100):** 62
- **Average number of reviews per paper:** 2
- **Total number of reviewers involved:** 23
- **Contact person for queries:**
  Name: Sergey Krivoshchekov
  Email: krivoshchekov@gmail.com
  Affiliation: Perm National Research Polytechnic University

IOP Conf. Series: Earth and Environmental Science 1021 (2022) 012001 doi:10.1088/1755-1315/1021/1/012001

Assessment of the main rheological parameters of rocks during field development

M Ya Khabibullin

Ufa State Petroleum Technological University, Branch of the University in the City of Oktyabrsky, 54a, Devonskaya St., Oktyabrsky, Republic of Bashkortostan, 452607, Russian Federation

E-mail: m-hab@mail.ru, info@of.ugntu.ru

Abstract. The total rock pressure on a formation wing is greater than the lateral pressure, and on the fold, on the contrary, the lateral rock pressure is greater than the total pressure. As a result, the formation material flows from wing zones to the fold, as a result of which the formation thickness on the fold increases, as noted by the studies of Soviet scientists. The study of the mechanism of hydraulic fracturing during water injection revealed large fracturing pressures in wells located in reduced parts of the structure (fold wings), and smaller pressures – in elevated parts (fold). The study showed that if the rock of the well bore zone is destabilized during field development for well operation, fresh water causes hydration and swelling or dispersion of clays, which act as the rock cementing material. This can both lead to a decrease in permeability and to the destruction of the well bore zone. This effect during well development may be prevented by using mineralized reservoir water or fresh water with hydration inhibitor additives such as NaCl (5-10%), $CaCl_2$ (1-3%), KCl (1-3%). Another complication in well completion is the colmatation of the well bore zone. In order to ensure the stability of the fixed part of the formation behind tubing in the bottomhole zone, there is a need to ensure the static equilibrium in the porous medium. Rocks with low rheological properties (low relaxation rate) provide longer resistance of the well bore zone to destruction.

1. Introduction
In the study of the development of hydrocarbon fields it is believed that for young folded zones where tectonic processes are most intense, the lateral rock pressure in one of the directions may be even higher than the total rock pressure [1-3]. The total rock pressure on the wing of the formation is greater than the lateral pressure, and on the fold, on the contrary, the lateral rock pressure is greater than the total pressure.

2. Materials and methods
As a result, the formation material flows from wing zones to the fold, as a result of which the formation thickness on the fold increases, as noted by the studies of Soviet scientists. Studying the mechanism of hydraulic fracturing during water injection the authors of the work [3] revealed large fracturing pressures in wells located in reduced parts of the structure (fold wings), and smaller pressures – in elevated parts (fold). The rock pressure on the formation is determined primarily by the action of gravity of overlying rocks, which can be expressed by the formula [1]

$$q_r = \sum_{i=1}^{n} \rho_i \cdot g \cdot h_i \approx \rho_{cp} \cdot g \cdot H \tag{1}$$

where q_r – total rock pressure, Pa; ρ_i – rock density of *i* formation, kg/m^3; h_i – thickness of *i* formation, m; g – gravity acceleration, m/s^2; H – formation depth, m; ρ_{cp} – weighted average density of overlying rocks, kg/m^3.

According to OAO SurgutNIPIneft, the density of rocks increases as the depth of the production facility increases (Table 1).

Table 1. Dependence of rock density on formation depth

Formation depth, m	0	300	600	900	1200
Rock density, kg/m^3	2260	2410	2520	2590	2620

If there are no tectonic movements in the horizontal formation, the lateral rock pressure q_∞ is determined by the expression:

$$q_\infty = \frac{\nu}{1-\nu} g_r = \xi \cdot g_r \tag{2}$$

Where $\xi = \frac{\nu}{1-\nu}$ – lateral pressure coefficient; ν – Poisson's ratio.

It is widely agreed that the lateral rock pressure does not exceed the total rock pressure, i.e. $\xi \leq 1$.

However, as noted in work [2], the lateral pressure coefficient for rocks under conditions of their occurrence should be taken equal to one, since all rocks are prone to plastic flow, and under high temperatures, pressures, water saturation and loading duration, which subsoil rocks are exposed to, are factors contributing to viscous-plastic flow, leading to stress equalization, i.e. $q_\infty = q_r$.

Formula (2), obtained by A.N. Dinnik in 1925, is valid for determining stresses in an elastic isotropic half-space with a horizontal surface and does not reflect the natural conditions of rocks. However, due to the absence of other more reliable dependencies, this formula is widely used in solving problems where it is necessary to take into account the lateral rock pressure.

Work [2] concludes that with a low content of cementitious material in sandstone, its viscous-plastic flow into the fold part due to rheological properties, as well as the relative movement of formations, can lead to a significant decrease in the strength properties of sandstone until its complete loss of connectivity and transformation into an uncemented sand layer. So, as a result of large deformations the subthrust formations of Starogroznenskoye field turned out to be looser than the same formations in the overriding wing.

The authors of the work [2] claim that about a year after the well drilling is completed, the lateral and total rock pressures during further development of the field are leveled as a result of the viscoplastic flow of formation rocks.

The study showed that if the rock of the well bore zone is destabilized during field development for well operation, fresh water causes hydration and swelling or dispersion of clays, which act as the rock cementing material. This can both lead to a decrease in permeability and to the destruction of the well bore zone. This effect during well development may be prevented by using mineralized reservoir water or fresh water with hydration inhibitor additives such as NaCl (5-10%), $CaCl_2$ (1-3%), KCl (1-3%).

Another complication in well completion is the colmatation of the well bore zone. According to the available data [3], the mud acid treatment induces rock permeability by about 50% of the initial value, while conventional flushing – by not more than 30%.

Therefore, the fight against sand-related problems in wells should begin during their construction through correct selection of working fluids. The work [4] studied the stability of sand arches in various forms of sand and the saturation of sand with fluids during fluid filtration. Laboratory studies revealed that the stability of arches depends on the degree of sand abrasion and fluid saturation. The authors note that the presence of clay particles in sand increases the stability of arches, while the increase in water

IOP Conf. Series: Earth and Environmental Science 1021 (2022) 012001 doi:10.1088/1755-1315/1021/1/012001

saturation reduces it. This is confirmed by the results of the analysis of the efficiency of thermal methods used enhanced oil recovery of clay sandstones saturated with high viscous oil.

If the conditions for the removal of sand to the surface along the tubing string are sufficiently studied, then the mechanism for the destruction of sandstone under the influence of filtration flow has not yet been fully disclosed. Based on the ideas of non-homogeneity of the oil-saturated formation, the work [5] concludes on the existence of directions with the greatest permeability, along which the bulk of sand is carried out from the formation.

Laboratory studies showed that the erosion of cemented rocks occurs at pressure gradients defined for each rock [6].

In our opinion, the method [7] to determine the permissible value of rock pressure depression in weakly cemented formations, where sandstone strength is taken constant in time, is incorrect, since rheological properties of rocks are not taken into account. So, for example, if rock strength is constant [8], then with a constant depression of rock pressure, the destruction of sandstone should occur until the pressure gradient on the well wall cavities is equal or less than the value of rock strength. Hence, there must be a direct dependence of the size of the cavity on the depression of rock pressure and rock strength. Moreover, the intensity of sand supply from the formation should decrease in time.

However, in practice, along with this trend, the flow of sand from the formation into the well is very common when the deposit is further developed over time with a constant amount of rock pressure depression becoming more intense. As is known, the main type of plastic defect of grainy cemented material is inter-grain sliding, which with a slight content of contact cement causes a decrease in rock strength. Over time, a complete loss of communication between the sand and on the open surface of the filter zone is possible, which serves as the beginning of direct destruction of the formation under the influence of a pressure gradient. Naturally, the greater the depression of rock pressure, the more intense the processes of sandstone destruction.

A sharp increase in reservoir pressure depression can lead to the collapse of the already loose part of the well zone. Even if the withdrawal of fluid from the formation is completely stopped, the "loosening" processes of the well bore zone will continue as a result of the visco-plastic flow of the stressed material. Thus, during the operation of wells that opened loose sand beds, the formation of stable arches is possible.

The presence of clay particles in the rock results in certain rheological properties of the formation leading to the destruction of arches over time. As a result of the visco-plastic flow of sandstone containing plastic cementitious material its strength is reduced. With the long-term effect of the filtration flow on such a formation, it is possible to completely disrupt the connections between the rock particles and the entry of sand into the borehole. As the rock pressure depression increases, the rate of the visco-plastic flow and destruction of the formation in the well bore zone increases.

Table 2. Elastic parameters of some rocks

Rock	Elastic modulus 10^{-4} *E*, MPa	Poisson's ratio, *v*	Shear modulus 10^{-4} G, MPa	Comprehensive compression modulus, 10^{-4} K
Basalt	4.39	0.12	4.45	1.9
Even-granular dolomite	5.05	-	-	-
Limestone	2.25	0.29	1.8	1.8
Clay limestone	6.5	0.22	0.68	3.8
Sandstone	2.6	0.28	0.68	3.8
Quartz sandstone	4.5	0.21	1.85	2.6

If we consider rocks as absolutely elastic bodies at instantaneous loading speed, then according to the theory of elasticity [9] the relationship between stresses and deformations will be determined by elastic parameters of rocks: Young's modulus, Poisson's ratio *v*, shear modulus *G*, comprehensive compression modulus *K* and elastic limit σ_e. Some rock parameters are given in Tables 2 and 3.

Table 3. Elastic parameters of some rocks

Rock	Elastic modulus 10^{-4} E, MPa		Poisson's ratio, *v*		Compression strength $\sigma_{сж}$, MPa		Tensile strength σ_p, MPa	
	E_n	E_l	v_n	v_l	$\sigma_{сж\ n}$	$\sigma_{сж\ l}$	σ_{pn}	$\sigma_{p\ l}$
Siltstone	2.67	1.72	0.25	0.29	50.6	67.5	-	-
Anthracite	0.45	0.38	-	0.13	10.5	16.0	-	-
Limestone	6.36	7.25	0.28	0.30	114-170	104-202	-	-
Coarse sandstone	1.93	1.73	0.45	0.36	118.5	142.3	4.43	5.25
Fine sandstone	3.83	2.64	0.20	0.19	138-226	133-226	8.30	7.25
Coking coal	0.04-0.8	0.06	-	-	-	-	-	-
Petroleum coke	-	-	-	-	0.2-1.3	0.2-1.3	-	-

3. Results and Discussion

In the field of plastic deformations, the linear relationship between the deformations and the load is broken, and the deformations become partially irreversible. As the load on the rock increases, the strain rate increases. The strength of cemented rocks is determined by the strength of cement. So, for clay shale $\sigma_{сж}$=14-57 MPa, while for sandy shale $\sigma_{сж}$=35-83 MPa.

The strength of coals depends on the degree of metamorphism and varies in the range from 1 MPa (for coking coals) to 35 MPa (for anthracites). When compressing petroleum cokes, its strength limits vary from 2 to 13 MPa. For most rocks, the value of tensile strength is in the range of up to 20 MPa and makes (0.1-0.2) $\sigma_{СЖ}$.

Rock strength in some section may be represented as linear dependence of tangent stresses in the sliding plane against normal stress [10]. The adhesion of clays is $2.5\cdot10^4$–$3.3\cdot10^5$ Pa, brown coals – $2\cdot10^4$–$2.5\cdot10^5$ Pa. The internal friction angle for fatty clays is 16-30°, for brown coals – 30-35°, for sand – 30-32°. The work [11] presents such data on the value of adhesion of weakly cemented sandstones of Azerbaijan fields: KS suite $4.9\cdot10^4$–$5.49\cdot10^5$ Pa and PS suite $9.8\cdot10^5$–$1.47\cdot10^6$ Pa.

Production wells typically operate for decades. With such a long time of the pressure drop effect on the formation near the bottom of the well, the stress state of rocks will also be determined by rheological properties in addition to these properties.

Besides, creep is another major rheological property of rocks. Significant creep is typical for coherent rocks (clays, argillites, clay shales). Krivorozhsky rocks, for example, quartzericite shale, arkosic sandstone have a creep of an average of 20-35% of the initial elasticity of deformation, with the greatest deformations occurring in the first 1.5-2 days of loading.

The phenomenon of reverse creep is relaxation, which is characterized by the fact that elastic deformations turn into plastic at a deformation rate of zero. There is a general pattern of changing properties over time: the longer the load, the smaller the elastic and strength properties of rocks. According to the work [12], the long-term strength of clays and marls ($\sigma_{сж\ g}$) is twice as low as the standard ($\sigma_{сж}$). For other rocks $\sigma_{сж\ др} = (0.7–0.8)\sigma_{сж}$. The Young's modulus also decreases with the time of load action and is equal to $E_{др}$ = = (0.67–0.95) *E*.

4. Conclusion

Thus, in order to ensure the stability of the fixed part of the formation behind tubing in the bottomhole zone, there is a need to ensure the static equilibrium in the porous medium. Rocks with low rheological properties (low relaxation rate) provide longer resistance of the well bore zone to destruction.

References

[1] Zaynagalina L Z 2020 On the development of an experimental design and field test of an upper bit tool *IOP Conference Series: Materials Science and Engineering* **905** 012093 DOI:10.1088/1757-899X/905/1/012093

[2] Zainagalina L Z 2019 Application of vibration diagnostics methods and programs to assess the remaining lifetime of oil equipment *Journal of Physics: Conference Series* **1333** 062029. DOI:10.1088/1742-6596/1333/6/062029

[3] Zaynagalina L Z 2020 Dynamics of the oscillatory system of the near-bit slurry grinder *IOP Conference Series: Materials Science and Engineering* **905** 012092 DOI:10.1088/1757-899X/905/1/012092

[4] Akhmetov R T, Mukhametshin V V, Kuleshova L S, Grezina O A, and Malyshev P M 2020 The generalized correlating function of capillary curves and the relationship of the filtration-capacitive parameters of reservoirs in Western Siberia with the size distribution of pore channels *Journal of Physics: Conference Series* **1661** 012016 DOI: 10.1088/1742-6596/1661/1/012016

[5] Haoran Zh, Yongtu L and Xingyuan Zh 2017 Sensitivity analysis and optimal operation control for large-scale waterflooding pipeline network of oilfield *Journal of petroleum science and engineering* **154** 38-48

[6] Sun W and Mun-Hong H 2017 Forecasting and uncertainty quantification for naturally fractured reservoirs using a new data-space inversion procedure *European Assoc Geoscientists & Engineers Computational geosciences 15th Conference on the Mathematics of Oil Recovery (ECMOR) (Amsterdam, Netherlands)*, 21(5-6), pp.1443-1458

[7] Khabibullin M Ya 2019 Managing the processes accompanying fluid motion inside oil field converging-diverging pipes *Journal of Physics: Conference Series* **1333** 042012 DOI: 10.1088/1742- 6596/1333/4/042012

[8] Khabibullin M Ya 2019 Managing the reliability of the tubing string in impulse non-slationary flooding *Journal of Physics: Conference Series* **1333** 052012 DOI:10.1088/1742-6596/1333/5/052012

[9] Soloviev N N, Mukhametshin V Sh and Safiullina A R 2020 Developing the efficiency of low-productivity oil deposits via internal flooding *IOP Conf. Ser.: Mater. Sci. Eng.* **952** 012064 DOI: 10.1088/1757-899X/952/1/012064

[10] Mukhametshin V Sh 2020 Rationale for the production of hard-to-recover deposits in carbonate reservoirs *IOP Conf. Ser.: Earth Env.* **579(1)** 012012 DOI: 10.1088/1755-1315/579/1/012012

[11] Galimullin M L, Khabibullin M Ya 2020 Experience with sucker-rod plunger pumps and the latest technology for repair of such pumps *Chemical and Petroleum Engineering* **55(**11-12) 896-901. DOI: 10.1007/s10556-020-00710-1

[12] Zeigman Yu V, Mukhametshin V V, and Kuleshova L S 2020 Management of flooding of low-production deposits according to geological and field data *IOP: Earth and Environmental Science (EES)* **579** 012019 DOI: 10.1088/1755-1315/579/1/012019

Problems of realization of low-productivity oil wells continuous operation

P N Tsylev[1], I N Shchapova[1], V A Shchapov[2]

[1] Perm National Research Polytechnic University, 29, Komsomolsky prospect, Perm, 614990, Russia
[2] Institute of Continuous Media Mechanics of the Ural Branch of the Russian Academy of Sciences, 1, Academika Koroleva street, Perm, 614013, Russia

E-mail: Irina.shchapova@gmail.com

Abstract. The expediency of low-productivity oil wells continuous operation has been specified. The analysis of possible ways of transferring oil wells with low flow rate to continuous operation is given. We noted the most rational use of low-speed asynchronous electric motors in the electric drive of plunger pumps. The problems of low-speed asynchronous electric motors development and production have been considered. The emphasis is placed on high weight, size and cost parameters of low-speed asynchronous electric motors and low values of their efficiency and power factor. A number of possible solutions to these problems are proposed. Thus, it is recommended to use a stator winding with a shortened pole pitch length to improve the weight, size and cost parameters. Reduction of electrical and magnetic loads and the use of copper instead of aluminum as a stator and rotor windings material contribute to the increase of the efficiency factor. The use of internal compensation of magnetizing current and reactive power is aimed at achieving the power factor values close to one.

1. Introduction
At present the number of oil wells with a low flow rate (less than 5 tons per day) is more than 2 thousand only in Perm region. There is a tendency to increase the number of such wells. This is largely explained by the depletion of oil fields in operation. Thus, the problem of optimal operation of low-productivity oil wells is relevant.

Numerous experiments on low-productivity oil wells show that oil production can be significantly increased if the well is transferred from the cyclic operation mode to the continuous mode. The transfer of low-productivity oil wells into the continuous operation mode also makes possible to reduce the water cut of oil and the installed power of oil wells electrical equipment.

2. Practical realization of low-productivity oil wells continuous operation
Plunger pumps with an asynchronous electric drive are used mostly for lifting oil from wells with a low flow rate. To realize a continuous operation mode of such wells the frequency of the pump plunger stroke has to be reduced to one stroke per minute [1]. Further reduction of the plunger stroke frequency may be impracticable due to the increased influence of the extracted liquid leak through the gap between the pump body and the plunger. When the plunger stroke frequency decreases the theoretical pump capacity decreases proportionally but the level of fluid leak continues to be the same.

Under these conditions the actual pump capacity may decrease to zero.

The stroke frequency of the pump plunger is determined by the speed of the asynchronous electric motor rotor and the gear ratios of the belt drive and the two-stage reducer.

In asynchronous electric motors the synchronous number of the rotor shaft turns at a voltage frequency of 50 Hz is determined by the number poles pairs of the three-phase winding laid in the stator slots. In asynchronous electric motors of rocking machines of domestic production [2], three-phase windings with a number of pole pairs 2, 3, 4 are placed in the stator slots, providing rotor rotation with a synchronous speed of 1500, 1000, 750 min^{-1}.

The gear ratio of the belt drive depends on the pulleys diameter mounted on the rotor shaft of the electric motor and on the input shaft of the two-stage reducer. By replacing the diameter of the pulley on the rotor shaft of an asynchronous electric motor. The gear ratio of the belt drive can be changed stepwise in the range of 2.8 – 4.5 by replacing the diameter of the pulley on the rotor shaft of an asynchronous electric motor.

The two-stage reducer of pumpjack has a gear ratio of 40. During the pumpjack operation, the gear ratio of two-stage reducer remains constant.

The maximum value of the total gear ratio the belt drive and two-stage reducer is 180.

So, the minimum stroke frequency of the pump plunger which can be obtained on pumpjack with standard mechanical and electrical equipment is 4 min^{-1}. This frequency of the pump plunger stroke is usually higher than the values that are required for the implementation of a continuous mode of fluid withdrawal from low-productivity oil wells.

The solution of an important practical problem of transferring low-productivity oil wells to a continuous operation mode can be achieved by increasing the gear ratio of pumpjack kinematic chain links or by reducing the shaft speed of asynchronous electric motors. The latter direction is associated with the use of frequency converters or the use of low-speed asynchronous electric motors.

Technical solutions allow changing the gear ratio of pumpjack kinematic chain in the direction of its increase [3]. Such solutions include: the use of additional belt drive, the use of gear motors, the replacement of the existing belt drive and two-stage reducer with a three-stage reducer articulated with the rotor shaft of a multi-speed asynchronous electric motor. However, these design solutions were not widely used in practice due to the necessity of pumpjack reconstruction and additional costs for maintenance and repair of mechanical equipment.

The installation of frequency converters on pumpjack makes it possible to regulate the rotor shaft revolutions of asynchronous electric motors, hence, deep regulation of the plunger movement frequency of plunger pumps in a wide range. At the same time, conditions for implementation of the optimal continuous mode operation of low-productivity oil wells are created.

The adjustable frequency electric drive of the plunger of sucker-rod pumping units (SRPU) has a number of disadvantages, the principle of which is: the high cost of frequency converters, significant operating costs for maintenance, the appearance of three-phase alternating voltage of higher time harmonics in the electric power network. These disadvantages of the frequency-controlled asynchronous electric drive of the SRPU plunger hinder the widespread introduction of frequency converters to optimize the process of oil production from low-productivity oil wells.

A large-scale experiment on the use of low-speed asynchronous electric motors with a synchronous rotor speed of 500 min^{-1} in the electric drive of sucker-rod pumping units was carried out at the enterprises of PJSC Tatneft. The results of experiment were published in [1] and confirmed the advantages of continuous operation of low-productivity oil wells, compared to the previously used cyclic mode. In particular, there is an increase in the volume of oil produced, efficiency factor of asynchronous electric drive, and significant savings in electrical energy.

The presence of low-productivity oil wells in oil-production enterprises with a wide range of flow rates required the development of asynchronous electric motors with a synchronous rotor speed less than 500 min^{-1} for the electric drive of plunger pumps. Among the enterprises working on this problem solution it is necessary to distinguish Vladimir Electric Motor Plant, Perm National Research Polytechnic University, and JSC "Privod" (Lysva, Perm Krai). Specialists of these enterprises developed asynchronous motors with a synchronous rotor rotation frequency of 375 and 200 min^{-1} with a power of 4 and 3 kW, respectively.

However, low-speed asynchronous electric motors in the drive of plunger pumps which lift oil from low-productivity wells have not been widely used untill now.

3. Reasons hindering the introduction of low-speed asynchronous motors in the plunger pumps drive

One of the reasons for hindering the introduction of low-speed asynchronous electric motors in the plunger pumps drive of low-productivity oil wells is the significant weight and size indicators of low-speed engines and their high cost.

Indeed, to obtain a low synchronous speed of the rotor shaft asynchronous motors, at a AC frequency 50 Hz, three-phase windings have to be placed in the slots of the stator, in which during operation electric currents excite a rotating magnetic field with the number of pairs of poles p , calculated according to the well-known formula:

$$p = \frac{60f}{n_0},$$

where f is the AC power frequency; n_0 is the synchronous speed of the electric motor rotor shaft.

For example, the synchronous speed of the electric motor rotor shaft 200 min^{-1} is achieved by placing on the stator a three-phase winding with a number of pole pairs of 15.

From a number of pole pairs of the three-phase stator winding of the asynchronous electric motor there is the diameter of its inner bore, the value of which, for the accepted length of pole pitch τ, is determined by the expression:

$$D_{в.с.} = \frac{2p\tau}{\pi}.$$

Thus, in an engine with a number of pole pairs equal to 15 and a pole pitch length of 10 cm the diameter of stator inner bore will be 95.5 cm. The tooth-slot zone of the stator core, the core yoke and the housing cause further increase in the dimensions, weight and cost of the stator.

The outer diameter of rotor asynchronous electric motors $D_{н.р.}$ differs from the diameter of the stator inner bore twice the amount of the air gap between the stator and the rotor. For low-power asynchronous electric motors the air gap does not exceed 0.5 – 0.7 mm and therefore $D_{н.р.} \approx D_{в.с.}$

Consequently, the outer diameter of rotor in low-speed asynchronous electric motors is also large. The height of rotor yoke will also be significant and it is defined as the difference between the radius of rotor, the radius of the rotor shaft and the height of rotor slots. The consequence of the above is that the magnetic induction in the rotor yoke is repeatedly underestimated in comparison with the permissible values for the yoke. As a result of the "hidden waste of steel" [4], the consumption of active materials increases and the cost of rotor and engine is getting higher as a whole.

4. Ways to reduce the weight and size parameters of low-speed electric motors and increase their energy efficiency

A certain improvement in the weight-size and cost parameters of low-speed asynchronous electric motors driving plunger pumps is achieved by placing a winding with a shortened pole pitch length [5] in the stator slots. The original design of the stator optimizes the basic geometric dimensions of motors and eliminates a number of disadvantages inherent in engines with a traditional stator design.

The development of low-speed asynchronous electric motors for the drive of SRPU plunger lifting oil from low-productivity oil wells requires solutions of the second most important problem, namely, ensuring high values of energy indicators, which include the power factor and efficiency factor.

Analysis of catalog and reference data on asynchronous electric motors of various series shows that as the pairs number of motor poles increases and the rotor synchronous speed decreases, the energy indicators of asynchronous energy converters decrease. Thus, in asynchronous motors developed by Vladimir Electric Motor Plant specifically for driving plunger pumps of pumpjack that have a winding on the stator that creates eight pairs of poles, the nominal values of energy indicators are 66 % and 0.48, respectively, against 85 % and 0.80 for electric motors with three pairs of poles.

Increasing the efficiency factor of low-speed asynchronous energy converters can be achieved by implementing measures aimed at reducing electrical, magnetic and mechanical losses. The most important one of these activities should include: electrical and magnetic loads reduction, the use of thinner and higher-quality electrical steel, the use of copper windings as a material instead of aluminum. These measures are successfully used by the world's leading electrical companies in the production of high-speed electromechanical energy converters of various capacities and designs.

In asynchronous motors, the conversion of alternating-current electrical energy into mechanical energy on the rotor shaft is carried out by a rotating magnetic field. For its creation a reactive magnetizing current is consumed from the power supply network. An indicator characterizing the influence of reactive magnetizing current on the operation of asynchronous motors is the power factor. The size and cost of asynchronous electromechanical energy converters depend on the value of power factor: with an increase in the power factor the size and cost of converters decrease and vice versa.

Reactive magnetizing current is generated by synchronous generators of power stations. The passage of magnetizing current through the current-carrying parts of the power supply systems elements of oil-producing enterprises causes losses of electric energy and voltage, which negatively affects the quality of electric energy, energy indicators of power supply systems, the operating characteristics of electric energy receivers including electric motors of pumpjack.

To reduce the reactive magnetizing current in switchgears with the voltage of 6 – 10 kV of the main step-down substations of oil-producing enterprises static capacitors batteries are installed in power supply systems. With the help of the same batteries of static capacitors the reactive magnetizing current of the transformers of the main step-down substations is compensated. However, the distribution electric networks of oil-producing enterprises connected to these networks and consumers of high and low voltage electric energy, including electric motors of rocking machines, remain loaded with reactive magnetizing current.

The development of low-speed asynchronous electromechanical energy converters for the electric drive of plunger pumps of low-flow oil wells required new design solutions aimed at reducing the consumption of reactive magnetizing current by asynchronous converters. One of the original solutions in this area should be considered the asynchronous energy converters with internal compensation of the reactive magnetizing current [6]. Theoretical and experimental studies in this field have allowed us to establish that the operation of asynchronous converters is carried out with a power factor the value of which is close to one and is invariant with respect to the load of the converters.

5. Conclusion

Serial production of low-speed asynchronous electromechanical energy converters of low power with a shortened length of the stator winding pole division, reduced electrical, magnetic and mechanical losses and internal reactive power compensation creates good prerequisites for their mass introduction at oil-producing enterprises for transferring low-productivity oil wells into continuous operation.

References

[1] Charonov V Ya 1996 *Oil industry* **12** 46–48

[2] *Catalog of oil equipment, automation equipment, devices and special materials* 1994 (Moscow: VNIIOENG Press)

[3] Ivanovsky V N, Darishchev V I, Sabirov A A, Kashtanov V S, Pekin S S 2002 *Borehole pumping units for oil production* (Moscow: Publishing House "Oil and gas" Gubkin Russian State University of Oil and Gas)

[4] Yuferov F M 1988 *Electrical machinery of automatic devices* (Moscow: High school Press)

IOP Conf. Series: Earth and Environmental Science 1021 (2022) 012002 doi:10.1088/1755-1315/1021/1/012002

[5] Belyaev E F, Tsylev P N, Vlasov E A 2010 RF Patent No 2402141 *Low-speed asynchronous energy converter.*
[6] Tsylev P N, Shchapova I N 2013 *Oil industry* **4** 62–65

Application of statistical dependencies to evaluate economic indicators of production efficiency by the example of oil fields in Azerbaijan

Q A Safarov[1], G Vagif gizi[1], S V Galkin[2]

[1] Azerbaijan State University of Oil and Industry, 34, Liberty ave., Baku, 050631, Azerbaijan
[2] Perm National Research Polytechnic University, 29, Komsomolsky prospect, Perm, 614990, Russia

E-mail: doc_galkin@mail.ru

Abstract. Improving the use of basic funds and production capacity in oil and gas recovery is not only the introduction of new techniques and technology, but also the improvement of labor, management and organization of work. The process of renewal of the fixed capital and production forces is considered intensive at the time when the vast majority of the newly incorporated basic funds and forces are aimed at replacement of morally obsolete and physically worn out parts. This time, their compliance with the economically justified life expectancy is of interest. All these are elements of recycling. Oil and gas wells are important elements of the active part of the main production funds and they determine the future development of oil and gas recovery. Productive use of production wells for oil and gas recovery is considered as a source of improvement of production efficiency in terms of production-resource potential (PRP). In order to increase the efficient use of oil and gas wells as an active element of the main production funds, special attention is paid to both extensive and intensive factors at oil and gas production enterprises.

1. Introduction

The direction of the exstensive use of the main production funds is to attract all existing well funds to operation, reduce the number of inactive wells, as well as increase the productive working time of each well. The intensive direction consists in the wide implementation of methods of influence on the well zone in order to ensure optimal oil production of each well and increase in the productivity of the wells in accordance with the given processing regimes. Here, real opportunities arise from the PRP's point of view, first of all, as a result of the implementation of various geological and technical measures based on the well fund.

The study of the use of investment bounties (means of production) among different directions of increasing the efficiency of public production is of particular importance. Exploration of production facilities for the oil and gas recovery industry, the main funds (well fund) that is a part of it, is necessary. Thus, the share of oil production is significant in the formation of the internal part of the country's budget and the incoming currency. In our view, increasing the efficiency of using the main funds in the oil industry is associated in two ways: a high pace of extensive reproduction and a decrease in the volume of products per manat of the production fund.

The main funds and, naturally, the production forces concentrate the values of scientific and technical progress, under its influence add new «color» to the development of productive forces. Intensification of public recovery puts the main fund and renewal of production power to the fore.

The study of the age structure of wells active in oil-gas recovery shows that there is a negative trend, that is, the specific weight of wells operated for a long time is large. This slows the further development of recovery volumes. In general, it has a negative impact on the level of the PRP at the oil and gas recovery enterprise.

Updating the well fund differs from the development project of the fields in a number of cases, and all this reduces the armament with funds and PRP in the field.

2. Methodology and methods

In the research work, grouping, comparative analysis, systematic approach, probability theory, methods of mathematical-statistical modeling and statistical criteria verification of adequacy of models, economic interpretation of models were used. Practical examples of conducting a geological and economic assessment are given in the articles [1].

Extensive recycling is the process of production of material goods, certain production relations that are constantly in motion and change. The general law in the conditions of scientific and technical promotion is the law of increasing the yield of means of production. This law manifests itself for the economy as a whole with the pace of being ahead of the growth and value of the public product and the means of production applied. The revised law on recycling in relation to production capacity sounds like this: the growth rate of production capacity compared to the growth rate of the cost of production capacity is defined as a preventive «growth rate» of production capacity.

The structure of the main funds of different oil and gas recovery enterprises is characterized by known diversity. They are determined by numerous factors, including the characteristics of the location and depth of the deposit, the technical level of the well construction, advanced technology used in oil, field development, optimal project decision making in the improvement of the mine, etc.

Improvement of technological regime of well operation and liquidation of their discharge lead to increase of oil recovery and implementation, increase of labor productivy, increase of profit and profitability without additional capital investment in oil and gas industry.

3. Current state of the oil fields of Azerbaijan

The following activities are performed on the extension of the operation time of the well [2–5]:

1. Introduction of inactive wells into operation.

Inactive wells determine the full use of the enterprise's production capacities, oil and gas production losses, deterioration of economic indicators and the PRP of oil and gas production as a whole.

Reduction of the number of inactive wells is provided by obtaining the following: prevention of decommissioning of wells; acceleration of the mastering of new wells; improvement of the technical level of repair work; determination and maintenance of optimal technological modes of operation of wells, proper selection of operating equipment, effective fight against sand blasting, paraffin deposition; elimination of accidents and discharges in the well operation process.

2. Effective increase of the inter-repair period of well operation.
3. Extension of the service life of oil wells.
4. Timely identification of surplus machinery and equipment.
5. Efficient extension of the service life of oil equipment.

These characteristics can be evaluated on the basis of forecast quick evaluation of the indices of of the oil fields [6]. All this increases PRP opportunities in oil and gas production by reducing capital investment in well equipment.

There is increase in the productivity of the use of the main funds in the unit of time (intensive resources), in other words, increasing the intensity of the use of wells, layers and deposits as a whole.

In our opinion, their implementation may be possible in the following ways: expansion and improvement of artificial impact on oil deposits, so that, along with the introduction of a network of sparse wells maintenance of lay pressure, during the period of its application allowed drilling of a large number of wells, and in terms of PRP, other benefits were realized. They are the application of methods of organization of the bottomhole zone; separate operation of two or more layers at the same time with one well; complex automation of the oil and gas production enterprise. So that makes it possible to achieve PRP improvement by solving the issue of «smart» well (mining) using the elements of modern digital economy; elimination of well product losses.

At the same time, it is very important to use them effectively in terms of formation of PRP and expansion of its opportunities in oil and gas production.

In our opinion, the important directions of increasing the effective use of the main production funds during the time unit can be the following: improvement of machinery and equipment for oil production; application of modern systems of the collection and transportation of oil and gas; preparation of its efficient processing, including dewatering and desalination of oil, in accordance with the established rule of the norms of regulations: drilling wells and lowering the cost of construction of oil fields.

As a result of the continuous increase of the main funds in oil and gas production, the amount of repair work carried out at these enterprises also sharply increased. Repair of the main funds is the condition created for their partial recycling. Effective organization of operation and repair of the main funds, preventing their premature physical wear, prolongs the state of readiness for operation, improves the use of labor tools that have a positive impact on the volume of production, labor productivity, profitability of production and other quality indicators of the enterprise's work.

The annual costs of the main funds in the oil and gas industry amounted to about 50 million manats. In this regard, the large amount of savings obtained from repair in these areas, among other conditions, is one of the important sources of the increase of the PRP in oil production.

The oil production trend is observed in the fields at the last stage of development. The study shows that this decrease is more felt in the fields exploited in the onshore area of the Azneft Production Association (PA). The current state of the oil fields on the onshore is characterized by a very small pace of oil production, which is explained by the complexity of the geological and energy characteristics of the fields and the limitation of the possibilities of the processing methods used in this regard. Development of oil fields takes into account maintenance of reservoir pressure, application of various methods of impact on the reservoirs and they are applied to different layers of the reservoir depending on geological, geophysical, physical, chemical and other conditions of the layer.

The development of the oil industry, stabilization and increase of production, among other conditions, depend on the effective use of the well Fund. So, thousands of wells have been drilled in the fields that have been operated for many years, and keeping them in working condition requires great financial resources. For certain reasons, every year hundreds of wells enter the inactive fund. Repairs to be carried out in these wells, their complexity, the necessary equipment and materials, financial resource to be spent on repairs and so on are very diverse and are formed mainly by a variety of natural conditions. From this point of view, it is important to study the dependence between the newly drilled well Fund and the daily oil production.

4. Application of statistical dependencies to evaluate economic indicators of production efficiency

Below, this dependence based on the actual indicators of «Azneft AP» using the smallest squares method on the modern computer was implemented on the «Statistics» program, and the following one-factor mathematical-statistical model was proposed:

$$NDW = 87.64 - 0.958X \tag{1}$$

where NDW is the number of newly drilled wells; X is the daily oil production, thousand ton.

Analysis of statistical indicators of the model shows that the dependence between daily oil production and the fund of newly drilled wells is not very strong, that is, the coefficient of correlation is r=-0.361. This is due to the fact that the daily debit from newly drilled wells is less than intended for economic design of deposits.

The analysis shows that the values of both F-Fisher and t-Student coefficients are less than their critical (table) values, with a probability of 95%.

There are several systems of main indicators (stock-yield, armament with fund) in oil production, which characterize the efficiency of the main funds. The most important of these is the stock yield, which is defined as the cost of the product per one manat of the main production funds.

Since the same type of products are produced in oil and gas production enterprises, the stock yield indicator is expressed both in nominal (ton / man.), and cost worth (man./man.). Stock yield in the oil and gas industry determines the level of the PRP of the enterprise better, and therefore it is advisable to study this indicator separately. Many factors affect the stock yield and this affects the volume of the product or the value of the main funds, or both of them.

In our opinion, the most basic factors are naturals conditions (geographical location, natural-chemical properties of oil and gas fields, physical conditions of oil and gas fields, depth of oil layers, yield of clay and wells, natural-chemical properties of oil and gas fields). The technical level of production; management, labor and production levels is related to the relocation and location of oil and gas production in remote, sparsely populated areas, to development of new layers and deposits, and to the drilling of wells and the construction of oil fields.

Taking all this into account, we proposed mathematical-statistical modeling of the stock yield indicator.

In order to improve the value of the indicator, the method «factory-year» was used, that is, to increase the length of the research period, the indicators of each oil extraction department in the last five years were summarized, the number of observations was n=40. And this satisfies the condition that the number of observations in mathematical-statistical models is 3-4 times more than the number of factors [7].

After the apriori analysis of factors to the level of oil stock yield (SY) three factors were maintained: average daily debit of oil well (X_1, t/day); average cost of oil well (X_2, thousand dollars); operating coefficient (X_3, part of the unit).

The implementation of the problem with the "statistical" (regression) program on the computer allowed one to obtain a three-factor model of the stock yield:

$$FV = 0.267 - 0.000515X_1 - 0.00012X_2 + 1.900X_3. \quad (2)$$

If we look at the statistical data of the model, we can see that the production values of both F-Fisher and t-Student coefficients are less than their critical (table) values with 95% probability. On the other hand with 95% probability all the reliability ratios of the indicators included in the model probably meet the requirements. The indicator of importance on all factors satisfies the criterion of importance P>0.05.

All this can be applied to the mathematical-statistical model of the stock yield indicator and to the rational analysis of the main funds, the capabilities of the PRP. It is natural that the efficient use of the main production funds contributes to the improvement of the stockyield, as well as the improvement of the PRP in oil and gas production as a whole. This also determines the level of labor productivity at the enterprise.

In order to measure the impact of working time losses on labor productivity and expenditure of salary fund funds from the point of view of PRP in oil and gas production, we propose the following methodical approach:

- we get the number of conditional workers not used during the analysis period by dividing the total of unplanned working time losses by the plan continuation time of the working period (quarter, year) ;
- multiplying the number of these workes by the the productivity of a worker's plan, we determine the value of this unused product;
- by dividing the volume of the unsold product by the actual average number of workers, we

determine the size of the worker's labor productivity decrease.

It is known that, since labor productivity is calculated according to the industrial production staff (IPS), it is necessary to multiply the determined indicator by the specific weight of workers in the total number of these people. The result will characterize "missed opportunities" of increasing labor productivity, which directly characterizes the PRP of oil and gas production;

The total of excess labor costs include the payment of daily and extraordinary idle work, the payment of Engineer-Technician workers (specialists), servers and part time in the cost of actually produced product. In order to measure these costs, we calculate the specific weight (in%) of the salary of engineering workers, servers and part time workers in the total product, multiply this indicator by the value of the unsold product. The result will characterize «missed savings» on the salary fund at the expense of conditional-permanent expenses in terms of payment of labor.

The level of labor productivity in oil-gas production and its subsequent growth have a high impact on: geological conditions (impact factors), technical and technological conditions of well operation, efficient use of basic funds, level of labor and production.

The increase in labor productivity affects the level of the cost of the unit of output, and this indicator is determined both in the natural and in the value terms. We have proposed a mathematical-statistical model among the factors affecting labor productivity and its impact on oil production. As factors affecting labor productivity (LP) we considered the flow rate (X_1 - ton/Day), specific weight of oil extraction by mechanized method (X_2 - %), katering of production (X_3 - %). The natural expression of this indicator (tons / person) was used, the problem of "factory-year" was solved, the number of observations was n=27.

The analysis showed that the double correlation coefficient between the indicators R_{x1x3} is $R_{x1x3}=0.672$. The mathematical-statistical model of labor productivity (LP) is as follows:

$$LP = -493.48 + 23.0X_1 + 1.048X_2 + 11.98X_3. \quad (3)$$

The calculated value of the F-Fisher coefficient with 95% probability is greater than its critical (Table) value.

The reliability coefficient of the model is within the norm. Therefore, the proposed mathematical-statistical model of labor productivity can be used in practical calculations.

Thus, we have seen a good source of reserve in improving the PRP in oil and gas production. With proper use of these sources, the oil and gas production enterprise can not only improve its own IRP, but also can produce competitive products.

5. Conclusion

Summarizing the above, we can note the following. Low oil production from new wells can be explained by the following:

- gas wells drilling in separate fields was realized in small productive areas, these areas were well landscaped, the depth of the layer is very large. In the field development project, the area with bad geographical relief (forest Strip), but productive field was intended;
- due to the lack of drilling equipment in the deep-lying layers, drilling of the well was implemented in a shallow place, which is not intended in the project, was carried out, and this resulted in the commissioning of small-debit wells;
- number of operation days of the new wells intended by the project was not complied;
- implementation of watering is less than the planned one in the project.

At the same time, it should be noted that the main reason for the fall in the production of fields operated in both onshore and offshore areas of "Azneft" PA in recent years is less renovation of the well fund in relatively new fields and unsatisfactory use of the existing well fund in the old fields, if other conditions are the same.

One of the main problems is to increase the efficiency of the use of production wells. The work should be carried out in several directions: ensuring optimal oil production from each well; commissioning of inactive wells with the implementation of geological and technical measures, increasing the period of repair of wells; using digital, "smart" technologies in the operation of oil fields.

References

[1] Nosov M A, Galkin V I, Krivoshchekov S N, Melkishev O A, Kochnev A A 2013 On substantiation of license area economic-geological evaluation by hydrocarbon resources and reserves *Oil Industry* **3** 14–17
[2] Aliyev T N, Babayev M T 2003 *Economic potential of oil gas production industry enterprises* (Baku: Azerneshr)
[3] Hajizade E M 2002 *Economical development model of oil gas production complex* (Baku: Elm)
[4] Alekseeva V A 2004 Management of production resource potential of oil gas producing enterprise *Neft, gas, biznes* **6** 24-32
[5] Guzhnovskii L P 2012 Method of system analysis of efficient functioning of oil gas producing industry *Problemy economik i upravlenia neftgasovim kompleksom* **5** 17-22.
[6] Poplygln V V, Galkin S V 2011 Forecast quick evaluation of the indices of the development of the oil deposits *Oil Industry* **3** 112–115
[7] Melkishev O A, Galkin V I, Galkin S V, Gurbanov V S, Koshkin K A 2018 Application of cluster analysis in assessing the initial total in place oil resources density for highly explored areas *SOCAR Proceedings* **3** 16–23

Processing and use of waste from the mining industry

V I Golik [1,2], S I Polovneva[3], R R Turluyev[4]

[1]North Caucasian Mining and Metallurgical Institute, 44 Nikolaeva St., Vladikavkaz, 362021, Russian Federation
[2]Moscow Polytechnic University, 38b Semenovskaya St., Moscow, 107023, Russian Federation
[3] Irkutsk National Research Technical University, 83, Lermontova str., Irkutsk, 664074, Russia
[4]Chechen State University named after A. H. Kadyrov, 17a, Dudaev Boulevard, Groznyy, 364015, Russia

E-mail: v.i.golik@mail.ru

Abstract. An increase in the extraction of resources is accompanied by their losses and dilution of ores due to the technogenic impact on ore-bearing massifs. The possibility of preserving the earth's surface is provided by technologies with filling technological voids with hardening mixtures based on leach tailings, the insufficient activity of which is increased by processing in disintegrators under the influence of high energy. Experimental proof of the possibility of controlling the state of the hardening mixture was obtained in the course of many years of use of the disintegrator as part of the filling complex. The possibility of replacing cement with activated tailings has been proven not only as inert fillers but also as binders, which makes it possible to replace commercial cement with cheaper production of commercial products. Utilization of metals from enrichment tailings of various types of minerals allows the use of secondary tailings without restrictions on sanitary requirements. The combined effect of mechanical and chemical energies allows waste-free disposal of enrichment tailings and creates, in essence, a new base of metal raw materials to meet the needs of industry.

1. Introduction

The use of intensive technologies for the extraction of mineral raw materials has brought the mining industry to the leaders of nature-destroying technologies [1–2].

The negative impact of technologies on the environment reduces the filling of the developed space with hardening mixtures [3–4].

Industrial enterprises have accumulated production waste, which is similar in chemical and mineralogical composition to the natural raw materials used for the preparation of the bookmark.

Work on the use of processing tailings for the manufacture of binders is carried out on a large scale [26, 27].

In the development of rock deposits, the property of discrete rocks to create stable structures with the preservation of bearing capacity has been known for a long time but is not always used. Strengthening the rigidity of rock structures is provided by the use of solid mixtures.

Mixtures based on low-activity tailings of processing have lower strength and can be used in certain conditions that are created by dividing the orefield with the help of tselikov or arrays from the bookmark into areas where the stresses do not exceed critical values [28].

The utilization of technogenic raw materials is carried out after the removal of metals to the sanitary level in two stages: the extraction of metals and the increase in the activity of ingredients that should be investigated simultaneously. The fate of extracted ores when used in metal–consuming industries depends on the success of measures to reduce dilution [5–6].

2. Methods and materials

The research methodology includes a set of tasks solved with the use of the Du-65 disintegrator for the first time at the Shokpak deposit laying complex in Northern Kazakhstan (Figure 1).

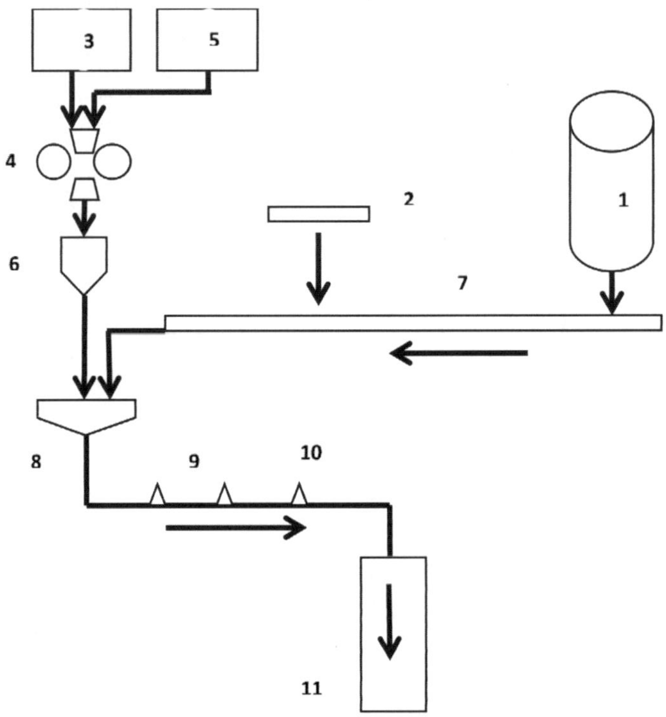

Figure 1. Workshop for preparing a hardening mixture: 1 – cement warehouse; 2 – classifier of inert aggregates; 3 – slag warehouse; 4 – disintegrator; 5 – water; 6 – vibrating mill; 7 – conveyor; 8 – mixer; 9 – laying pipeline; 9, 10 – vibrators; 11– blast chamber.

The parameters of metal extraction by metal leaching were investigated by comparing the indicators of metal extraction during agitation leaching and during leaching in a disintegrator under different processing modes.

The criterion for evaluating the studied options is the extraction of metals, established by the processing of samples taken after the disintegrator, and the strength under uniaxial compression of solidifying mixtures made on the basis of tailings of ferruginous quartzite enrichment [14].

To study the processes of metal leaching, a DU-11 disintegrator was used.

Experimental data on obtaining, accumulating, and storing the results of studies of the technogenic cycle of mining and metallurgical enterprises is a data bank and is improved upon receipt of new data [7, 20].

3. Results and discussion

Leaching of the tailings of the enrichment of polymetallic ores of Sadon deposits in heavy suspensions containing, along with other components (%) of the same – 4.4; zinc – 0.95, lead – 0.84, compared with leaching in agitators increased the extraction of metals and reduced the duration of the process.

Similar results were obtained by leaching the tailings of ferruginous quartzite enrichment containing 8% iron, 5*10–3% copper, 4*10–3% nickel, 5*10-4% zinc, and other elements – (30...50)*10–5%.

Mechanical activation of the tails in the disintegrator increases the strength of the hardening mixture. Data on the strength of the mixture with the addition of 100 kg/m3 of cement are shown in Figure 2.

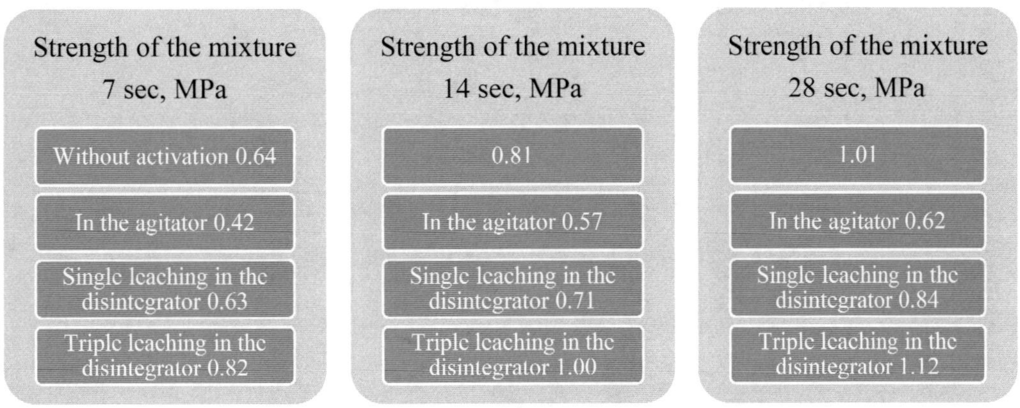

Figure 2. Dependence of the strength of hardening mixtures on the method of activation of the binding tails at the same cement consumption.

The process of metal leaching reduces the strength of the hardening mixture due to an increase in humidity, but repeated activation of the enrichment tails in the disintegrator during storage of other experimental conditions increases the strength of the mixture (Figure 3).

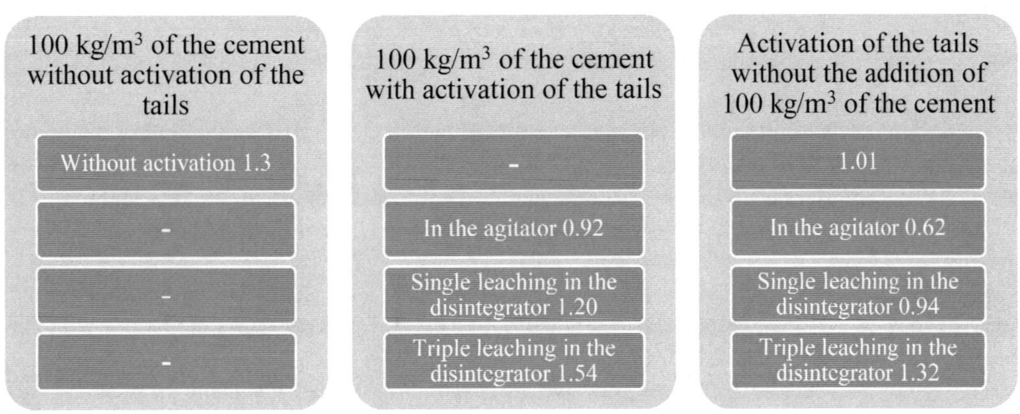

Figure 3. Dependence of the strength of hardening mixtures on the activation parameters of the tails at the same cement consumption.

The increase in the strength of hardening mixtures based on enrichment tailings activated in the disintegrator is estimated at 20–30%.

IOP Conf. Series: Earth and Environmental Science 1021 (2022) 012004 doi:10.1088/1755-1315/1021/1/012004

Hardening mixtures based on activated binder tailings of enrichment, even without the addition of commercial cement, can be used to lay the worked-out space in the areas of deposits unloaded from high stresses.

Technologies with the laying of voids with hardening mixtures, most often, a two-stage process of dredging stocks: first, the extraction of stocks of cameras of the first stage, and then cameras of the second stage.

The effectiveness of the technology can be determined from the expression [8]:

$$E=\{(M_aP_m+Q_eЦ_{rs})\}-\sum_{n=1}^{n}[K(1+E_r)+E_l+E_e]-[(M_eP_{lm}+QN_{le})+N_{ed}C_m]\sum_{0}^{n}K_n$$

where M_a is the amount of metals extracted from the tailings;
P_m – the price of metals;
Q_e – the number of recovered effects;
P_{rs} – the price of recycled substances;
E_r – environmental remediation interest rate ratio;
E_l – loan interest rate ratio;
E_e – the interest rate coefficient on environmental restoration;
M_e – the amount of lost metals;
P_{lm} – the price of lost metals;
N_{le} – the number of lost effects;
N_{ed} - the number of effects of environmental damage;
C_g the cost of compensation for global damage factors;
C_m – the cost of management;
K – storage management costs;
K_c – the sum of coefficients for unaccounted factors.

Compositions of mixtures based on 0.5...1.0 MPa enrichment tailings activated in the disintegrator can be used as hardening mixtures when filling man-made voids under appropriate geomechanical conditions [21-25].

The basis of resource-saving technologies are variants of chamber development systems with the laying of the developed space, the basic design principles of which are as follows:

- complete filling of the cleaning chambers with a filling mixture without leaving an open cleaning space;
- differentiation of solutions for managing the geomechanics of the ore-containing array, based on the stability of reserves and minimizing the costs of laying work.

4. Conclusion

If the developed space is protected by jammed structural defects, it is possible to use a bookmark composition of different strengths from different materials [13, 15-17].

The supply of rocks to the developed space being laid is carried out from the workings of the ventilation-laying horizon. In rare cases, a rock bookmark is used, extracted by blasting the host rocks with delivery by scrapers or the force of an explosion.

Having filled the required volume of the chamber with a rock bookmark, they proceed to lay the remaining part with hardening mixtures to ensure the penetration of sand-cement mortar into the rocks.

The economic efficiency of the new technology is determined by the volume of replacement of commodity components of hardening mixtures with secondary tailings of mechanochemical processing and sale of simultaneously extracted metals.

Aspects of the topic under consideration have been studied by the works of foreign specialists [9–12].

References

[1] Valiev N G, Propp V D and Vandyshev A M 2020 The 100th anniversary of the department of mining engineering of UrSMU *News of Higher Educational Institutions. Mining Magazine* **8** 130–143
[2] Dushin A V, Valiev N G, Lagunova Yu A and Shorin A G 2018 Ural Mining University and Moscow Mining University: interaction of higher education institutions *Mining Journal* **4** 4–10
[3] Golik V I, Dmitrak Yu V, Gabaraev O Z and Kozhiev Kh Kh 2018 Minimizing the impact of mining on the environment *Ecology and Industry of Russia* **6** 26–29
[4] Golik V I, Razorenov Yu I, Puzin V S and Stas G V 2021 Differentiated assessment of stability of exposed rock surfaces in sublevel stoping with backfill *Physico-Technical Problems of Mineral Development* **5** 85–93
[5] Panov V S, Eremeeva Zh V, Sharipzyanova G H, Skorikov R A, Mikheev G V and Ageev E V 2014 The effect of the nature of nanoscale particles and method of mixing on the tribological properties of powder steel 70P *Proceedings of the Southwest State University* **6 (57)** 8–14
[6] Bosikov I I, Klyuev R V, Aimbetova I O and Makhosheva S A 2021 Development of methods and management tools aerogasdynamics processes at mining sites *Sustainable Development of Mountain Territories* **3** 397–405
[7] Shchurov N I, Myatezh S V, Malozyomov B V, Shtang A A, Martyushev N V, Klyuev R V and Dedov S I 2021 Determination of inactive powers in a single-phase ac network *Energies* **14 (16)** 4814
[8] Khasheva Z M and Golik V I 2015 The ways of recovery in economy of the depressed mining enterprises of the Russian Caucasus *International Business Management* **9 (6)** 1210–1216
[9] Ben-Avua E, Richter O, Elkington T and Purrahimyan Yu 2016 Strategic mining options optimization: open pit mining, underground mining or both *International Journal of Mining Science and Technology* **26 (6)** 1065–1071
[10] Dominguez M S K, Baptista A L F and Tato D M 2017 Engineering complex systems applied to risk management in the mining industry *International Journal of Mining Science and Technology* **27** 611–616
[11] Shchurov N I, Dedov S I, Malozyomov B V, Shtang A A, Martyushev N V, Klyuev R V and Andriashin S N 2021 Degradation of lithium-ion batteries in an electric transport complex *Energies* **14 (23)** 8072
[12] Rybak J, Kongar-Suryun Ch, Tyulyaeva Yu, Khairutdinov A and Akinshin I 2021 Geomechanical substantiation of parameters of technology for mining salt deposits with a backfill *Mining Science* **28** 19–32
[13] Bolobov V I, Chupin S A, Bochkov V S, Akhmerov E V, Plaschinskiy V A 2020 The effect of finely divided martensite of austenitic high manganese steel on the wear resistance of the excavator buckets teeth. *Key Engineering Materials* **854** 3-9
[14] Karlina A I, Ershov V A, Kargapoltsev S K, Gozbenko V E, Karlina Y I 2021 Algorithm for calculating the time of thermal pulse processing of small-sized high-precision parts. *IOP Conference Series: Materials Science and Engineering* **1064(1)** 012018
[15] Konyuhov V Yu, Gladkih A M, Galyautdinov I I, Shchadova E I 2020 Calculations of efficiency in implementing progressive mold forming methods. *IOP Conference Series: Materials Science and Engineering* **760** 012027
[16] Salamanova M, Murtazaev S-A, Bataev D, Nakhaev M, Sarsenbayev B 2021 Recipes and properties of multicomponent systems "Aspiration dust - mineral powder - Na2Sio3. *Materials*

Science Forum **1017** 143–152

[17] Lombardi P, Styczynski Z, Sokolnikova T, Suslov K 2014 *Use of energy storage in Isolated Micro Grids.* Proceedings - 2014 Power Systems Computation Conference, PSCC 2014, 7038361

[18] Konyuhov V Yu, Gladkih A M, Chemezov A V 2019 Evaluation of the economic feasibility of the introduction of plasma hardening technologies in the Far North enterprises. *IOP Conference Series: Materials Science and Engineering* **560** 012147

[19] Karlina Y I, Kargapoltsev S K, Gozbenko V E, Karlina A I 2020 Removal of burrs from small-size high-precise parts for coaxial radio components. *IOP Conference Series: Materials Science and Engineering* **832(1)** 012089

[20] Gorbunov R N, Gorbunova Z V, Gorbunova L N, Kolchin V S, Karlina A I 2021 Evaluation of the efficiency of the road network functioning. *IOP Conference Series: Materials Science and Engineering* **1061(1)** 012018

[21] Konyuhov V Y, Gladkih A M, Zott R S 2020 Accelerator as an effective replacement of a business incubator in the Irkutsk region. *Journal of Physics: Conference Series* **1582** 012045

[22] Voropai N, Ukolova E, Gerasimov D, Suslov K, Lombardi P, Komarnicki P 2019 *Simulation approach to integrated energy systems study based on energy hub concept.* 2019 IEEE Milan PowerTech, PowerTech 2019 8810666

[23] Lombardi P, Sokolnikova T, Suslov K V, Styczynski Z 2012 *Optimal storage capacity within an autonomous micro grid with a high penetration of Renewable Energy Sources.* IEEE PES Innovative Smart Grid Technologies Conference Europe, 6465706

[24] Stotzer M, Gronstedt P, Styczynski Z, Buchholz B M, Glaunsinger W, Suslov K V 2012 *Demand side integration-A potential analysis for the German power system.* IEEE Power and Energy Society General Meeting, 6345090

[25] Konyuhov V Yu, Gladkih A M, Galyautdinov I I, Kiseleva T Yu 2019 Ecological architecture: The green roofs. *IOP Conference Series: Earth and Environmental Science* **350** 012035

[26] Perelygin D N, Boichuk I P, Gorlov A S 2019 Rational design of low-tonnage technological complexes for fine and ultrafine grinding of materials. *IOP Conference Series: Materials Science and Engineering* **560(1)** 012118

[27] Sevostyanov V S, Uralsky V I, Boichuk I P, Ermilov R A, Lukyanova E A 2019 Analysis of separator operating conditions in a technological module with vibration-centrifugal closed-cycle grinding unit for obtaining high-dispersity materials. *IOP Conference Series: Materials Science and Engineering* **552(1)** 012040

[28] Maksarov V V, Keksin A I, Filipenko I A 2022 The effect of the magnetic abrasive finishing process on the roughness of flat products made of aluminum alloy of the AMc brand. *Tsvetnye metally* **7** 34-41

Evaluation of the variability of the oil viscosity of the Permian-Carboniferous reservoir Usinskoye field

S N Kultysheva[1], **A V Raspopov**[1], **G P Khizhnyak**[2]

[1] Branch of LLC «LUKOIL-Engineering» «PermNIPIneft» in Perm, 3a, Permskaya street, 614000, Perm, Russia
[2] Perm National Research University, 29, Komsomolsky ave., Perm, 614990, Russia

E-mail: Svetlana.Kultysheva@pnn.lukoil.com

Abstract. The article describes the results of assessing the variability of the viscosity of reservoir oil from the Permian-Carboniferous reservoir of the Usinsk field, obtained on the basis of sampling, research and analysis of downhole samples. Emphasis is placed on the study of oil properties (viscosity, gas content, molar mass, resin content, etc.) in the edge of the reservoir; a number of regularities in the change in viscosity over the area and section of the reservoir have been identified, which can later be used to optimize technologies for the development of super-viscous oil.

1. Relevance of the development of heavy oils

As the reserves of light oil are depleted, the issue of involving unconventional hydrocarbons, including heavy oils, in the development becomes more and more urgent. Their involvement in industrial turnover can support production and cover the forecasted oil and gas deficit in the near future. Heavy oils are characterized by low mobility in reservoir conditions, and therefore require special methods of extraction from the reservoir, which increases their cost.

2. The need to take into account the variability of oil viscosity when determining oil recovery factor and choosing a technology for the development of heavy oils

Viscosity is the most important technological property of oil, because it determines the degree of oil mobility in the process of its production from the reservoir and transportation through trunk pipelines. The viscosity value is used to assess the filtration rate in the reservoir, to select the type of displacing agent and its parameters, to justify the development technology, to calculate the capacity of downhole pumping equipment in oil production and to solve other problems.

A reliable assessment of the oil recovery factor when designing a development and determining an effective technology for the development of heavy oils directly depends on the degree of knowledge and completeness of accounting for the geological structure of the reservoir, the nature of oil occurrence in the reservoir, physical and chemical properties and the conditions of their filtration in a porous medium. Often, when designing a development, only a minimal set of parameters is taken into account: heterogeneity of the physical properties of the reservoir (porosity, permeability, initial oil saturation and displacement coefficient), the difference in the mobility of oil and water, and the presence of initial water-oil zones. Meanwhile, the nature of the development of reserves is significantly influenced by the magnitude and variability of viscosity.

The works of many researchers are devoted to the study of the regularities of the distribution of the physicochemical properties of reservoir oils in Russia and abroad. So, in the dissertation of Voskresensky V V [1] based on the study of the variability of oil properties in the Ural-Volga region, it was found that oil properties change significantly with approaching the oil-water contact: the density and viscosity of oil, the content of resins and asphaltenes increases, the coefficient of gas solubility in oil decreases - the gas-oil ratio decreases and the saturation pressure, the proportion of nitrogen and heavy hydrocarbons in the composition of the dissolved gas increases, the proportion of methane decreases.

The results of multivariate analysis of the regularities in the distribution of physical properties of reservoir oils, performed in [1], showed that:

- viscosity and density increase from the crest of the reservoir to the wings and with the depth of occurrence;
- volumetric ratio and gas factor decrease;
- in the presence of disjunctive disturbances and significant lithological discontinuity of the reservoir, the variability of physical properties is revealed in individual areas or blocks of the reservoir.

3. Features of the Permian-Carboniferous reservoir of the Usinsk field

In the Timan-Pechora region, PJSC LUKOIL is engaged in commercial development of heavy anomalously viscous oil at two large fields - Yaregskoye and Usinskoye [2]. Both fields are a priority for the Company and require increased attention due to the complexity, versatility and science intensity of the tasks being solved.

The Permian-Carboniferous heavy oil reservoir of the Usinskoye field lies at a depth of 1.0-1.5 km and contains super-viscous oil. A massive reservoir with an oil-bearing floor up to 350 m, underlain by bottom water. Carbonate deposits are characterized by fractured-vuggy-porous reservoir type. The cracks form two mutually perpendicular systems. In addition, tectonic fractures are distinguished, cutting through a significant thickness of the deposits and covering a significant area of the deposit.

Filtration of super-viscous oil during development in natural mode and during cold waterflooding is difficult in practice [2]. The main filtration channels are intervals of abnormally high permeability. The possibility of fluid movement in the formation is significantly influenced by the relaxation properties of oils, which consist in the fact that there is a delay in the change in shear stresses with respect to the change in the shear rate [3].

Laboratory studies have shown that an increase in temperature contributes to the weakening of the relaxation properties. Thus, when Usinsk oil is heated to 150°C, its viscosity decreases by almost 140 times. This fact is a prerequisite for the use of thermal methods of enhanced oil recovery (EOR), which make it possible to multiply the oil recovery factor [2].

The main methods to increase oil recovery in the reservoir are cyclic steam treatment of producing wells and stationary / cyclic thermal steam treatment through injection wells.

4. Chronology of the study of oil viscosity at the Usinskoye field

On the area of the Permian-Carboniferous reservoir of the Usinskoye field, there are zones: inner (central) in the uplifted part of the structure and marginal (peripheral) in the wing sections of the structure, significantly differing in both oil-saturated thicknesses and filtration-capacity properties. The selection of depth samples of oil to study the patterns of distribution of its properties was carried out in stages, which is associated with the periods of development of the reservoir (Table 1):

- **stage 1 (1977-2016)**: production drilling and active development were carried out mainly in the inner zone of the reservoir, therefore, the main part of deep oil samples was confined to it, and the edge zone was poorly illuminated. The average viscosity of the reservoir oil according to the studied samples was 710 mPa·s and was taken on the balance sheet;

- **stage 2 (2019-2020)**: the beginning of active drilling of the marginal zone since 2016 made it necessary to study the properties of oil in the areas being commissioned. As the project fund was formed, depth samples were taken from the wells of the edge zone, the results of which changed the idea of the viscosity distribution within the reservoir.

Table 1. Chronology of the study of oil viscosity

Period	Stage 1 (inner zone)	Stage 2 (marginal zone)	TOTAL
	1977-2016	2019-2020	1977-2020
Total samples taken,	43	18	61
of which air onditioned	*33*	*16*	*49*

The proportion of wells in which sampling was carried out against the background of the total drilled stock is low (49 samples with a stock of more than 1500 wells), however, due to the wide coverage in the area and section, the results obtained allow us to judge the variability of physical and chemical properties.

5. Identification of the main regularities of viscosity change in The Permian-Carboniferous reservoir of the Usinsk deposit

Based on the analysis of the results of the study of deep oil samples taken from 1977 to 2020, carried out by the authors of the article, it was revealed that with an increase in the depth of occurrence, the viscosity of reservoir oil increases (Figure 1). At the same time, in the upper part of the section (above absolute depth -1150 m), the increase in viscosity is insignificant (544-737 mPa·s), and in the lower part (below absolute depth -1150 m), a multiple increase is noted (up to 2623 mPa·s).

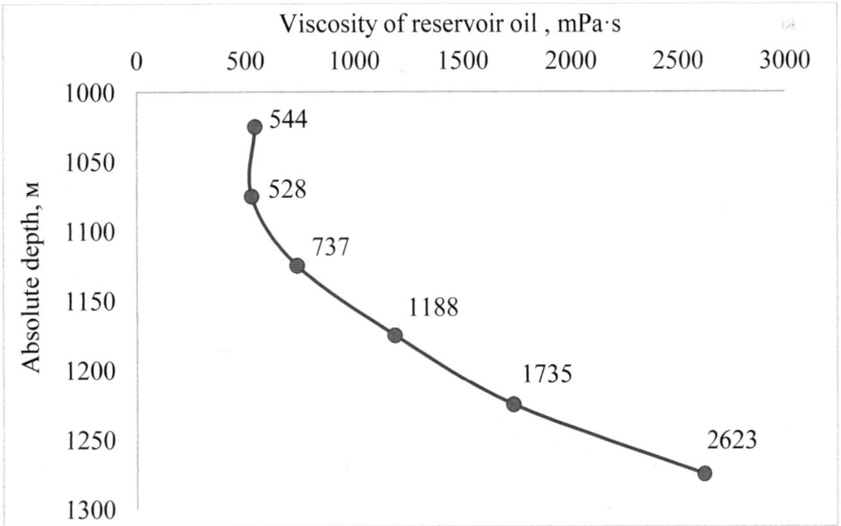

Figure 1. Dependence of the viscosity of the reservoir oil of the Permian-Carboniferous reservoir Usinsk deposit from depth (absolute depth)

The investigated oil samples were taken at a pressure higher than the bubble point pressure, which indicates a single-phase state of the fluid (associated petroleum gas was in a dissolved state).

IOP Conf. Series: Earth and Environmental Science 1021 (2022) 012005 doi:10.1088/1755-1315/1021/1/012005

It was revealed that with an increase in the gas content of oil, its viscosity decreases (Figure 2). When gas dissolves in oil, its volume increases due to an increase in intermolecular distances and, consequently, with a decrease in the forces of intermolecular interaction, the viscosity decreases [4].

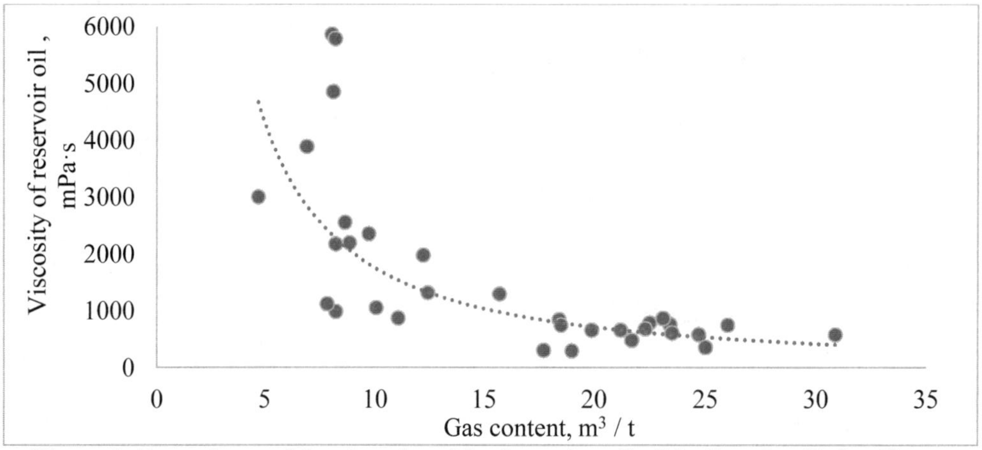

Figure 2. Dependence of the viscosity of the formation oil of the Permian-Carboniferous reservoir Usinskoye field from gas content

Correlation of the dependence obtained above along the section of the deposit has been carried out (Figure 3). With increasing depth and approaching the edges of the structure, a decrease in gas content is noted, which leads to an increase in viscosity - in the edge part of the deposit, the average value is 2720 mPa·s, while in the inner part it is 710 mPa·s.

The saturation pressure of oil with gas and the volumetric coefficient decrease with increasing depth. In the crest of the structure, oil is less viscous and is characterized by the highest values of saturation pressure and volumetric coefficient, and with approaching the bottom of the formation, the saturation pressure and volumetric coefficient decrease, at the same time, the oil viscosity increases.

6. Summary characteristics of oil properties of The Permian-Carboniferous reservoir of the Usinskoye field

The dependencies described above are presented in table 2.

The oil in the marginal zone of the reservoir is more viscous than the inner oil (the difference is 4 times), and is characterized by a higher molar mass, i.e. contains more heavy hydrocarbon components and resins, has a lower (2 times) gas content, is characterized by a lower saturation pressure (2 times).

The distribution of the viscosity of reservoir oil by the cardinal points shows that:

- the central part (inner zone) is characterized by the lowest viscosity (average ~ 700 mPa·s);
- the eastern and southern parts of the marginal zone have high viscosity (average ~ 2000-2500 mPa·s);
- the western and northern parts of the marginal zone are characterized by the highest viscosity (average ~ 3000-4000 mPa·s), while the maximum values for the wells in the northern part are due to high fracturing and the presence of disjunctive disturbances that promote gas migration outside the reservoir.

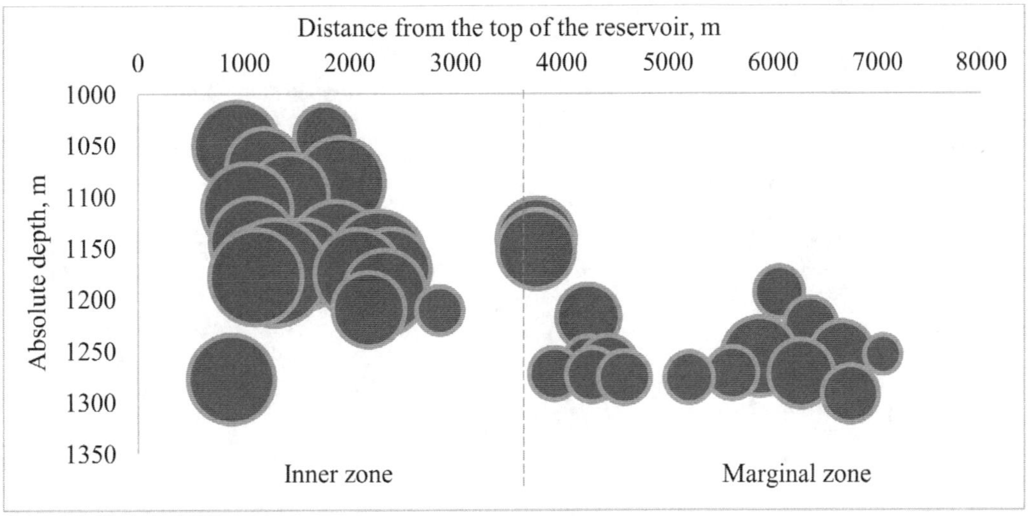

(a)

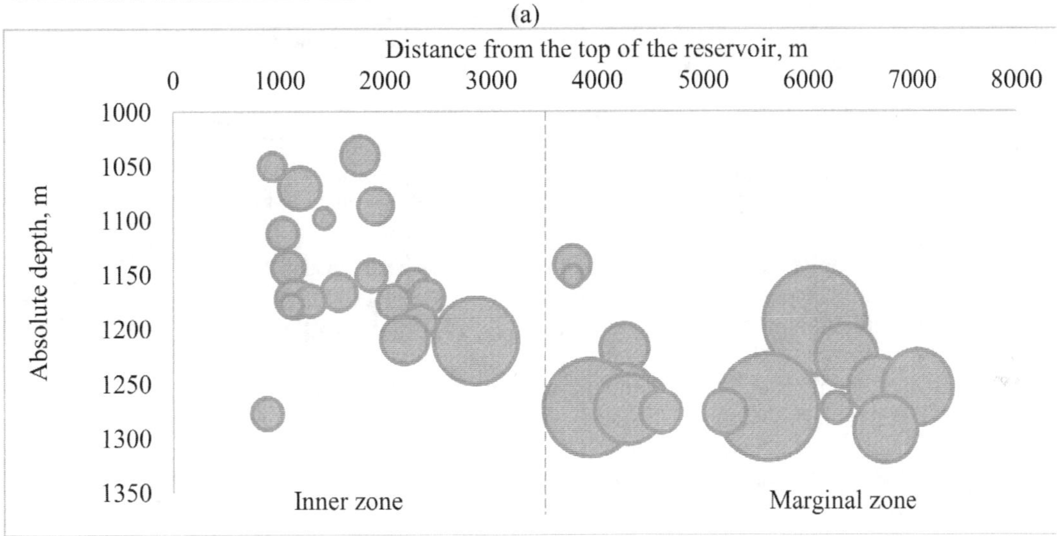

(b)

Figure 3. Dependence of gas content (a) and viscosity (b) of reservoir oil on depth and distance from the top of the reservoir (the size of the circle characterizes the value of the gas content / viscosity)

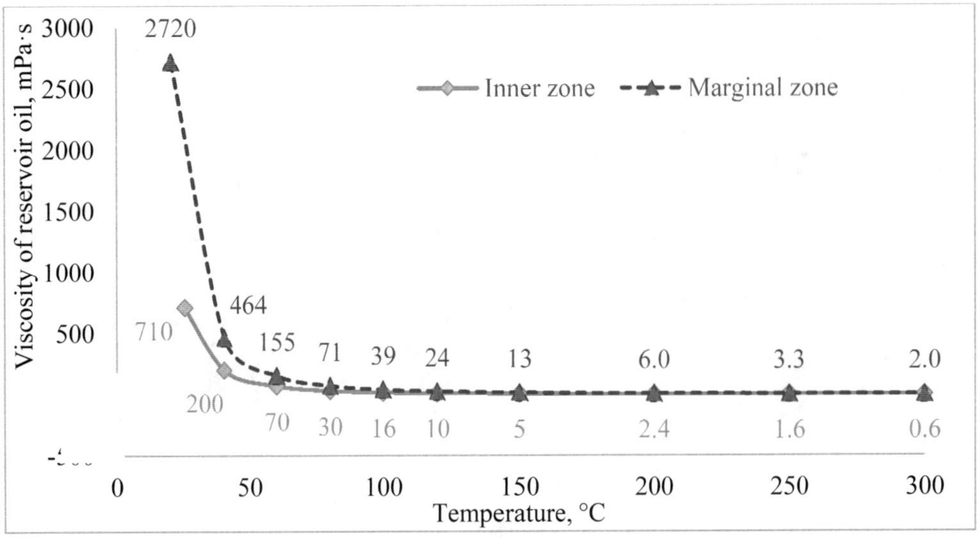

Figure 4 Dependence of the viscosity of the reservoir oil of the Permian-Carboniferous reservoir Usinskoye field on temperature

With isobaric heating of oil, a significant decrease in its viscosity is observed (Figure 4). At the same time, the main changes (a decrease in viscosity by 44-70 times - from 710-2720 to 16-39 mPa·s) are characteristic up to a temperature of 100°C, and further heating only slightly reduces the viscosity. Heating increases the vibration of molecules (or the speed of their movement), which in the absence of a limiting pressure capable of maintaining a constant volume, leads to an increase in intermolecular distances and volume of the substance (expansion). With an increase in intermolecular distances, the force of mutual attraction of molecules and friction caused by the collision of the latter decrease, as a result of which the viscosity decreases.

The use of thermal EOR (cyclic steam treatment, thermal steam treatment) in the Permian-Carboniferous reservoir of the Usinskoye field, involving heating of the formation and the oil contained in it, allows to obtain a significant increase in efficiency compared to the natural regime (a multiple increase in well production rates), and in the long term - to increase the oil recovery factor.

In addition to temperature and gas content, oil viscosity is influenced by its component composition. Under laboratory conditions, the determination of hydrocarbon (HC) components is carried out up to C6 (light fractions), and all other fractions (including heavy ones) belong to C7 + and are not differentiated. Indirectly, the presence of heavy components can be judged by the value of the molar mass of oil and the content of resins in it, because the manifestation of anomalously viscous properties by oils is due to high molecular weight components (resins, asphaltenes, paraffins), which are structure-forming compounds. The dependences shown in Figure 5 show an increase in the content of heavy components and resins with depth.

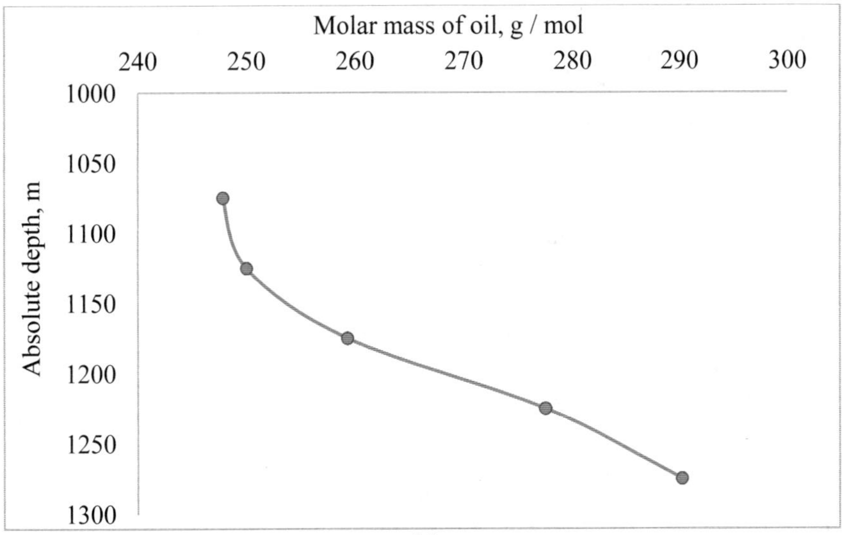

(a)

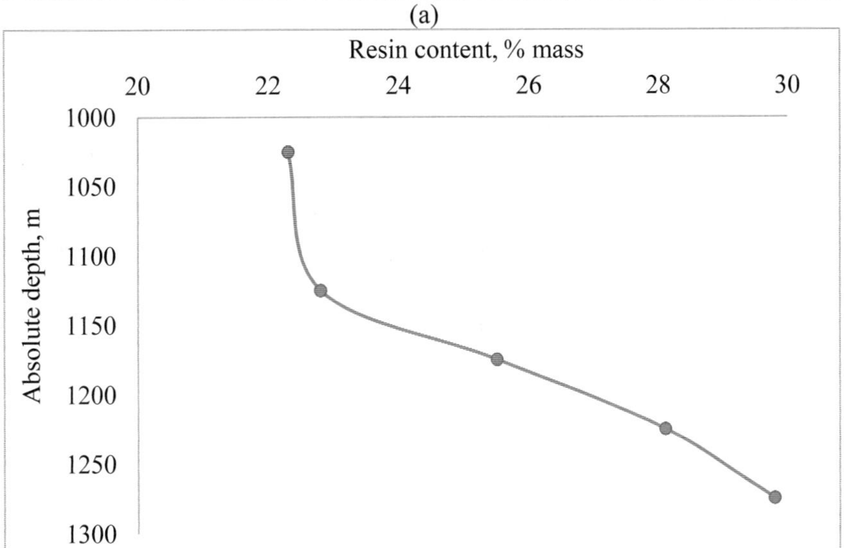

(b)

Figure 5. Dependence of molar mass of oil (a) and resin content (b) in reservoir oil on depth

Table 2. Summary characteristics of the physical and chemical properties of oil

Deposit zone	Parameter	Viscosity of reservoir oil, mPa·s	Molar mass of oil, g / mol	Gas content, m3 / t	Saturation pressure, MPa	Resin content, % mass	Paraffin content, % mass
Inner zone	Range of values	282-1295	219.5-291.0	11.1-26.9	4.6-9.5	15.1-33.0	0.02-1.8
	The average	**710**	**251.6**	**21.2**	**6.9**	**23.3**	**0.3**
Marginal zone	Range of values	984-5861	259.4-314.5	4.7-17.5	1.4-7.2	17.2-36.0	0.2-3.3
	The average	**2720**	**282.9**	**10.3**	**3.8**	**28.2**	**1.7**

IOP Conf. Series: Earth and Environmental Science 1021 (2022) 012005 doi:10.1088/1755-1315/1021/1/012005

7. Conclusion

In conditions of depletion of light oil reserves, the issue of involving unconventional oil in the development, incl. heavy oils is gaining special relevance.

In the Timan-Pechora region, PJSC LUKOIL is producing heavy oil at two large fields - Yaregskoye and Usinskoye - using thermal EOR, the implementation of which is associated with high complexity, versatility and science intensity of the tasks being solved.

Significant depths and wide areas of heavy oil distribution in The Permian-Carboniferous reservoir of the Usinskoye field require a differentiated approach to its development.

The authors of the article carried out research work to study the distribution of the physicochemical properties of oil and to identify the main patterns of viscosity change.

- The research results revealed the following patterns:
- with an increase in the depth of occurrence, the viscosity of reservoir oil, its molar mass, the content of resins increases, and the saturation pressure of oil with gas and the volumetric coefficient decrease;
- with an increase in the gas content of oil, its viscosity decreases;
- when approaching the edges of the structure, a decrease in gas content is noted, which leads to an increase in viscosity - in the edge part of the deposit, the average value is 2720 mPa·s, while in the inner part it is 710 mPa·s;
- with an increase in temperature, a significant decrease in oil viscosity is observed, while the main changes (a decrease in viscosity by 44-70 times - from 710-2720 to 16-39 mPa·s) are characteristic up to a temperature of 100°C, and further heating only slightly decreases the viscosity;
- the northern part of the edge zone of the reservoir, complicated by fracturing and the presence of disjunctive disturbances, is characterized by the highest viscosity of reservoir oil (~ 3500 mPa·s), which is probably associated with gas migration through decompressed zones.

The new information on the distribution of oil properties obtained in the course of research work creates the basis for further optimization of technologies used in the development of the Permian-Carboniferous reservoir of the Usinskoye field (cyclic steam treatment, thermal steam treatment), selection of new effective technologies, as well as effective operation of downhole pumping equipment based on the specific properties of oil in a given part of the reservoir.

References

[1] Voskresensky V V *Influence of variability of oil viscosity on the technological indicators of the development of oil reservoirs* 1975: author. dis. Cand. tech. Sciences: 05.15.06 / (Ufa) pp 5, 7

[2] Ruzin L M, Chuprov I F, ed. Tskhaday N D 2007 *Technological principles for the development of deposits of abnormally viscous oils and bitumen* (Ukhta: USTU)

[3] Ametov I M, Baidikov Yu N, Ruzin L M, Spiridonov Yu A 1985 *Extraction of heavy and high-viscosity oils* (Moscow: Nedra)

[4] Evdokimov I N, Eliseev N Yu, ed. prof. Nagaev V B 2005 *Molecular mechanisms of liquid and gas viscosity* **Part 1.** *Basic concepts* (Moscow: Russian State University of Oil and Gas named after THEM. Gubkina)

Calculation and Forecast of Resource Extraction during Exploration

V Sh Mukhametshin

Ufa State Petroleum Technological University, Branch of the University in the City of Oktyabrsky, 54a, Devonskaya St., Oktyabrsky, Republic of Bashkortostan, 452607, Russian Federation

E-mail: vsh@of.ugntu.ru

Abstract. Vast experience accumulated in the development of various groups of deposits in the carbonate reservoirs of the Volga-Ural oil and gas province, gave a set of geological and statistical relationships enabling to calculate and forecast resource extraction, provided that a representative volume of geological and field data is not available at the stage of geological exploration. The findings show that developing deposits in carbonate reservoirs requires a differentiated approach due to some specific features of oil recovery in these deposits.

1. Introduction

Studies of oil production and oil recovery patterns were carried out separately for the wells in the conditions of 14 groups of deposits in the carbonate depletion-drive reservoirs of the Volga-Ural oil and gas province [1]. This approach is due to the fact that the formations are generally drilled out for quite a long time and new wells being commissioned significantly affect a depletion drive form. Forms constructed individually for each well are devoid of this serious drawback. However, in this case, it is necessary to know that some factors have a certain effect can lead to a distorted depletion drive form [1–4].

A current flow rate of wells, being pivotal for studying oil production, can be influenced by: changes in the operating mode of nearby production wells, waterflooding or changes in the operating mode of injection wells, bottom-hole interventions, forced fluid withdrawal through more productive equipment [5–10].

As per the said deposits, production wells either have little or no interact with each other, which is attributed to low reservoir properties, their lenticular structure with a relatively sparse spacing. Therefore, changes in operating conditions of nearby wells does not affect a depletion drive form for a single well. The effect of this factor is not further considered, so is more efficient equipment installed, since the oil production rate (with a few exceptions for some deposits), is primarily determined by production capacity of the well itself, by which the production equipment is immediately selected [11–12].

The wells were chosen located outside the pilot areas where water injection was organized or no, or of little effect, measures were taken to stimulate oil production to get rid of the effect of waterflooding and bottom-hole treatments.

2. Materials and methods

The above rationale suggests that oil production was studied along with defining depletion drive forms for the wells of the selected groups of formations in the current annual oil production/time coordinates. A time series of flow rates for each of the 394 wells were approximated by various functions in the range from (oil production in the first year of well operation) to (minimum profitable annual oil production for the last year of operation). The minimum profitable annual oil production for the target formations was taken at the level of 72 t/year.

For approximation, the functions presented in the Table were used, where Q_o is the current annual oil production, t/year; t – time since the beginning of well operation, year.

Table 1. Average relative errors in approximation time series of well flow rates by various functions

No.	Approximating function	By wells of groups of formations													
		1	2	3	4	5	6	7	8	9	10	11	12	13	14
1	$Q_o = a + bt$	23	34	22	11	37	21	58	33	47	18	34	20	11	16
2	$Q_o=a+b/t$	29	27	31	30	20	35	26	28	15	31	23	18	25	27
3	$Q_o = a + b \cdot ln(t)$	3	12	9	9	15	8	28	9	10	14	16	13	10	12
4	$Q_o = \frac{1}{b(t+a/2)}$	723	632	145	142	116	146	908	170	340	200	22	19	17	34
5	$Q_o = \frac{1}{a} - \frac{b}{a^2} \cdot \frac{1}{t+b/a}$	900	815	232	174	139	338	433	321	471	198	360	401	204	180
6	$Q_o = \frac{1}{b(\ln(t)+a/b)}$	308	380	312	990	212	340	442	292	803	900	193	404	49	90
7	$Q_o = ae^{bt}$	4	17	9	14	29	14	22	9	И	13	24	9	6	8
8	$Q_o = ae^{b/t}$	53	44	52	49	44	58	75	57	32	52	39	18	32	30
9	$Q_o = at^b$	26	39	26	30	19	31	36	25	18	28	20	21	17	22

3. Results and Discussion

The function that best describes the dynamics of oil production over time was determined based on the smallest relative error of actual and calculated flow rates. As can be seen from the Table, for the groups 1-6, 8, 9, 11, the time series are best approximated by function (3). A relative well flow error is from 3 to 16%, and on average – 10%. For the groups 7, 10, 12-14, the best function is (7), by which the error values vary from 6 to 22%, and on average account for 12%.

Given that these errors are within a well production error measured in field conditions, it is possible to recommend using formulas (3) and (7) to predict the annual well production rates of the corresponding groups 4–5 years after the wells were put into operation.

The annual flow rates (Q_o^i), recoverable reserves (Q_{rec}) were determined by summing up the annual flow rates for the period from t_1 to t_{tot}, where t_1 is the flow rate for the first year of operation, and t_{tot} is the total lifetime until the flow rate decreases to the minimum profitable and total lifetime of wells depending on the primary productivity of wells. Some sufficiently close significant linear relationships are available (the correlations for different groups varied from 0.645 to 0.994), thereby indicating a significant contribution of this factor to the oil recovery in depletion-drive reservoirs.

In view of the productivity, variations in annual oil production rates for the conditions of various groups is as follows:

$$Q_H^i = 701.7 I_{prod} - 224.3 I_{prod} \ln t_i \quad (1)$$

(for wells of groups 1, 2);

$$Q_o^i = 971.0 I_{prod} - 294.1 I_{prod} \ln t_i \quad (2)$$

(for wells of group 3);

$$Q_o^i = 843.0I_{prod} - 265.6I_{prod} \ln t_i \tag{3}$$

(for wells of group 4);

$$Q_o^i = 1032.7I_{prod} - 317.8I_{prod} \ln t_i \tag{4}$$

(for wells of group 5);

$$Q_o^i = 878.9I_{prod} - 277.3I_{prod} \ln t_i \tag{5}$$

(for wells of group 6);

$$Q_o^i = 882.0I_{prod}\, e^{-0,238t_i} \tag{6}$$

(for wells of group 7);

$$Q_o^i = 977.3I_{prod} - 337.2I_{prod} \ln t_i \tag{7}$$

(for wells of group 8);

$$Q_o^i = 557.9I_{prod} - 166.9I_{prod} \ln t_i \tag{8}$$

(for wells of group 9);

$$Q_o^i = 1119I_{prod}\, e^{-0,156t_i} \tag{9}$$

(for wells of group 10);

$$Q_o^i = 751.6I_{prod} - 248.1I_{prod} \ln t_i \tag{10}$$

(for wells of groups 11);

$$Q_o^i = 524.0I_{prod}\, e^{-0,139t_i} \tag{11}$$

(for wells of groups 12);

$$Q_o^i = 441.0I_{prod}\, e^{-0,116t_i} \tag{12}$$

(for wells of groups 13, 14).

The correlation coefficients ranged from -0.934 to -0.990. Based on the correlations obtained that look like dynamic models it is possible to forecast the current annual oil production by the productivity index since a well started to operate.

The recoverable reserves can be estimated by the following formulas:

$$Q_{rec} = I_{prod} \sum_{t=1}^{t_{tot}} 701.7 - 224.3 \ln t_i \tag{13}$$

(for wells of groups 1, 2);

$$Q_{rec} = I_{prod} \sum_{t=1}^{t_{tot}} 971.0 - 294.1 \ln t_i \tag{14}$$

(for wells of group 3);

$$Q_{rec} = I_{prod} \sum_{t=1}^{t_{tot}} 843.0 - 265.6 \ln t_i \tag{15}$$

(for wells of group 4);

$$Q_{rec} = I_{prod} \sum_{t=1}^{t_{tot}} 1032.7 - 317.8 \ln t_i \tag{16}$$

(for wells of group 5);

$$Q_{rec} = I_{prod} \sum_{t=1}^{t_{tot}} 878.9 - 277.3 \ln t_i \tag{17}$$

(for wells of group 6);

$$Q_{rec} = 882\ I_{prod} \sum_{t=1}^{t_{tot}} e^{-0.238 t_i} \quad (18)$$

(for wells of group 7);

$$Q_{rec} = I_{prod} \sum_{t=1}^{t_{tot}} 977.3 - 337.2 \ln t_i \quad (19)$$

(for wells of group 8);

$$Q_{rec} = I_{prod} \sum_{t=1}^{t_{tot}} 557.9 - 166.9 \ln t_i \quad (20)$$

(for wells of group 9);

$$Q_{rec} = 1119\ I_{pod} \sum_{t=1}^{t_{tot}} e^{-0.156 t_i} \quad (21)$$

(for wells of group 10);

$$Q_{rec} = I_{prod} \sum_{t=1}^{t_{tot}} 751.6 - 248.1 \ln t_i \quad (22)$$

(for wells of group 11);

$$Q_{rec} = 524\ I_{prod} \sum_{t=1}^{t_{tot}} e^{-0.139 t_i} \quad (23)$$

(for wells of group 12);

$$Q_{rec} = 441\ I_{prod} \sum_{t=1}^{t_{tot}} e^{-0.116 t_i} \quad (24)$$

(for wells of groups 13, 14),
where the total lifetime of wells (t_{tot}), subject to productivity and minimum profitable annual oil production ($Q_o^{\min}$), is determined from the equations:

$$t_{tot} = 22.8 e^{-44.6 \cdot 10^{-4} Q_o^{\min} / I_{prod}} \quad (25)$$

(for wells of groups 1, 2);

$$t_{tot} = 27.1 e^{-34{,}0 \cdot 10^{-4} Q_o^{\min} / I_{prod}} \quad (26)$$

(for wells of group 3);

$$t_{tot} = 23.9 e^{-37{,}7 \cdot 10^{-4} Q_o^{\min} / I_{prod}} \quad (27)$$

(for wells of group 4);

$$t_{tot} = 25.8 e^{-31{,}5 \cdot 10^{-4} Q_o^{\min} / I_{prod}} \quad (28)$$

(for wells of group 5);

$$t_{tot} = 23.8 e^{-36{,}1 \cdot 10^{-4} Q_o^{\min} / I_{prod}} \quad (29)$$

(for wells of group 6);

$$t_{tot} = 28.5 - 4.188 \ln\left(Q_o^{\min} / I_{prod}\right) \quad (30)$$

(for wells of group 7);

$$t_{tot} = 18.2 e^{-29{,}7 \cdot 10^{-4} Q_o^{\min} / I_{prod}} \quad (31)$$

(for wells of group 8);

$$t_{tot} = 28.2 e^{-60{,}0 \cdot 10^{-4} Q_o^{\min} / I_{prod}} \quad (32)$$

(for wells of group 9);

$$t_{tot} = 45.0 - 6.41 \ln\left(Q_o^{\min} / I_{prod}\right) \tag{33}$$

(for wells of group 10);

$$t_{tot} = 20.7 e^{-40,3 \cdot 10^{-4} Q_o^{\min} / I_{prod}} \tag{34}$$

(for wells of group 11);

$$t_{tot} = 45.0 - 7.194 \ln\left(Q_o^{\min} / I_{prod}\right) \tag{35}$$

(for wells of group 12);

$$t_{tot} = 52.5 - 8.62 \ln\left(Q_o^{\min} / I_{prod}\right) \tag{36}$$

(for wells of groups 13, 14).

A relationship between productivity, current annual oil production, recoverable reserves, and the total lifetime of wells is shown as nomograms in the figure. The graphs shows that for various groups with fixed equal primary productivity index, the primary and current annual well rates, recoverable reserves, operation life and production decline rates differ, which signifies geological heterogeneity, type of rocks-collectors, activity of bottom and marginal waters, etc.

The wells operating the deposits of the Tournaisian and Famennian stages (groups 1-8) have primary annual oil production rates and recoverable reserves per unit of productivity higher than the wells operating the deposits of the Kashirian, Podolsky, Vereysky horizons and the Bashkirian stage (groups 10-14). The difference between these groups is largely due to the relative bottom water activity. Over time, in wells producing oil from the Tournaisian and Famennian deposits, as the reservoir pressure decreases and the pressure drop between the peripheral and the oil-saturated reservoir increases, water penetrating through a sealing layer partially compensates for a decrease in reservoir pressure in the recovery zone and displaces oil. The bottom water in the formations of groups 10-14 is much lower due to a denser and less permeable sealing layer at the oil-water interaction. The deposits of the Asselian-Sakmaro-Artinskian age of group 9 have intermediate values.

By recoverable reserves and primary annual oil production rates, the deposits of the Lower Carboniferous and Upper Devonian systems are arranged in the following order: 5 → 3 → 8→ 6 → 4 → 1, 2 → 7. Based on the formation pressure varying during operation, bottom water is most active in the Tournaisian deposits of the South-Eastern slope of the Russian platform (group 5), and least active in Tournaisian deposits of the Birsk saddle (except for the Shchelkanovskoe field). The distribution confirms this fact. The volumes of the Middle Carboniferous system are distributed by the values Q_{rec} и Q_o^i in the following order: 10 → 11 → 12 → 13 → 14. The difference is also attributed to the relative activity of bottom and edge waters.

According to the graph, most efficient wells within each group of formations have high current flow rates and recoverable reserves, which is quite reasonable. What is more, these wells have longer operation lives. This indicates a better recovery of reserves and higher drainage efficiency by more efficient wells. However, the influence of productivity on these indicators varies within groups.

In general, the formations of the Tournaisian and Famennian stages (groups 1-8) are characterized by a faster oil production decline at large primary flow rates (all other things being equal) and approximately 1.5 times shorter operation life (once minimum profitable flow rates are reached) compared to the formations of the Podolsk, Kashirskiy, Vereiskiy horizons and the Bashkirian stage (groups 10-14), which is also due to the hydrodynamic features of the production. Under these conditions, the impact of productivity on the lifetime of wells within the groups of formations 1-8 is much less than within the groups (10-14). For example, enhanced efficiency of wells of group 5 from 2 to 10 t/day·MPa will make the operation time from 23.0 to 25.2 years, while enhanced efficiency of wells of groups 13 and 14 from 2 to 10 t/day·MPa will make the operation time from 21.6 to 35.5 years.

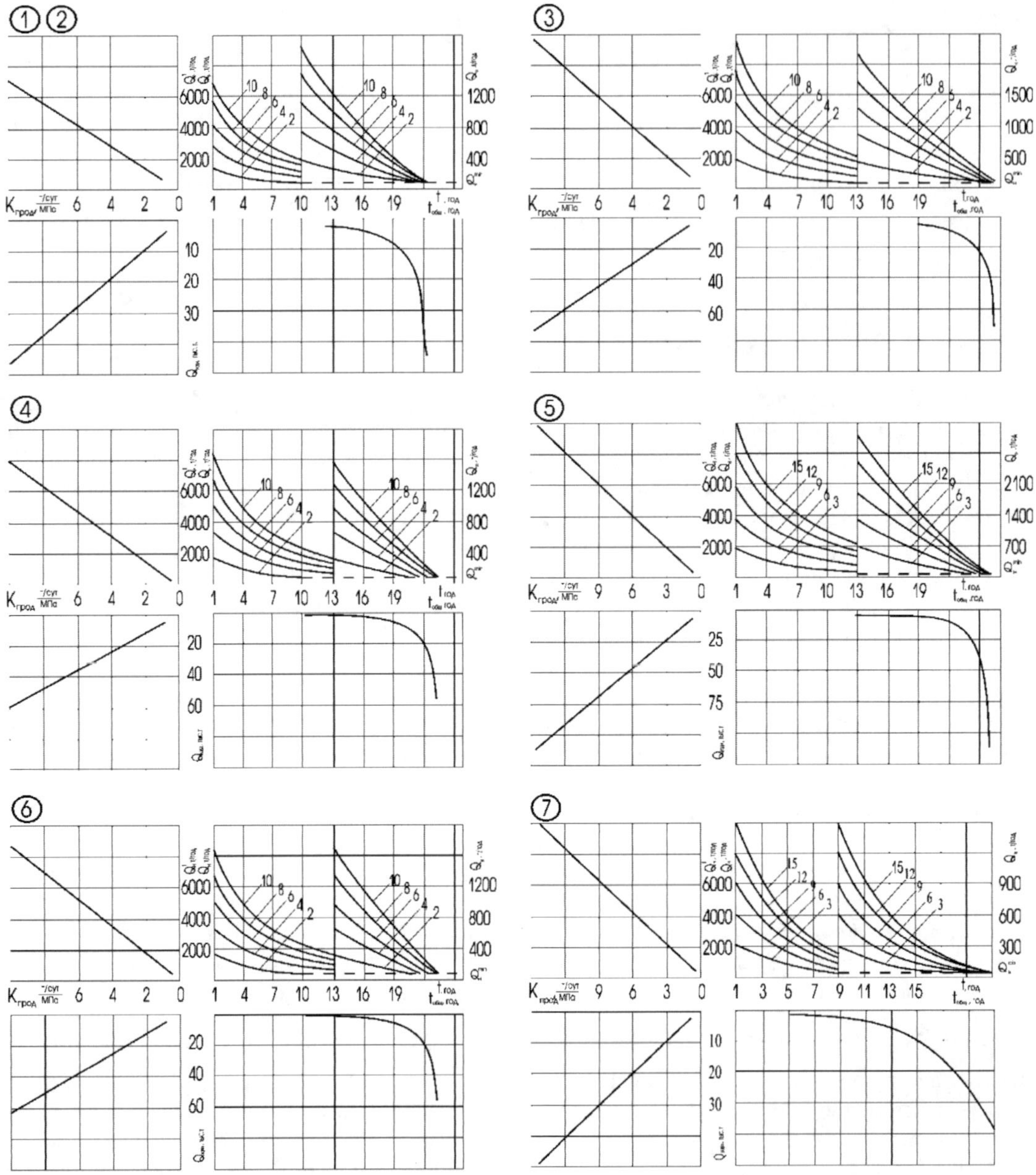

Fig 1. Nomograms for annual oil production, recoverable reserves and operation life: ⑩ – number of group of formations; Q_o^{min} – minimum profitable well production rate; I_{prod} – cipher of curves

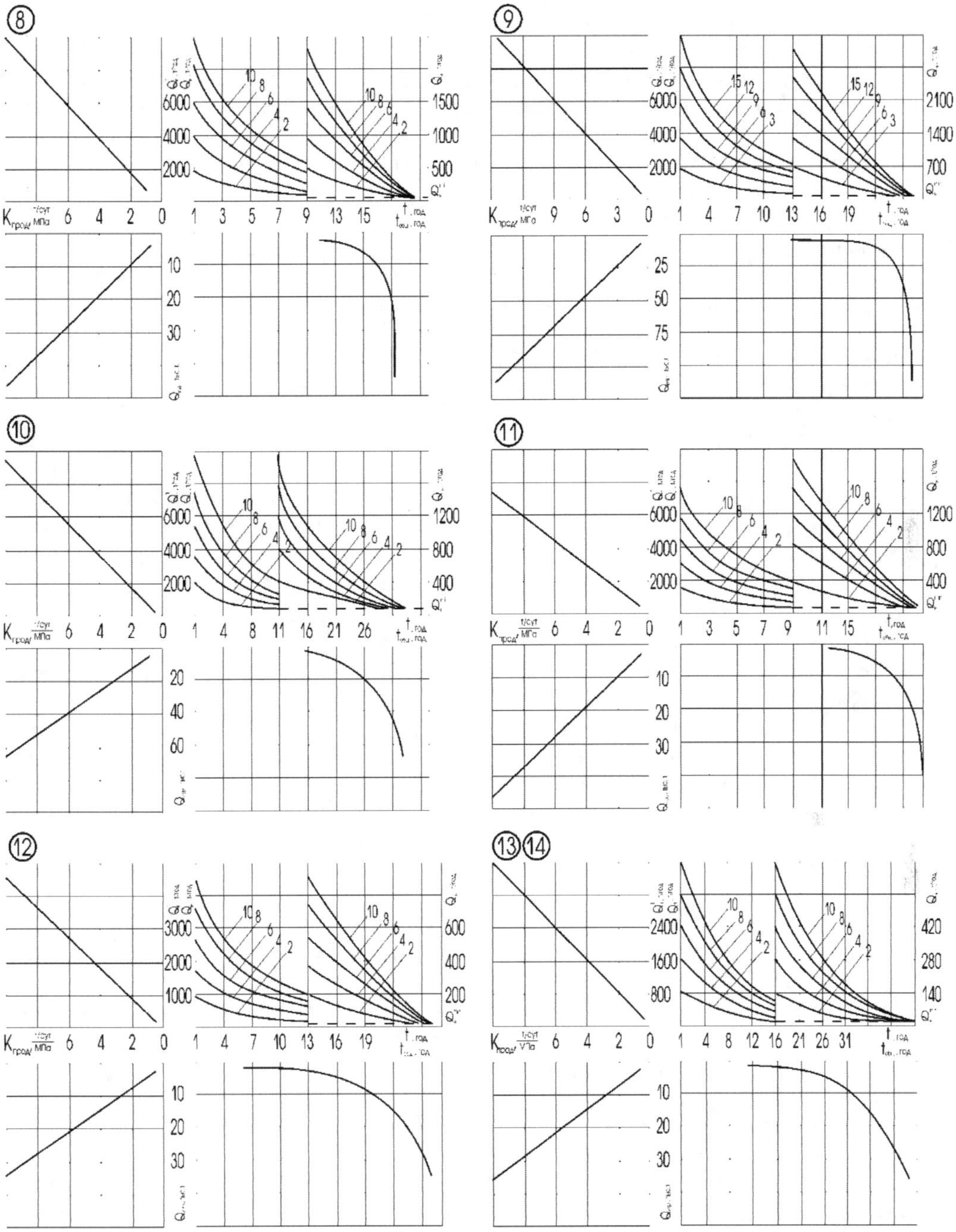

Fig 1 (continued). Nomograms for annual oil production, recoverable reserves and operation life: (10) – number of group of formations; Q_o^{min} – minimum profitable well production rate; I_{prod} – cipher of curves

4. Conclusion

The findings show that developing deposits in carbonate reservoirs requires a differentiated approach due to some specific features of oil recovery in these deposits.

The graphs allow the following solutions:

- based on the primary rating, to predict primary annual production rates (upper left graph), changes in the current annual oil production (upper right), recoverable reserves (lower left), operation life up to the economic profitability limit (lower right). The dependencies can be used for express calculations of technological indicators of small and medium-sized deposits during primary production when there is no representative geological production material;
- to evaluate the effectiveness of methods to influence the bottom-hole zone and the formation as a whole;
- to monitor the oil recovery;
- to find deposits and areas that will be economically feasible to drill (through the dependence of recoverable reserves and operation time on productivity).

References

[1] Mukhametshin V Sh 2003 *Differentiation and grouping of complex deposits in solving development problems* (Ufa: USPTU)

[2] Economides J M and Nolte K I 2000 *Reservoir stimulation* (West Sussex, England: John Wiley and Sons)

[3] Akhmetov R T, and Mukhametshin V V 2018 Estimation of displacement coefficient with due account for hydrophobization of reservoir using geophysical data of wells *IOP Conference Series: Earth and Environmental Science* **194** 062001 DOI: 10.1088/1755-1315/194/6/062001

[4] Zeigman Yu V, Mukhametshin V Sh, Khafizov A R, Kharina S B, Abutalipova E M, and Avrenyuk A N 2017 Peculiarities of selecting well-killng fluids composition for difficult conditions *Oil Industry* **1** 66–69

[5] Sun S Q and Wan J C 2002 Geological analogs usage rates high in global survey *Oil & Gas Journal* **100(46)** 49-50

[6] Batalov D A, Mukhametshin V V, Dubinskiy G S, and Andreev V E 2018 Laboratory grounding of waterproofing sealant based on acrylic polymers *IOP Conference Series: Earth and Environmental Science* **194** 042003 DOI: 10.1088/1755-1315/194/4/042003

[7] Yaskin S A, Mukhametshin V V, Andreev V E and Dubinskiy G S 2018 Forecasting the parameters of non-stationary water flooding of oil deposits *IOP Conference Series: Earth and Environmental Science* **194** 032037 1-5 DOI: 10.1088/1755-1315/194/6/062037

[8] Watson R W, and Boukadi F H 1990 Effect of surface area and pore-size distribution on oil recovery at breakthrough, on oil recovery at infinite water-oil ratio, on residual oil saturation and on wettability (Paper SCA 9007) *Proceedings of the Society of Core Analysts, Fourth Annual Technical Conference (Dallas, Texas, USA, 1990)*

[9] Mukhametshin V V 2017 The need for creation of a unified comprehensive method of geological and field analysis and integration of data on effective influence on the bottom-hole formation zone *Oil Industry* **4** 80–84 DOI: 10.24887/0028-2448-2017-4-80-84

[10] Yakupov R F, and Mukhametshin V Sh 2013 Problem of efficiency of low-productivity carbonate reservoir development on example of Turnaisian stage of Tuymazinskoye field *Oil Industry* **12** 106–110

[11] Mukhametshin V V and Andreev V E 2017 Search and argumentation of decisions aimed at increasing the efficiency of bottom-hole zone stimulation in oil accumulations with challenged reserves *SPE Russian Petroleum Technology Conference (16-18 October 2017, Moscow)* DOI: 10.2118/187785-MS

[12] Byrnes A P, and Bhattacharya S 2006 Influence of Initial and Residual Oil Saturation and Relative Permeability on Recovery From Transition Zone Reservoirs in Shallow-Shelf Carbonates *SPE/DOE Symposium on Improved Oil Recovery (Tulsa, Oklahoma, USA, 22-26 April 2006)* DOI: 10.2118/99736-MS

IOP Conf. Series: Earth and Environmental Science 1021 (2022) 012007 doi:10.1088/1755-1315/1021/1/012007

The dependence of the efficiency of hydraulic fracturing on the geological factors of the Tourne-Famene deposit at the Gagarinsky oilfield

V I Galkin, P O Chalova

Perm National Research Polytechnic University, Komsomolsky Prospekt 29, Perm, 614990, Russia

Email: chalovapolina@yandex.ru

Abstract. In this article, a factor analysis of well flow rate increase after hydraulic fracturing depending on facies conditions is carried out using the example of the Turne-Famenskoye reservoir of the Gagarinskoye field as an example. By constructing the equations of multiple regression, the influence of geological factors on the productivity coefficient and, consequently, on the additional oil production was determined. Based on the results of factor analysis and construction of statistical models, high significance of bottom hole and formation pressure values and their change depending on facial environment has been established. It is recommended to select candidate wells for hydraulic fracturing operations taking into account their facial affiliation, specific features of wells where hydraulic fracturing operations were carried out quite effectively, taking into account residual reserves density, water cut of well production, proximity to oil-bearing contour and to injection wells.

1. Introduction

During the period of 2010-2020, 53 geological and technical events took place at the investigated operational facility. 28 units of them were conducted using the hydraulic fracturing technology (53 % of the total number of geological and technical events) with a total additional oil production of 33.1 thousand tons (30 % of the total additional oil production). The specific additional oil production per one hydraulic fracturing operation is 1.2 thousand tons that corresponds to the minimum value in comparison with other types of geological and technical events that were carried out in the wells of the Gagarinsky oilfield for the same period of time.

It is worth noting that the conditions for the development of deposits become more complicated over time. The water content of well production is increasing, and reservoir and bottom-hole pressures are decreasing. This leads to an annual reduction in additional oil production due to hydraulic fracturing.

Due to the low efficiency of hydraulic fracturing, it is advisable to perform a factor analysis of the increase of the well flow rate for liquid and oil after geological and technical events and determine the influence of geological factors on the additional oil production.

IOP Conf. Series: Earth and Environmental Science 1021 (2022) 012007 doi:10.1088/1755-1315/1021/1/012007

2. The dependence of the efficiency of hydraulic fracturing on the geological factors of the Tourne-Famene deposit at the Gagarinsky oilfield

The Turne-Famen deposit of the Gagarin oilfield is characterized by a complex geological structure. Within the limits of the deposit, due to different conditions of sedimentation, there are zones with different filtration and reservoir properties.

The Gagarin oilfield has a typical structure for the reef massifs of the Upper Devonian-Tournaisian paleoshelf [1, 2]. According to the results of lithological-facies analysis, within the deposit, there are four facies zones that characterize the development of organogenic-carbonate structures: the biohermic core (reef ridge), the upper and lower part of the rear plume of the reef, and the slope of the reef (Figure 4).

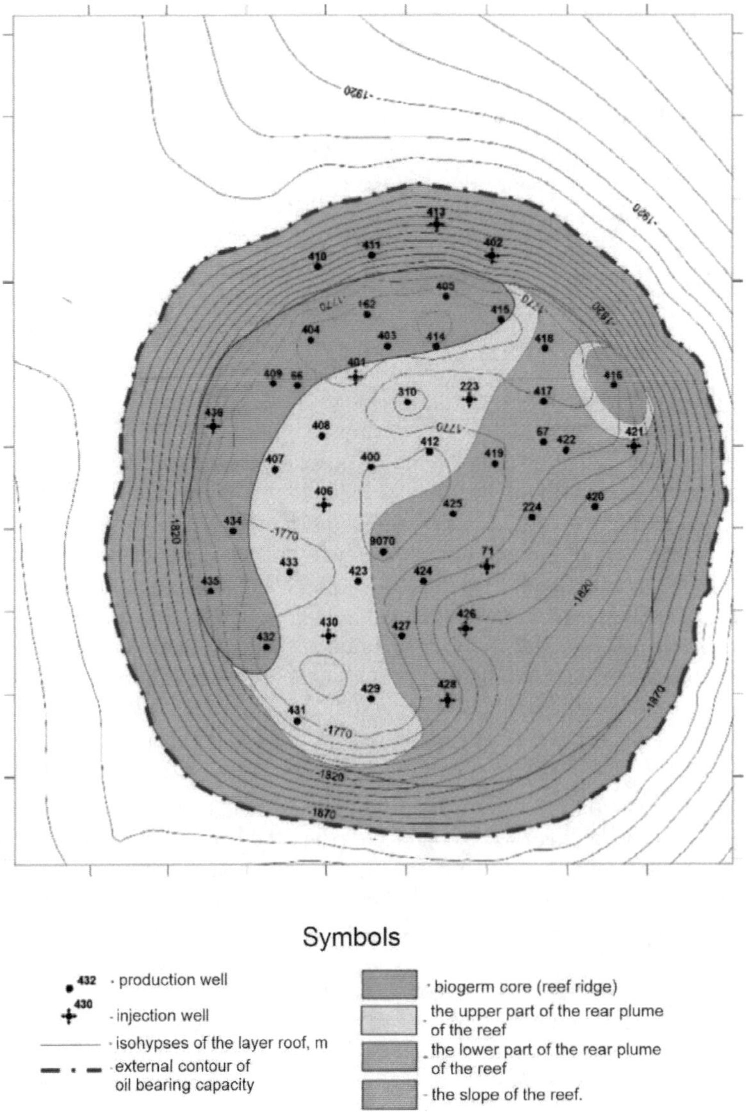

Figure 1. Schematic lithological-facial map of the Tourne-Famene productive sediments of the Gagarinskoye oilfield

IOP Conf. Series: Earth and Environmental Science 1021 (2022) 012007 doi:10.1088/1755-1315/1021/1/012007

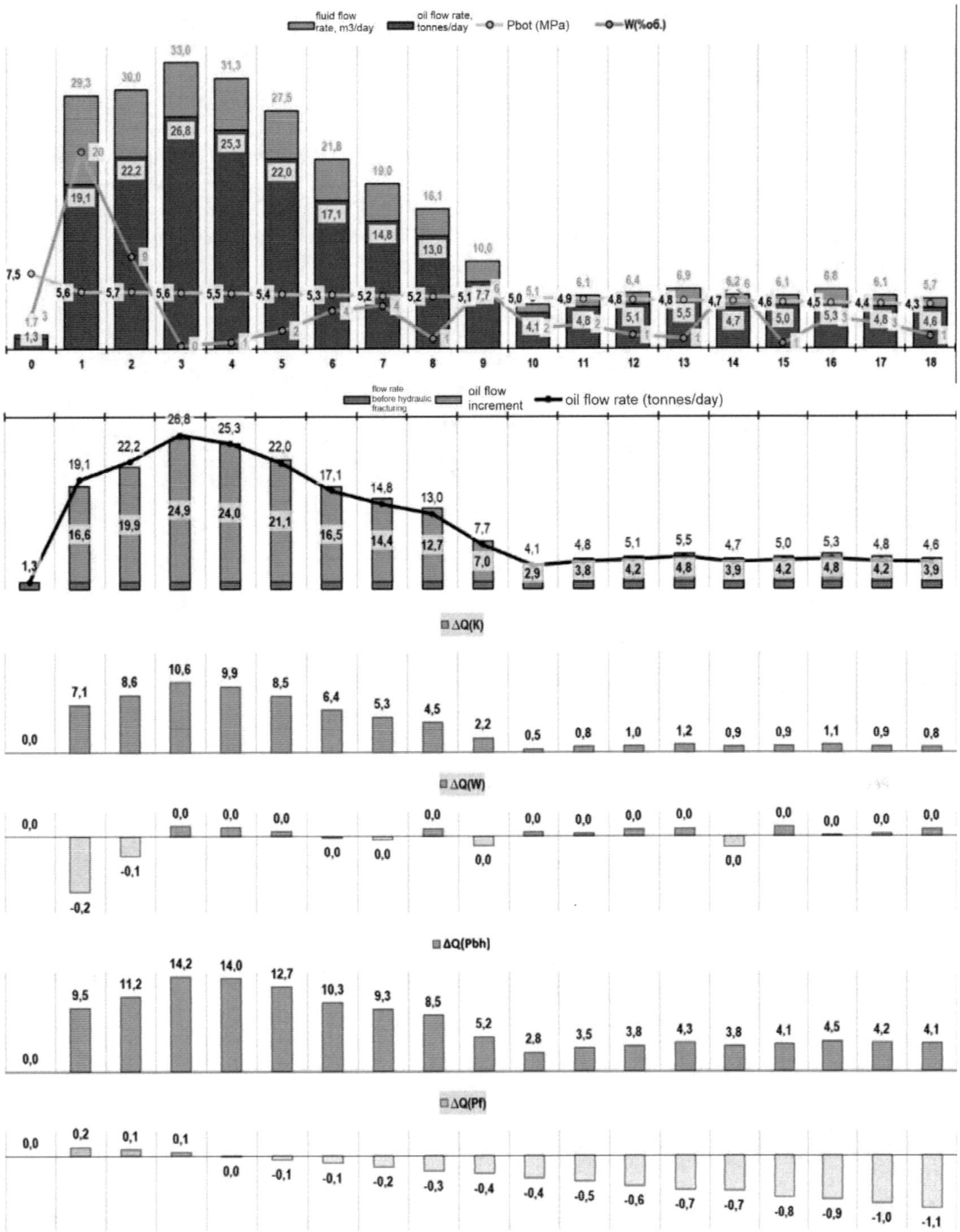

Figure 2. Graphical representation of the factor analysis of flow rate increase after hydraulic fracturing. Well No. 407. The Gagarinskoye oilfield

The filtration and reservoir properties of the deposit vary in a wide range, most of the reservoirs are low-porous and low-permeable [3]. Hydraulic fracturing technology improves well productivity by creating high-conductivity fractures in the target formation to allow the produced fluid to flow to the bottomhole. But, often, as a result of hydraulic fracturing, the flow rate increases several times in some wells, while in some wells the effect cannot be achieved.

Failure to achieve the planned indicators is caused by various factors, both geological and technological.

To assess the effectiveness of hydraulic fracturing, a factor analysis was performed to estimate the increase in oil flow rate from changes in the productivity factor (K), bottomhole pressure (Pbh), reservoir pressure (Pf) and water cut of the well product (W). The integral deterministic factor analysis method was used for this purpose. A graphic example of the result of factor analysis of the production rate increment of well No. 407 after well interventions is shown in Figure 2.

According to the results of factor analysis of hydraulic fracturing efficiency of well No. 407, the change of productivity coefficient provides less than 50 % of the total incremental oil production rate.

It is worth paying attention to the section from 3 to 10 months of well operation after hydraulic fracturing. In this section, there is a simultaneous decrease in flow rate increment as a result of both productivity factor and bottomhole pressure (Pbh) changes, while Pbh in this section decreases relative to the pressure prior to hydraulic fracturing, which should result in an increase in the oil flow rate accordingly. This behavior of the parameters in this method of analysis indicates the relationship between downhole pressure and production rate. Figure 3 shows the dependence of the productivity coefficient on bottomhole pressure.

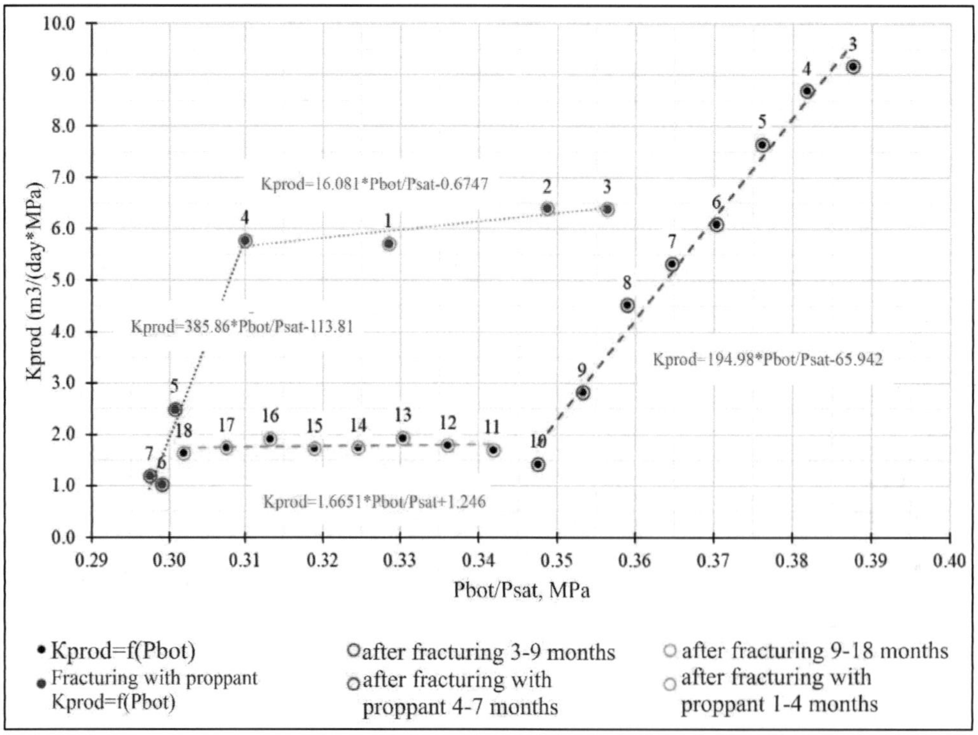

Figure 3. Dependence of the productivity coefficient on downhole pressure for hydraulic fracturing and fracturing with proppant fixation. Well No. 407. The Gagarinskoye oilfield

During the period from 3 to 10 months after fracturing, there is a more than a 6-fold reduction in the productivity factor. At the same time, the bottomhole pressure decreased by 0.6 MPa to 5 MPa. This decrease in the production rate is most likely due to fracture closure as a result of reduced downhole pressure. In addition, the effect of downhole pressure on the productivity coefficient has decreased, as indicated by the angular coefficient in the trend equation. Similar dependences of the productivity coefficient on downhole pressure were obtained for other facies. At the end of 2020, the second hydraulic fracturing with proppant fixation of the fracture was carried out in well 407. This method of fracturing reduced the influence of downhole pressure on the productivity coefficient and shifted the dependence to the left along the abscissa, so that the value of downhole pressure, at which fracture clamping occurs, decreased.

The factor analysis shows that, among all facies, the maximum value of the ratio of specific oil flow rate increases due to changes in productivity coefficient in the first month after hydraulic fracturing to the specific oil flow rate before hydraulic fracturing corresponds to the lower back reef facies and is 7.3 p.u. The minimum duration of the effect due to changes in the productivity coefficient and, consequently, the least value of additional oil production corresponds to this same facies. Biohermal core facies corresponds to minimum flow rate increment, but at the same time maximum additional production and effect time are achieved due to high values of its energy characteristics. It should also be noted that wells with ineffective hydraulic fracturing operations (low flow rate increase and effect time) are characterized by low bottomhole pressure (less than 5 MPa). In this case the fractures in the reservoir are closing and the productivity coefficient is decreasing. For those wells where hydraulic fracturing operations were successful (production rate increase of more than 1000 tons), the average bottomhole pressure value is 5.9 MPa.

In summary, the results of the factor analysis show:

- the presence of relationship of the productivity coefficient with bottomhole pressure and formation pressure;
- there is significant decrease of the productivity coefficient when bottomhole pressure is lower than 5 MPa;
- proppant fixing of the fracture helps to decrease the downhole pressure effect on the well productivity coefficient;
- fracturing is advisable in wells located within the biohermal core facies due to a better energy condition compared to other facies.

3. Assessment of hydraulic fracturing efficiency depending on facies

The flow rate of a fluid is directly related to the productivity factor and pressure differential. The productivity factor is a complex quantity that includes both fluid properties and reservoir properties, and it is influenced by a variety of factors. As noted earlier, pressure has a significant impact on the coefficient of productivity, with significant changes in the filtration pattern and fracture closure. In order to account for total influence of the studied parameters on well productivity after hydraulic fracturing operation, we build a probabilistic statistical model for Turne-Famen reservoir based on results of 28 hydraulic fracturing measures. The technique of developing individual probability-statistical models is given in [4, 5, 6].

Multivariate regression equations for the reef crest, upper and lower reef plume facies were derived from the values of geological and process parameters to estimate the post-fracture productivity factor. The reef slope facies was excluded from the analysis at this stage due to the small amount of accumulated fracture data (there was only one performed fracture operation).

The final multiple regression equations include the variables that have the greatest influence on the value of productivity coefficient (K) - effective oil-saturated thickness (Tef), relative values - the ratio of bottomhole and formation pressure to saturation pressure (Pbot/Psat, Pform/Psat). The equations are as follows.

The facies zone of the reef ridge:

$$K = 3.523 + 111.229 * (\frac{P_{bot}}{P_{sat}}) - 4.012 * (\frac{P_{form}}{P_{sat}}), \text{ (r=0.89)}.$$

The facies zone of the upper part of the reef rear plume:

$$К = -0.044 + 9.485 * (\frac{P_{bot}}{P_{sat}}) - 0.089 * T_{ef}, \text{ (r=0.83).}$$

The facies zone of the lower part of the reef rear plume:

$$К = -8.76 + 0.2887 * T_{ef} + 92.862 * \left(\frac{P_{bot}}{P_{sat}}\right) - 40.873 * (\frac{P_{form}}{P_{sat}}), \text{ (r=0.74).}$$

A stepwise regression analysis was used to construct the models. The sequence of parameters in the above equations corresponds to the order of their inclusion in the regression equation. The first included variable has the greatest impact on the productivity coefficient.

Correlation fields between the factual productivity coefficient (Kprod_f (m^3/(day*MPa)) and the model (Kprod_m (m^3/(day*MPa)) by facies are shown in Figure 4.

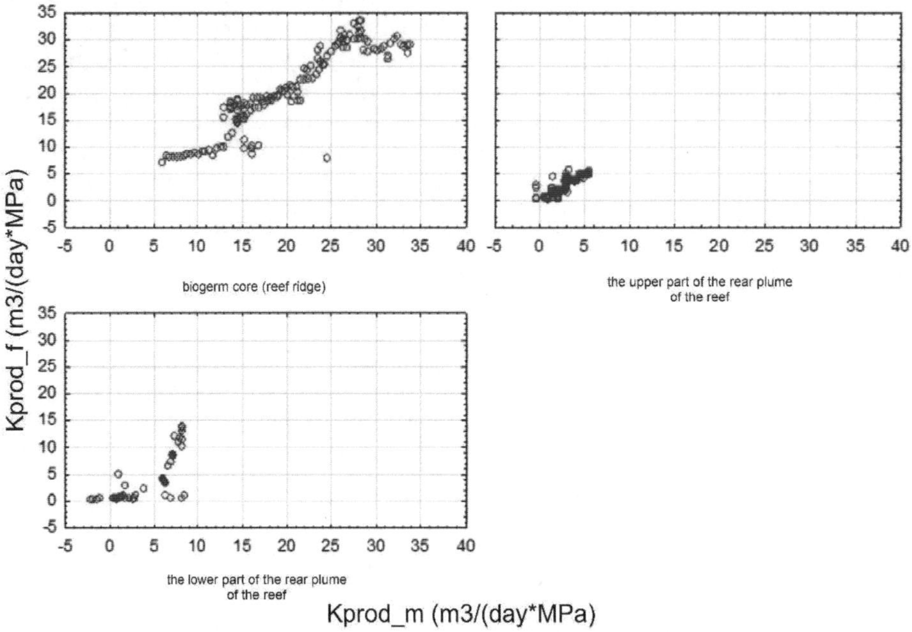

Figure 4. Correlation fields for factual and modelled productivity coefficients by facies. Gagarinskoye oilfield.

Figure 4 shows that for each facies there is quite a close correlation between factual productivity coefficient and model productivity coefficient. Ranges of productivity coefficient values for each facies are different.

The biohermal core facies has a greater variation than the other two facies. It is characterized by the highest productivity factor after hydraulic fracturing due to its high permeability properties and increased energy potential (high formation pressure, which allows one to operate well with high bottomhole pressure). The minimum scatter of productivity coefficient values is observed for the facies of the upper part of the reef rear plume. Productivity coefficient values for lower back reef facies are in the area of facies 1 and 3, overlap and are characterized by the smaller correlation coefficient (0.74). It should be also noted that in the area of zero values the regression model works worse. For all facies the values of the correlation coefficient are statistically significant.

Thus, these models show the high importance of considering downhole and reservoir pressures and their variation depending on the facies of the well when planning geological and technological measures.

4. Conclusion

Based on this analysis, it is recommended that candidate wells for fracturing operations should be selected based on their facies, the characteristics of wells where fracturing operations have been performed efficiently, residual reserve density, water cut of well production, proximity to the oil-bearing zone and to injection wells.

References

[1] Krivoshchekov S N, Kozlova I A 2012 Geodynamic characteristics of the conditions of french-famennian strata dispersed organic matter immersion and catagenesis in perm region *Neftyanoe Khozyaystvo - Oil Industry* **7** 82-85

[2] Krivoshchekov S N, Kozlova I A, Sannikov I V 2014 Estimate of the petroleum potential of the western solikamsk depression based on geochemical and geodynamic data *Neftyanoe Khozyaystvo - Oil Industry.* **6** 12-15

[3] Mengaliev A G, Martyushev D A 2020 Accounting the parameter of anisotropy of permeability in geological and hydrodynamic models of carbonate objects (on the example of the Gagarin deposit) *Proceedings of Tomsk Polytechnic University. Engineering of Georesources* **331(5)** 7-17

[4] Mengaliev A G, Martyushev D A 2020 Assessment of technological and economic efficiency of fracturing formations using geological and hydrodynamic models, considering features of carbonate reservoir structure *Proceedings of Tomsk Polytechnic University. Engineering of Georesources* **331(7)** 37-48

[5] Galkin V I, Rastegaev A V, Kozlova I A and Krivoschekov S N 2014 Probability evaluation of a formation hydraulic fracturing based on mining-geological features (deposits of Kogalym region are taken as an example) *Geology, Geophysics and Development of Oil and Gas Fields* **7** 100-104

[6] Galkin V I, Ponomareva I N and Koltyrin A N 2018 Development of probabilistic and statistical models for evaluation of the effectiveness of proppant hydraulic fracturing (by the example of the tl-bb reservoir of the Batyrbayskoe field) *Geology. Oil and gas and mining* **17(1)** 37-49

Differentiation of exploration and production wells to prevent the breakthrough of water injected into the reservoir

V V Mukhametshin[1], L S Kuleshova[2]

[1] Ufa State Petroleum Technological University, 1, Kosmonavtov st., Ufa, Republic of Bashkortostan, 450062, Russian Federation
[2] Ufa State Petroleum Technological University, Branch of the University in the City of Oktyabrsky, 54a, Devonskaya St., Oktyabrsky, Republic of Bashkortostan, 452607, Russian Federation

E-mail: vv@of.ugntu.ru

Abstract. The equations of canonical discriminant functions and distributions were derived for deposits of the Tournaisian age of the Volga-Ural oil and gas province. This enables solution to the problems of increasing the efficiency of implementation of various contour waterflood systems, choosing exploration and production wells for focal waterflood, and estimation of the well grid density at the stage of preparation of the design documents, pressure and injection volumes, as well as making scientifically grounded management decisions for regulating the exploitation process. The method has been proposed to prevent a catastrophic breakthrough of water injected into the reservoir, which is the reason for a significant decrease in the degree of oil reserves recovery.

1. Introduction

The development of deposits in carbonate reservoirs shows that in some cases, in some areas, catastrophic water breakthrough into reservoirs occurs through fracture systems and individual high-permeability interlayers, which does not always have a favorable effect on technical and economic indicators [1–8]. The vast majority of these cases are due to high discharge pressures. At injection pressure comparable to vertical mining (P_m), fractures open up in the reservoir and water breaks through individual highly permeable fractures into the reservoir and quickly floods it, which is undesirable.

At the same time, at injection pressures below 0.9 P_m, water breakthrough occurs in individual high-permeability interlayers, which significantly reduces the sweep efficiency in thickness and reduces the degree of reserves recovery. For example, in the Sakmar-Assel-Artinsky limestones of the Ishimbay group of fields, these facts did not allow maintaining reservoir pressure and oil displacement by water. Similar rapid water breakthrough took place in the analyzed Tournaisian fields of the Volga-Ural oil and gas province. To prevent these undesirable phenomena, the zones of deposits with the highest probability of rapid water breakthrough in individual layers should be identified. This will enable making necessary control decisions in advance in order to prevent a decrease in the degree of reserves recovery as a result of rapid watering of reservoirs [9–11].

2. Materials and methods

Since water breakthrough events are most often observed in cases of very good hydrodynamic interaction of production and injection wells, the analyzed wells of the group where the values of the cross-correlation function (*R*) are greater than or equal to 0.5[12] were divided into two more groups: 1 – highly responding to injection ($R > 0.7$) and 2 – weakly responding to injection ($0.5 \le R \le 0.7$). Group 3 includes wells not responding to injection ($R < 0.5$) that remained unchanged.

Calculations were carried out using the method of canonical discriminant functions for six options using a wide scope of geological information.

3. Results and Discussion

The representation of wells in the axes of the first two canonical discriminant functions established the boundaries separating the objects according to the values of the cross-correlation functions (Figure 1). The obtained distributions can be used for a qualitative assessment of the values of the cross-correlation functions. Based on the values of the geological and technological parameters of the production and injection wells, it is necessary to calculate the values of the canonical discriminant functions (1)–(24) and determine the group to which the object is assigned.

Comparison of the results of falling objects into two [12] and three groups (Table 1) shows that in the second case, the accuracy of assigning wells to a particular group is lower than that in the first case. As can be seen, the number of wells in the zone of uncertainty on average for all six options (parameter Z_1) is 2.1 fold greater in the second case compared to the first one, and the relative value of the area of the zone of uncertainty (parameter Z_2) is 1.4 fold greater in the second case compared to the first one. In this case, the projection accuracy is reduced, but deeper differentiation of wells can be performed. Analysis reveals no need to use the entire set of the considered parameters. The parameters that are significant according to the Kullback criterion (Table 1) are reasonably sufficient, and the results for options 2, 4 and 6 should be applied, although the use of option 6 with no regard to the peculiarities of the geological structure of the deposits can lead to significant errors.

Table 1. Relative values of the areas of uncertainty zones and the number of wells in them incorrectly grouped in the axes of the canonical discriminant functions y_1' and y_2' according to options for objects divided into 2 and 3 groups

Number of groups	Parameter	Option					
		1	2	3	4	5	6
3	Z_1,%	7	7	9	9	18	18
	Z_2,%	10	10	22	22	30	36
2	Z_1,%	4	4	4	4	7	10
	Z_2,%	8	10	9	11	25	30

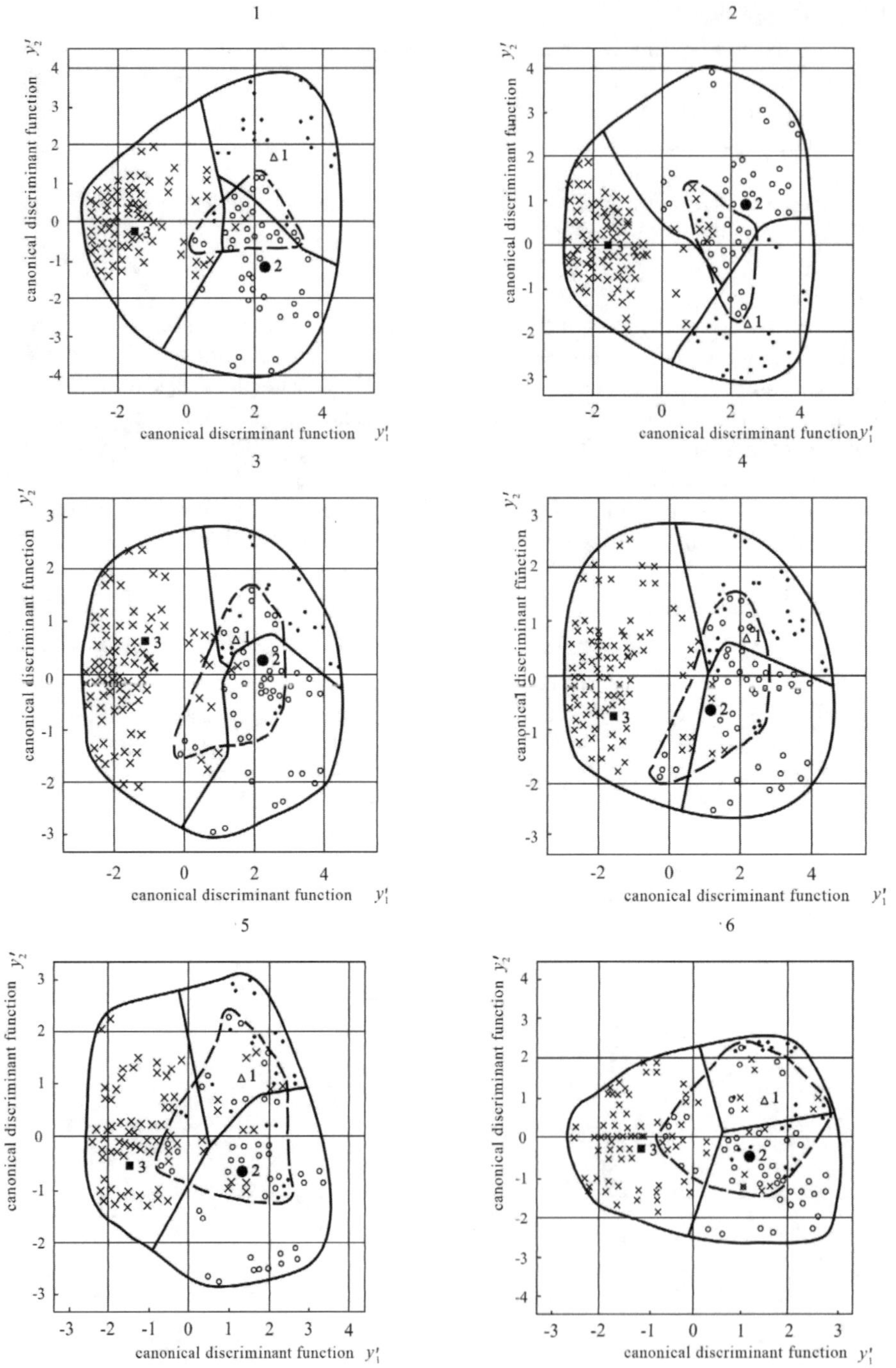

Δ1, ●2, ■3 – centroid and number of the well group that reacted (R > 0.7), slightly reacted (0.7 ≤ R ≤ 0.5), and did not react (R <0.5) to water injection, respectively; ○, ●, × – wells of groups 1, 2, 3, respectively; – well group boundary; ○, – the concentration zone of all wells and the zone of uncertainty, respectively

Figure 1. Distribution of wells in the axes of canonical discriminant functions

IOP Conf. Series: Earth and Environmental Science 1021 (2022) 012008 doi:10.1088/1755-1315/1021/1/012008

The equations of the canonical discriminant functions are as follows:

- option 1:

$$\begin{aligned} y_1 = {} & -0{,}04F - 0{,}476P_i / P_m + 0{,}002Q_i - 0{,}71t - 0{,}13Q^P_{O\max} - 0{,}20\mu_0 - \\ & -0{,}03\rho_H + 0{,}31K_{heter} + 0{,}76K_{prod} + 2{,}71T^I_e + 1{,}40T^I_i + 0{,}52n^i - \\ & -0{,}53M^i_r + 0{,}39T^p_e + 0{,}88T^p_i - 0{,}13n^p + 0{,}07M^p_r - 3{,}25K^i_r - \\ & -1{,}47T^i_{total} + 0{,}19K^p_r + 0{,}14T^p_{total} - 0{,}17Q^p_o + 0{,}37Q^p_o + 0{,}33Q^p_w - 0{,}30f_1, \end{aligned} \quad (1)$$

$$\begin{aligned} y_2 = {} & -0{,}03F - 0{,}57P_i / P_m + 0{,}004Q_i - 2{,}32t + 1{,}55Q^p_{o\max} + 1{,}66\mu_0 - \\ & -0{,}79\rho_o + 0{,}07K_{heter} - 0{,}31K_{prod} + 0{,}30T^i_e + 2{,}61T^i_i + 0{,}68n^i - \\ & -0{,}03M^i_r - 0{,}07T^p_e + 0{,}42T^i_e + 0{,}39n^p + 0{,}18M^p_r - 3{,}03K^i_r - \\ & -0{,}86H^i_{total} - 0{,}04K^p_r - 0{,}07T^p_{total} - 0{,}42Q^p_o + 0{,}58Q^p_{cum} + 0{,}23Q^p_w + 0{,}26f_1, \end{aligned} \quad (2)$$

$$\begin{aligned} y'_1 = {} & 13{,}6 + 0{,}0001F - 5{,}25P_i / P_m + 0{,}0001Q_i - 0{,}008t - 0{,}001Q^p_{o\max} - \\ & -0{,}10\mu_0 - 0{,}007\rho_o + 0{,}05K_{heter} + 1{,}07K_{prod} + 0{,}86T^i_e + 1{,}28T^i_i + 0{,}35n^i - \\ & -0{,}40M^i_r + 0{,}13T^p_e + 0{,}85T^p_i - 0{,}08n^p + 0{,}05M^p_r - 22{,}4K^i_r - 0{,}27T^i_{total} + \\ & +1{,}27K^p_i + 0{,}02T^p_{total} - 0{,}001Q^p_o + 0{,}0001Q^p_{cum} + 0{,}003Q^p_w - 1{,}24f_1, \end{aligned} \quad (3)$$

$$\begin{aligned} y'_2 = {} & 180 + 0{,}0001F - 6{,}63P_i / P_m + 0{,}0001Q_i - 0{,}03t + 0{,}01Q^p_{o\max} + 0{,}77\mu_0 - \\ & -0{,}21\rho_o + 0{,}01K_{heter} - 0{,}44K_{prod} + 0{,}10T^i_e + 2{,}36T^i_i + 0{,}46n^i - \\ & -0{,}02M^i_r - 0{,}02T^p_e + 0{,}40T^p_i + 0{,}24n^p + 0{,}13M^p_r - 20{,}9K^i_r - \\ & -0{,}16T^i_{total} - 0{,}24K^p_r - 0{,}01T^p_{total} - 0{,}004Q^p_o + 0{,}0001Q^p_{cum} + 0{,}002Q^p_w + 1{,}13f_1; \end{aligned} \quad (4)$$

- option 2:

$$\begin{aligned} y_1 = {} & -0{,}98T_i + 1{,}71t_s + 0{,}72K_{prod} + 2{,}78T^i_e + 0{,}72T^i_i - 0{,}39M^i_r + \\ & +0{,}34T^p_e + 0{,}83T^p_i + 0{,}06M^p_r - 2{,}62K^i_r + 0{,}25K^p_r - 0{,}02F - \\ & -0{,}27P_i / P_m - 0{,}04Q_i - 1{,}14T^i_{total} - 0{,}13t - 0{,}09Q^p_{o\max}, \end{aligned} \quad (5)$$

$$\begin{aligned} y_2 = {} & 3{,}38T_i - 2{,}57t_s + 0{,}28K_{prod} - 0{,}24T^i_e - 1{,}47T^i_i - 0{,}07M^i_r - \\ & -0{,}45T^p_e + 0{,}06T^p_t - 0{,}37M^p_e + 1{,}96K^i_r + 0{,}10K^p_r - 0{,}02F + \\ & +0{,}49P_i / P_m - 0{,}13Q_i + 0{,}24T^i_{total} + 1{,}19t - 1{,}18Q^p_{o\max}, \end{aligned} \quad (6)$$

$$\begin{aligned} y'_1 = {} & -190 - 0{,}01T_i + 0{,}70t_s + 1{,}0K_{prod} + 0{,}88T^i_e + 0{,}65T^i_i - 0{,}30M^i_r + \\ & +0{,}11T^p_e + 0{,}81T^p_e + 0{,}04M^p_r - 18{,}0K^i_r + 1{,}64K^p_r + 0{,}0001F - \\ & -3{,}11P_i / P_m + 0{,}0001Q_i - 0{,}21T^i_{total} - 0{,}001t - 0{,}001Q^p_{o\max}, \end{aligned} \quad (7)$$

$$\begin{aligned} y'_2 = {} & 265 + 0{,}03T_i - 1{,}06t_s + 0{,}40K_{prod} - 0{,}08T^i_e - 1{,}34T^i_i - 0{,}05M^i_r - \\ & -0{,}14T^p_e + 0{,}06T^p_i - 0{,}27M^p_r + 13{,}5K^i_r + 0{,}65K^p_r + 0{,}0001F + \\ & +5{,}7P_i / P_m + 0{,}0001Q_i + 0{,}05T^i_{total} + 0{,}01t - 0{,}01Q^p_{o\max}; \end{aligned} \quad (8)$$

- option 3:

$$y_1 = -0{,}08F - 0{,}47P_i / P_m - 0{,}01Q_i - 0{,}50\mu_0 + 0{,}04\rho_o + 0{,}24K_{heter} + \\ +0{,}76K_{prod} + 2{,}18T_e^i + 0{,}94T_i^i + 0{,}37n^i - 0{,}39M_r^i + 0{,}63T_e^p + \\ +0{,}63T_i^p - 0{,}19n^p + 0{,}04M_r^p - 2{,}14K_r^i - 1{,}08T_{total}^i + 0{,}03K_r^p + 0{,}11T_{total}^p, \quad (9)$$

$$y_2 = 0{,}44F - 0{,}61P_i / P_m - 0{,}06Q_i + 0{,}62\mu_0 - 0{,}63\rho_o + 0{,}19K_{heter} - \\ -0{,}43K_{prod} - 1{,}78T_e^i + 1{,}27T_i^i + 0{,}27n^i + 0{,}62M_r^i + 0{,}31T_e^p - \\ -0{,}08T_i^p - 0{,}45n^p + 0{,}84M_r^p + 0{,}44K_r^i + 0{,}74T_{total}^i + 0{,}32K_r^p + 0{,}38T_{total}^p, \quad (10)$$

$$y_1' = -2{,}70 + 0{,}0001F - 5{,}44P_i / P_m + 0{,}0001Q_i - 0{,}23\mu_0 + 0{,}01\rho_o + \\ +0{,}04K_{heter} + 1{,}06K_{prod} + 0{,}69T_e^i + 0{,}85T_i^i + 0{,}25n^i - 0{,}30M_r^i + 0{,}20T_o^p + \\ +0{,}61T_i^p - 0{,}12n^p + 0{,}03M_r^p - 14{,}7K_r^i - 0{,}20T_{total}^i + 0{,}17K_r^p + 0{,}02T_{total}^p, \quad (11)$$

$$y_2' = 130 + 0{,}002F - 7{,}01P_i / P_m + 0{,}0001Q_i + 0{,}29\mu_0 - 0{,}17\rho_o + 0{,}03K_{heter} - \\ -0{,}60K_{prod} - 0{,}56T_e^i + 1{,}15T_i^i + 0{,}18n^i + 0{,}47M_r^i + 0{,}10T_e^p - \\ -0{,}07T_i^p - 0{,}28n^p + 0{,}61M_r^p + 3{,}03K_r^i + 0{,}14T_{total}^i + 2{,}09K_r^p + 0{,}07T_{total}^p, \quad (12)$$

- option 4:

$$y_1 = -0{,}97T_i + 1{,}71t_s + 0{,}71K_{prod} + 2{,}54T_e^i + 0{,}63T_i^i - 0{,}35M_r^i + \\ +0{,}41T_e^p + 0{,}75T_i^p + 0{,}04M_r^p - 2{,}21K_r^i + 0{,}13K_r^p - 0{,}06F - \\ -0{,}30P_i / P_m - 0{,}02Q_i - 1{,}02T_{total}^i, \quad (13)$$

$$y_2 = -3{,}84T_i + 3{,}90t_s - 0{,}49K_{prod} - 1{,}72T_e^i + 1{,}13T_i^i + 0{,}73M_r^i + \\ +0{,}15T_r^p + 0{,}27T_i^p + 0{,}80M_r^p + 0{,}54K_r^i + 0{,}30K_r^p + 0{,}44F - \\ -0{,}47P_i / P_m - 0{,}12Q_i + 0{,}94T_{total}^i, \quad (14)$$

$$y_1' = -191 - 0{,}01T_i + 0{,}70ts + 1{,}00K_{prod} + 0{,}80T_e^i + 0{,}57T_i^i - 0{,}26M_r^i + \\ +0{,}13T_e^p + 0{,}73T_i^p + 0{,}03M_r^p - 15{,}2K_r^i + 0{,}87K_r^p + 0{,}0001F - \\ -3{,}50P_i / P_m + 0{,}0001Q_i - 0{,}19T_{total}^i, \quad (15)$$

$$y_2' = -436 - 0{,}04T_i + 1{,}60t_s - 0{,}68K_{prod} - 0{,}54T_e^i + 1{,}03T_i^i + 0{,}56M_r^i + \\ +0{,}05T_e^p + 0{,}26T_i^p + 0{,}58M_r^p + 3{,}70K_r^i + 1{,}96K_r^p + 0{,}002F - \\ -5{,}43P_i / P_m + 0{,}0001Q_i + 0{,}17T_{total}^i; \quad (16)$$

- option 5:

$$y_1 = 0{,}93P_i / P_m + 0{,}20Q_o^p + 0{,}12Q_{cum} + 0{,}20Q_w^p - 0{,}30f_1, +0{,}15Q_i + 0{,}15t - \\ -0{,}34Q_{o\max}^p - 0{,}18F, \quad (17)$$

$$y_2 = 0{,}13P_i / P_m - 0{,}11Q_o^p + 0{,}53Q_{cum} - 0{,}04Q_w^p + 0{,}46f_1, +0{,}01Q_i - 1{,}11t + \\ +1{,}22Q_{o\max}^p + 0{,}06F, \quad (18)$$

$$y_1' = -4{,}80 + 10{,}8P_i / P_m + 0{,}002Q_o^p + 0{,}0001Q_{cum} + 0{,}002Q_w^p - 1{,}31f_1, + \\ +0{,}0001Q_i + 0{,}002t - 0{,}002Q_{o\max}^p - 0{,}001F, \quad (19)$$

$$y_2' = 2,99+1,47P_i/P_m-0,001Q_o^p+0,0001Q_{cum}+0,0001Q_w^p+2,05f_1,+$$
$$+0,0001Q_i-0,01t+0,007Q_{o\max}^p+0,0001F; \quad (20)$$

- option 6:

$$y_1 = 0,95P_i/P_m+0,07Q_i+0,04t+0,008Q_{o\max}^p-0,19F, \quad (21)$$
$$y_2 = 0,01P_i/P_m+0,17Q_i-0,66t+1,20Q_{o\max}^p+0,004F, \quad (22)$$
$$y_1' = -5,04+11,0P_i/P_m+0,0001Q_i+0,0001t+0,0001Q_{o\max}^p-0,001F, \quad (23)$$
$$y_2' = -2,28+0,14P_i/P_m+0,0001Q_i-0,007t+0,007Q_{o\max}^p+0,0001F. \quad (24)$$

The interpretation of the parameters is provided in [13].

In equations (1), (2), (5), (6), (9), (10), (13), (14), (17), (18), (21), and (22), the coefficients of canonical discriminant functions are normalized. Table 2 presents the values of the functions in the centroids of the chosen groups.

Table 2. Values of discriminant functions in the group centroids

Option	Function	Well group number		
		1	2	3
1	y_1'	2.58	2.13	–1.60
	y_2'	1.64	–1.00	0.05
2	y_1'	5.18	1.41	–1.01
	y_2'	1.09	–0.62	0.01
3	y_1'	2.45	2.06	–1.54
	y_2'	0.89	–0.53	0.02
4	y_1'	2.38	2.04	–1.51
	y_2'	0.87	–0.51	0.02
5	y_1'	2.34	2.10	–1.52
	y_2'	–1.47	0.85	–0.02
6	y_1'	1.51	1.41	–1.01
	y_2'	1.09	–0.62	0.01

4. Conclusion

Thus, the equations of canonical discriminant functions and distributions were derived for deposits in carbonate reservoirs of the Tournaisian age of the Volga-Ural oil and gas province. This enables solution to the problems of increasing the efficiency of implementation of various contour waterflood systems, choosing wells for focal waterflood, and estimation of the grid density at the stage of preparation of the design documents, pressure and injection volumes, as well as taking scientifically grounded management decisions in regulating the exploitation process.

The method has been proposed to prevent a catastrophic breakthrough of water injected into the reservoir, which is the reason for a significant decrease in the degree of oil reserves recovery.

References

[1] Batalov S A, Andreev V E, Lobankov V M and Mukhametshin V Sh 2019 Numerical simulation of oil formation with regulated disturbances. Oil recovery quality simulation *Journal of Physics: Conference Series* **1333** 032006 DOI: 10.1088/1742-6596/1333/3/032006

[2] Shuchart C E and Buster D C 1995 Determination of the Chemistry of HF Acidizing with the Use of F NMR Spectroscopy *SPE International Symposium on Oilfield Chemistry (San Antonio, Texas, 14–17 February 1995)* 12 p DOI: 10.2118/28975-MS
[3] Batalov S A, Andreev V E, Lobankov V M and Mukhametshin V Sh 2019 Numerical simulation of the oil reservoir with regulated disturbances. Oil recovery stability simulation *Journal of Physics: Conference Series* **1333** 032007 DOI:10.1088/1742-6596/1333/3/032007
[4] Andreev V E, Chizhov A P, Chibisov A V and Mukhametshin V Sh 2019 Forecasting the use of enhanced oil recovery methods in oilfields of Bashkortostan *IOP Conference Series: Earth and Environmental Science* **350** 012025 DOI: 10.1088/1755-1315/350/1/012025
[5] Polyakov V N, Chizhov A P, Kotenev Yu A and Mukhametshin V Sh 2019 Results of System Drilling Techniques and Completion of Oil and Gas Wells *IOP Conference Series: Earth and Environmental Science* **378(1)** 012119 DOI: 10.1088/1755-1315/378/1/012119
[6] Brian C and Metcalf S 2009 Intensification of oil production at the San Andreas site thanks to the use of weak acid *Oil and Gas technologies* **12** 21–24
[7] Malyarenko A M, Bogdan V A, Kotenev Yu A, Mukhametshin V Sh, Umetbaev V G 2019 Wettability and formation conditions of reservoirs *IOP Conference Series: Earth and Environmental Science* **378(1)** 012040 DOI: 10.1088/1755-1315/378/1/012040
[8] Khokhlov V I, Galimov Sh S, Devyatkova S G, Kotenev Yu A, Sultanov Sh Kh and Mukhametshin V Sh 2019 Justification of impact and planning of technology efficiency on the basis of limy-emulsion formulation in low-permeability highly-rugged reservoirs of Tyumen deposits *IOP Conference Series: Earth and Environmental Science***378(1)** 012114 DOI: 10.1088/1755-1315/378/1/012114
[9] Lungwitz B, Fredd C, Brady M, Miller M, Ali S, and Hughes K 2004 Diversion and Cleanup Studies of Viscoelastic Surfactant-Based Self-Diverting Acid *SPE International Symposium and Exhibition on Formation Damage Control (Lafayette, 18–20 February 2004)* 10 p DOI: 10.2118/86504-MS
[10] Valeev A S, Kotenev Yu A, and Mukhametshin V Sh 2018 Evaluation of Water-Alternating-Gas Efficiency when Using Wide Range of Gas Composition *IOP Conference Series: Earth and Environmental Science* **194** 082042 DOI: 10.1088/1755-1315/194/8/082042
[11] Nasr-El-Din H A, Van Domelen M S, Sierra L, and Welton Th D 2007 Optimization of Surfactant-based Fluids for Acid Diversion *European Formation Damage Conference (Scheveningen, The Netherlands, 30 May-1 June 2007)* 11 p DOI: 10.2118/107687-MS
[12] Batalov D A, Mukhametshin V V, Dubinskiy G S, and Andreev V E 2018 Laboratory grounding of waterproofing sealant based on acrylic polymers *IOP Conference Series: Earth and Environmental Science* **194(4)** 042003 DOI: 10.1088/1755-1315/194/4/042003
[13] Zeigman Yu V and Mukhametshin V V 2009 Flooding systems correspondence to formation geological structure features *Oilfield Engineering* **5** 10–12

IOP Conf. Series: Earth and Environmental Science 1021 (2022) 012009 doi:10.1088/1755-1315/1021/1/012009

The mathematical layer structure of the digital twin of hazardous production facilities at a gas transmission enterprise

A E Kirilov[1, 2], **K A Chernyi**[1]

[1] Perm National Research Polytechnic University, 29, Komsomolsky ave., Perm, 614990, Russia
[2] JSC "Gazprom transgaz Tchaikovsky", 30, Primorsky blvd., Tchaikovsky, Perm Region, 617760, Russia

E-mail: cever78@yandex.ru

Abstract. The article presents the structure of the mathematical layer for digital twins (DT) of hazardous production facilities (HPF) operated by gas transportation enterprises. Mathematical models of technical device destructions, models to control parameters of technological processes are presented, the types of affecting factors which can influence the personnel of HPF are described, as well as a safety assessment model for the evacuation of personnel in the event of an accident or a fire. The presented structure of DT and models make it possible to adequately describe the features of HPF and optimise operational safety. The areas of application of digital twins in the production activities of a gas pipeline enterprise have been determined. To determine the most possible scenarios of accidents at HPF the presented models have been tested and implemented in the operational activities of JSC "Gazprom transgaz Tchaikovsky".

1. Introduction

Digital twins have become a core component of Industry 4.0, they enable hazardous enterprises, included pipeline industry, to effectively understand their data, optimise complex processes and facilitate operational activities [1–9].

The concept of realization of the «digital twins» (DT) technologies within the single information space of the Gazprom Group defines the structure of DT, which includes the following components: reference data layer, the electronic design and estimate, engineering and executive documentation layer, graphical data representation layer, engineering data layer, mathematical model layer, and «on-line» data layer. Each of these components has specific functionality to serve the specific needs of industrial analytics. The mathematical model layer is one of the key layers, as it is used for equipment technical state forecast, forecasting, and technological processes management. If the digital twin is a hazardous production facilities (HPF), then a layer of mathematical models can forecast such technogenic events as an accident, a fire, failures, evacuation of HPF personnel, and other incidents. Certainly the basic concepts defined by the safety requirements should form the basis for the structure of the DT mathematical model layer of HPF. These include an accident, an incident, HPF technical devices, occupational health and safety management system, HPF personnel training, etc.

To make it possible to adequately describe the features of HPF when creating DT the authors proposed a structure and presented models that make it possible to adequately describe the features of HPF when creating DT.

2. Methods

The authors presume that the structure of the mathematical model layer is formed according to this principle, which includes the following components:

- models of technical devices destruction used on HPF;
- models of accidents affecting factors and their consequences;
- model of personnel evacuation in case of HPF accidents.

The structure of the DT mathematical model layer of HPF is shown in Figure 1. The main element of the layer is the function that describes the state of HPF:

$$C_n = f\left(b_i, b_{ERi}, Q_{Di}, N_{HSi}, Q_P, Q_E, \tau_{kr}, \mathrm{Pr}_i\right), \quad (1)$$

where C_n is HPF state, $n=\overline{1...3}$; b_i – safety index of i-th parameter, which characterizes the process carried out on HPF; b_{ERi} is the safety index of i-th escape route; Q_{Di} is the destruction probability of the i-th technical device in the HPF. N_{HSi} is the number of hazardous substances from the i-th technical device or facility as a result of an accident or an incident, kg. Q_P is the probability of personnel violating labour and technological discipline (we assumed that the criterion by which the level of work and technological discipline is assessed is the probability of errors by the employee or personnel in the operation of the HPF). Q_E is the probability of explosion at HPF; τ_{kr} is the critical duration of a fire, min; Pr_i is the value of probit function of probabilistic injury of personnel, buildings, equipment by a i-th damaging factor in case of HPF accident. In (1) $i=\overline{1...N}$, where N – number of measured parameters.

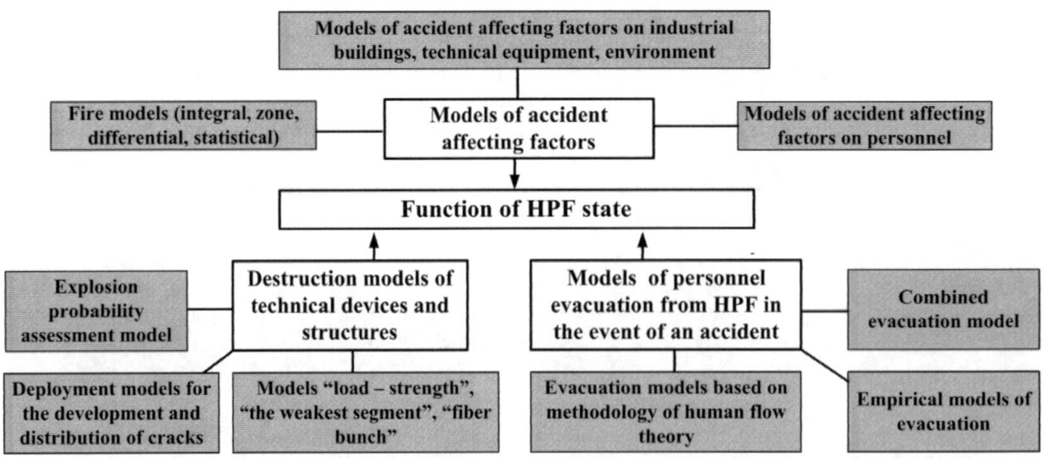

Figure 1. The structure of the DT mathematical model layer of HPF.

We suppose that conditionally HPF can be in three states: C_1 is an HPF safety state, C_2 is an HPF hazard state and C_3 is an accident (incident) state. The C_1 state is characterized by: 1) the technological mode of HPF operation without exceeding the maximum allowable values; 2) the low probability of personnel error in the HPF operation; 3) the absence of fire and accident affecting factors; 4) the possibility of the timely evacuation of personnel.

The HPF can be transferred into C_2 only if: 1) the HPF technology mode parameters exceed the allowable values; 2) the probability of HPF personnel error increases.

IOP Conf. Series: Earth and Environmental Science 1021 (2022) 012009 doi:10.1088/1755-1315/1021/1/012009

The principle of the HPF conditional state-to-state transition is shown by the directed graph (Figure 2). Typically HPF is in a state of C_1 most of the time. The HPF state C_2 can be seconds, minutes, months, or even years. Everything depends on the criteria by which the state and conditions of transition are determined.

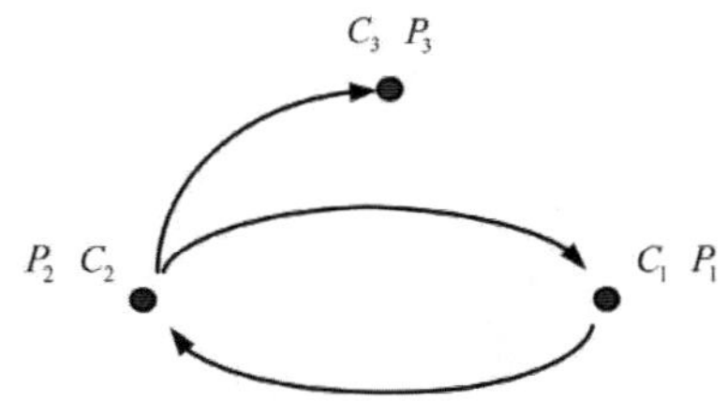

Figure 2. Directed graph of HPF states.

The HPF will transfer into C_3 state when one or more or all of the following events occur: the destruction of one or more technical devices, the release of hazardous substances from technical devices, explosion, accident factors, a fire that will block escape routes, and (or) harm HPF personnel, equipment, buildings, and the environment. The transition conditions are represented in the following expressions:

$$C_1 = \begin{Bmatrix} b_i \geq 0 \\ b_{ERi} > 0 \\ Q_{Di} \leq 0{,}1 \\ N_{HSi} \leq N_{\text{H}} \\ Q_P \leq 0{,}1 \\ Q_E \leq 0{,}01 \\ \Pr_i = 0 \end{Bmatrix}, \quad C_2 = \begin{Bmatrix} b_i < 0 \\ b_{ERi} > 0 \\ 0{,}1 < Q_{Di} \leq 0{,}95 \\ N_{HSi} \leq N_{norm} \\ Q_P > 0{,}1 \\ 0{,}1 < Q_E \leq 0{,}95 \\ \Pr_i = 0 \end{Bmatrix}, \quad C_3 = \begin{Bmatrix} b_{ERi} \leq 0 \\ Q_{Di} > 0{,}95 \\ N_{HSi} > N_{norm} \\ Q_{\text{E}} > 0{,}95 \\ \Pr_i > 0 \end{Bmatrix}, \tag{2}$$

where N_{norm} is the threshold value of the N_{HSi} determining the following [10].

To determine HPF in this or that state the following probabilities are used: P_1 is the probability of HPF in C_1 state, P_2 is the probability of HPF in C_2 state, P_3 is the probability of HPF in C_3 state. The calculation of probabilities is based on the statistical principle:

$$P_1 = \frac{N(C_1)}{N_{test}}, \quad P_2 = \frac{N(C_2)}{N_{test}}, \quad P_3 = \frac{N(C_3)}{N_{test}}, \tag{3}$$

where $N(C_1)$ is the number of iterations when the HPF was conditionally in C_1 state; $N(C_2)$, $N(C_3)$ is the number of iterations when the HPF was conditionally in C_2 и C_3 states respectively. The required total number of iterations N_{test} shall be determined from the Chebyshev inequality [11, 12]. Calculation (3) shall be carried out provided that:

$$\begin{gathered} P_1 + P_2 + P_3 = 1, \\ N(C_1) + N(C_1) + N(C_1) = N_{test}. \end{gathered} \tag{4}$$

The criteria and conditions necessary for the assessment of the HPF state are determined on the basis of the technological characteristics of the specific HPF, so the quantitative values of the parameters and values specified in (2) may change. It should be clarified that HPF can only be in one of the three states.

3. Results

Depending on the structural features there are different mathematical models to describe the destruction of technical devices, linking the strength parameters of the technical devices structure to the parameters of destructive effects. The "load–strength" model is widely accepted:

$$Q_P = 1 - \Phi\left\{\frac{M_{[\varepsilon]} - M_\varepsilon}{\sqrt{S_{[\varepsilon]}^2 - S_\varepsilon^2}}\right\}, \tag{5}$$

where [ε] is a random value denoting the strength of the structure (material); ε is a random value denoting the equivalent stress from external (internal) factors on the structure (material); $M_{[\varepsilon]}$ – mathematical expectation [ε]; M_ε – mathematical expectation ε; $S_{[\varepsilon]}$ – r.m.s. deviation [ε]; S_ε – r.m.s. deviation ε. It must be borne in mind that the load and strength must be physically adequate for each other, which is also taken into account when choosing the dimensions for the model (5).

The probability of Q_P can also be computed by other probability distributions (Weibull, Gamma, normal, etc.) that fairly accurately describe the destruction process of a particular technical device [13].

It is known that any process takes place under the influence of many different factors, some of which may be accidental in origin. Therefore, to calculate b_i we will use the principles that are based on the truncated normal distribution which is the basis for the technological process and workplace safety models [11–13]:

$$b_i = \frac{x_d - x(t)_i\left(1 + \frac{C_{xi}}{\sigma_{xi}\sqrt{2\pi}}\exp\left[-\frac{\left(x(t)_i - M_{xi}\right)^2}{2\sigma_{xi}^2}\right]Q_x\right)}{x_d}, \tag{6}$$

where x_d is the minimum permissible value $x(t)_i$; $x(t)_i$ is the current value of i-th technological process parameter; C_{xi} is the truncation coefficient of the truncated normal distribution of all possible values $x(t)_i$, shall be taken to be 1.0028 [11–13]; σ_{xi} is the r.m.s. deviation of the truncated normal distribution of the dangerous value $x(t)_i$; M_{xi} is the mathematical expectation of the dangerous value $x(t)_i$; Q_x is the probability of an accidental variation event $x(t)_i$. The value of $x(t)_i$ is better to be determined in real-time mode from the automatic process control systems currently used on virtually any HPF. Model (6) works as follows. The expression $\frac{C_{xi}}{\sigma_{xi}\sqrt{2\pi}}\exp\left[-\frac{\left(x(t)_i - M_{xi}\right)^2}{2\sigma_{xi}^2}\right]Q_x$ describes random deviations of $x(t)_i$ in the form of jumps of the controlled parameter, changing the value of b_i, which is then taken into account in (2).

Model (6) can be used to describe the process of pipeline destruction due to corrosion of pipe walls. In this case $x(t)_i$ will be the function that characterizes the process of changing the structure and wall thickness of the pipe, x_d will be the minimum allowable wall thickness of the pipe [14]. The principle of connectivity among the current, random, and allowable values of $x(t)_i$ will also change:

$$b_i = \frac{\left(ae^{-v} - ae^{-v}\frac{1.0028}{\sigma_r\sqrt{2\pi}}\exp\left[-\frac{ae^{-v} - M_r}{2\sigma_r^2}\right]e^{-\lambda_r}\right)}{S_d},$$

where a is the initial (designed) thickness of the tube wall, mm; ν is the corrosion rate of the wall, mm·year^{-1}; t is the time the dimension of which is determined in each case. σ_r is r.m.s. deviation of the distribution of the tube wall unacceptable thickness, mm; M_r is the mathematical expectation of the tube wall unacceptable thickness, mm. λ_r is the rate of pipeline failure due to the pipe wall thinning, h^{-1}; S_d is the minimum thickness of pipe wall allowed, mm. The method of ν calculation is given in [14], the parameter λ_r is calculated based on the statistical data available in the organization.

The model for assessing the personnel safety in the event of their evacuation from an area affected by an accident or a fire is similar:

$$b_{ERi} = \frac{1}{\tau_b}\left[\tau_b - t_{ei}\left(1 + \frac{C_U}{\sigma_U\sqrt{2\pi}}\exp\left[-\frac{(U_{ei} - M_U)^2}{2\sigma_U^2}\right]P_U\right)\right],$$

where τ_b is the time when the escape routes are blocked by fire hazards or their accompanying manifestations, s; t_{ei} is the personnel evacuation time from i-th production premises which is determined on an experimental or calculated basis, s; U_{ei} is the event of an accidental change in the time of personnel evacuation (evacuation under a tight technological arrangement, exit from limited enclosed spaces, descending from safety system transverses, lifting gears, etc.); σ_U is r.m.s. deviation U_{ei}, s; M_U is mathematical expectation U_{ei}, s; P_U is the probability of the event of evacuation time accidental change U_{ei}.

The parameters C_U, σ_U, M_U, and U_{ei} are determined based on the HPF specifics, as well as the characteristics of the organization and the performance of the work process. The probability P_U is determined by the expression $P_U = t_w / t_y$, where t_w is the time that the employee(s) are at work during a year, hour; t_y is the number of working hours per calendar year. The value of P_U in the personnel's operational activities before evacuation shall be equal to one if the stated actions are the responsibility of the employees. The possibility of an accidental change in the evacuation time due to erroneous actions of personnel is not unlikely. For example, making the decision to used emergency firefighting equipment when these activities are not effective due to exceeding the critical values of hazardous fire factors for humans. The model considered (4) refers to a combined method of assessing the safety of personnel evacuation, as it provides both an empirical way of determining evacuation time and a calculated way based on the theory of human flows.

The HPF explosion model foresees an explosive atmosphere capable of ignition from an ignition source:

$$Q_E = 1 - \prod_{j=1}^{M}\left(1 - Q_{afj}Q_{sj}\right), \quad (7)$$

where Q_E is explosion probability; Q_{afj} is the probability of an explosive atmosphere formation during the depressurization (destruction) of j-th technical device; Q_{sj} is the probability of j-th ignition source appearance capable of triggering a combustion reaction of an explosive mixture; j is the number of technical device, M – quantity of the technical device. Methods to determine Q_{afj} and Q_{sj} are given in the standard [15]. If there is a direct or correlative relationship between Q_{afj} and Q_{sj}, then the latter can be substituted in (7) instead of Q_{afj}.

The value of gas leakage (emissions) from communications, apparatus, containers, and other technical devices is determined according to the methods developed by the organizations taking into account their technological characteristics.

The impact of personnel on the HPF security is hard to overstate, as personnel has different specific influences at all stages of the HPF life cycle and this must be taken into account in assessing the reliability of HPF. It is proposed to use the "Conditional Fault Tree" method and the probability function polynomial formation rule. The probability of the personnel error can be determined by the dependencies indicated in the source [16]:

- the dependency of the error probability on the time allowed for the task decision;
- the dependency of human's reliability on the amount of workload (ratio of total work time to total shift time);
- the dependency of human's reliability on the degree of the task suddenness;
- the dependency of human error probability on the number of signals received that require a selective response to each signal;
- the dependency of incorrect identification probability of adjacent anomalous events on the time since their occurrence.

The handbook [16] provides data on the reliability of a human while working with the administration.

Almost any accident is associated with the formation of affecting factors that may adversely affect HPF personnel, HPF components (buildings, structures, technological equipment), and the environment. We shall divide the affecting factors into two groups: 1) HPF accident factors; 2) fire hazards. The first group of factors includes [17]: fragments of destroyed gas pipeline, containers, and other technological equipment; dynamic pressure of gas streams; gas contamination; heat radiation from a fire. In the second group, there are flames and sparks; thermal radiation; increased concentration of toxic combustion and thermal decomposition products; reduced concentration of oxygen; reduced visibility in smoke [18]. Mathematical models to determine the level of affecting factors exposure on HPF personnel, HPF components (buildings, structures, technological equipment), and the environment are discussed in detail in legislative rules [17, 18].

4. Conclusion

Thus, the article examines aspects of the development of the HPF DT mathematical model layer of the gas transportation enterprise. The models and principles characterizing the destruction models of technical devices used at HPF are presented. The mathematical principle of change in the state of the HPF is shown. There are explanations about the models describing the effects of affecting factors on personnel, technological equipment, buildings, and structures which may occur as a result of accidents (fires) on HPF. The model allowing to assess the possibility of employees' timely evacuation in case of an accident or a fire at HPF is considered. The presented models have been tested and implemented in the operational activities of JSC "Gazprom transgaz Tchaikovsky" to determine the most possible scenarios of accidents at hazardous facilities.

References

[1] Hlady J, Glanzer M and Fugate L 2018 Automated creation of the pipeline digital twin during construction: improvement to construction quality and pipeline integrity *Proc. 12th Int. Pipeline Conf.* (September 24–28, Calgary, Alberta, Canada) vol. 2

[2] Winning K 2018 Lessons learned: creating the digital twin. Concurrently for pipeline projects *Proc. Esri European Petroleum GIS Conf.* (October 30–31, London)

[3] Iureva R A, Kremlev A S, Subbotin V, Kolesnikova D V and Andreev Y S 2020 Digital twin technology for pipeline inspection *Intelligent Decision Technologies. Proc. 12th KES Int. Conf. on Intelligent Decision Technologies (KES-IDT 2020)* (June 17–19, Split, Croatia) ed I Czarnowski, R Howlet et al (Springer, Singapore) pp. 329–39

[4] Fryer T 2019 *Eng. Technol.* **14(1)** 28–9

[5] Stark R and Damerau T 2019 Digital twin. *CIRP Encyclopedia of Production Engineering* ed Int. Academy for Production Engineering, Chatti S et al (Berlin, Springer, Heidelberg) 5–18

[6] Adjei P and Montasari R A 2020 *Int. J. Strateg. Eng.* **3(1)** 51–61

[7] Harper E, Ganz Ch and Malakuti S 2019 *IIC J. Innov.* **November**

[8] Janda P, Hajicek Z and Bernardin P 2019 Implementation of the digital twin methodology *Proc. 30th Int. DAAAM Symp. "Intelligent Manufacturing & Automation"*(October 23–26, Zadar, Croatia) ed. B Katalinic (Vienna: DAAAM International) 533–8

[9] Dobrynin A, Chernykh K, Kupriyanovsky V, Kupriyanovsky P and Sinyagov S 2016 *Int. J. Open Inf. Technol.* **4(1)** 4–11

[10] Rostekhnadzor OrderNo. 29. *Methodological Recommendations on the Classification of Technogenic Events in the Field of Industrial Safety at Hazardous Facilities of the Oil and Gas Complex: Safety Guidelines*: Retrieved from: http://docs.cntd.ru/document/556353126

[11] Kirilov A E 2017 *Fire Safety* **3** 54–60

[12] Kirilov A E and Chernyi K A 2018 *Perm J. of Petroleum and Mining Eng.* **18(2)** 190–200

[13] Kirilov A E and Chernyi K A 2020 *Occupational Safety in Industry* **8** 37–47

[14] Ostreikovsky V A 2003 *Reliability Theory* (Moscow: Vyssh. shk.)

[15] GOST 12.1.004–91 *Occupational Safety Standards System. Fire Safety. General Requirements* (Moskow: Standardinform)
[16] Ireson W G 1966 *Reliability Handbook* (New York: McGraw-Hill Book Company) p 702
[17] Rostekhnadzor Order from December 26 2018 No. 647 *Accidents Risk Assessing Methodology on Hazardous Production Facilities of Main Pipeline Gas Transportation*: Retrieved from https://docs.cntd.ru/document/552083261, 27 June 2021
[18] *Methodology for Determining the Fire Risk Calculated Values in Buildings, Constructions and Facilities with Various Functional Fire Hazard Classes*: Order of EMERCOM of Russia from June 30 2009 No. 382. Retrieved from https://docs.cntd.ru/document/902167776, 27 June 2021

IOP Conf. Series: Earth and Environmental Science 1021 (2022) 012010 doi:10.1088/1755-1315/1021/1/012010

A study of the effect of the limiting shear gradient on the pressure recovery curve during hydraulic fracturing

L F Yusupova, R R Khusnutdinova, R A Gilyazetdinov, and E M Almukhametova

Ufa State Petroleum Technological University, Branch of the University in the City of Oktyabrsky, 54a, Devonskaya St., Oktyabrsky, Republic of Bashkortostan, 452607, Russian Federation

E-mail: xusnutdinova.1986@mail.ru, info@of.ugntu.ru

Abstract. The article analyzes the effect of the limiting shear gradient of viscoplastic oil on the results of interpretation of well testing by the pressure recovery curve (PRC) method following the methods of stimulation by hydraulic fracturing. The lack of mobile water in the collector was analyzed. An analysis of the effect of the limiting shear gradient on the pressure build-up curve and interpretation results was conducted. According to the results of interpretation by the pressure recovery curve method in the presence of a limiting shear gradient in oil, curves were constructed for various dependences and deviations were revealed.

1. Introduction

Well testing is the most accurate and reliable method for assessing the effectiveness of most geological and technological measures, including hydraulic fracturing. Well testing is a set of measures performed at wells, which invovles measuring changes in the flow rate, temperature, and pressure with downhole or wellhead tools and their subsequent processing to determine the formation parameters. Well testing is used to study productive formations, during the development and operation of a well stock in order to obtain information on the reservoir properties, to determine reservoir pressure, reservoir boundaries, heterogeneities, etc. [1].

The well testing's goals are as follows: 1) Industrial exploration : collection of information about properties for calculating reserves and determining possible macroinhomogeneities; 2) Trial / commercial operation: further determination of the geological characteristics of the field to control the development; analysis of the energy state of the formation; determination of the effectiveness of technological operations aimed at increasing oil production (hydraulic fracturing, treatment of the bottomhole zone (BHZ), heat treatments) and identification of possible candidates for performing these operations. The research methodology involves finding parameters from known input and output signals. The input signal is the response of the well operation mode, the output signal is the pressure change. This problem is called the "inverse problem". When analyzing the data, various possible models of the system are used, 34 ones are associated with the pressure response to flow rate changes. The system model and parameters are set to values that must coincide with the real data from the study.

The types of hydrodynamic studies of wells differ depending on the purpose of the well, the reservoir properties, data obtained. Consider one of the main types of research [2-4].

IOP Conf. Series: Earth and Environmental Science 1021 (2022) 012010 doi:10.1088/1755-1315/1021/1/012010

Well testing using the pressure recovery curve (PRC) method : the well operates for a certain period of time t with a constant flow rate q, after which it is stopped and the pressure change is measured (Figure 1).

This type of research involves the following paameters: the constant flow rate through the well before it is closed for research; the operation time of the well before stopping should be sufficient to reach a steady state of operation.

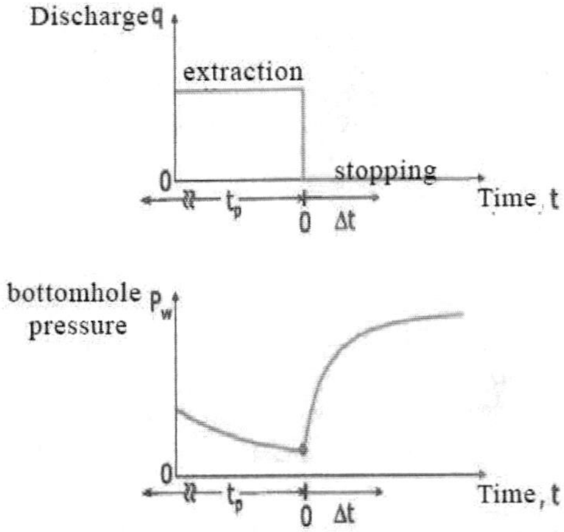

Figure 1. Research of a production well by the pressure recovery method.

2. Materials and methods

To study the effect of the limiting shear gradient on the pressure recovery process, the mathematical model of three-phase filtration was built. As a research tool, the package of hydrodynamic modeling "Tempest-More" version 6.7.1 was used. It make it possible to develop an analytical model taking into account the limiting shear gradient. A radial formation with a radius of 10,000 m and a height of 20 m was used as a model; a production well is located in the center. Assume that the reservoir is homogeneous both along the section and laterally. The reservoir model is shown in Figure 2.

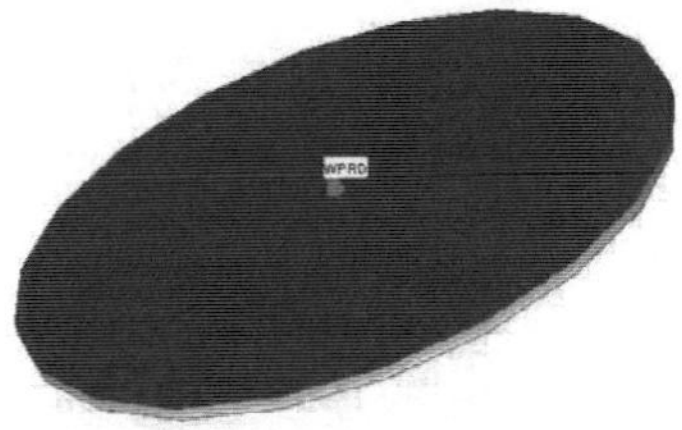

Figure 2. The model of the reservoir used in the calculations.

The depth is 470-490 m. The initial reservoir pressure and temperature are P_0 = 5.8 MPa, T_0 = 31°C. The density and viscosity of water in reservoir conditions are 1.04 g/cm^3 and 0.83 mPa·s, respectively. The density of oil in 54 surface conditions is 0.920 g / cm^3. In reservoir conditions, oil viscosity is 406 mPa·s, gas content is 25 m^3/m^3. Oil saturation pressure with gas is 2.29 MPa. The relative phase permeabilities (RP) are shown in Figure 3.

IOP Conf. Series: Earth and Environmental Science 1021 (2022) 012010 doi:10.1088/1755-1315/1021/1/012010

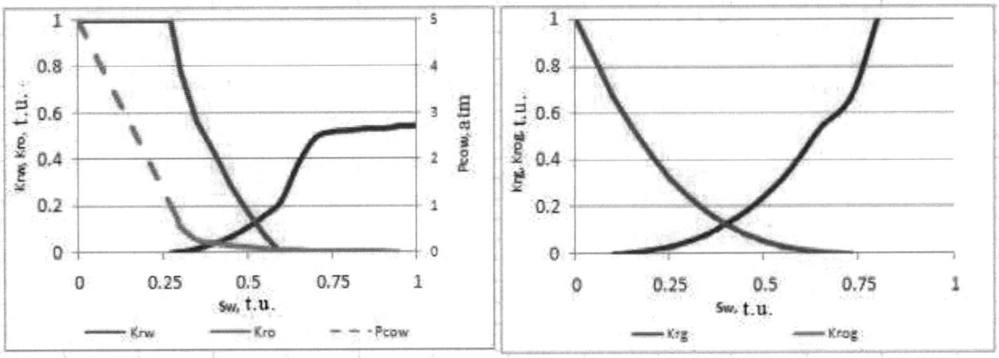

a) RP in the oil-water system, b) RP in the oil-gas system.
Figure3. Relative phase permeabilities of the model formation for oil, water and gas.

The technological conditions for the development of the deposit area are as follows. The production well is put into operation for 365 days. This production interval is chosen for the well to reach a pseudo-steady state, after which the well is stopped with an analysis of the pressure recovery curve (PRC). The flow rate threshold was set equal to 10 m^3 / day. The purpose of this work is to study the difference between the pressure recovery curve in a formation with a viscoplastic fluid and the parameters obtained during the interpretation from the pressure recovery curve in a formation with a fluid obeying Newton's law. Viscoplastic fluid is characterized by ultimate shear stress. In Tempest-More, the non-Newtonian oil flow is modeled by a multiplier that changes the dynamic viscosity of oil within the specified boundaries of the reservoir pressure gradient [5-7]. Below the value of the pressure gradient D_P, the limiting factor 55 is equal to zero - the case of infinite oil viscosity, i.e. the lack of flow. Above D_{prel} and up to $D_{Pnewton}$, the multiplier varies from 0 to 1 in a linear relationship. Above $D_{Pnewton}$, the oil flow has a Newtonian character.

Fix the upper boundary of the area of non-Newtonian oil flow in reservoir conditions $D_{Pnewton}$ = 0/02 MPa / m. We will vary the D_{prel} value from 0.001 to 0.019 MPa/m. Figure 4 shows the dependence of the change in factor F on the pressure gradient [9-11].

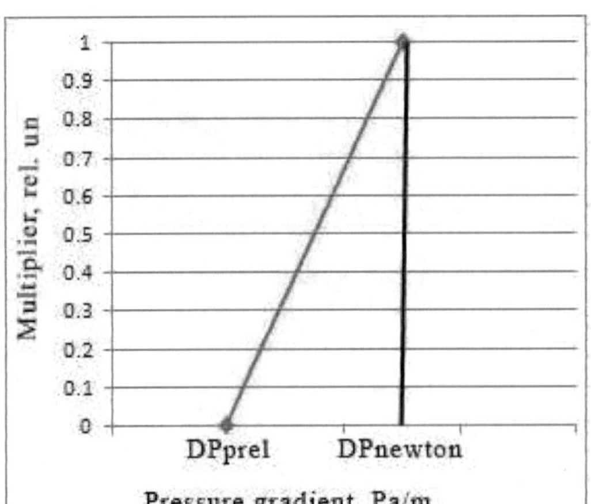

Figure4. Dependence of coefficient F, which determines the non-Newtonian nature of oil flow, in different ranges of pressure gradient values

The calculations were performed at water cut values of 0 and 2%. The analyzes were carried out using the tangent method.

3. Results

The results of interpretation by the pressure recovery curve method in the presence of a limiting shear gradient in oil and in the absence of mobile formation water [8] are presented here.

This calculation assumes that there is no mobile water in the formation. Consider the pressure build-up curve for Newtonian oil in the first calculation. The results were processed by the tangent method. The following are the data for performing calculations. Q_0 = 11.6 10-5m^3/s, Pd_0 = 4.34 MPa, βn pl = 0.406 Ns/m^2, N_{ef} = 20 m, dex = 0.073 m, m = 0.34, βn = 1.45 10-10 (N/ m^2) -1, βw = 3.75 10-10 (N / m2) -1, βp = 4.35 10-10 (N / m2) -1, water cut - 0%. T = 31536000 sec. According to the measurements obtained, plot the dependence ΔPd (t) - logt (Figure 5).

Coefficient A is calculated as the point of intersection of the straight-line section of the curve with the ordinate, coefficient i is determined as the tangent of the slope of the curve. Drawing a tangent along the straight section, the following values of the coefficients can be obtained: A = 0.198 MPa, i = 0.165 MPa.

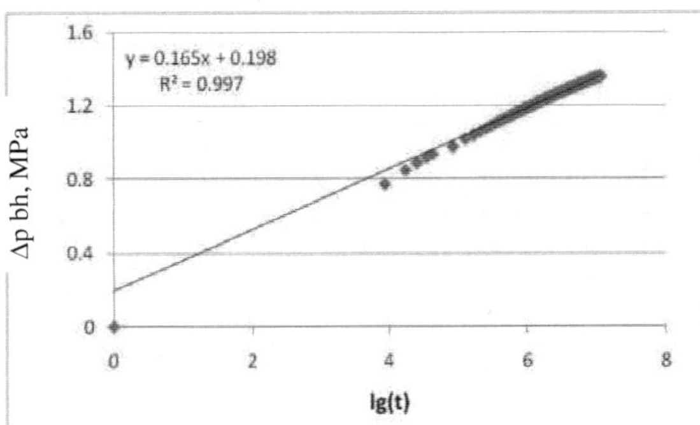

Figure5. Dependence Pd (t) - lgt for Newtonian oil

Determine the coefficient of reservoir conductivity:

$$\varepsilon = \frac{0.183 \cdot 116 \cdot 10^{-6}}{165000} = 1.28 \cdot 10^{-10} \frac{m^5}{N \cdot sec}. \quad (1)$$

Calculate χ (piezo conductivity coefficient):

$$\chi = \frac{1.28 \cdot 10^{-10}}{20 \cdot (0{,}34 \cdot 1.45 \cdot 10^{-10} + 4.35 \cdot 10^{-10}} = 1.32 \cdot 10^{-2} \frac{m^2}{sec}. \quad (2)$$

Calculate the reduced radius of the well:

$$r_{wr} = \sqrt{\frac{1.32 \cdot 10^{-2} \cdot 2.25}{10^{\frac{198\,000}{165\,000}}}} = 4.3 \cdot 10^{-2} \text{m}. \quad (3)$$

Determine productivity factor K of the well

$$K = \frac{2 \cdot 3.14 \cdot 4.4 \cdot 10^{-10}}{\ln \frac{3500}{4.3 \cdot 10^{-2}}} = 0.71 \cdot 10^{-10} \frac{m^3}{MPa \cdot sec} \quad (4)$$

Determine the formation permeability:

$$k = \frac{1.28 \cdot 10^{-10} \cdot 0.406}{20} = 2.61 \cdot 10^{-12} m^2 \quad (5)$$

Interpret the results obtained with the limiting shear gradient. During the simulation, the value $D_{Pnewton}$ = 0.02 MPa / m; it did not change, the D_{prel} parameter varied from 0.001 to 0.0175 MPa / m.

Consider the values $D_{Pnewton}$ = 0.02 MPa / m and D_{prel} = 0.001 MPa/m. Similarly to the interpretation with Newtonian oil, build the dependence ΔPd (t) - logt (Figure 6).

As can be seen from Figure (6), when choosing a straight section and constructing the continuation of this line to the intersection with the ordinate curve, the values of 58 coefficients were obtained: i = 0.519 MPa, A = –1.868 MPa. It is impossible to interpret this study due to the negative value of coefficient A, since the negative value indicates the non-physical nature of the results obtained.

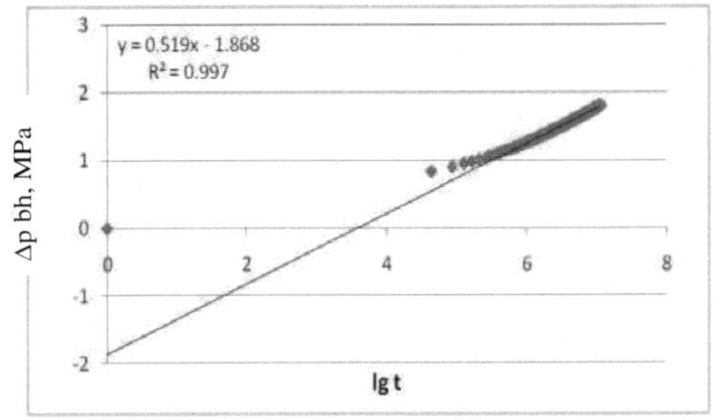

Figure6. Dependence Pd (t) - logt at $D_{Pnewton}$ = 0.02 MPa / m, D_{prel} = 0.001 MPa/m.

At Dprel = 0.015 MPa / m and more, "kinks" are observed when plotting the Pd(t) - lgt graph, which introduces additional uncertainty in the interpretation process when choosing a straight-line section of the curve. This can be seen from Figure 7.

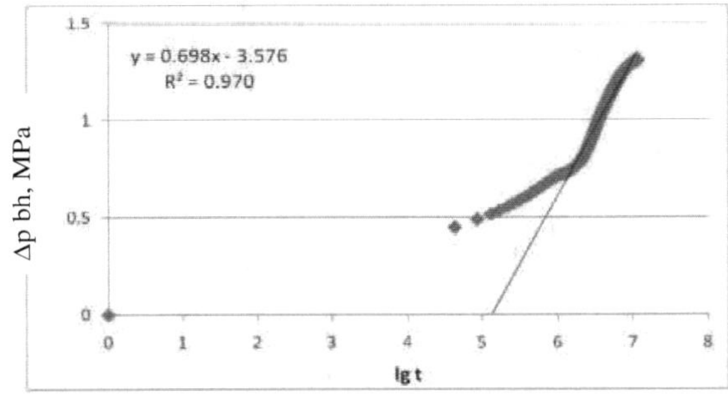

Figure7. Dependence Pd (t) - logt at $D_{Pnewton}$ = 0.02 MPa / m, D_{prel} = 0.015 MPa/m.

Figure 8 shows that the greater the value of coefficient DPprel, the greater the noise level is observed on the graph Pd(t) - lgt.

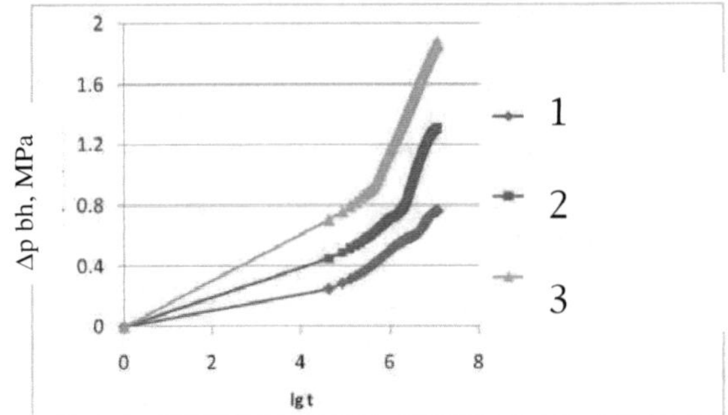

Figure8. Dependence Pd (t) - logt at $D_{Pnewton}$ = 0.02 MPa / m,
D_{prel} =0.015 MPa/m (1)
D_{prel} 0.0175 MPa/m (2)
D_{prel} 0.0125 MPa/m (3)

Consider the study at Dprel = 0.0175 MPa/m (Figure 9-12)

With a given value of the parameter DPprel, it is difficult to distinguish a straight-line section on the Pd (t) - lgt graph. Four different interpretations are possible. Sections that were taken as a straight line segment to find the coefficients A and i are marked red.

Option 1

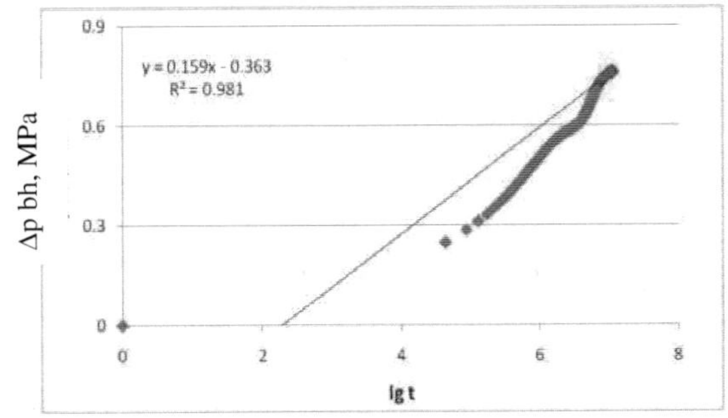

Figure 9. Dependence Pd (t) - logt at $D_{Pnewton}$ = 0.02 MPa / m, D_{prel} =0.0175 MPa/m.

Option 2

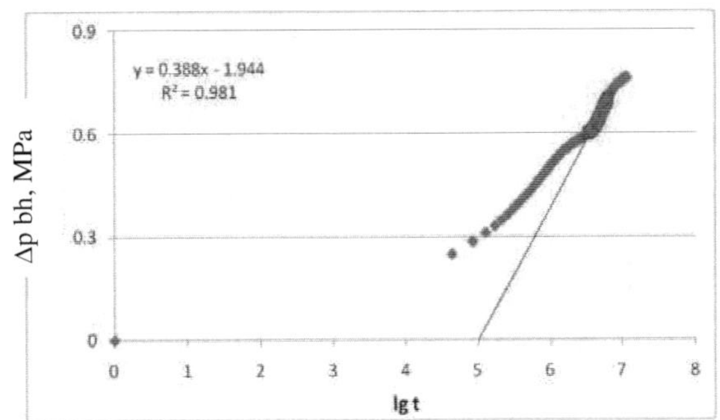

Figure10. Dependence Pd (t) - logt at $D_{Pnewton}$ = 0.02 MPa / m, D_{prel} =0.0175 MPa/m.

Option 3

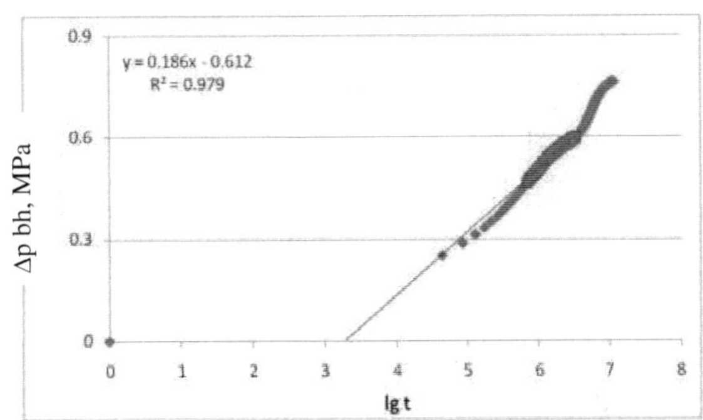

Figure 11. Dependence Pd (t) - logt at $D_{Pnewton}$ = 0.02 MPa / m, D_{prel} =0.0175 MPa/m.

Option 4

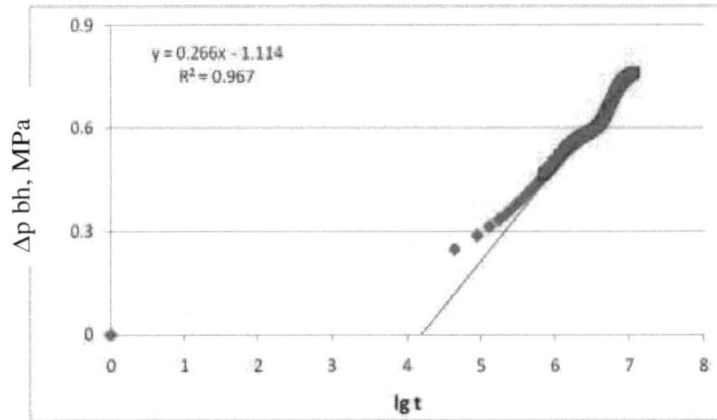

Figure12. Dependence Pd (t) - logt at $D_{Pnewton}$ = 0.02 MPa / m, D_{prel} =0.0175 MPa/m.

The results are presented in Table 1.

Table 1. Results of data interpretation at $D_{Pnewton}$ = 0.02 MPa/m, D_{prel}= 0.0175 MPa/m

No of option	A, MPa	i, MPa
Option 1	-0.363	0.159
Option 2	-1.944	0.388
Option 3	-0.612	0.186
Option 4	-1.114	0.266

As can be seen from Table 1, when choosing different sites, we obtain different values of coefficients A and i.

Present all the values of coefficients A and i obtained when calculating the flow of viscoplastic fluid in the formation in the absence of mobile formation water, except for $D_{Pnewton}$ = 0.02 MPa / m, D_{prel} = 0.0175 MPa / m (due to the strong uncertainty in the choice of a final result).

Table 2. The results of data interpretation

D_{Prel},MPa/m	A, MPa	i, MPa
0.001	-1.868	0.519
0.002	-1.9	0.523
0.005	-2.103	0.552
0.0075	-2.234	0.579
0.01	-2.365	0.605
0.0125	-2.899	0.679
0.015	-3.179	0.698

Construct the dependences of coefficients A and i on D_{prel} (Figures 13, 14).

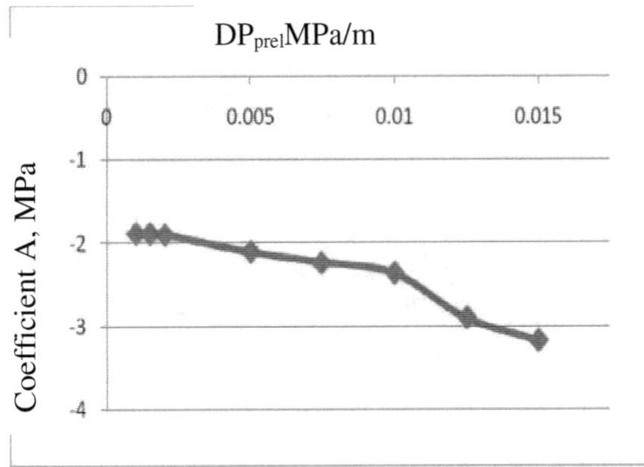

Figure13. Dependence of parameter A on DP_{prel}.

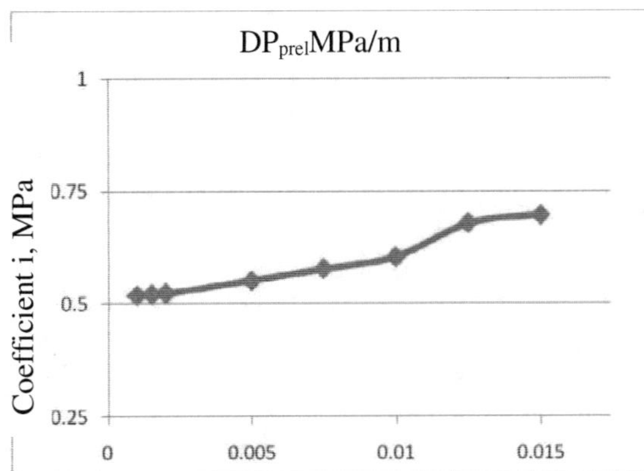

Figure14. Dependence of parameter i on DP_{prel}.

On both graphs, there is no linear dependence on the DPprel coefficient. With an increase in the Dprel, the value of parameter A decreases, and of parameter i increases. On both graphs there is a deviation from the linear dependence at Dprel ~ 0.01 MPa/m. In a homogeneous reservoir in the absence of external factors 64 (the presence of faults, the influence of the neighboring fund, etc.) and mobile water, the negative value of coefficient A indicates the presence of a limiting shear gradient.

4. Conclusion

The influence of the limiting shear gradient of viscoplastic oil on the results of interpretation of the well test by the pressure recovery curve (PRC) method was studied. Well test resutls were interpreted using the tangent method. There was no mobile water in the collector [12-14].

Based on the results of interpretation of the studies using the pressure recovery curve method in the presence of the limiting shear gradient in oil and in the absence of mobile formation water, the following conclusions can be drawn:

- in the absence of the limiting shear gradient (Newtonian oil flow), the positive value of coefficient A was obtained;
- in the presence of the limiting shear gradient (non-Newtonian oil flow), the value of coefficient A was negative; thenegative value of coefficient A indicates the presence of the limiting shear gradient;
- when plotting the dependence of coefficients A and i on D_{prel}, it can be seen that with an increase in coefficient D_{prel}, the value of parameter A decreases, and of parameter i increases.

The graphs show a deviation from the linear dependence at D_{prel} ~ 0.01 MPa / m.

Thus, this well testing method can be used to assess the effectiveness of technological operations aimed at increasing the volume of oil production (hydraulic fracturing, etc.)

5. References

[1] Fei H, Cheng Hassan L S, Hou J, Liu C Z, Feng J D 2008 Threshold Pressure Gradient in Ultra-low Permeability Reservoirs *Petroleum Science and Technology* **26** 10241035
[2] Baoquan Z, Linsong C, Chunlan L 2011 Threshold Pressure Gradient in Ultra-low Permeability Reservoirs *Petroleum Science and Technology* **80** 1-6
[3] Yu R, Bian Y, Zhou S, Wang K, Lu Q, Chen Z 2012 Nonlinear flow numerical simulation of low-permeability reservoir *Journal of Central South University* **19** 1980-1987
[4] Stone L 2009 Herbert Iterative solution of implicit approximations of multidimensional partial differential equations. *SIAM Journal on Numerical Analysis***5** 530-558
[5] Grezina O A 2018 Downhole device design and results of its utilization under acid-implosion action *Advances in Engineering Research* **157** 203–206 DOI: 10.2991/aime-18.2018.39
[6] Akhmetov R T, Mukhametshin V V, Kuleshova L S 2019 Simulation of the absolute permeability based on the capillary pressure curves using the dumbbell model, *Journal of Physics: Conference Series* **1333(3)** 032001 DOI: 10.1088/1742-6596/1333/3/032001
[7] Almukhametova E M, Fattakhov D I, Zakirov A I 2018 The efficiency analysis of applying hydraulic fracturing of formation at AB1-2 facility of Potochnoye field *IOP Conference Series: Earth and Environmental Science* **194 082002** DOI:10.1088/1755-1315/194/8/082002
[8] Izmailova G R 2018 Research on the high-frequency electromagnetic acoustic heating of an oil reservoir with further pumping of a solvent *High Temperature* **56(6)** 910–914
[9] Tyncherov K T, Mukhametshin V Sh, Paderin M G, Selivanova M V, Shokurov I V, Almukhametova E M 2018 Thermoacoustic inductor for heavy oil extraction, *IOP Conference Series: Materials Science and Engineering* **327(4)** 042111 DOI:10.1088/1757-899X/327/4/042111
[10] Izmaylova G R 2019 Mathematical modeling of high-viscosity index oil formation with a sonic field at various frequencies *Journal of Physics: Conference Series* **1333(3)** 032030 doi:10.1088/1742-6596/1333/3/032030
[11] Petrova L V, Yarullin, D R 2019 Evaluation of the effect of asphalt resin paraffin deposits on oil well performance *IOP Conference Series: Materials Science and Engineering* **560** 012084 doi:10.1088/1757-899X/560/1/012084
[12] Zainagalina L Z, Petrova L V, Petrov V A 2019 Analysis of the efficiency of telemetric systems for drilling wells *IOP Conference Series: Materials Science and Engineering,* **560** 012096 doi:10.1088/1757-899X/560/1/012096
[13] Yusupova L F, Khalikova K M, Khusnutdinova R R 2021 Technological feature of water shutoff operations. *IOP Conf. Series: Earth and Environmental Science* **868** 012086 DOI:10.1088/1755-1315/868/1/012086
[14] Yusupova L F, Almukhametova E M Faizullin I A 2021 Sources of oily waste affecting the Earth's crust interior. *AIP Conference Proceedings* **2410** 020034 https://doi.org/10.1063/5.0067642

Evaluation of hydrocarbon resources of the Devonian terrigenous rock basin of Perm Region

I A Kozlova and D D Borisova

Perm National Research Polytechnic University, 29 Komsomolsky prospekt, Perm, 614990, Russia

Email: ikozlova@pstu.ru

Abstract. The geological structure of the Devonian terrigenous complex within the Bashkirian arch located in the well-explored southern part of Perm Krai has been considered in the paper. Due to the limited extent and complex structure of the bulk of reservoir rock in the prospective horizons, these deposits are explored unevenly, usually alongside with the overlying oil and gas bearing complexes. On the basis of the performed analysis of systematic data, the authors justify quantitative parameters for evaluation of non-localized resources after the method of comparative geological analogies in the Devonian terrigenous deposits. The estimated quantity of hydrocarbon resources in the territory under study unequivocally defines it as prospective for exploration and discovery of reserves in the Devonian oil and gas bearing complex. Therefore, the authors have identified areas for priority prospect evaluation survey taking into account the extent of the most promising facies zones of deltaic channels and beach bars.

1. Introduction

In the territory of the Bashkirian arch (BA), the surface of the reflecting horizon (RH) III has relatively flat and shallow geometry from -1,780 m to -2,020 m. Against the monoclinal northward submersion, the RH structural geometry has minor positive structures: Kuedinsky and Moskudyinsky bars, Dubovogorsgaya structural terrace, Tatyshlinsky ledge, Kapkanskaya and Chernushinskaya uplift zones, complicated with local uplifts. Structural geometries of the overlying RHs – II^p, II^k, I^p, I^k are more dissected and have increasing amplitudes due to occurrence of reef structures in the Frasnian-Famennian stage in the section side parts. In the BA sedimentary cover, commercial oil and gas bearing potential is typical of Devonian terrigenous, Upper Devonian-Tournaisian Carboniferous, Lower-Middle-Visean Terrigenous, Upper-Visean-Bashkirian Carboniferous and Moskovian Terrigenous Carboniferous oil and gas bearing complexes (OGC). Development of reserves is related to the extent of the arch, axial and southeast side parts of the Kama-Kinel trough system.

In the Devonian OGC, three beds are commercially oil-bearing. Bed D_o is in the lower part of the Timanian horizon, bed D_1 – associated with Pashiy deposits, and bed D_2 – associated with Mullino reservoirs. Bed D_2 mostly consists of coarse siltstones with subordinate interbeds of fine sandstones, it is not continuous and often replaced with dense rock, especially in the east of the territory under study. Bed D_1 is composed of quartz fine and medium-grain sandstones. In the east of the territory under study, the reservoir is often replaced with impermeable rock or siltstone with low reservoir properties. Bed D_0 occurs in the form of irregular interbeds or lenses of sandy siltstone grading to sandstone. The bed is made of sandstones that are medium- or fine-grain, silty, less often uneven-grained, clayey, or loamy-carbonaceous, and siltstones that are coarse-grain, sandy, or clayey-calcareous. The terrigenous bulk of Devonian deposits occurred in the highly variable sedimentation environment, when transgressions alternated with regressions, and subsidence with low-amplitude rising and wash-out of previously accumulated sediments. By the beginning of the Givetian tansgression, the territory under study had an uplifted plain of the Perm-Bashkir paleoarch, which served as a source of terrigenous material migration. Therefore, thinning of Givetian deposits occurs from northwest (Krasnokamsk-Chusovoy paleotrough) to southeast (Perm-Bashkir paleoarch). These conditions resulted in development of cyclic deposits having significant lithological inhomogeneity. The lower regions of Mullino and Timanian horizons accumulated thick bulks of sandstones; deposits in the upper regions of the horizons mostly consist of clayey siltstones, and the cycle is completed with clayey structures formed in deep water conditions [1,2].

Such cyclicity is typical of deltaic plain deposits which are a good hydrocarbon reservoir. However, frequent replacement of reservoirs with impermeable rock in conditions of complex facies structure substantially complicates the process of exploration and survey of such deposits.

Commercial oil bearing potential of beds D_0, D_1 and D_2 has been established for Alnyashskoe, Barykinskoe, Vinnikovskoe, Lesnoe, Chikulaevskoe, Aspinskoe, Sofyinskoe, Dubovogorskoe, Kapkanskoe, Karamorskoe, Kamyshlovskoe, Kalmyarskoe, South Chernushinskoe, Khatymskoe,Odinovskoe, Moskudyinskoe, Shagirtsko-Gozhanskoe, Krasnoyarsko-Kuedinskoe fields. Net pays of the reservoirs are from 0.4 to 4.7 m. Reservoir properties widely vary: porosity from 13% to 21%, permeability from 0.011 to 0.544 μm^2, oil saturation 0.53-0.93 unit fractions. D_0 bed has the best mean properties. The complex is regionally sealed by argillites and clayey limestones of the Timanian and Sargayevo horizons. Reservoirs in the terrigenous Devonian deposits are inherently caused by manifestation of structural, lithological and tectonic factors.

Thus, Devonian terrigenous oil and gas bearing complex deposits are a promising source of hydrocarbons in Perm Krai, but they are unevenly and insufficiently studied due to the complex geological structure compared to overlying deposits [3]. Therefore, the authors have made an attempt to evaluate non-localized resources with justification of specific volumetric parameters and development of recommendations for exploration of priority areas taking into account favorable facies zones of the Devonian terrigenous OGC.

2. Materials and methods

Re-evaluation of total initial extractable reserves (TIRR) of oil and gas bearing regions is performed periodically to obtain a qualitative assessment and establish the profile of reserves across the sedimentary cover section [4,5]. TIRR includes production output and reserves in the discovered fields, as well as localized and non-localized resources within the province (its part) and all OGCs. The study evaluates non-localized resources (part of TIRR) with a focus on the Devonian terrigenous OGC for BA using the method of comparative geological analogies by specific density per unit area. The method consists in using a single measure of similarity between the reference region (RR) and evaluation region (ER) – analogy coefficient which takes into account the change in volumetric parameters. Forecast parameters are selected using expert judgement and can include diverse criteria groups taking into account structural, lithological, morphological, hydrogeological, geochemical properties, thickness, porosity and permeability, and other features of the rock sequence under consideration.

The priority task of evaluation is to justify areas of the reference and evaluation regions. As a rule, RR and ER area is justified on the basis of the extent of geological and geophysical knowledge and field development. Apart from this important attribute, the authors have justified the area boundaries on the basis of horizontal and vertical extent of favorable facies zones. In terms of geological and geophysical knowledge, Bashkirian arch is one of the well-studied regions of Perm Krai. However, territory coverage by 2D seismic exploration is uneven: seismic exploration density varies within less than 1 km/km^2 for BA southeast and more than 5 km/km^2 for Chernushinskaya swell-like zone and

Dubovogorskaya structural terrace. 3D seismic exploration covers most of the territory, including areas with low 2D coverage, which generally suggests high extent of seismic exploration coverage of the territory. State of exploration of the territory by deep drilling is more differentiated – the highest deep well spacing density (0.075 – 0.125 wells/km^2 and more) developing the Devonian terrigenous OGC, is mostly associated with regions of already discovered wells with proven oil and gas bearing potential of the complex. The fewest number of deep wells is concentrated in BA east and southeast (0.025-0.075 wells/km^2). Thus, on the basis of available geological and geophysical knowledge, the reference region was identified in the more explored western part of BA, whereas the evaluation region was identified in the less explored (eastern) part. Apart from the traditional criteria of the extent of knowledge about the territory, the authors have based their area justification on the parameter of horizontal extent of more than 2 meters thick reservoirs, taking into account the most promising facies. The territory under study, as mentioned above, is associated with the coastal-marine facies type with accumulative formations. The extent area of the most promising reservoirs is associated with sandy structures of deltaic channels, made up of more than 4 m thick sandstones and sandy formations of bars, spits and coastal accumulative bodies more than 2 m thick, made up of sandstones with thin interbeds of siltstones and argillites. Thus, the area of the identified evaluation region in the southeast was reduced taking into account the extent of facies zones (Figure 1).

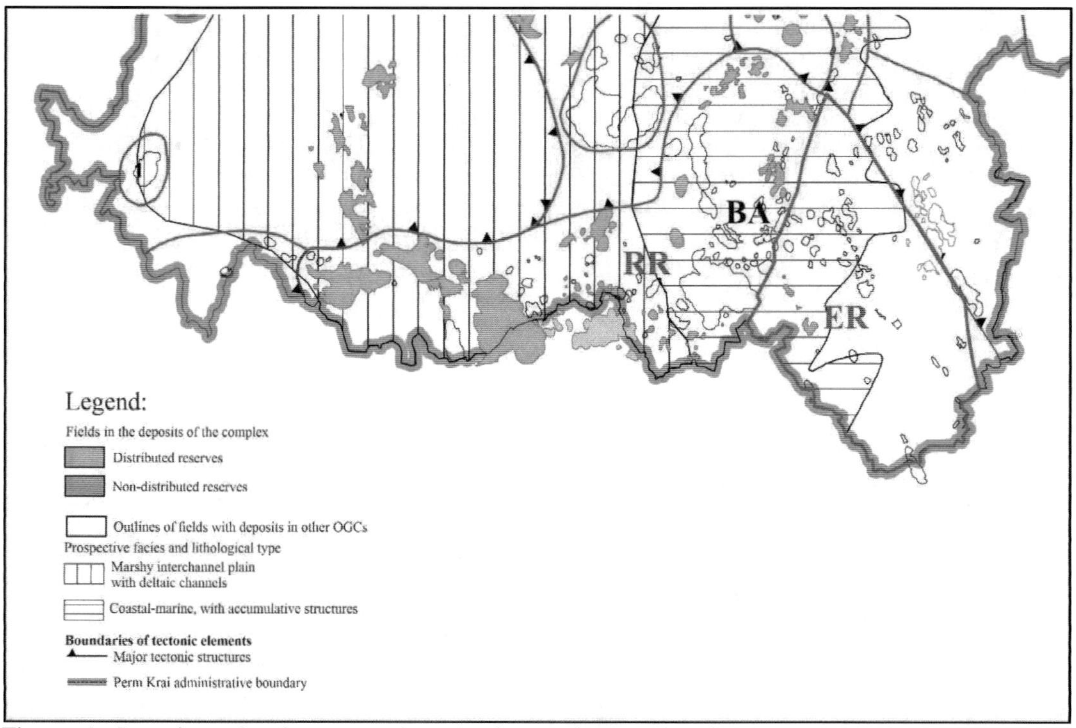

Figure 1. Location of reference and evaluation regions taking into account facies zones of the Devonian terrigenous OGC

The analogy coefficient which is the desired output of the method of comparative geological analogies takes into account the degree of similarity of the OGC under study within the RR and ER against a set of geological and geophysical, geochemical, hydrogeological and other characteristics of the object under study, defining generation, migration and accumulation of hydrocarbons. Therefore, the reference and evaluation regions were used to collect and analyze the characteristics of the Devonian terrigenous bulk in general, productive strata, reservoir and seal properties. In order to determine the most insightful volumetric parameters with maximum degree of similarity, mathematical statistic methods were used. Analysis of the two datasets (for the reference and evaluation regions) was based on the test of hypotheses using Student (t) criterion for comparison of samples by mean values, and using Fisher (F) criterion for comparison of samples by their dispersions. Based on the computation results, the following volumetric parameters have been identified as the most similar: mean seal thickness OGC (Os), m; mean thickness of identified reservoirs in OGC (Or), m; mean content of organic carbon (Corg), %; knowledge coefficient (Ckn), unit fractions; mean coefficient of discovered porosity (Kdp), unit fractions.

Mean thickness of Timanian seal (Os) was determined on the basis of geophysical logging data, taking into account the double gamma-ray index. This index is calculated on the basis of gamma ray curve and shows the actual percentage content of the clayey component in the total rock volume. The lower limit value for calculation was taken from dense limestones of Sargayevo or Mendym stage, where geophysical logging data is minimal, and the upper limit value from where the geophysical logging data reached its maximum – the cleanest Devonian and Vendian clays.

Mean thickness of Devonian reservoirs within the territory under study for RR and ER was obtained on the basis of well log interpretation data (WLID) of the wells under development in the area. The analysis used cumulative reservoir thicknesses of Givetian stage (D_2), Pashiy (D_1) and Timanian (D_0) horizons.

Mean Corg value was based on the analysis of self-potential of oil and gas source rock of the Devonian terrigenous bulk on the basis of available geochemical research data in the Devonian terrigenous OGC strata (based on the data by JSC KamNIIKIGS). For RR, the analysis was based on the study of 62 samples; for ER, on 11 samples.

Knowledge extent for RR and ER was determined on the basis of knowledge coefficient (Ckn). Knowledge coefficient (Ckn) is a computed value and is determined for RR and ER based on the data for reserves, prepared and localized resources in the regions under study.

The porosity coefficient (Cdp) is the basic parameter descriptive of the reservoir properties of rock. Analysis of the Cdp change for RR and ER used WLID for the wells under development within the Bashkirian arch.

3. Results

On the basis of the parameters set out by the authors, specific analogy coefficients were determined for RR and ER, as shown in table 1.

Table 1. Mean parameter values and specific analogy coefficients

Parameter	RR	ER	Specific Ca
Os, m	10.2	5.2	0.51
Or, m	7.3	3.8	0.52
Corg, %	0.45	0.74	1.64
Ckn, u.fr.	2.9	2.1	0.72
Cdp, u. fr	0.156	0.151	0.97

Multiplied specific analogy coefficients (Ca) provide a total analogy coefficient equal to 0.3 u. fr. Taking into account ER area, resource density in RR and obtained Ca, prospective D_1 category resources in ER amount to 2,495 fuel equiv. un.

To evaluate prospective resources in RR, the underexploration extent of the region was determined and amounted to 0.07 u. fr., and prospective D_1 category resources in RR were found equal to 3,305 fuel equiv. un.

Thus, in the south of Perm Krai, promising geological resources of D_1 category amount to 5,800 fuel equiv. units, which suggests feasibility of further geological and geophysical exploration.

In the course of development of promising GGE areas in the south of Perm Krai, the authors recommend to take into account the limited horizontal extent of the reservoirs due to facies nature of the complex, and existence of local uplifts identified on the surface of the reflecting horizon across the roof of the Timanian terrigenous deposits. The superimposed map of facies zones of Pashiy deposits and structural map for RH III is shown in Figure 2.

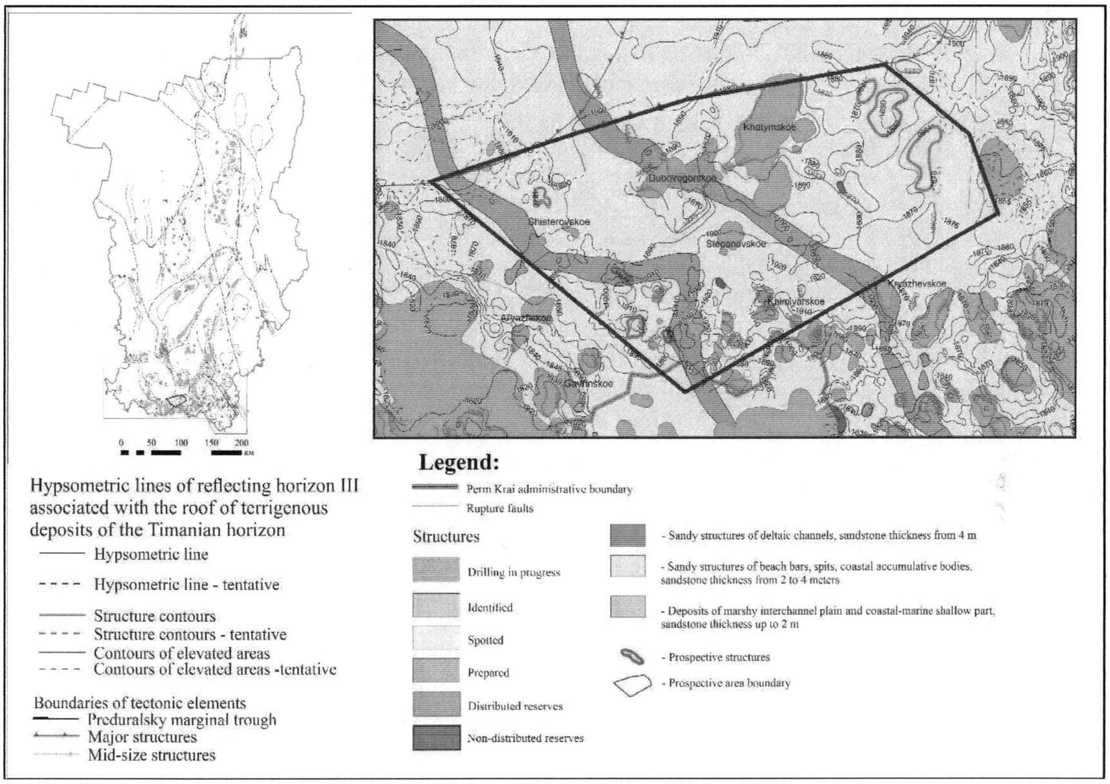

Figure 2. Forecast map of prospective area location

Therefore, one of the identified priority areas for prospect evaluation survey is the promising area in the north of the Dubovogorskaya structural terrace. Occurrences of sandy deltaic channel structures are associated with 2 local uplifts (Nos. 1, 2) stretching north-westward. Occurrences of sandy bars, spits and coastal accumulative bodies are 5 structures (Nos. 3, 4, 5, 6, 7) stretching north-eastward.

4. Conclusion

Evaluation of prospective resources of the Devonian terrigenous OGC in Perm Krai is based on the ad hoc approach due to the complex structure and lesser degree of knowledge compared to the overlying OGCs. The main peculiarity of the OGC under study is its limited horizontal extent across the reservoir area, therefore evaluation of D_1 category resources took into account the area of their most probable occurrence. The authors determined the analogy coefficient through statistical justification of the most insightful parameters descriptive of the degree of similarity between geological structures of RR and ER identified in the Devonian terrigenous OGC. Based on the performed computations, it was proven that the complex under study is promising for geological exploration. The most prospective area was determined taking into account the facies and structure peculiarities, specifically the presence of local uplifts within facies zones of deltaic channels and sandy bars. Therefore, it is recommended to perform increased density 3D seismic exploration in the prospective area, build detailed seismic facial schemes, and perform thorough exploration of the identified local objects.

References

[1] Melkishev O A, Galkin V I, Kozhevnikova E E, Karaseva T V 2014 Ediction of zonal hydrocarbon potentials devonian clastic sediments on the south of perm region *Neftyanoe khozyaystvo – Oil Industry* **6** 4-8

[2] Kozlova I A, Krivoschekov S N, Zykova L Yu, Shadrina M A, Bashkova S E 2014 Geological and geochemical assessment of oil and gas in the upper proterozoic possibility sediments in the perm region *Neftyanoe khozyaystvo – Oil Industry* **5** 55-59

[3] Galkin V I, Karaseva T V, Kozlova I A, Nosov M A, Krivoshchekov S N 2014 Differentiated probabilistic assessment of the generation processes in domanic sediments of perm region *Neftyanoe khozyaystvo – Oil Industry* **12** 103-105

[4] Melkishev O A, Galkin V I, Galkin S V, Gurbanov V Sh, Koshkin K A 2018 Application of cluster analysis in assessing the initial total in place oil resources' density for highly explored areas *Scientific works of NIPI Neftegaz SOCAR* **3** 16-23

[5] Galkin V I, Kozlova I A, Nosov M A, Krivoshchekov S N 2015 Solutions to regional problems of forecasting oil bearing according to geological and geochemical analysis of dispersed organic matter of domanic type rocks *Neftyanoe khozyaystvo – Oil Industry* **1** 21-23

Impact of hydraulic fracturing on development indicators at BS10, BS11, US11 zones of Yuzhno-Yagunskoye field

N Kh Gabdrakhmanov, E M Almukhametova, A A Karimova and R R Stepanova

Ufa State Petroleum Technological University, Branch of the University in the City of Oktyabrsky, 54a, Devonskaya St., Oktyabrsky, Republic of Bashkortostan, 452607, Russian Federation

E-mail: rrngm_304@mail.ru, info@of.ugntu.ru

Abstract. The study considers three production zones of Yuzhno-Yagunskoye field. The study analyzes one method of oil intensification from the formation, namely Hydraulic Fracture Treatment (HFT), and provides the results of its influence on development indicators within three zones of Yuzhno-Yagunskoye field. A series of hydraulic fracturing operations was carried out in the period from 1995 to 2017. The methods for analyzing the bottomhole zone of the formation were used to process field data. However, it should be noted that in comparison with previous periods of hydraulic fracturing technology, the effect is reduced, which implies further modification of the used hydraulic fracturing technology. The purpose of this study is to evaluate the process efficiency of hydraulic fracturing at Yuzhno-Yagunskoye field. The above results were used during hydraulic fracturing at Yuzhno-Yagunskoye field.

1. Introduction

Hydraulic fracturing technologies have been actively used for many years to intensify production and increase the permeability of the bottom-hole formation zone. Hydraulic fracturing in producing and injection wells is one of the effective tools for enhanced oil recovery that involves low-permeability zones and interlayers, which allows transferring part of off-balance reserves to industrial analogs [1].

The main reason for the low productivity of wells along with poor natural permeability of the formation and poor-quality perforation is a decrease in the bottomhole zone permeability [2].

The main task of the development of any field is the maximum possible production of oil and gas at a low level of technological and economic costs per unit of production.

The purpose of this study is to consolidate and deepen the theoretical training of students and their acquisition of practical skills and competencies in the field of production activities, develop the ability to independently carry out research work related to complex professional problems in innovative conditions, as well as to evaluate the technological efficiency of HFT at Yuzhno-Yagunskoye field.

This purpose implies the following main objectives:

1) To analyze the prospects and possibilities of using the achievements of scientific and technological progress in the innovative development of the industry.

2) To carry out analytical, simulation and experimental studies, evaluate the obtained data.

For HFT, preference should be given to wells meeting the following criteria. The latter as a whole make it possible to ensure the intensification of oil production with high probability.

IOP Conf. Series: Earth and Environmental Science 1021 (2022) 012012 doi:10.1088/1755-1315/1021/1/012012

2. Materials and methods

Incremental oil production from HFT in injection wells is 30% higher than in production wells. This is caused by the stronger influence of the resulting increase in the injection well flow rate on the drainage mode of the site at the same productivity increase ratios as production wells.

The reduction of bottomhole zone permeability occurs during well operation, which is accompanied by violation of thermobaric equilibrium in the formation system and extraction of free gas, paraffin and asphalt-resinous substances from oil that clog the pore space of the reservoir. Intense contamination of the bottomhole zone is noted as a result of penetration of working fluids into it during various repair works in wells.

Hydraulic fracturing in production and injection wells is one of the effective tools for enhanced oil recovery that involves low-permeability zones and interlayers, which allows transferring part of off-balance reserves to industrial analogs [3].

There are some factors that should be considered for HFT.

There are various lithological characteristics of the formation, namely type of the reservoir, degree of grain consolidation, degree of fracturing and cavern porosity, degree of clay content. The HFT experience in Russia shows that the greatest effect of HFT operations is obtained in carbonates or strongly cemented sandstones with low clay content and low degree of fracturing.

The unsuccessful HFT were marked by some signs and one of the first is the destruction of shale barriers and, as a result, increased water production. The presence of cracks in the formation jeopardizes the HFT operations, since the fracturing fluid may be lost in natural cracks and we will not get any effect. Lithological heterogeneity is characterized by the net-to-gross sand ratio, number of permeable intervals, anisotropy coefficient. A large effect is obtained when exposed to a homogeneous formation with low permeability anisotropy coefficient and other factors.

The novelty of the study results was evaluated as part of the state of the oil production complex of the Republic of Bashkortostan and the Russian Federation, where the "intellectualization" of oil production is quite low. The performed study might bring the state of oil production of the Republic of Bashkortostan and the Russian Federation to a new level.

3. Results

The HFT at Yuzhno-Yagunskoye field began in 1995. As of 01.01.2017 in total 474 wells were put into operation after HFT, of which 385 production wells of the operating well stock, 78 – put into operation with the HFT from drilling and 11 – were put into the injection well stock after the HFT. In addition, 4 wells were not put into operation after the HFT. Most of the HFT within the operating well stock was made on operating wells (167 HFT), 34 HFT – on non-operating wells, 24 wells were transferred from another horizon and 4 wells were returned to the object with HFT [4].

During the joint operation of objects, regardless of how the processing was carried out, the fracturing process parameters were similar (proppant weight – 13.2 and 14.5 tons, fracturing fluid flow – about 2.8 m^3/min). In both cases, after the HFT there was an increase in water content (up to 73.0 and 72.2%, respectively). At the same time higher fluid flow rate was obtained with combined formation stimulation – 50.6 t/day, oil flow rate – 14.1 t/day [1].

The average HFT efficiency by well category is shown in Fig. 1. Well treatment after drilling showed slightly higher results compared to hydraulic fracturing concerning the operating well stock by an average of 13-16% in fluid and oil production rate. Water production after the HFT on average reaches quite similar values: 49.1% for the drilling well stock and 47.7% for the operating well stock.

In particular, the higher HFT effects on drilling wells were caused by higher net thickness (13.3 m versus 11.0 m), as well as the use of 2 times the weight of proppant and the higher rates of the fracturing fluid injection.

It is noted that 6 months after the fracturing operations at drilling wells, the fluid flow rate decreases by an average of 20%, at production wells – by 30%. The achieved effects on oil in the first month for both categories of wells were quite close (19.7 t/day for the drilling well stock).

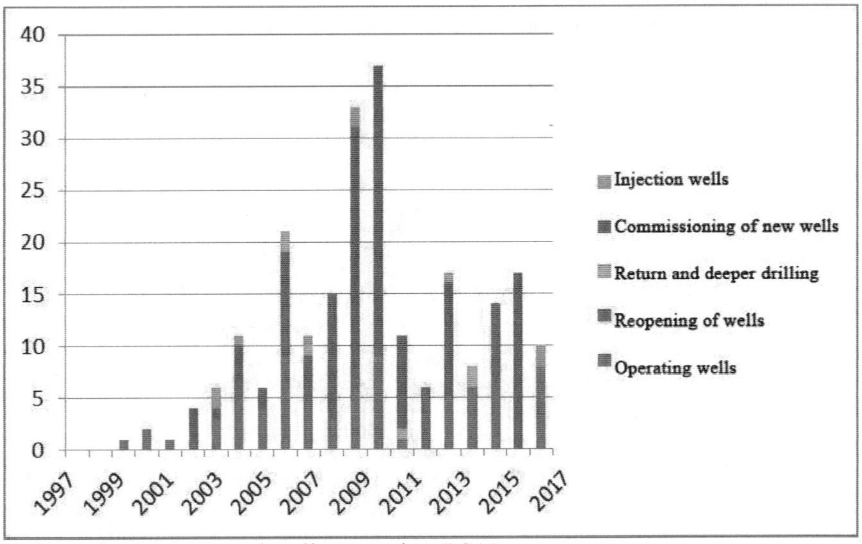

Fig. 1. HFT dynamics by years and well categories. BS11

The treatment on the group of BS11 formations was carried out under the following geological conditions: effective thickness in the range of HFT averaged 2.7 m (variation interval from 0.4 to 4.2 m), permeability – 30.9 mD (variation interval from 4.2 mD to 58.4 mD), oil and gas saturation – 0.46 unit fraction.

The HFT at BS11 was carried out in the 2005-2010 period. During this time, 6 HFT operations were performed, all at the wells of the current operating well stock. When estimating the average oil and fluid production rates before and after HFT for 3 months of well operation, the following results were obtained: on average, before HFT, the production rate of wells was 3.4 t/day for oil and 7.4 t/day for fluid, with water production of 54.1%.

After treatment, average oil and fluid rate reached 7.9 t/day and 54.1 t/day, respectively, while watering increased to 85.4%. The reasons for high water content in wells are considered below. Incremental oil production due to HFT in the group of BS10 objects amounted to 1598.6 thousand tons (of which 27.1 thousand tons accounts for by 5 drilling wells) [3].

160 HFT operations were performed at SUS11 from 1999 to 2017, of which 73 were the performed on the operating well stock, 81 – drilling wells and 6 – wells that were later put into the FPM system. The largest volume of treatment was carried out in 2010 (37 HFT), with 73% performed in drilling wells.

After carrying out two large-volume HFT at BS10, the oil flow rate was 6.0 t/day – well No. 1300 (proppant weight 46 t) and 25.7 t/day – well No. 1536 (proppant weight 30 t), fluid – 91.5 t/day and 42.5 t/day, respectively. For well No. 1536, this HFT was a repeated operation; the effect obtained corresponds to the results of the first treatment performed with a proppant weight of 10.5 tons. At well No. 1300 after HFT there was a jump in water production from 34 to 94%; due to the fluid withdrawal forcing, the effect lasts for more than two years, the current incremental oil production is 4.1 thousand tons, fluid – 74.9 thousand tons.

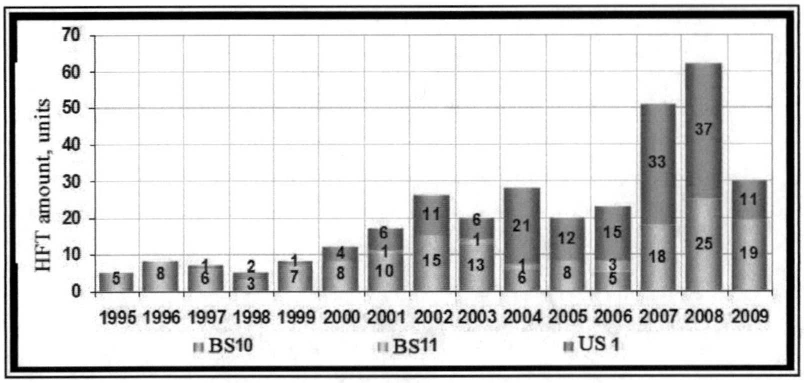

Fig. 2. HFT dynamics by years and objects of impact

As a result of hydraulic fracturing with injection of 10 40.3 tons of proppant into the formation with subsequent combined operation at well No. 258, the initial oil flow rate was 3.2 t/day, fluid flow rate – 25.1 t/day, water production – 87.4% (before HFT – 20-30%), incremental oil production – 1.3 thousand tons.

After HFT during combined operation of BS11 formations, oil production increased to 10.6 t/day, fluid – to 17.5 t/day, water – to 15.9%. As of 1.01.2017, incremental oil production amounted to 5.9 thousand tons, the average annual increase in oil production – 9 t/day.

Incremental oil production due to HFT at SU11 amounted to 1565.6 thousand tons, on average 10.2 thousand tons are accounted per one job, including 678.2 thousand tons and 9.3 thousand tons/well, for drilling wells – 887.3 thousand tons and 11.0 thousand tons/well. Thus, during formation stimulation with an effective capacity of less than 12 m, the average fluid flow rate reaches 29.4 t/day, oil – 13.8 t/day [1]. In addition to modeling the main process in hydrodynamic simulators, it is recommended to perform mini-HFT to increase the overall HFT efficiency [5].

Mini-HFT is the most important field test before the main HFT and is defined as relatively short in time injection at the speed of the major job using relatively large volumes of fluid and a small amount of proppant. It is performed before the main hydraulic fracturing to minimize differences, to obtain reliable data on the geomechanical properties of rocks.

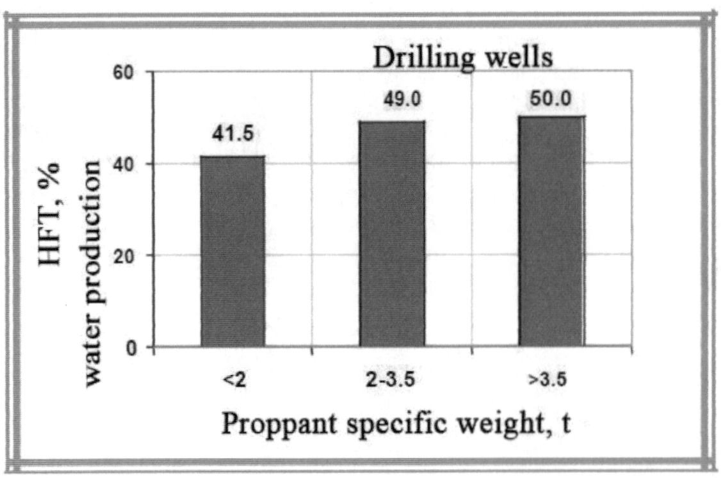

Fig. 3. Distribution of water production after HFT according to proppant specific gravity ranges. BS11

Besides, it is worth considering the history of these events at the site to conduct a more effective HFT. Since, in view of the similar geological peculiarity of the reservoir and physicochemical properties of its saturating fluids, it is possible with a high degree of probability to predict the success of HFT at a particular well or its impracticability [5].

The well was operated jointly on two formations BS11 and BS10 with flow rates of 0.86 t/day for oil and 0.90 t/day for fluid and with water content of 5.1% prior to HFT. In June 2012, HFT was performed in the formation interval BS11 with an effective thickness of 4.2 m and a permeability factor of 33.8 mD. During treatment, 43.4 tons of proppant were injected into the formation at an aqueous gel flow rate of 2.8 m^3/min, with a maximum proppant concentration of 910 kg/m^3 and an average concentration of 552 kg/m^3.

After HFT, during the joint operation of BS11 and BS10 formations, the fluid flow rate increased to 48.2 t/day, water production – to 97.4%, incremental oil rate amounted to 0.4 t/day. Well No. 767 is located near highly watered wells (No. 750, 751, 752). The effect continues due to a further decrease in water production to 91-92%, and the well additionally produced 3.8 thousand tons of oil.

Table 1. Comparison of the main geological, technological parameters and efficiency indicators of the first and second HFT at BS11

Parameter	Well No. 3362	Well No. 3116	Well No. 3268
Net thickness, m	4.2	6.0	4.9
Oil saturation, fraction units	0.49	0.52	0.52
Alpha SP, fraction units	0.56	0.64	0.64
Permeability, mD	33.8	31.5	40.0
Fluid flow rate prior to HFT, t/day	0.9	2.5	2.7
Oil flow rate prior to HFT, t/day	0.86	1.4	2.9
Water production prior to HFT, %	5.1	6.1	9.2
Fluid flow rate after HFT, t/day	80.1	86.1	67.3
Oil flow rate after HFT, т/сут	48.2	47.8	35.4
Water production after HFT, %	68.2	70.2	77
Initial incremental oil rate, t/day	12.0	6.2	7.1
Incremental oil rate, тыс.	10.4	6.5	3.8
Working hours for oil, days	1274	1353	566
Max. proppant concentration, kg/m^3	853	637	737
Average proppant concentration, kg/m^3	470	315	420

Table 1 shows that the oil production rate will be higher in wells where three HFT were carried out than in wells with seven HFT. The more fracturing was carried out at the wells, the lower the efficiency of oil production and the higher water cut.

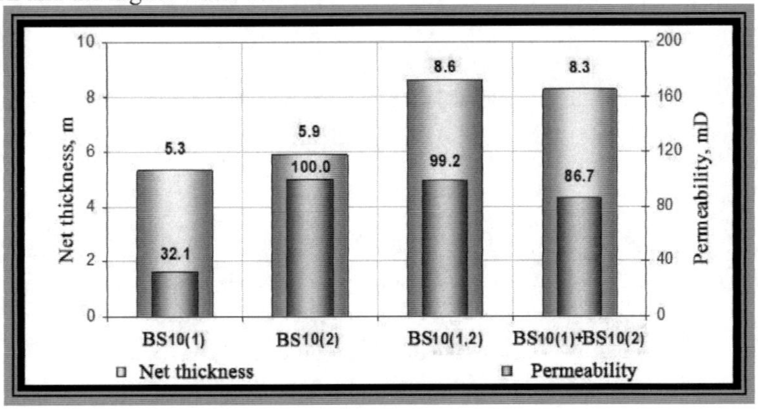

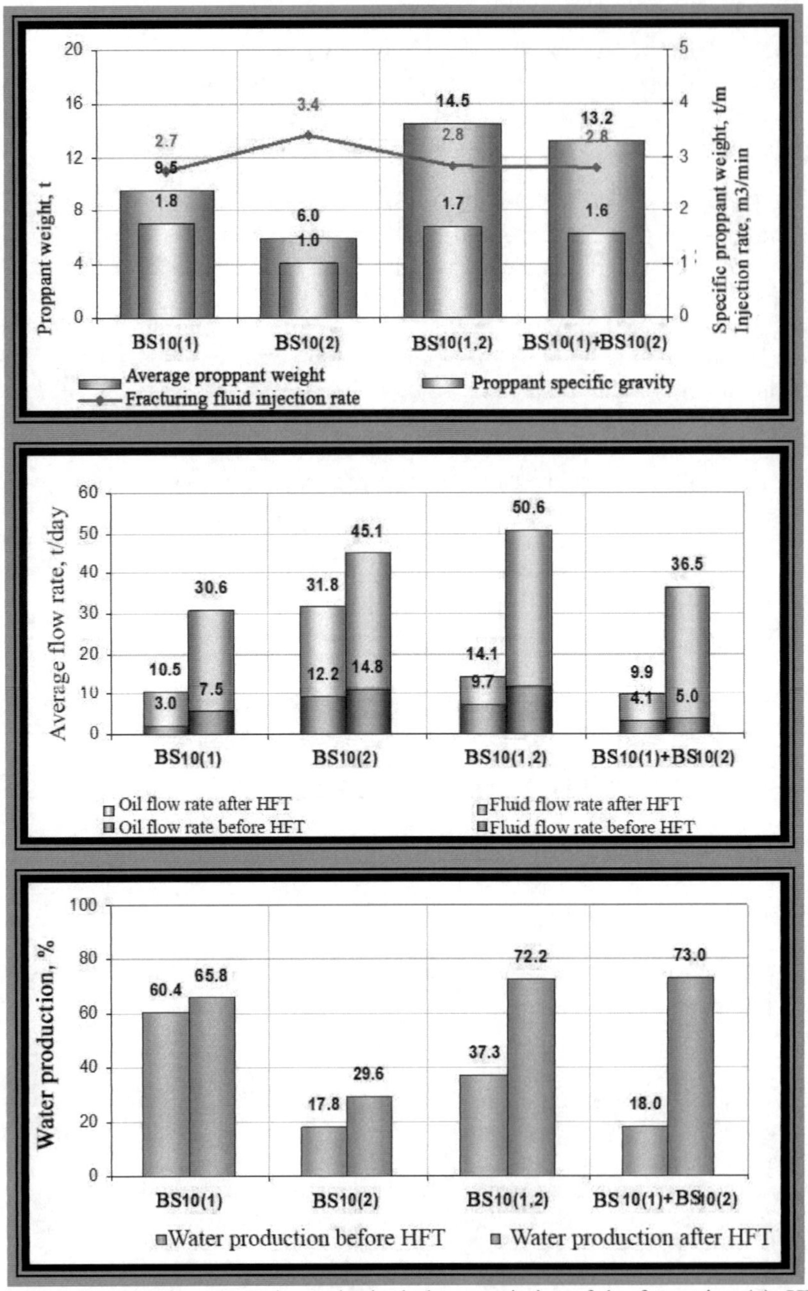

Fig. 4. Average values or main geological characteristics of the formation (a), HFT process parameters (b) and treatment efficiency (c, d) for wells with HFT by objects of impact BS10

It should be noted that in general in 2012-2017, the HFT at BS10 were carried out with higher proppant weight compared to the HFT in 1997-2017 (12.8 versus 7.0 tons) and lower fracturing fluid injection rate (2.2 versus 3.4 m^3/min). At the same time, oil efficiency after the HFT in 2012-2017 was lower relative to the HFT in 1997-2017 by 44% (9 versus 16 t/day), with similar indicators of fluid flow rates after the HFT (32.3 and 35.6 t/day, respectively).

To analyze the results of the performed HFT, paired dependencies of efficiency indicators were built for the main influencing factors (net thickness, proppant weight and proppant specific gravity), which also indicate the range of variation of these values.

No significant correlations between the influencing parameters and performance indicators were revealed at BS10 in any of the graphs. It may only be noted that there is a weak tendency to increase the specific fluid flow rate while increasing the specific gravity of the proppant. The major share of the HFT is performed with proppant weight up to 15 tons and specific gravity up to 3 t/m. There is also a tendency to oil production decrease after the HFT with an increase in proppant weight.

4. Conclusion

Physical, chemical and physicochemical methods of enhanced oil recovery are being applied at the field in order to improve the development and increase the ultimate oil recovery factor [4].

The comparison of incremental oil production from the methods used revealed that at present one of the most effective methods for enhanced oil recovery is hydraulic fracturing. The HFT does not only intensify the production of reserves located in the drainage zone, but also under certain conditions allows significantly expanding this zone by adding weakly drained zones and interlayers to the production, and, therefore, achieving higher ultimate oil recovery.

For HFT, preference should be given to wells meeting the following criteria. The latter as a whole make it possible to ensure the intensification of oil production with high probability.

When performing the HFT at BS10 with subsequent joint operation with BS10, the flow rates after the HFT are slightly inferior to the values at BS10 – 36.5 t/day for fluid and 9.9 t/day for oil.

The dynamics of flow rates given at the date of the HFT shows that during the year the oil flow rates remain stable – at the level of 29-32 t/day and 15-17 t/day, respectively for two areas (Fig.1.3). For BS10 and BS10 (1) + BS10 (2), on the contrary, over the same period of time, there is a decrease in both fluid (by 25.7 and 37.8%, respectively) and oil flow rates (by 26.5 and 28.0%, respectively).

The reservoirs are low permeable (HFT ensures the increase of the filtration surface), so the following criteria should be met:

1. Net thickness – not less than 5 m;
2. Absence of gas from the gas cap in well production, as well as inject or edge water;
3. The productive formation subjected to the HFT is separated from other permeable formations by impermeable sections more than 8-10 m thick;
4. The distance of the well from gas-oil contact and water-oil contact shall exceed the distance between production wells;
5. Cumulative oil uptake from the well shall not exceed 20% of specific recoverable reserves;
6. The compartmentalization of the productive interval (subject to the HFT) – not more than 3-5;
7. The well shall be technically serviceable, both the state of the production string and the adhesion of cement stone to the column and rock shall be satisfactory in the interval above and below the filter – by 50 m;
8. Formation permeability shall be not more than 0.03 μm^2 at oil viscosity under formation conditions – not more than 5 MPa/s.

Hence, the above criteria make it possible to carry out a versatile preliminary expert assessment of each well from technical, technological and geological-field perspective.

Thus, at the production wells, the best oil efficiency, both initial and average annual, as a result of the HFT was obtained at US11. For other objects it is much lower due to high water production.

For most wells of BS10, there is an increase in water production after the HFT.

In general, after the HFT smaller effects were obtained at BS11 compared to BS10. So, on average, after six treatments, the fluid flow rate was 54.1 t/day, oil flow rate – 7.9 t/day, and water content – 85.4%. At BS11, repeated HFT operations were not performed.

For US11, higher efficiency of hydraulic fracturing was obtained during the introduction of new wells than for the operating well stock: the average fluid flow rate after hydraulic fracturing was 37.1 and 32.0 t/day, oil flow rate – 18.9 and 16.7 t/day, respectively. In particular, this is caused by higher net thicknesses (13.3 vs. 11.0 m), as well as the use of 2 times the weight of proppant (41.1 vs. 24.1 t). Water production after the HFT on average reaches close values: 49.1% for drilling wells and 47.7% for the operating well stock.

US11 is also characterized by an increase in water content after the HFT. So, if the water production of 18% of wells was more than 50% before the HFT, then after the HFT their number increased to 31%.

For drilling wells on average the level of water content amounted to 51.4%, for 48% of wells the water content exceeded 50%. As a rule, high water content after operations is caused by the introduction of drilling wells with the HFT in the edge zones.

Depending on the initial permeability of the formation and the condition of the downhole zone, the criteria are grouped into the following two categories.

The reservoirs are low permeable (HFT ensures the increase of the filtration surface).

Hydraulic fracturing in medium and low permeability reservoirs is used for enhanced oil recovery due to elimination of increased filtration resistances in the bottomhole zone.

Between the formations of the BS10 group the best oil efficiency, both initial and average annual, was observed for BS10 as a result of the HFT (average annual increase – 21.2 t/day). For other objects of impact, it is significantly lower, which is caused by high water production (average annual increase – 6.5-11.4 t/day).

It is noted that 6 months after fracturing operations at drilling wells, the fluid flow rate decreases by an average of 20%, at production wells – by 30%. The achieved effects for oil in the first month for both categories of wells are quite similar (19.7 t/day for drilling wells, 21.7 t/day for the operating well stock). However, in the following months, oil production from drilling wells increased on average for 6 months by 4%. The production wells, on the contrary, are characterized by a decrease in oil production – on average over the same period by 26%.

The following results were achieved during this study.

Knowledge on decision-making in conditions of uncertainty and multi-criteria in relation to oil and gas industry.

Knowledge on the design and development of new innovative technological processes and equipment for hydrocarbon transport.

References

[1] Almukhametova E M 2018 Recommendations on thermal non-stationary waterflooding when modeling oil field development *IOP Conference Series: Earth and Environmental Science* **6** 062002 DOI:10.1088/1755-1315/194/6/062002

[2] Almukhametova E M, Zakirov A I, Fattakhov D I, Faizullin A A, Safiullina A R 2018 Analysis of the technology of intensifying oil production through the bottomhole formation zone treatment in the Potochnoe field *IOP Conference Series: Earth and Environmental Science* **8** 082005 DOI:10.1088/1755-1315/194/8/082005

[3] Kotenev Yu A, Mukhametshin V Sh, Sultanov Sh Kh, Kotenev A Yu 2020 Cyclic impact on productive formations of high-viscosity oil deposits *IOP Conference Series: Materials Science and Engineering* **905** 012082 DOI: 10.1088/1757-899X/905/1/012082.

[4] Mukhametshin V V 2020 Oil Production Facilities Management Improving Using the Analogy Method *SOCAR Proceedings* **4** 42-50 DOI: 10.5510/OGP20200400464

[5] Gidley J L 1989 R*ecent advances in hydraulic fracturing* (SPE)

[6] Mukhametshin V V, Kuleshova L S 2019 Justification of Low-Productive Oil Deposits Flooding Systems in the Conditions of Limited Information Amount *SOCAR Proceedings* **2** 16–22 DOI: 10.5510/OGP20190200384

[7] Terracina J M, Turner J M, Collins D H, Spillars S 2010 *SPE Annual Technical Conference and Exhibition,* Florence, Italy, 19-22 September

[8] Brannon H D, Malone M R, Rickards A R, Wood W D, Edgeman J R, Bryant J L 2004 SPE Annual Technical Conference and Exhibition, Florence, Italy, 19-22 September, pp. 1-23

[9] T.T. Palisch, M.C. Vincent, P.J. Handren, *SPE Annual Technical Conference and Exhibition,* pp 1-20 (2008)
[10] Mukhametshin V Sh, Andreev V E, Yaskin S A 2020 Designing measures to increase oil recovery based on the identification and grouping of deposits,IOP: Earth and Environmental Science **579** 012013, DOI: 10.1088/1755-1315/579/1/012013
[11] Mukhametshin V V, Kuleshova L S 2020 Analogies method application to increase the involvement degree in the deposits development of the Jurassic oil and gas complex in Western Siberia, Journal of Physics: Conference Series **1582**, 012065 DOI: 10.1088/1742-6596/1582/1/012065
[12] Kadyrov R R, Mukhametshin V V, Galiullina I F, Kuleshova L S, Safiullina A R 2020 Prospects of applying formation water and heavy brines derived therefrom in oil production and national economy *IOP Conference Series: Materials Science and Engineering* **905** 012081 DOI: 10.1088/1757-899X/905/1/012081

IOP Conf. Series: Earth and Environmental Science 1021 (2022) 012013 doi:10.1088/1755-1315/1021/1/012013

Use of geotechnical monitoring data in the design of the Serginsky field development

Alsu F Shakurova, Aygul F Shakurova

Ufa State Petroleum Technological University, Branch of the University in the City of Oktyabrsky, Oktyabrsky, Republic of Bashkortostan, Russian Federation

E-mail: alsu0017@mail.ru, info@of.ugntu.ru

Abstract. The paper deals with the use of geotechnical monitoring data to substantiate commissioning of production facilities in the Serginsky field. At the first stage, the geological and physical features of the reservoirs of the Serginsky field were considered; at the second stage, the results of substantiation were adjusted with regard to technological and economic factors. In addition, the experience of deposit exploitation gained during the period of pilot industrial development of the field was considered to substantiate commissioning of production facilities. The main part of the oil reserves of the Serginsky field (85%) is accumulated in two reservoirs J_{2-3} and J_{10}. Geological oil reserves of J_1 and J_4 reservoirs account for about 15.2% of the total oil reserves of the field.

1. Introduction

Tectonically, the Serginsky field is located in the Serginsky dome of the West Siberian platform and is confined to the structure of the same name.

Commercial oil deposits were found in the terrigenous deposits of the Bazhen-Abalak complex of the Upper Jurassic, and in the deposits of the Tyumen and Sherkala series of the Middle and Lower Jurassic.

2. Methods and materials

The software IRAP RMS (Roxar) was used as a research tool.

3. Results

The commercial oil content in the Serginsky field is associated with the deposits of the Abalak, Tyumen and Sherkala series of J_1, J_{2-3}, J_4, J_{5-1}, J_{5-2}, J_9, J_{10} reservoirs.

The location of oil deposits of the field mainly coincides along the oil-water contact (Figure 1, Table 1).

When commissioning production facilities, the following conditions must be taken into account:

- reservoirs can be combined for joint development only if they belong to one oil-bearing column, that is, the reservoirs must be located at close depths and be characterized by a slight difference in the values of the initial reservoir pressure and temperature;
- natural reservoir regimes must be similar;
- reservoirs must be similar in lithology and reservoir type in order to avoid differences in the type of filtration of reservoir fluids in reservoirs with different structures of the void space [1];

- combined reservoirs must not fundamentally differ in permeability and heterogeneity to ensure uniform reservoir production at similar bottomhole pressure;
- viscosity of oil in reservoir conditions must be of the same order in the combined reservoirs, which ensures the general patterns of oil displacement;
- oil from reservoirs must be of the same commercial qualities to avoid combining oils that require different technologies for field preparation and processing;
- production facility must exhibit significant reserves per unit of area (specific reserves) to ensure the long-term well operation [2].

The commissioning of production facilities was substantiated in two stages. At the first stage, the geological and physical features of the reservoirs of the Serginsky field were considered; at the second stage, the results of substantiation were adjusted with regard to technological and economic factors. In addition, when commissioning of production facilities was substantiated, the experience of deposit exploitation accumulated during the period of pilot development of the field was taken into account [3, 4].

The Serginsky field is multi-layer. Each reservoir comprises several deposits.

The geological oil reserves were estimated for 11 deposits confined to 7 reservoirs.

Table 2 presents the ratio of oil reserves by reservoirs.

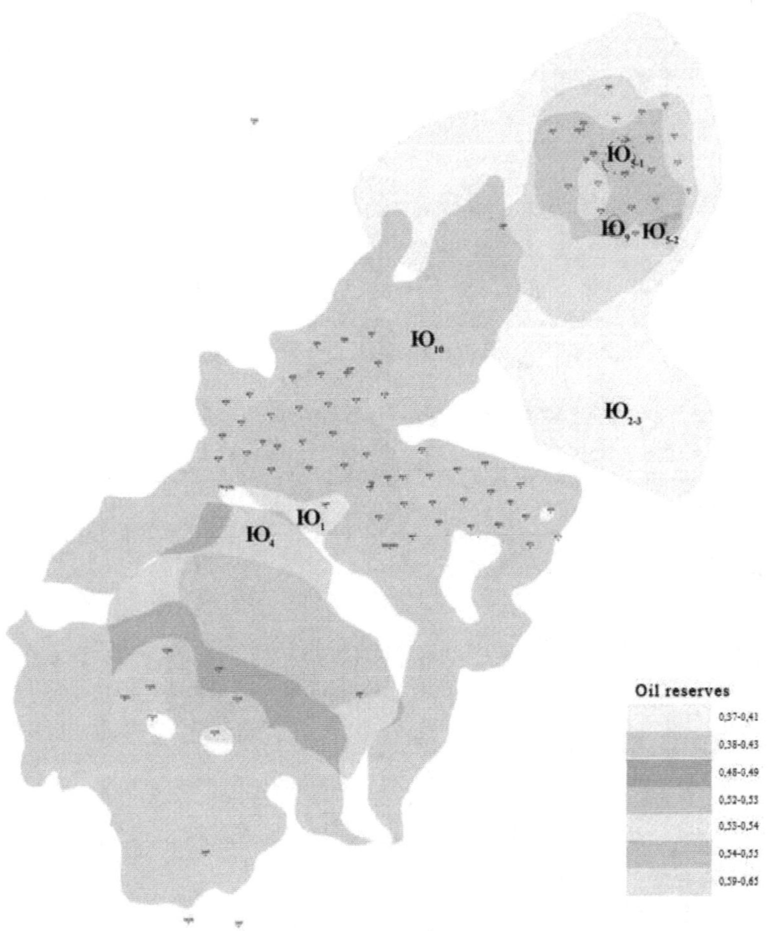

Figure 1. Map of the combined oil-water contact s in the Serginsky field

Table 1. analysis of the coincidence of production facilities in design in the Serginsky field

Reservoir	Degree of the coincidence of reservoir in design,%						
	J_1	J_{2-3}	J_4	J_{5-1}	J_{5-2}	J_9	J_{10}
J_1	-	41	26.6	1.0	0.9	0	53.9
J_{2-3}	35	-	0	0.9	0.8	0.3	23.5
J_4	57.8	0	-	0	0	0	28.5
J_{5-1}	100	100	0	-	0	0	100.0
J_{5-2}	100	100	0	0	-	0	21.0
J_9	100	100	0	0	0	-	97.4
J_{10}	26.2	13.3	6.4	0.5	0.1	0.2	-

Table 2. Distribution of oil reserves in reservoirs

Reservoir	Initial geological oil reserves (IGR), thousand tons			Share of the total volume of IGR,%	Oil productive area, thousand m^2
	POZ	*OWZ*	*Total*		
J_{10}	14919	7217	22136	53	68776
J_{2-3}	9804	3516	13320	32	38923
J_1	316	434	750	1.8	34119
J_4	4640	980	5620	13.4	15590
J_{5-2}	17	18	35	0.08	308
J_{5-1}	0	24	24	0.06	350
J_9	9	0	9	0.02	130
Total	29705	12189	41894	100	158196

The above data show that the main part of oil reserves of the Serginsky field (85%) is concentrated in two reservoirs J_{2-3} and J_{10}. Geological oil reserves of J_1 and J_4 reservoirs are about 15.2% of the total oil reserves of the field. Formations J_{5-1}, J_{5-2} and J_9 make up an extremely insignificant part of the total reserves (0.16%). In addition, the oil-water contacts of these objects in design completely coincide with the oil-water contacts of J_{2-3} and J_{10} reservoirs. The oil-water contact of J_1 reservoir almost completely coincide in design with the deposits of J_{2-3}, J_4 and J_{10} reservoirs.

The deposits are distributed over the area at a small distance from each other, in places where the deposits coincide in design in the section, from two to four deposits can be concentrated. Deposits differ in structure from lenticular and partially lithologically shielded to fully underlain by water, that is, without an internal oil-water contact, which indicates different periods for oil reserves exploitation, individual systems of development and placement of wells. These factors indicate that for better oil reserves exploitation and higher oil production rates, separate exploitation of reservoirs is preferable [5–7].

The need for commissioning of each of production facilities separately is substantiated by the scattered depths of the reservoirs, which differ from each other by tens and hundreds of meters, as well as the coincidence in the pure oil zone of one reservoir and water-oil zone of another one, for example, the deposits of J_{5-1} and J_{2-3} reservoirs.

All this necessitates further development of each reservoir as an independent production facility [8, 9].

The analysis of the fields of the current oil saturation is carried out both in the area and in the section, and allows determination of:

- weighted average oil saturation values;
- zones of low and high oil saturation;
- pattern of the distribution of oil saturation along the section;
- level of oil reserves production by elements of heterogeneity.

The values of the weighted average oil saturation of reservoirs by the area were determined based on the design oil saturation fields (Ko.s.cur.> Ko.s.res.) within the outer contour of the OWC.

The degree of production of mobile oil reserves can be clearly estimated using the so-called "production factor" (1), which includes comparative data on the relative change in mobile oil reserves and serves as a very flexible and convenient tool in the form of a map:

$$K_{prod.} = 1 - \frac{K_{o.s.t.} - K_{o.s.res.}}{K_{o.s.init.} - K_{o.s.res.}}, \quad (1)$$

where $K_{o.s.cur.}$ is the current value of the oil saturation coefficient,

$K_{o.s.init.}$ is the initial value of the oil saturation coefficient,

$K_{o.s.res.}$ is the value of the residual oil saturation coefficient determined for the reservoirs in the study objects.

The production coefficient changes over time, since the front of the water entering the reservoir captures all new sections of the reservoir and interlayers as it moves, and when the direction of filtration flows changes, it captures stagnant and dead-end zones.

Thus, the value of the production factor equal to "0" indicates the absence of changes in saturation at a given point in the reservoir and, in fact, identifies the zones that are not affected. The value of the production factor equal to "1", on the contrary, indicates a decrease in saturation at a given point in the reservoir to the residual level and shows the full depletion of the covered reserves. The value of the production factor at the end of production will show the degree of incomplete production of mobile drained oil reserves, i.e. loss of oil due to the production termination as a result of technical and economic restrictions: when the maximum permissible water cut or minimum oil production rate are attained using the existing methods and technologies. In fact, the value of the production factor at the end of production is the waterflooding factor [10–12].

To date, the following method of commissioning production facilities is the most optimal in terms of production experience, and current geological-physical and technical-economic information.

Commissioning of seven production facilities: J_1, J_{2-3}, J_4, J_{5-1}, J_{5-2}, J_9, J_{10} reservoirs.

Analysis of the distribution of the amount and density of geological reserves in these objects is presented in Figure 2.

In production facilities J_{10} and J_{2-3}, reserves production takes place fairly uniformly with average rates of oil withdrawal. Comparative analysis of the operation indicators (chronology of changes in production rates, water cut, injection volumes, bottomhole and reservoir pressures) with the results of 3D exploration seismology showed that a section of wells located in the north-west of the main deposit of J_{10} reservoir – well #43 – requires control and production management. Analysis of the dynamics of the well performance in this area indicates the need to change the operating parameters of injection wells, in particular, their transfer to a more radical mode of cyclic injection with alternating changes in the direction of filtration flows. As of the date of the analysis, the current compensation in this area is 138%, and the accumulated compensation is of about 120%. Almost all the wells operate at pressures corresponding to the saturation pressure level. The production of initial oil reserves in the section of J_{10} reservoir in the area of well #43 is 40% with the current water cut of 42%. At present, the intensification of fluid production in this area will not lead to an increase in oil production. Therefore, it is possible to stabilize the levels of oil production or reduce the rate of its loss in J_{10}, well #43, only by regulating water injection and ensuring rational bottomhole pressures in the production stock.

J_{10} reservoir is distinguished by the amount of reserves as the basic production facility. The largest concentration of oil reserves is observed in J_{10} in net oil thicknesses of less than 8 m (about 69% of all reservoir reserves), in J_{2-3}, about 65% of the geological oil reserves are concentrated in the thickness ranging from 4 to 12 m.

Increased permeability of the reservoirs increases the production factor. In this regard, two groups of reservoirs characterized by the degree of production can be distinguished in terms of permeability. These are actively drained reservoirs with permeability of more than 50 μm, and weakly drained

IOP Conf. Series: Earth and Environmental Science 1021 (2022) 012013 doi:10.1088/1755-1315/1021/1/012013

reservoirs with permeability of less than 50 μm. Most of the mobile reserves are concentrated in weakly drained reservoirs.

The location of the objects in terms of the area of the licensed site (Figure 1), and the characteristics of the size and density of reserves (Figure 2) indicate that it is reasonable to design two independent systems in the zones of joint occurrence of $J_{10} - J_{2\text{-}3}$ and $J_{10} - J_4$.

At the same time, it is possible to develop all three objects with a single grid of wells and use the technology of simultaneous-separate exploitation (SSE). in the zones of combined oil-water contacts. This technology will reduce capital investments by reducing the number of wells for drilling and ensuring separate operation of each facility.

J_1, $J_{5\text{-}1}$, $J_{5\text{-}2}$ and J_9 are return production facilities, and the start of their exploitation will correspond to the moment of commissioning of wells that fulfilled their tasks in J_{10}, $J_{2\text{-}3}$, and J_4.

In the section of $J_{2\text{-}3}$, mobile oil reserves are uniformly distributed with a shift towards the roof, and the production occurs evenly. In the section of J_{10}, mobile oil reserves are mainly concentrated in the roof and central parts of the reservoir, the highest production of reserves falls on the central and bottom part.

Field exploitation options were created based on the analysis of the geological structure of deposits, the current state of exploitation with regard to the degree of exploitation of each deposit in the field, previously adopted and partially or completely implemented solutions for systems of well placement and impact on reservoirs, with regard to the actual results of geological and technical measures taken to intensify and increase oil recovery, performed at the wells in the Serginsky field.

Currently, the main deposits of the Serginsky field include the central reservoir J_{10} and the deposit of $J_{2\text{-}3}$, which are at the final stage of drilling. The rest of the deposits of the field have not been previously exploited. The current design provides the location of wells according to two inscribed nine-point systems with a grid density of 32 ha/well – one for the exploitation of the Sherkala oil deposits, and the other one for the exploitation of oil deposits in the reservoirs of the Tyumen suite.

For production facilities ($J_{5\text{-}1}$, $J_{5\text{-}2}$ and J_9), the area and reserves of which do not allow the creation of even one exploitation element, the options were calculated using reservoir pressure maintenance ($J_{5\text{-}1}$ and $J_{5\text{-}2}$) and without it (J_9), hydraulic fracturing is provided at all the facilities. The exploitation of J1 is carried out in natural mode for all options and with the use of hydraulic fracturing in option 3 [5, 6].

The average values of the operating factors for both production and injection wells are taken equal to 0.452 unit fractions in the first year after drilling and 0.951 unit fractions in subsequent years.

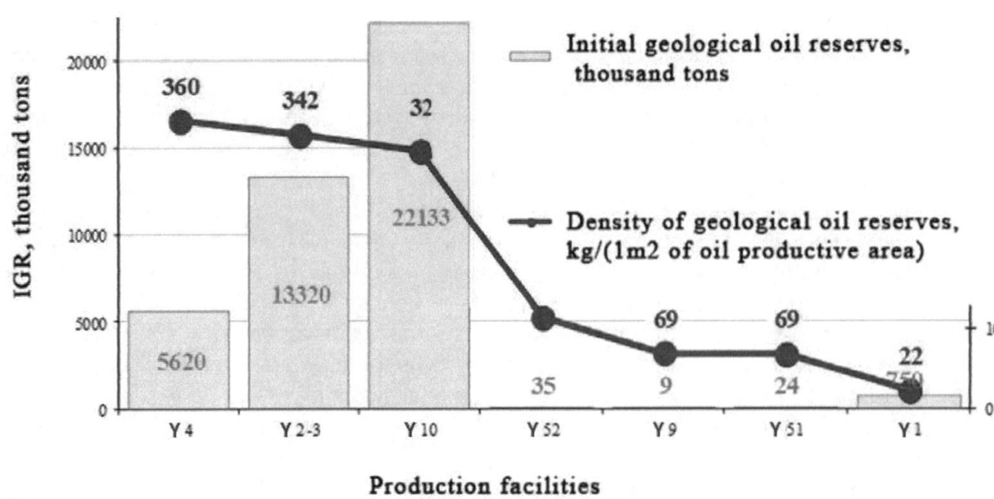

Figure 2. Characteristics of the production potential and density of oil reserves at the production facilities in the Serginsky field

In all the design options for production wells, a restriction is set on the bottomhole pressure equal to the saturation pressure. Bottomhole pressures of injection wells are limited by the value of the hydraulic fracturing pressure. Withdrawal compensation by injection was maintained at 115%. The shutdown of production wells in the prognostic estimation was carried out when the water cut attained 98%.

To achieve the highest oil recovery during the exploitation of the Serginsky field, with regard to the current economic conditions, a complex of innovative technologies is proposed at the wells of the field: injection of a water-gas mixture (using surplus associated gas), drilling horizontal wells, hydraulic fracturing, drilling horizontal wells using interval hydraulic fracturing, simultaneously-separate operation of facilities, sidetracking at a later stage of exploitation, the implementation of which will allow achieving the required oil recovery factor [7].

When choosing design options for the exploitation of production facilities in the Serginsky field, the following general principles were considered:

- Application of selective adaptive oilfield exploitation system, which includes dispersed waterflooding, originally designed as areal waterflooding according to the inverted 9-point pattern, is converted into selective waterflooding during drilling, including near-contour selective waterflooding. The selective adaptive system includes the principle of drilling from the center to the periphery, from the known to the unknown, from the best to the worst; accounts data obtained during drilling and well survey; allows combination of the industrial exploitation of oil reservoirs with additional survey;
- Provision of individual water injection into injection wells and optimal injection pressure by plunger pumps located directly on the well clusters, which optimizes the joint operation of injection and surrounding production wells; uniformly maintain reservoir pressure, preventing oil outflow into the aquiferous zone, thereby increasing oil production rates of wells and oil recovery of reservoirs;
- The use of deep intensive perforation of reservoirs in underbalanced conditions with a depth of perforation channels of 50–100 cm and breakup of the zone of penetration of drilling fluid using physicochemical methods for restoring productivity, including ultrasound;
- Drilling and well exploitation is carried out with high quality; the primary opening of the reservoir is carried out at equilibrium using modern drilling fluids and high-quality resistant bits, which allow passing the largest interval of productive deposits in one hole advancing;
- Continuous control over the operation of production and injection wells (control over the values of fluid and oil production rates, water cut, water injection, bottomhole and reservoir pressures and other parameters). The control accuracy is high and allows for engineering decisions to optimize the operation of production and injection wells, their repair in a timely manner, and shut out of operation.

At all the facilities of the Serginsky field, an areal inverse nine-point adaptive exploitation system is being implemented.

In the first (basic) option, the development of the exploitation system is considered according to the approved version of the previous design document, with the drilling of the remaining number of directional wells with independent grids in the zones where J_{2-3} and J_{10} contours, and J_{10} and J_4 contours coincide. Drilling of production wells in the field is carried out along two inscribed square grids with a density of 32 ha/well each. The well clusters of J_{10} facility are drilled first. Drilling of reservoirs of the Sherkalinskaya suite resumes in 2012. Production drilling in J_4 begins in 2020. Then, starting from 2023, drilling of J_{2-3} continues. The total density of the grid of production drilling of the field (in the areas of joint exploitation in design of production facilities) varies from 16 to 32 ha/well.

To attain the highest oil recovery, a set of technologies is proposed: injection of a water-gas mixture, drilling horizontal wells, hydraulic fracturing, and acid treatment.

The second option implies the development of exploitation system with a denser grid of 16 ha/well (400x400 m) at a thickness of more than 2 m is considered, with the drilling of the remaining number of vertical and directional wells with independent grids in the zones, where J_{2-3} and J_{10} contours, and J_{10} and J_4 contours are combined. Drilling of production wells in the field is carried out along two square grids (16 ha/well) inscribed in each other. The well clusters of J_{10} are drilled first. Drilling of reservoirs of the Sherkalinskaya suite resumes in 2012. Production drilling of J_4 is planned for 2027. Then, starting from 2031, drilling of J_{2-3} continues. The total density of the grid of production drilling of the field (in the areas of joint exploitation in design of production facilities) varies from 8 to 16 ha/well.

To attain the highest oil recovery, a set of technologies is proposed: injection of a water-gas mixture, drilling horizontal wells, hydraulic fracturing, and acid treatment.

The third (recommended) option considers the development of exploitation system with a well grid density of 32 ha/well (566x566 m).

Drilling of wells is provided for by a single grid of wells using the technology of simultaneous-separate exploitation of objects (in the areas of joint exploitation in design of production facilities). Production drilling of reservoirs resumed in 2012.

To attain the highest oil recovery, a set of technologies is proposed: Simultaneous-separate operation of facilities, injection of a water-gas mixture, hydraulic fracturing, drilling horizontal wells using interval hydraulic fracturing, acid treatment, sidetracking at a later stage of exploitation.

With regard to the practice of developing similar low-permeability Jurassic objects, hydraulic fracturing is planned for 95% of the production wells.

4. Conclusion

A comprehensive assessment of geological field data and analysis of the current state of exploitation in the section of deposits in the Serginsky field, as well as analysis of the substantiation of the choice of production facilities, yield the following conclusions:

1. It is proposed to retain seven production facilities in the Serginsky field: J_1, $J_{2\text{-}3}$, J_4, $J_{5\text{-}1}$, $J_{5\text{-}2}$, J_9 and J_{10}.

2. The technology of simultaneous-separate exploitation of wells in the zones where $J_{2\text{-}3}$, J_4 and J_{10} contours are combined is proposed to reduce capital investments by reducing the stock of wells for drilling.

3. J_1, $J_{5\text{-}1}$, $J_{5\text{-}2}$ and J_9 are return production facilities, and the start of their development will correspond to the moment of commissioning of wells that fulfilled their tasks at facilities $J_{2\text{-}3}$, J_4 and J_{10}.

4. For J_1, $J_{5\text{-}1}$, $J_{5\text{-}2}$ and J_9, the main task is to further study them in the process of pilot operation.

References

[1] Economides J M, Nolte K I 2000 *Reservoir stimulation* (West Sussex, England: John Wiley and Sons)

[2] Mukhametshin V Sh 2020 Justification for increasing the performance of hydrochloric acid treatment in wells of fields with carbonate reservoir, *IOP Conference Series: Materials Science and Engineering* **905** 012083 DOI: 10.1088/1757-899X/905/1/012083.

[3] Nasr-El-Din H A, Van Domelen M S, Sierra L, Welton T D 2007 Optimization of Surfactant-based Fluids for Acid Diversion, Europ. Format. Damage Conf. (Scheveningen, The Netherlands, 30 May – 1 June 2007) pp. 1–11. DOI: 10.2118/107687-MS

[4] Malyarenko A M,. Kotenev Yu A, Mukhametshin V Sh, Bogdan V A, Kotenev A Yu, Blinov S A, Umetbaev V G 2020 The influence of the rocks material composition on the reservoir properties of the reservoir *Journal of Physics: Conference Series* **1582** 012061 DOI: 10.1088/1742-6596/1582/1/012061.

[5] Shakurova Al F, Shakurova Ay F 2019 Influence of orientation of vertical crack of hydraulic fracturing treatment on the efficiency of reserves recovery *Journal of Physics: Conference Series* **1333** 032076 doi:10.1088/1742-6596/1333/3/032076

[6] Shakurova Al F, Shakurova Ay F 2019 Study of geological and technological factors on the efficiency of hydraulic fracturing, *Journal of Physics: Conference Series* **1333** 062025 doi:10.1088/1742-6596/1333/6/062025

[7] Petrova L V, Buravov V V 2019 Evaluating the technological effect of a horizontal hydraulic fracturing Journal of Physics: Conference Series **1333** 072019 doi:10.1088/1742-6596/1333/7/072019

[8] Bakhtizin R N, Fattakhov I G, Kadyrov R R, Safiullina A R 2016 Integral analysis aimed at identification and analytical solution of issues on oil recovery efficiency enhancement,

International Journal of Applied Engineering Research **11 (3)** 1612–1621.

[9] Almukhametova E M, Shamsutdinova G F, Sadvakasov A A, Tyncherov K T, Petrova L V, Stepanova R R 2018 Modeling development of Fyodorovsky deposit *IOP Conference Series: Materials Science and Engineering* **327** 042100

[10] Tiab D, Donaldson E C 1999 *Petrophysics: Theory and Practice of Measuring Reservoir Rock and Fluid Transport Properties* (Houston, TX: Gulf Publ.)

[11] Akhmetov R T, Mukhametshin V V, Kuleshova L S 2019 Simulation of the absolute permeability based on the capillary pressure curves using the dumbbell model *Journal of Physics: Conference Series* **1333 (3)** 032001 DOI: 10.1088/1742-6596/1333/3/032001.

[12] Webb K J, Black C. JJ, Tjetland G A 2005 Laboratory Study Investigating Methods for Improving Oil Recovery in Carbonates, Int. Petrol. Technol. Conf. (Doha, Qatar, 21–23 November 2005) pp. 1–6. DOI: 10.2523/IPTC-10506-MS

The structure of the oil-bearing strata and possibility of extracting oil from them

L B Akhmetianova, R N Bagmanov, M M Raymanov, D I Kobishcha, and Z N Sagitova

Ufa State Petroleum Technological University, Branch of the University in the City of Oktyabrsky, 54a, Devonskaya St., Oktyabrsky, Republic of Bashkortostan, 452607, Russian Federation

E-mail: lena21vek18@mail.ru, info@of.ugntu.ru

Abstract. It is important to study the structure of oil-bearing strata is. In many ways, the structure of the Earth's layers affects the possibility of intensifying oil production. This is one of the urgent problems of the oil industry. Obviously, at various stages of oil field development, methods of intensifying oil production, depending on the structure of oil-bearing rocks and geological and physical characteristics have their own features, which are manifested when they are used in the field conditions. Determination of the oil recovery factor (ORF) causes great difficulty in the operational assessment of oil reserves. At the stage of initial assessment, this indicator is taken either by analogy, or approximately. In order to make this parameter more valid, the main factors influencing its formation were studied. The formula for calculating oil recovery factor takes into account three main coefficients: grids, displacement and waterflooding.

1. Introduction

The problem of intensifying oil production has become more crucial with the introduction of oil fields with low-productive formations. Experience shows that the problem of intensification arises when opening productive formations in the process of drilling wells. It can be solved with the use of high-quality drilling fluids, appropriate equipment and technological methods of well casing, which ensure reliable separation of productive formations and exclude clogging [1–7].

Another task is to restore the productivity of wells, which decreases during their operation for a variety of reasons, for example, the deposition of inorganic salts, high molecular weight oil components, the formation of stable emulsions, etc. The well productivity decreases due to a violation of the technology of underground workover and during repair and isolation works.

The third task is to increase the productivity index of the well. This task is of particular relevance due to the need to develop fields with hard-to-recover oil reserves. It is solved by the use of hydraulic fracturing, hydrosand-jet perforation, and acid treatment. Other methods are being developed in order to increase the productivity factor of wells [8–12].

2. Materials and methods

The sweep efficiency is usually determined after drilling, taking into account geological information and density of the well pattern.

IOP Conf. Series: Earth and Environmental Science 1021 (2022) 012014 doi:10.1088/1755-1315/1021/1/012014

When calculating the grid factor in terrigenous reservoirs, it should be taken into account that, since the spreading factor is related to the average thickness of the reservoirs, the grid factor should also be determined by their distribution over the area of the reservoir. In this case, the mesh factor should depend on the drilling density. The ratio of the volume of reserves covered by this grid to the total reserves is the grid coefficient. At low values of the average thickness of formations, it is recommended to take the into account features of the formation structure.

There is no data on the displacement factor in the rapid assessment of reserves, especially for small deposits. Therefore, they use statistical dependencies determined in the analysis of a large amount of factual material. Using the average value of porosity, these relationships can be used to determine the displacement factor.

No dependencies were found to estimate the reservoir sweep efficiency. It is recommended to take its value by analogy with closely located reservoirs with similar average values of oil permeability and viscosity. In this case, it is impossible to ensure its high accuracy.

3. Results and Discussion

At the fields characterized by water pressure regimes (Tula, Bobrikov, Askyn horizons and Trans-Volga above-horizon), the oil recovery factor (ORF) was determined by the ratio

$$\eta = K_{dis} K_{fl} K_s E,$$

где K_{dis} – oil water displacement factor; K_{fl} – flooding factor; K_s – sweep efficiency; E – displacement geometric efficiency factor.

The dependences of the displacement coefficient on the permeability (or porosity) for many fields in Bashkiria have been investigated. For the Tula horizon, there are 12 core permeability determinations (0.106 μm2); for Bobrikov horizon - formation CVI1 - 79 determinations (0.169 μm2), for formation CVI2 - 44 determinations (average value 0.323 μm2); along the Kizelov horizon - 221 determinations (0.044 μm2); no core was taken for the Trans-Volga over-horizon and the Askyn horizon; along the Kynovsky horizon, there were 7 determinations (0.148 μm2). To determine the displacement factor for the Lower Carboniferous sandstones (Tula and Bobrikov horizons), the dependences of the displacement factor on the permeability obtained for the terrigenous deposits of the Lower Carboniferous of the Voyadinskoye field were used; for the Askyn horizon, the dependence of the displacement factor on the porosity obtained for the carbonates of the Famennian stage of the Uzybashevskoye field was used. The maximum possible value of the displacement factor obtained from the experimental data was taken for the Trans-Volga superhorizon - 0.57. The values of the displacement factors are presented in Table 1.

Table 1. Displacement Factor Values

Objects	Permeability, μm ^{2}	Displacement factor, units
Tula horizon	0.106	0.60
Bobrikov horizon (CVI1)	0.169	0.62
Bobrikov horizon (CVI2)	0.323	0.65
Tran-Volga over-horizon	0.061	0.57
Askyn horizon	0.332; m=0.11	0.50

The sweep efficiency depends on the heterogeneity parameter a/k_0, the mobility coefficient μ_0 and the limiting water cut of wells, usually taken equal to 99%.

For a quantitative assessment of the reservoir heterogeneity parameter, quantile diagrams of the permeability distribution were constructed. The heterogeneity parameter for the CVI1 formation was 0.16, and for the CVI2 formation - 0.3. Due to the paucity of core data for the Tula horizon, this parameter was taken by analogy with the CVI1 formation (0.16) (the net-to-gross ratio and compartmentalization of these objects were close to each other). Due to the absence of core material for the Trans-Volga over-horizon and the Askyn horizon, this parameter was taken equal to 0.9 by analogy with the carbonate reservoirs of the neighboring Ardatovs (Kizelov horizon, Trans-Volga over-horizon) and Tuimazin (Kizelov horizon) fields.

The mobility coefficient μ_0 was determined from the relation:

$$\mu_0 = \frac{\mu_o}{\mu_d} \overline{K}_p$$

where μ_o – oil viscosity in reservoir conditions, mPa·s; μ_d – displacement fluid viscosity, mPa·s; $\overline{K}_p$ – relative phase permeability for water in the flooded zone.

According to the curves of phase permeabilities for analogue reservoirs, the relative permeability for water in the flooded volume can be taken equal to 0.3 for terrigenous reservoirs of the Tula and Bobrikov horizons; for the Trans-Volga overhorizon - 0.1 and 0.15 for the Askyn horizon.

The viscosity of formation water was estimated by its salinity and was equal to 1.3 mPa·s.

The values of the flooding efficiency are presented in Table. 2.

The sweep factor reflects the proportion of oil reserves covered by a grid of wells and participating in development. This coefficient was estimated for each oil reservoir by the ratio:

$$K_c = 1 - \frac{Q'_{res}}{Q_{res}}, \text{ c = g}$$

where Q'_{res} – geological oil reserves not covered by the well grid, Q_{res} – initial geological reserves of oil deposits.

Table 2. Sweep efficiency values when oil is displaced by formation water

Object	Deposit	a/c$_0$	$\mu_н$, mPa·s	$\mu_в$, mPa·s	K_f, units	μ_0	K_s units
Tula horizon		0.16	62.3	1.3	0.3	14.4	0.61
Bobrikov horizon	1	0.16	62.3	1.3	0.3	14.4	0.61
	2	0.16	62.3	1.3	0.3	14.4	0.61
	3	0.16	62.3	1.3	0.3	14.4	0.61
	3a	0.3	62.3	1.3	0.3	14.4	0.57
	4	0.16	62.3	1.3	0.3	14.4	0.61
	5	0.16	62.3	1.3	0.3	14.4	0.61
	6	0.16	62.3	1.3	0.3	14.4	0.61
	7	0.16	62.3	1.3	0.3	14.4	0.61
Trans-Volga overhorizon		0.9	17.8	1.3	0.1	1.4	0.48
Askyn horizon		0.9	19.6	1.3	0.15	2.3	0.46

When determining Q_{res}, it was assumed that undeveloped oil reserves are contained in zones adjacent to the replacement boundaries at a distance of 100 m from the replacement boundary. The values of this coefficient were calculated taking into account all measures for drilling new wells and sidetracks, returning wells from other facilities, using idle well stock. The values are presented in Table 3.

Table 3. Values of the sweep efficiency factor

Object	Deposit	Not covered by the grid of wells				K_g
		Oil-bearing area, thousand m ^{2}	Average oil-saturated thickness, m	Reservoir volume, thousand m ^{3}	Q', Thousand tons	
Tula horizon		435	0.7	305	49.7	0.84
Bobrikov horizon	1	-	-	-	0	1
	2	270	0.3	81	10	0.84
	3	-	-	-	-	1
	3a	225	1.0	225	20	0.94
	4	18	1.0	18	3	0.92
	5		the deposit is depleted			
	6	-	-	-	-	1
	7		the deposit is depleted			
Trans-Volga overhorizon	-	-	-	-	-	1
Askyn horizon	-	-	-	-	-	1

The geometric displacement efficiency factor is applicable only for reservoirs with water-oil zones. It takes into account the deterioration of oil displacement by water due to coning and depends on the ratio of water-saturated and oil-saturated thickness and the ratio of viscosities of formation oil and formation water.

The value of this coefficient by Sattarov for a purely water-oil zone is determined from the ratio:

$$E_0 = 1 - \left(\frac{h_{os}}{h_{os0}}\right)^2,$$

where h_{os} and h_{os0} – current and initial oil-saturated formation thickness,

$$\frac{h_o}{h_{o0}} = \frac{(1-f_w)\frac{\mu_o}{\mu_w}}{f_w + (1-f_w)\frac{\mu_o}{\mu_w}} \cdot \left(1 + \frac{h_{w0}}{h_{o0}}\right)$$

where h_{w0} – initial water-saturated thickness of the oil-water zone; f_w – limiting water cut of well shutdown, fraction of units.

If there is a purely oil zone, the geometric efficiency factor of displacement is estimated from the ratio:

$$E = \frac{Q_{por} + Q_{wor} E_0}{Q_{por} + Q_{wor}},$$

where Q_{por} and Q_{wor} – oil reserves in pure oil and water-oil zones.

When calculating this coefficient, the limiting water cut of wells for deposits with terrigenous reservoirs was taken equal to 99%, with carbonate reservoirs - 0.98, and the ratio h_{w0} / h_{o0} was taken equal to 1 based on the condition of the horizon position of the oil-water contact. The values of the coefficient E for deposits of production facilities are presented in Table 4.

Table 4. Displacement Geometric Efficiency Factor Values

Object	Deposit	μ_o, mPa·s	μ_{w}, mPa·s	H_o/h_{o0}	E_0	Q_{por}, thousand tons	Q_{wor}, thousand tons	E
Tula horizon		62.3	1.3	0.652	0.57	290	27	0.96
Bobrikov horizon	1	62.3	1.3	0.652	0.57	0	108	0.57
	2	62.3	1.3	0.652	0.57	0	63	0.57
	3	62.3	1.3	0.652	0.57	43	38	0.80
	3a	62.3	1.3	0.652	0.57	110	230	0.71
	4	62.3	1.3	0.652	0.57	26	10	0.88
	6	62.3	1.3	0.652	0.57	57	8	0.95
Trans-Volga overhorizon		17.8	1.3	0.437	0.81	7	5	0.92
Askyn overhorizon		19.6	1.3	0.471	0.78	0	48	0.78

The values of oil recovery factors for the objects and deposits are presented in Table 5.

Table 5. Factor values

Object	Dposit	K_{rec}, units	K_{sw} units	K_g	E	η
Tula horizon		0.60	0.61	0.84	0.96	0,295
Bobrikov horizon	1	0.62	0.61	1	0.57	0,216
	2	0.62	0.61	0.84	0.57	0,181
	3	0.62	0.61	1	0.80	0,303
	3a	0.62	0.57	0.94	0.71	0,236
	4	0.65	0.61	0.92	0.88	0,321
	6	0.65	0.61	1	0.95	0,377
Trans-Volga overhorizon		0.57	0.48	1	0.92	0,252
Askyn overhorizon		0.50	0.46	1	0.78	0,179

For the deposits of the Tula horizon, which is operated by well No. 1, recoverable reserves were estimated using the Kambarov method in accordance with the equation:

$$V_o = \beta - \alpha \frac{1}{V_l},$$

where V_o и V_l – cumulative oil and liquid production;

According to the previously investigated dependence $V_o = f(V_l)$, built on the basis of actual well operation data, the values of empirical factors were: α = 72.6190; β = 15.8796, with a correlation coefficient of 0.9979.

Recoverable oil reserves are determined from the ratio:

$$Q_{rec} = \beta - \sqrt{\alpha \cdot f_o},$$

They are equal to 14.7 thousand tons (with a final oil fraction of 0.02).

For reservoir 3 of the Bobrikov horizon, where well No. 2 is operated, recoverable reserves are estimated from the equation:

$$V_o = \beta - \frac{\alpha}{t},$$

where t – number of operation years.

According to the previously investigated dependence $V_o = f(1/t)$, built according to the actual well operation data, the values of the empirical factors were: α = 41.1732; β = 8.1724, with a correlation coefficient of 0.9982.

Recoverable oil reserves are determined from the ratio:

$$Q_{rec} = \beta - \sqrt{a \cdot q_{\lim}},$$

where $q_{\lim}$ – the limiting annual rate of oil withdrawal calculated from the condition of the limiting production rate of wells 0.2 t/day ($q_{\lim} = 0.008$). Recoverable reserves amounted to 7.6 thousand tons.

For the Kynov horizon, which is operated by well No. 3, recoverable oil reserves were determined using the Kopytov method. Dependence $V_o = f(1/t)$. The values of the empirical coefficients of this well were: α = 49.0571; β = 14.7492, with a correlation coefficient of 0.9983. Recoverable oil reserves calculated under the same conditions as in the previous case amounted to 13.9 thousand tons.

4. Conclusion

After analyzing a significant volume of field materials, the main development indicators such as the oil displacement factor, the sweep efficiency, the flooding factor, the geometric displacement efficiency factor, the oil recovery factor for the Tula, Bobrikov and Askyn horizons and the Trans-Volga overhorizon were calculated.

Based on the previously investigated relationships, empirical relationships were determined and recoverable reserves were estimated using various methods.

References

[1] Gabzalilova A Kh, Batalov D A, Mukhametshin V Sh, and Andreev V E 2021 Geological and technological justification of the parameters of acid-clay treatment of wells *IOP Conference Series: Materials Science and Engineering* **1064** 012058 DOI: 10.1088/1757-899X/1064/1/012058

[2] Mukhametshin V V, and Kuleshova L S 2021 Using the method of canonical discriminant functions for a qualitative assessment of the response degree of producing wells to water injection during the development of carbonate deposits *IOP Conference Series: Materials Science and Engineering* **1064** 012069 DOI: 10.1088/1757-899X/1064/1/012069

[3] Andreev V E, Chizhov A P, Chibisov A V and Mukhametshin V Sh 2019 Forecasting the use of enhanced oil recovery methods in oilfields of Bashkortostan *IOP Conference Series: Earth and Environmental Science* **350** 012025 DOI: 10.1088/1755-1315/350/1/012025

[4] Batalov D A, Andreev V E, Mukhametshin V V, and Kuleshova L S 2021 Development regulation of oil and gas reservoirs based on effective geological and geophysical information *IOP Conference Series: Materials Science and Engineering* **1064** 012055 DOI: 10.1088/1757-899X/1064/1/012055

[5] Malyarenko A M, Bogdan V A, Blinov S A, Kotenev Yu A, Mukhametshin V Sh, and Kotenev M Yu 2021 Improving the reliability of determining physical properties of heterogeneous clay reservoir rocks using a set of techniques *Journal of Physics: Conference Series* **1753** 012074 DOI: 10.1088/1742-6596/1753/1/012074

[6] Hodgin J E and Harrell D R 2006 The Selection, Application, and Misapplication of Reservoir Analogs for the Estimation of Petroleum Reserves *SPE Annual Technical Conference and Exhibition (San Antonio, Texas, USA, 24-27 September 2006)* DOI: 10.2118/102505-MS

[7] Beliveau D 2008 Waterflooding Viscous Oil Reservoirs *SPE Indian Oil and Gas Technical Conference and Exhibition (Mumbai, India, 4-6 March 2008)* DOI: 10.2118/113132-MS

[8] Mukhametshin V G, Dubinskiy G S, Andreev V E, Mukhametshin V V, and Kuleshova L S

2021 Geological, technological and technical justification for choosing a design solution for drilling wells under different geological conditions *IOP Conference Series: Materials Science and Engineering* **1064** 012061 DOI: 10.1088/1757-899X/1064/1/012061

[9] Averyanov A P, Andreev V E, Chizhov A P, Mukhametshin V Sh, and Chibisov A V 2021 Theoretical and practical aspects of absorbing layers insulation in the conditions of hydraulically perfect modes of grouting solutions injection *IOP Conference Series: Materials Science and Engineering* **1064** 012054 DOI: 10.1088/1757-899X/1064/1/012054

[10] Efimov E R, Chizhov A P, Mukhametshin V Sh, Chibisov A V, and Andreev V E 2021 Process modeling of gas displacement of oil on the example of oil fields in Bashkortostan *IOP Conference Series: Materials Science and Engineering (MEACS 2020 – International Conference on Mechanical Engineering, Automation and Control Systems)* **1064** 012060 DOI: 10.1088/1757-899X/1064/1/012060

[11] Watson R W, and Boukadi F H 1990 Effect of surface area and pore-size distribution on oil recovery at breakthrough, on oil recovery at infinite water-oil ratio, on residual oil saturation and on wettability (Paper SCA 9007) *Proceedings of the Society of Core Analysts, Fourth Annual Technical Conference (Dallas, Texas, USA, 1990)*

[12] Allan J, and Sun S Q 2003 Controls on Recovery Factor in Fractured Reservoirs: Lessons Learned from 100 Fractured Fields *SPE Annual Technical Conference and Exhibition (Denver, Colorado, 5-8 October 2003)* DOI: 10.2118/84590-MS

Probabilistic and statistical justification of the forecast of oil and gas potential in the area with established industrial oil capacity

N N Krysin[1], **M V Sologubova**[2]

[1] Perm National Research Polytechnic University, 29, Komsomolsky ave., Perm, 614990, Russia
[2] Perm LUKOIL-Engineering LLC, PermNIPIneft Branch in Perm, 3A, Permskaya st., Perm, 614060, Russia

E-mail: nkrysin@gmail.com

Abstract. The article provides an example of an integrated approach to geological exploration in an area with an established industrial oil capacity. This approach is based on the analysis of the geological structure of the area under consideration and the stochastic assessment of the resources of the prospects using the Monte-Carlo simulation. Within the area under consideration, two regions were allocated for separate forecasting of the oil and gas potential of the prospects: south-western and north-eastern. Potentially oil and gas-bearing structures have been established in each region: in the south-western region, all structures are classified as unpromising, in the north-eastern region – as exploration targets. For each structure in the north-eastern region, possible oil-bearing areas were calculated, analog fields were selected for the probabilistic assessment of the prepared resources using the Monte-Carlo simulation. Based on the results of the evaluation, the most exploration target was selected.

1. Introduction

When conducting geological exploration, even within the territories with established oil and gas potential, negative results of work are expected, therefore, the issue of increasing their efficiency is particularly acute. To increase the probability of a positive result of geological exploration, various methods can be applied, one of which is a qualitative analysis of the geological structure of the area.

At the stage of preparing the structure for deep drilling, it is impossible to accurately assess the resources within it. This uncertainty leads to additional economic risks [1] associated with overestimation or underestimation of hydrocarbon resources. At the moment, the resources of the prospects can be estimated by deterministic and stochastic methods. The deterministic method allows you to get a specific amount of resources, the stochastic method – an interval estimate of resources. The use of the stochastic estimation method is more justified on prospects, since it allows one to level out the error that occurs when taking the calculated parameters as discrete values from analog fields.

Thus, by combining a qualitative analysis of the geological structure and a stochastic assessment of the resources of prospects, it is possible to increase the accuracy of geological exploration.

2. Analysis of the geological structure

To demonstrate this approach, an area was chosen in which the presence of both industrial fields and non-productive structures was established. Geographically, the region is located in the southern part of the Perm Krai. The area is tectonically confined to one large geostructural element, in terms of oil and gas geology, it is located in one oil and gas accumulation zone.

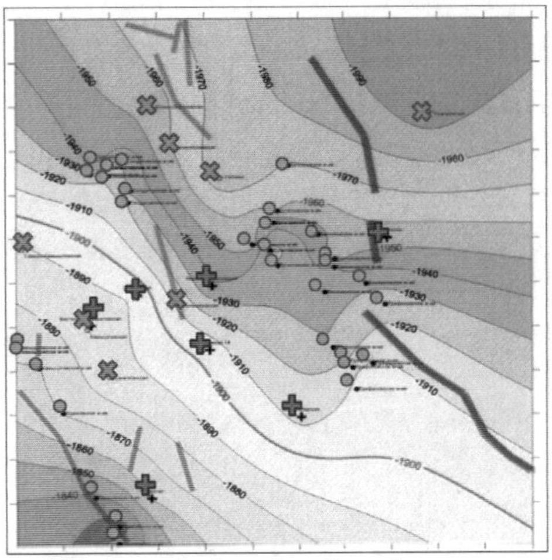

Figure 1. Caprock underground contour map D_3tm^{ter}.
There are 6 fields under development (indicated by orange circles), 7 prospects (indicated by blue crosses) and 8 non-productive structures (indicated by red crosses) in the work area.

When analyzing the oil and gas potential of productive fields, it was noted that the Devonian terrigenous oil and gas condensate field (OGCF) is distinguished only in the southwestern part of the area, the Upper Devonian-Tournaisian carbonate and Lower Middle Visean OGCF are represented throughout the area. Accordingly, in the south-western part of the area, search operations should be carried out for all the considered OGCFs, in the north-eastern part – for the Upper Devonian-Tournaisian carbonate and Lower Middle Visean OGCFs.

According to the drilling data, regional maps were constructed for five reflecting horizons, and their structure was analyzed. It was found that some of the fields under consideration are located hypsometrically below other developed fields and non-productive structures.

This observation raises the question of the need to divide the area into several parts for the possibility of predicting the oil and gas potential of exploration targets. The analysis was built by the graph of the absolute elevations of the caprock Tula terrigenous deposits (C_1tl^{ter}) from the absolute elevations of the caprock Timan terrigenous deposits (D_3tm^{ter}) shown in figure 2 which clearly stands out clear boundary separating the considered structure into two groups according to the absolute level of -1900 m on the caprock Timan terrigenous deposits.

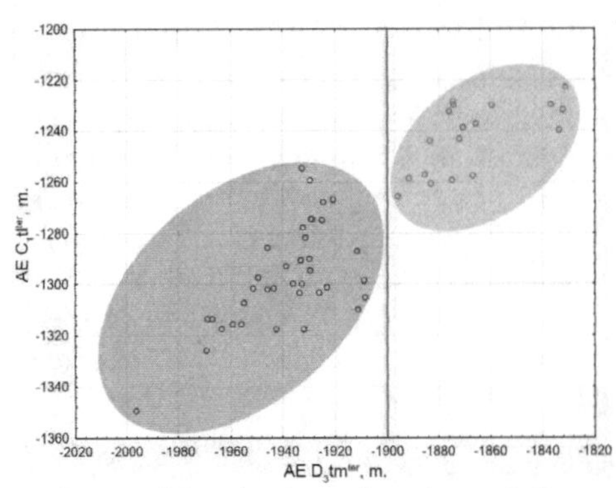

Figure 2. Graph of the dependence of the absolute elevations of the caprock C_1tl^{ter} on the absolute elevations of the caprock D_3tm^{ter}

The entire area was divided into two separate regions (Figure 1) – south-western and north-eastern. The prospects were assigned to one of these regions: 3 structures were assigned to the south-western region, 4 structures were assigned to the north-eastern region. The further forecast was carried out separately for each region.

3. Forecast of oil and gas potential on exploration target

The forecast of oil and gas potential on exploration target was carried out using discriminant analysis. To classify structures by productivity, the following parameters were used: absolute caprock elevations of Tula terrigenous deposits (AE_{C1tlT}), absolute caprock elevations of Tournaisian tier (AE_{C1t}), absolutecaprock elevations of Timan terrigenous deposits (AE_{D3tmT}), the thickness of the Visean deposits (T_{C1V}), the coordinates of the structures X and Y, the distance to the western faults (L_{WF}).

According to the initial data, the following discriminant equations were obtained for the southwestern (equation 1) and north-eastern (equation 2) parts of the area:

$D = 13090.7942 + 0.064 \times AE_{C1tlT} - 0.0543 \times AE_{D3tmT} - 0.0032 \times X - 0.0019 \times Y(1);$

$D = 13.2456 - 0.0429 \times AE_{C1t} + 0.1039 \times T_{C1V} - 0.0002 \times X + 0.0006 \times L_{WF}(2).$

Using equations 1 and 2, the roots of the discriminant function for productive and nonproductive structures were calculated, after which histograms of the distribution of these roots were constructed (Figures 3, 4) and boundary values were determined that allow the predictive structure to be classified as productive or non-productive. So, for the southwestern part of the area, the boundary value is 0: the root of the discriminant function of productive structures takes the value 0 and more (Figure 3). For the north-eastern part of the area, the boundary value is 2: the root of the discriminant function of productive structures takes the value of 2 and less (figure 4).

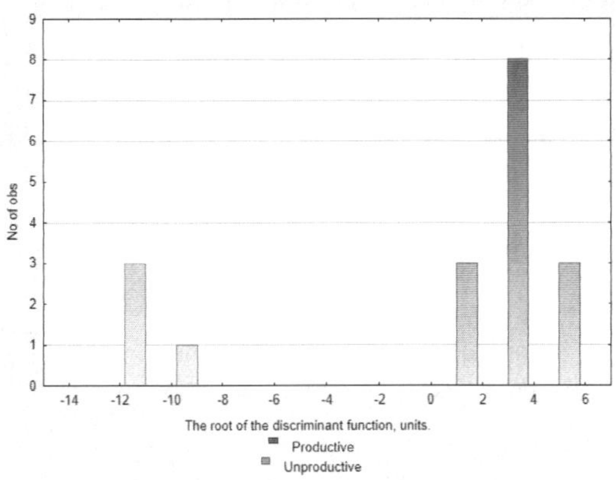

Figure 3. Histogram of the distribution of the roots of the discriminant function for the southwestern part of the area

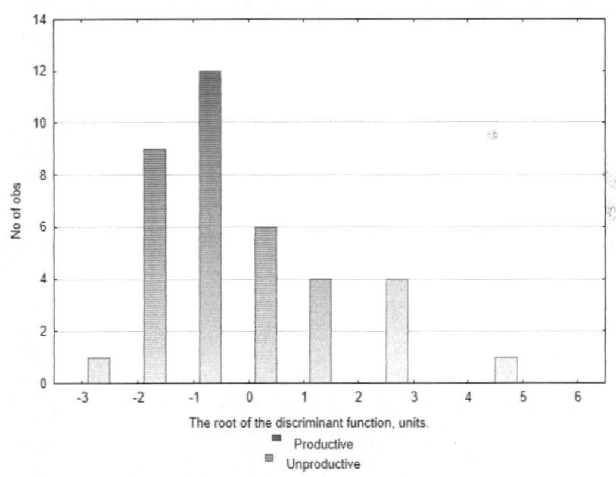

Figure 4. Histogram of the distribution of the roots of the discriminant function for the north-eastern part of the area

The values of the roots of the discriminant function were calculated for all the prospects. The roots of all three structures of the southwestern part of the area have a value less than 0. Accordingly, all prospects of the southwestern part of the area are classified as unpromising. The roots of the four prospects of the north-eastern part of the area have a value of less than 2, that is, all prospects of the north-eastern part of the area are classified as potentially productive.

Further resource assessment, respectively, will be carried out on four structures in the north-eastern part of the area.

IOP Conf. Series: Earth and Environmental Science 1021 (2022) 012015 doi:10.1088/1755-1315/1021/1/012015

4. Resource assessment on potentially productive structures

The resources on the prospects were estimated using the stochastic method. Stochastic resource estimation is based on the application of statistical distributions of calculation parameters and allows us to obtain an interval estimate of the resources of the structure [2]. In this case, a triangular distribution was used to model the calculation parameters. When using this approach, the probability of obtaining boundary values of the value of the counting parameter decreases.

By repeated random implementation, the distributions of each counting parameter are modeled, after which the obtained distributions are multiplied. As a result, you can get the distribution of hydrocarbon reserves on the structure. Then, using the Monte-Carlo simulation the proven resources (P90), probable resources (P50) and possible resources (P10) are estimated according to the PRMS classification [3].

To determine the boundary and modal values of the calculation parameters, it is necessary to select analogs for each prospect of the deposit. All deposits are located in relative proximity to each other, so the selection of an analog was carried out based on the common conditions of the deposit and the structure. As a parameter characterizing the geological structure of the structure, the absolute caprock elevation of the reflecting horizons were chosen, since many characteristics of both formations and deposits depend on the depth of the layers [4].

4.1. Determination of the deposit area

According to the maps of reflecting horizons, the area of the deposit in probably productive layers was predicted. On each structure, the areas of deposits were determined at the values of the trap-filling coefficient from a 0.1 unit fraction to 1.0 unit in increments of a 0.1 unit fraction. Table 1 shows an example of the values obtained for two of the four structures:

Table 1. The areas of prospects at different values of the trap fill factor

Structures		Structure 1		Structure 2	
Oil and gas complex		Lower Middle Visean	Upper Devonian-Tournaisian	Lower Middle Visean	Upper Devonian-Tournaisian
The area at the trap fill factor, thousand m^2	0.1	119	88.7	97.4	61.4
	0.2	210.9	198	292	173.2
	0.3	468.6	321	359.6	294.4
	0.4	519.8	480.4	513.8	468.5
	0.5	590.2	634.5	569.5	646.8
	0.6	798.1	875.9	780.1	781.4
	0.7	1064.7	1160.7	942.2	893.5
	0.8	1324.1	1465.0	1090.2	996.9
	0.9	1683.1	1687.4	1205.9	1155.3
	1	2038.0	1933.5	1348.7	1352.2

Linear equations constructed for each interval were used to determine the intermediate values of the areas by the value of the trap fill factor. Examples of the obtained dependency graphs are shown in Figures 5 and 6.

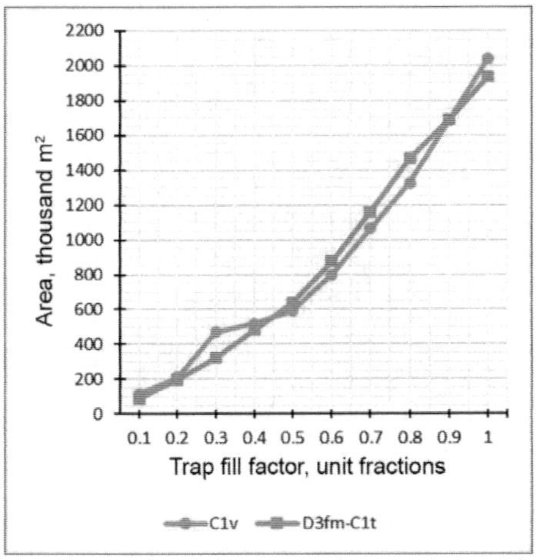

Figure 5. Graph of the dependence of the deposit area on the trap fill factor for Structure 1

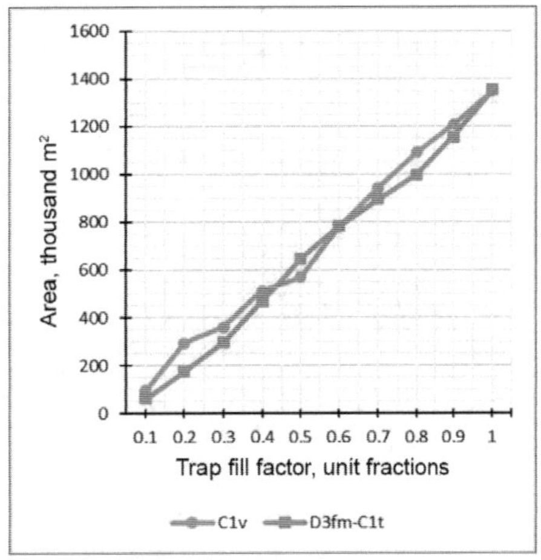

Figure 6. Graph of the dependence of the deposit area on the trap fill factor for Structure 2

When modeling the trap fill factor, a triangular distribution was used. The minimum, maximum and modal values of the trap fill factor were taken from analog fields. After that, the simulated values of the trap fill factors were substituted into linear equations. As a result, the values of the deposit areas were obtained with random values of the trap fill factor for each structure.

4.2. Modeling of calculation parameters and resource estimation

The calculation parameters were also modeled using a triangular distribution. The minimum, maximum and modal values of the calculated parameters were taken from analog fields. As a result, using the Monte-Carlo simulation [2] distributions of all calculated parameters for each productive formation were obtained.

To evaluate the prepared resources of the structure, it is necessary to pre-evaluate the resources for each OGCF. To estimate the resources within the OGCF, the calculated parameters were averaged: the effective oil-saturated thicknesses were calculated as a weighted average over the area of the oil and gas complex, and the remaining parameters were calculated as average values. After that, histograms of the distribution of calculation parameters for each oil and gas complex were obtained. The histogram data were multiplied, as a result, the values of geological and recoverable resources of both oil and dissolved gas for each OGCF were obtained. Histograms of the distribution of recoverable oil resources are shown in Figures 7-10.

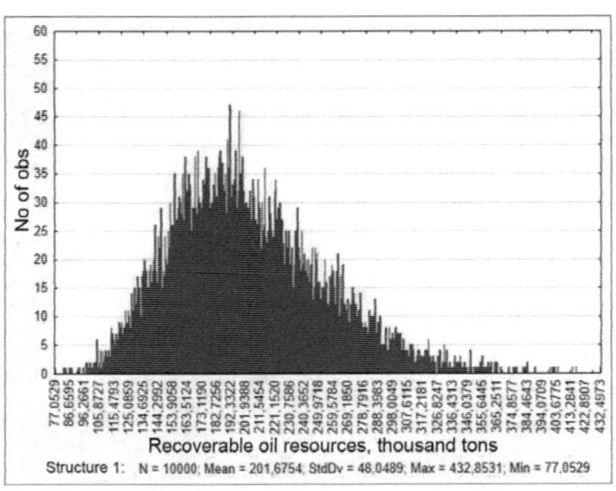

Figure 7. Histogram of the distribution of recoverable oil resources for Structure 1

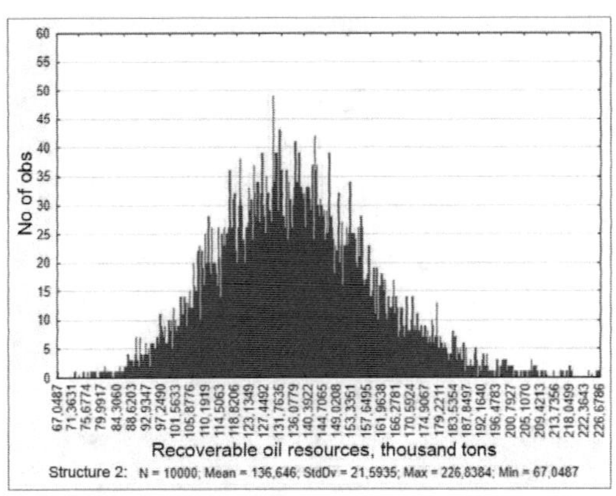

Figure 8. Histogram of the distribution of recoverable oil resources for Structure 2

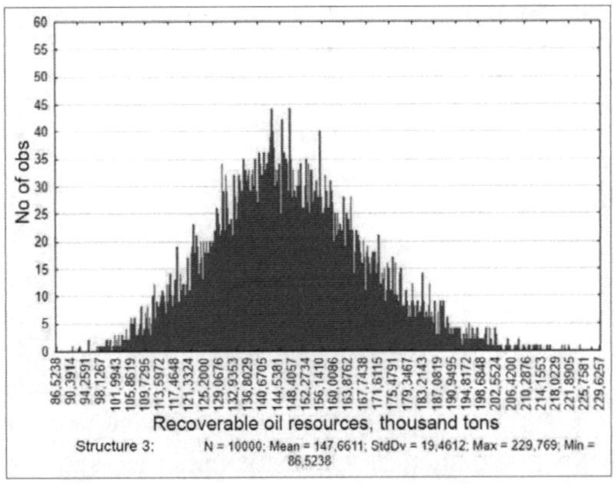

Figure 9. Histogram of the distribution of recoverable oil resources for Structure 3

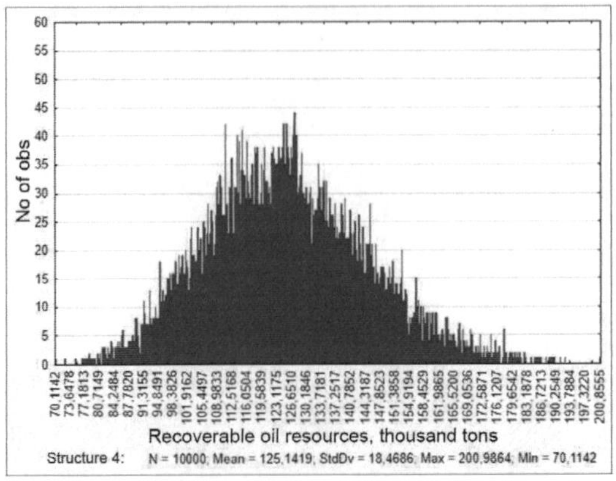

Figure 10. Histogram of the distribution of recoverable oil resources for Structure 4

The obtained histograms of the distribution of geological and recoverable resources were analyzed, resulting in the values of resource values corresponding to proven resources (P90), probable resources (P50) and possible resources (P10) according to the PRMS classification [3]. The resource values for each formation are shown in Table 2.

Table 2. Hydrocarbon resources on prospects

Structures	Resources											
	Geological resources of oil, thousand tons			Recoverable oil resources, thousand tons			Geological resources of gas, million m^3			Recoverable gas resources, million m^3		
	P90	P50	P10	P90	P50	P10	P90	P50	P10	P90	P50	P10
Structure 1	411	553	735	144	196	267	15	20	27	5	7	10
Structure 2	314	385	461	110	136	165	11	14	17	4	5	6
Structure 3	284	369	479	101	132	171	9	12	16	3	4	6
Structure 4	243	295	354	102	124	148	9	11	13	4	5	6

According to Table 2, it can be seen that the largest number of hydrocarbon resources is expected in Structure 1.

5. Conclusion

Thus, the paper demonstrates an integrated approach to solving search tasks in areas with established industrial oil capacity. This approach provides for a qualitative analysis of information about the geological structure of the area and a stochastic assessment of the resources of the prospects. Based on the conducted analysis, it is possible to make a reasonable choice of the most prospecting structure for further staging of search and evaluation works on it.

References

[1] Galkin S V, Ivanov A A *Probabilistic-statistical methodology for considering the risks of prospecting operations when evaluating alternative investment projects* pp. 29–28

[2] Melkishev O A, Krivoshchekov S N 2012 *Stochastic evaluation of oil resources forecast on the stage of geological exploration work*

[3] Kelliher C, Mahoney L *Using Monte Carlo Simulation to Improve Long-Term Investment Decisions* pp. 44-56.

[4] SPE/AAPG/WPC/SPEE/SEG *Guidelines for Application of the Petroleum Resources Management System*

IOP Conf. Series: Earth and Environmental Science 1021 (2022) 012015 doi:10.1088/1755-1315/1021/1/012015

[5] Proshlyakov B K 1987 *Reservoir properties of sedimentary rocks at great depths* (Moscow: Nedra) pp. 60-64

Development of an experimental set-up for studying the effect of elastic oscillations on fluid flow through rocks

E P Riabokon, A L Kazantsev, E A Gladkikh, M S Turbakov

Perm National Research Polytechnic University, 29, Komsomolsky ave., Perm, 614990, Russia

E-mail: riabokon.evgenii@gmail.com

Abstract. The article presents the results of experimental studies of the influence of elastic oscillations on the structure of paraffin oil of a Perm region field. Oil samples are treated by elastic oscillations at a frequency of 22 kHz. As a result of cavitation and an increase in oil temperature, a decrease in the concentration and a decrease in the size of high-molecular paraffin compounds in oil is observed. With an increase in the time of ultrasonic treatment, the intensity of dissolution of paraffins in oil increases. A model of an experimental setup is proposed to study the effect of elastic oscillations on the permeability of a rock during the flow of paraffin oil. The impact of permeability deterioration around the production well on its flow rate is assessed.

1. Introduction

Precipitation of paraffins in reservoirs is one of the factors that complicates oil production in Perm region since it deteriorates rock permeability and leads to a decrease in rate of production wells.

There are several reasons for paraffin precipitation such as decrease in reservoir temperature (as a result of cooling of a productive formation when using a waterflooding system with cold water) and pressure below the values of paraffin crystallization in oil [1, 2]. This also includes a decrease in reservoir pressure below the saturation pressure of oil with gas (appearance of free gas in the reservoir) [3].

As a result of a decrease in reservoir pressure as the formation fluid approaches the well (occurrence of a cone of depression) the paraffin precipitation is intensified. Considering that the decrease in the permeability of the near-wellbore zone has a significant impact on the well flow rate, it becomes necessary to clean (restore) the near-wellbore zone.

The main methods for restoration the near-wellbore zone from paraffin is treatment of wells using such types of impact as chemical (acid treatment) and physical (electromagnetic, heat, vibration treatment) [4, 5]. Despite during an acid treatment of the near-wellbore rock pore space become cleaner, destruction of the rock occurs as well; it is also required to shut down the well for the treatment. Vibration treatments contribute to the restoration of the pore space of the near-wellbore zone, increase the rate of capillary imbibition, improve the oil rheological properties and can be performed without shutting down the well.

The results of field [6, 7] and experimental [8, 9] studies suggest that a vibration treatment of the near-wellbore zone of production wells with elastic oscillations contributes to an increase in rock permeability and an increase in well productivity. The paper [10] presents the filed results of wells

Published under licence by IOP Publishing Ltd

treatment using oscillators of various operating principles. Due to the highest intensity of the observed effects, acoustic well treatment in the ultrasonic range became most widespread. An ultrasonic downhole oscillator operating at frequencies of 20-27 kHz and the mechanism of its effect on the fluid flow in the near-wellbore zone are presented in [11]. In order to achieve the maximum effect from the vibration treatment of the near-wellbore zone a mathematical model is proposed in [12] that allows one to select the optimal values of the oscillation of an vibration unit in a perforated well. Another mathematical model for evaluating the fluid flow in the field of elastic oscillation to a perforated well is given in [13]. In a recent field application review [14], the authors suggest using elastic oscillations influence on rock as one of the most promising EOR methods, combining vibration treatment with injection of gas, liquid and other chemicals into the near-wellbore zone. According to [15, 16], under the action of elastic oscillation on the rock, the effects are observed that are as follows: capillary effects, poroelastic motion, thermal effects, increased degassing, change in oil viscosity, mobility ratio and oil displacement by water as well as rock permeability. According to the review [17], the effect of cavitation is one of the most important effects influencing the stimulation of oil flow to the well. Exposure to elastic waves leads to a destruction of paraffin compounds and a decrease in oil viscosity. The authors [18] designed an experimental system simulating a borehole in the actual scale. They concluded that an increase in an oil flow rate was caused by the effective viscosity reduction. While acoustic cleaning the authors [19] built an experimental set-up containing a linear cell. They conducted tests at a frequecny of 20 kHz using Berea sandstone samples of the diameter of 7.5 cm and the length of 20 cm that contained a mud filter cake. The author obtained 100% permeability regained due to formation damage removal. Using the experimental set-up containing 7 systems and the ISCO pump, the authors [20] conducted tests on removing drilling fluid, polymer and paraffin deposition from man-made quartz cores of 2.5 cm in diameter and 8 cm in length by an ultrasonic wave. The authors found an optimal ultrasonic transducer for paraffin deposition plug removal which allowed them to recover permeability of cores by 30%.

The efficiency of elastic oscillation core treatment against paraffin precipitation is demonstrated in the experimental papers. However, the mechanism by which the effect of increasing the permeability of the rock is achieved remains not fully disclosed. In this regard, the purpose of this work is to experimentally study the effect of elastic oscillation on the properties of a fluid and to propose an experimental set-up for performing flow experiments. The paper is organized as follows. In Section 2, the impact of permeability degradation around the well on its production rate is assessed. The results of the experimental effect of elastic oscillation on the paraffin components of oil are presented in Section 3, followed by the model of a small experimental set-up proposed for studying the flow characteristics of a rock in the field of elastic oscillations in Section 4.

2. Evaluation of the impact of permeability degradation around the well on the flow rate

In order to assess the impact of the difference between permeabilities of a near-wellbore zone k_1 and a remote formation zone k_2 it is convenient to compare the flow rate of a well in a zonal heterogeneous reservoir with the flow rate of a well in a homogeneous reservoir. Assuming the plane radial inflow to a well, the production rate of a well located in a zonal heterogenious Q and homogeneous Q_2 reservoir can be estimated by the formulas:

$$Q = \frac{2\pi h(p_c - p_w)}{\mu\left(\frac{1}{k_1} ln\frac{R}{r_w} + \frac{1}{k_2} ln\frac{R_c}{R}\right)}, \quad (1)$$

$$Q_2 = \frac{2\pi k_2 h(p_c - p_w)}{\mu \, ln\frac{R_c}{r_w}}, \quad (2)$$

in which we set $r_w = 10$ cm, $R_c = 10$ km and R as a well radius, supply contour radius and a distance to a boundary of a change in permeabilities respectively; h, μ, p_c, p_w denote thickness of a certain reservoir, viscosity of a fluid, pressure on the supply contour and pressure in the well respectively.

Using formulas (1) and (2) of well flow rates in zonal heterogeneous and homogeneous formations, we compose the ratio:

$$\frac{Q}{Q_2} = \frac{\log\frac{R_c}{r_w}}{\log\frac{R_c}{R}+\frac{k_2}{k_1}\log\frac{R}{r_w}} \quad (3)$$

and in figure 1 we build the ratio of production rates depending on the radius of the near-wellbore zone R and the ratio of permeabilities k_1 and k_2.

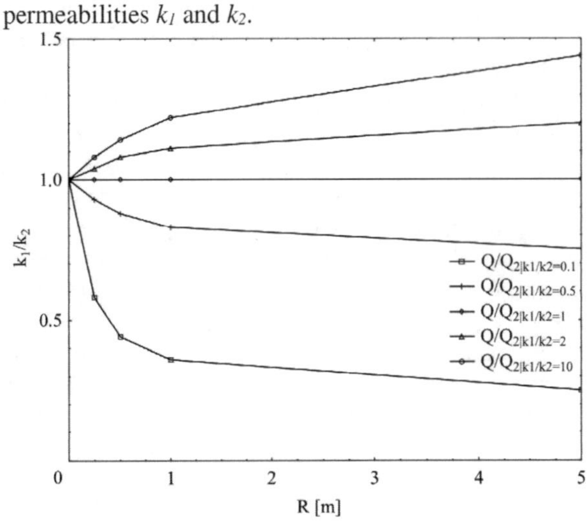

Figure 1. Dependence of the ratio of the flow rates of zonally heterogeneous Q and homogeneous Q_2 reservoirs on the radius R of the near-wellbore zone and different ratios of permeabilities of the near-wellbore zone k_1 and the remote formation zone k_2.

Figure 1 shows that the deterioration of the near-wellbore zone permeability causes a sharp decrease in well production rate; for example, a decrease in permeability by a factor of 10 (k_1=0.1k_2) in a radius of R = 1 m around the well reduces the well flow rate almost 3 times (by 64%). If the same change in permeability was inside the annular area with an outer radius of R = 0.5 m, then the well flow rate would also decrease by 56%. Comparing the last two examples leads to an important conclusion. Namely, with the mentioned tenfold deterioration of permeability at the first 40 cm from the borehole wall (from r_w = 10 cm to R = 50 cm), the flow rate decreases by 56%; with the same decrease in permeability for another 50 cm (from R = 50 cm to R = 100 cm), the well flow rate is further reduced by 8% only. Therefore, the flow rate of the well is most influenced by the change in permeability in the zone immediately adjacent to the wall of the well, where the pressure gradients are the greatest. On the other hand, a tenfold increase in permeability in the same annular zones with outer radii R = 50 cm and R = 100 cm leads to an increase in well production by 14% and 22% respectively. Hence it follows that the deterioration of the permeability in a certain annular region affects the well flow rate any number of times stronger than the increase in permeability in the same region in the same number of times [21].

In this regard, given that the greatest pressure drawdowns that lead to paraffin pressipitation and block the pore space are observed in the near-wellbore zone, it is necessary to restore the permeability of the near-wellbore zone to the initial level in order to increase the well production well rate. As noted earlier, treatment of the rock around the well helps to unblock the pores by breaking down the paraffin compounds of the oil. The next section evaluates the influence of elastic oscillations on the structure of oil.

3. Study of the influence of elastic oscillations on the structure of oil

Oil samples used as the test fluid were taken in the production well 1/1 which operates an oil reservoir A in clastic Bobrikovsky (C_1bb) deposits in the north of the Perm region. Operating parameters of the well were as following: liquid flow rate - 50.9 m^3/day, oil flow rate - 40.1 t/day, water cut - 5%.

Oil samples were treated by elastic oscillations at an ultrasonic technological apparatus USTA-0.8 / 22-ORV. The working tool of the apparatus contains a piezoelectric element that introduces high-intensity ultrasonic oscillations into the liquid medium (oil). As a result, a mode of developed cavitation is observed in the oil, during which energy is released and high-molecular paraffin compounds are destroyed.

The analysis of changes in the oil structure under the action of elastic oscillation was carried out on an Axio Imager light microscope with an M2M support.

It can be seen from figure 2 that in comparison with initial oil structure (figure 2*a*) treatment for 1 minute did not change the oil structure much (figure 2*b*). However, after exposure to elastic oscillations for 4 and 10 minutes (figures 2*c* and 2*d*), the large paraffin compounds in oil became destroyed, reduced in size, and oil structure became smoother. During the experiments, the longer was the treatment the higher was the temperature of an oil sample glass.

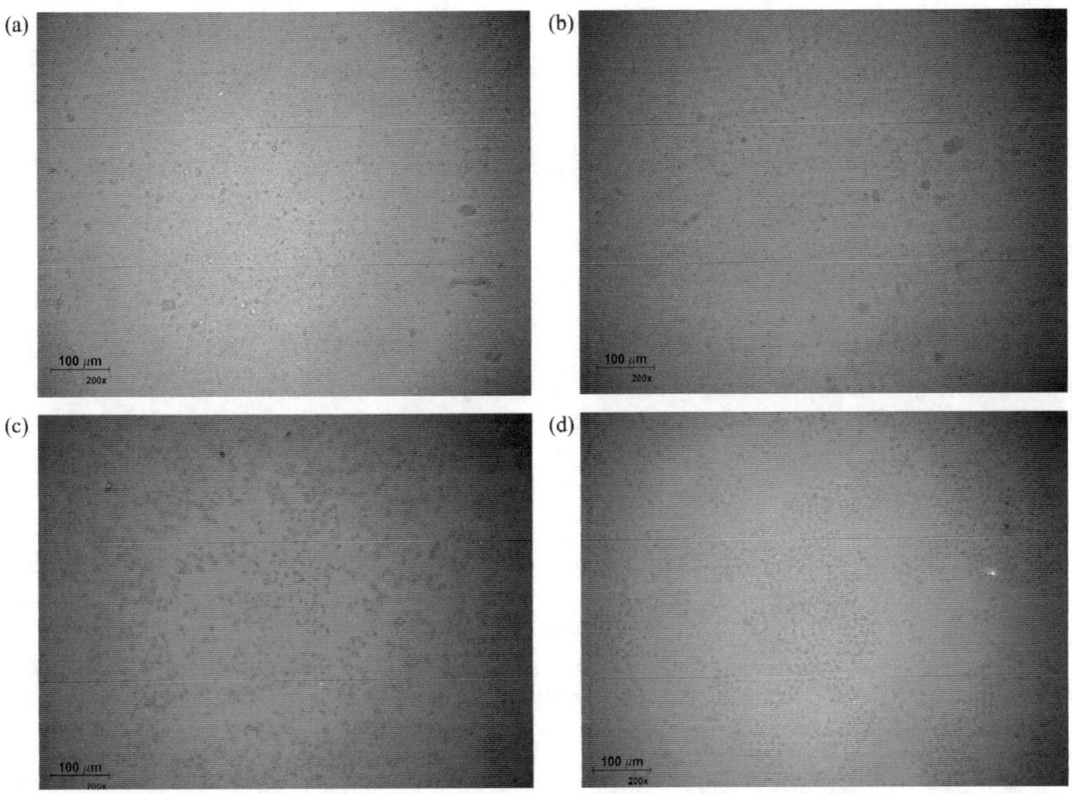

Figure 2. The structure of oil sample under a microscope at a 200x magnification after different periods of ultrasonic treatment with a frequency of 22 kHz; (a) without an ultrasonic treatment; (b) treatment for 1 minute; (c) treatment for 4 minutes; (d) treatment for 10 minutes.

It has to be said that during acoustic treament the temperature of the fluid can increase significantly. That can lead futher in flow experiments to a conclusion that rock permeability increases. Thus, the appropriate correction has to be implemented considering change in fluid viscosity.

4. Experimental set-up for studying the effect of elastic oscillations on oil flow through rocks

In order to study the influence of elastic oscillations on the flow of oil through the rock, a small experimental set-up was developed (figure 3). The set-up modells the field conditions in which a downhole oscillation generator is lowered to the bottom of a production well in fornt of the payzone on a tubing string and performs a treatment of the near-wellbore zone with elastic oscillations without shutting down the well.

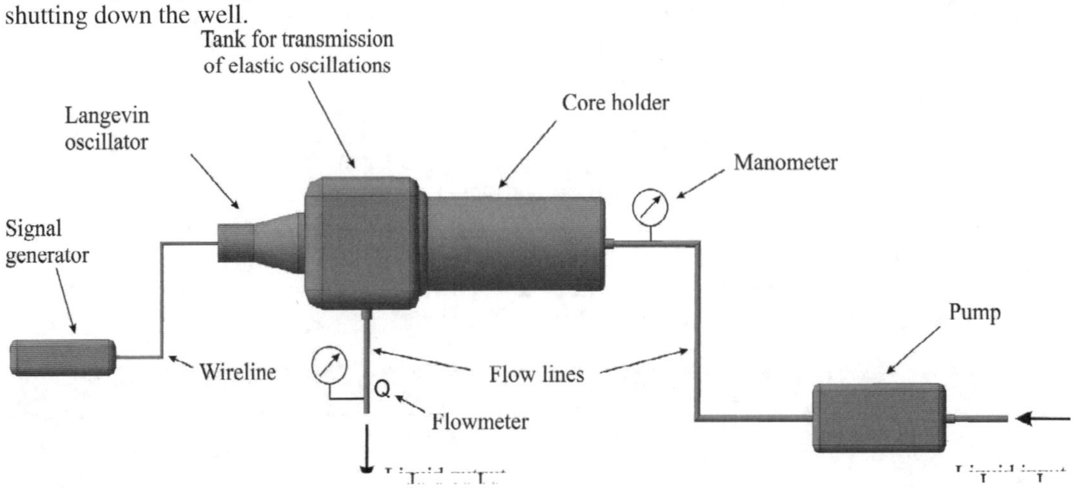

Figure 3. Schematics of the experimental set-up for flow studies under the influence of elastic oscillations.

During the flow set-up, a rock sample with a length $l = 3$ cm and a dimater $d = 2.5$ cm is installed in a core holder. In order to model the conditions of overburden pressure and to ensure the fluid flow in the axial direction, the sample is exposed to the pressure of the hydraulic crimp of the rubber cuff inside the core holder. After the hydraulic compression pressure is created, the pump is turned on, to the inlet of which the work fluid (oil) is supplied. The fluid passes through the rock sample, tank, and then directed to the outlet line. Pressure gauges at the inlet and outlet of the unit allow to determine the differential pressure ΔP across the sample. The flow meter allows to determine the volume of liquid Q passing through a rock sample per unit of time. Using the indicators of manometers and a flow meter, it is possible to determine the permeability k of the rock for the flowing liquid.

In order to carry out the studies of fluid flow under the influence of elastic oscillations, the set-up contains a Langevin oscillator. The oscillator is rigidly fixed to the end surface of the tank using epoxy glue. The oscillator operates in the ultrasonic frequency range. To carry out flow studies, different oscillators with a power of 100 W with a frequency range from 17 to 40 kHz can be used. The oscillator operates on the basis of the piezoelectric effect, which manifests itself in mechanical deformations of the material under the action of an electric field. The electrical signal to the oscillator comes from a signal generator, which forms a sinusoidal signal with a voltage from peak to peak up to 90 V.

The oscillator transmits elastic vibrations of the working fluid through the tank wall, which can withstand a load of up to 2 MPa, which in turn transfers them to the rock sample. Under the influence of elastic oscillations in the tank and the rock, the effect of cavitation is observed, which leads to the destruction of high-molecular paraffin compounds, reducing the dynamic viscosity μ of the fluid and cleaning the pore space of the rock from paraffin precipitations. As a result of these processes, the fluid permeability k of the rock increases.

Indirectly, the increase in the rock permeability k can be determined on the basis of instrument parameters using the Darcy formula:

$$k = \frac{Q\mu l}{S\Delta P}. \tag{4}$$

There is no much information on existing set-ups for evaluation of the effect of elastic oscillations on the flow characteristics of oil. Therefore, the proposed set-up can be usefull for modelling the effect of vibrations generated by a downhole tool to the oil inflow through porous permeable rock to a production well.

5. Conclusion

In this work the problem of paraffin precipitation in the near-wellbore zone is considered. Since the paraffin blocks the pore space, the permeability around the well is degraded resulting in a decrease in oil rate of a production well. Elastic oscillation treatment is considered as the most optimal method for rock permeability recovery due to its efficiency for permeability restoration and no negative impact on rock mechanical (strength) characteristics.

As a result of the influence of elastic oscillation on paraffin oil in the ultrasonic range, the destruction of high-molecular paraffin chains and a decrease in the viscosity of oil samples are observed. An experimental set-up was developed to study the effect of elastic oscillations on fluid flow through a rock. The small-scale experimental set-up simulates the field conditions in which a downhole oscillation generator is lowered to the bottom of a production well in front of the payzone on a tubing string and performs treatment of the near-wellbore zone with elastic oscillations without shutting down the well.

Acknowledgments

The support of the Russian Science Foundation (Project No. 19-19-00408) is gratefully acknowledged.

References

[1] Yonghai G, Kai1 L, Xinxin1 Zh, Hao L, Yanchun C, Guizhen X, Baojiang S 2018 *Pet. Explor. Dev.* **45(2)** 351-357
[2] Nie X, Yang S 2014 *Geosyst. Eng.* **17(4)** 219-225
[3] Brown T S, Niesen V G, Erickson D D 1994 69th SPE Annual Technical Conference and Exhibition SPE 28505
[4] El-Dalatony M M, Jeon B-H, Salama E-S, Eraky M, Kim W B, Wang J, Ahn T 2019 *Energies* **12** en12060967
[5] Sousa A L, Matos H A, Guerreiro L P 2019 *J. Pet. Explor. Prod. Technol.* **9** 2091-2107
[6] Wang Zh, Xu Yu 2015 *Energy* **89** 259-267
[7] Mullakaev M S, Abramov V O, Abramova AV 2017 *J. Petrol. Sci. Eng.* **159** 1-7
[8] Beresnev I A, Johnson P A 1994 *Geophysics Eng.* **59(6)** 1000–1017
[9] Ariadji T 2005 Asia Pacific Oil & Gas Conference and Exhibition SPE 93112
[10] Riabokon E P 2020 *Neftyanoe Khozyaystvo - Oil Industry*, **6** 76-79
[11] Mullakaev M S, Abramov V O, Pechkov A A 2009 *Chem. Pet. Eng.* **45**(3-4) 133-137
[12] Pérez-Arancibia C, Godoy E, Durán M 2018 *Wave Motion* **77** 214-228
[13] Prachkin V G, Mullakaev M S, Asylbaev D F 2014 *Chem. Pet. Eng.* **50**(9-10) 571-578
[14] Meribout M 2018 *IEEE Access* **6** 51110-51118
[15] Cidoncha J G 2007 International Oil Conference and Exhibition in Mexico SPE 108643
[16] Riabokon E P 2020 *Neftyanoe Khozyaystvo - Oil Industry*, **4** 54-57
[17] Avvaru B, Venkateswaran N, Uppara P, Iyengar S B, Katti S S 2018 *Ultrason. Sonochem.* **42** 493-507
[18] Mullakaev M S, Abramov O V, Abramov V O, Gradov O M, Pechkov A A 2009 *Chem. Pet. Eng.* **45**(3-4) 203-210
[19] Wong S-W, van der Bas F, Groenenboom J, Zuiderwijk P 2003 SPE European Formation Damage Conference SPE 82198
[20] Wang Zh, Xu Yu, Bajracharya S 2015 *Ultrason. Sonochem.* **27** 339-344
[21] Shchelkachev V N, Lapuk B B 1949 *Underground hydraulics* (Moscow: Gostoptekhizdat)

Ways to improve the development efficiency in the complex-structured carbonate deposits of the Tournaisian stage using hydraulic fracturing

V A Grishchenko[1,2], **V V Mukhametshin**[3], **L S Kuleshova**[1]

[1] Ufa State Petroleum Technological University, Branch of the University in the City of Oktyabrsky, 54a, Devonskaya St., Oktyabrsky, Republic of Bashkortostan, 452607, Russian Federation
[2] «RN-BashNIPIneft» LLC, 86/1, Lenin St., Ufa, 450006, Russia
[3] Ufa State Petroleum Technological University, 1, Kosmonavtov St., Ufa, Republic of Bashkortostan, 450064, Russian Federation

E-mail: vv@of.ugntu.ru

Abstract. Depletion of the main oil reserves within the Volga-Ural oil and gas province causes a constant shift in production volumes towards complex low-productive carbonate deposits. Clarification of the geological structure, which has a prevailing effect on the final result, is carried out in order to ensure an efficient process of developing reserves. The analysis of the influence of clay dam thickness between oil and water-saturated formations was done for various injection sites on the efficiency of hydraulic fracturing and treatment of the bottom-hole zone in production wells. The study showed that the critical value of the dam thickness is equal to three meters; the waterflooding efficiency is significantly reduced below this value.

1. Introduction
The development of carbonate objects is accompanied by complicating factors associated with the peculiarities of their geological structure: discontinuous reservoirs, low permeability of the matrix, significant variability of properties both along the section and in the area [1–6]. This requires the formation of differentiated approaches to the operation of wells, depending on the structural features of reservoirs in a specific subsoil area [7–12].

2. Materials and methods
One of the fields of the Volga-Ural oil and gas province is considered. The main object, both in terms of initial and residual reserves at the field, is the Tournaisian stage. The Tournaisian stage is represented by the Kizelovskiy, Cherepetskiy, and Upinskiy-Malevskiy horizons. Commercial oil-bearing capacity is established in the Kizelovskiy horizon, the rest are water-saturated. The Kizelovskiy horizon is represented by brownish-gray, organogenic-sludge, clayey, hard, unevenly porous, industrially oil-bearing limestones. The thickness of the Kizelovskiy horizon varies from 10 to 30 m. The productive unit of C1ksl.1 occurs in the top of the Tournaisian stage 3-5 m below the base of the Bobrikovskiy horizon. The productive unit C1ksl.2 is separated by carbonate varieties from the overlying unit of C1ksl.1. A map of the bulkhead thickness between C1ksl.1 and C1ksl.2, built on the basis of GIS, is shown in Figure 1; Figure 2 is an example of a correlation scheme.

The object is drilled along a triangular grid of directional wells of 400*400 m, a focal-selective waterflooding system is formed. The object is characterized by an increasing level of production due to intensive geological and technical measures (GTM) [13-15] (bottomhole treatment (BHT), acidic HFT) and drilling of horizontal wells.

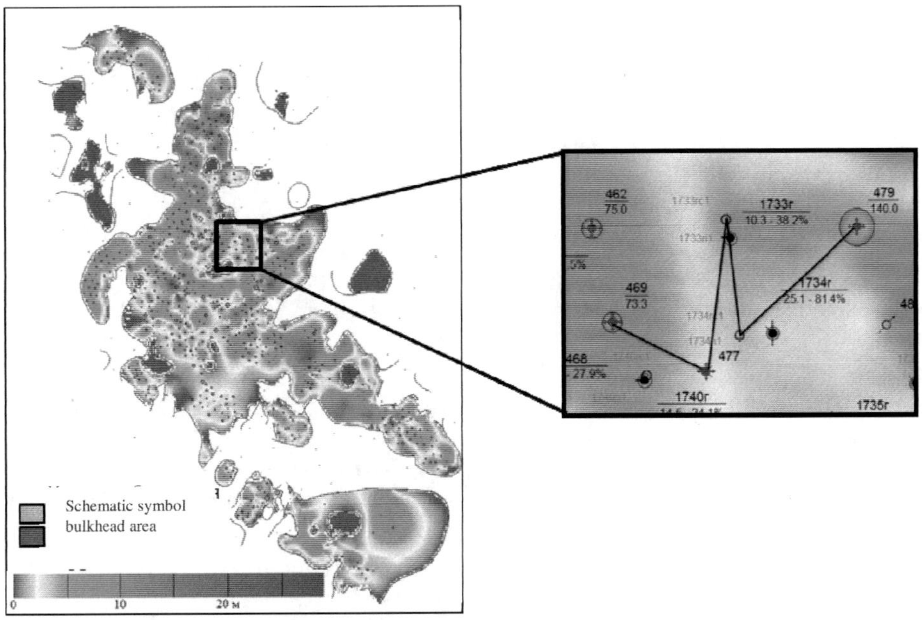

Figure 1. Map of the bulkhead thickness between C1ksl.1 and C1ksl.2

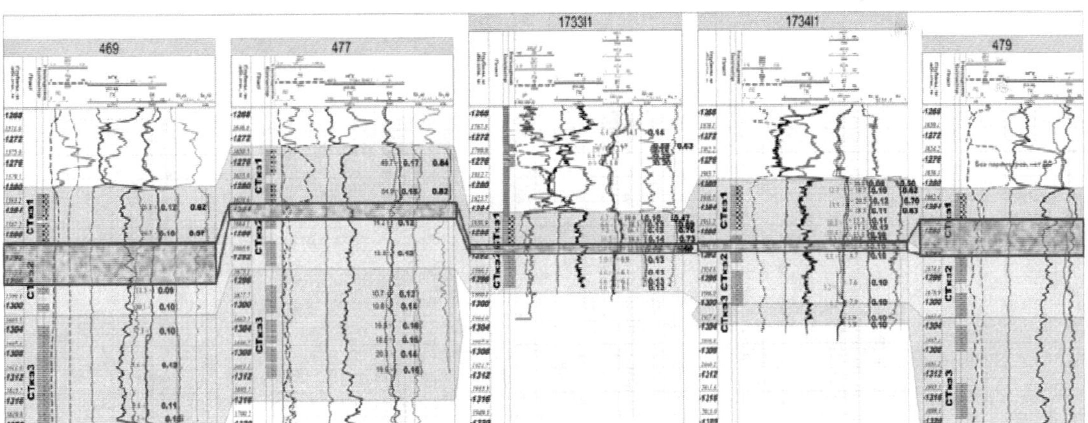

Figure 2. Correlation scheme for the Tournaisian stage

3. Results and Discussion

The presence of a hydrodynamic connection is established between the oil-saturated unit of C1ksl.1 and the underlying water-saturated unit C1ksl.2 during development. A significant number of results of production geophysical studies, which trace the presence of intra-reservoir flows (by thermometry and RGE) is a key source of information when diagnosing the relationship between the units. An example of an injectivity profile for an injection well is shown in Figure 3.

IOP Conf. Series: Earth and Environmental Science 1021 (2022) 012017 doi:10.1088/1755-1315/1021/1/012017

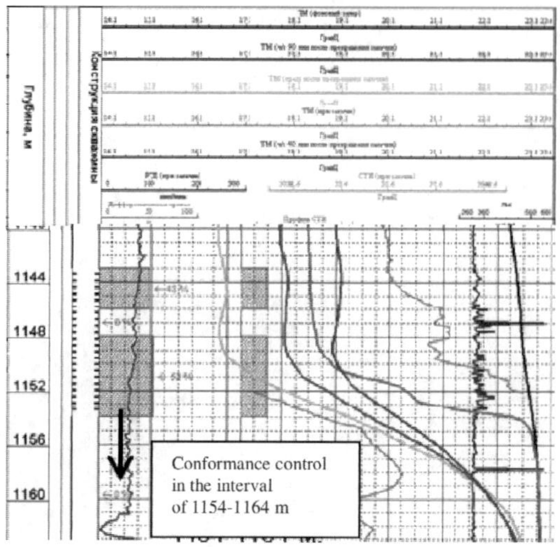

Figure 3. Injection profile with intra-reservoir flows

The GTM program is being actively pursued at the object. The difference in the dynamics of the wells operation is observed in terms of the liquid flow rate after the given measures. Figure 4 presents the dynamics of operation after HFT.

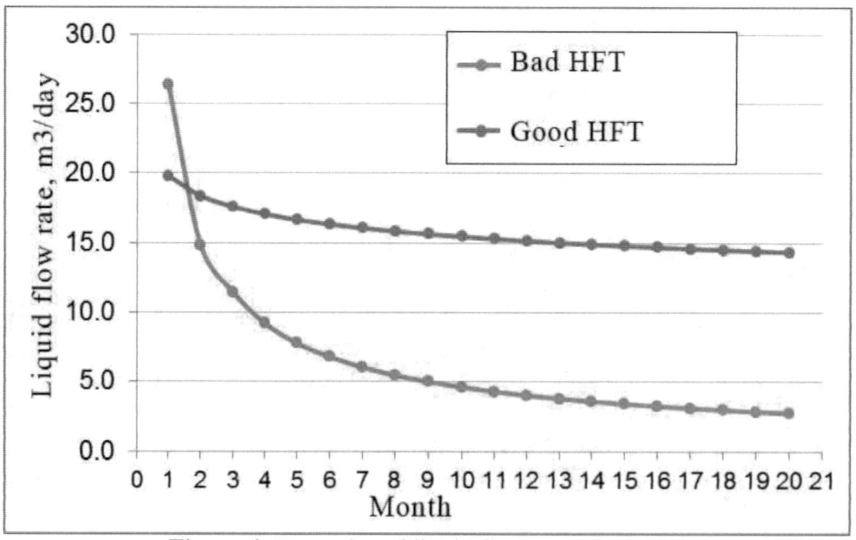

Figure 4. Dynamics of liquid flow rate after HFT

Every well is conditionally divided into two groups: "good" with a low rate of liquid decline and "bad" with an intensive decline. A similar picture is observed when conducting BHT in production wells. The main reason for the decrease in liquid flow rates is a drop in reservoir pressure, given that almost all sections are provided with waterflooding centers. It follows that in the sections with "good" and "bad" GTM, the PMM system efficiency (pressure maintenance method) is different, which depends on the amount of water outflow into the underlying reservoir C1ksl.2. A schematic representation is shown in Figure 5.

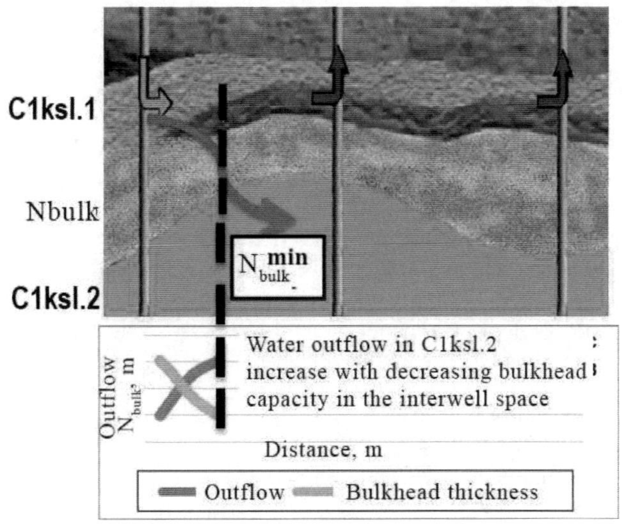

Figure 5. Schematic representation of the bulkhead capacity effect on water outflow into C1ksl.2 reservoir

Thus, the amount of water leaving in off-target intervals depends on the bulkhead power in the interwell space between the production and injection wells. The distribution of HFT and BHT depending on the minimum value of the bulkhead between the injection and production wells is shown in Figure 6.

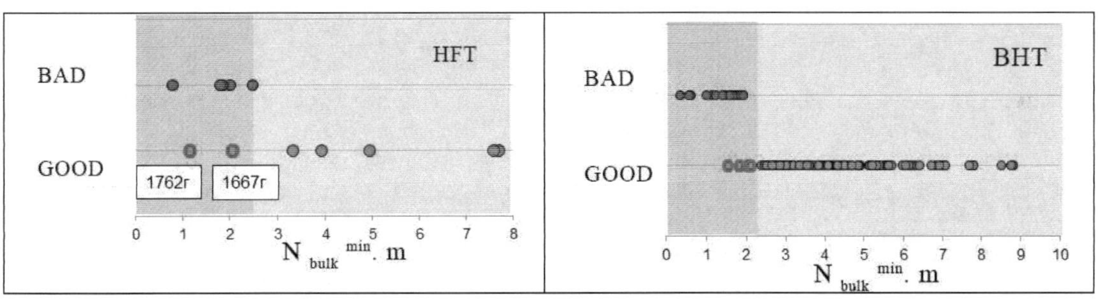

Figure 6. Distribution of GTM by bulkhead power

Figure 6 shows that most of the "good" GTM are located in the bulkhead thickness zone of more than three meters, the "bad" ones are less than three meters. According to HFT, two wells are an exception - 1762 and 1667, which belong to the category of highly productive, but at the same time they are located in an area with a bulkhead thickness of 1-2 m. For the well of 1762, the reason is the high compensation in the area (more than 500%), which makes it possible to provide production wells with the required volume of efficient compensation even with a large amount of outflow. For the well of 1667, the reason is that the injection in the area is organized in a horizontal well drilled along the top of the C1ksl.1 formation, which allows reducing outflow and significantly increase the injection efficiency.

4. Conclusion

Three meters is a critical minimum bulkhead thickness for efficient injection. Confirmation of withdrawal based on actual GTM results is: 100% for HFT and 97% for BHT. Further development planning is possible on the basis of the constructed map of the bulkhead capacity, it is feasible to conduct GTM without additional measures for zones with a bulkhead thickness of more than three meters, additionally plan an increase in compensation or organizing PMM in horizontal wells - for areas with a bulkhead which are less than three meters.

References

[1] Suchkov B M 2005 *Oil production from carbonate reservoirs* (Moscow: SIC "Regular and chaotic dynamics")

[2] Yakupov R F and Mukhametshin V Sh 2013 Problem of efficiency of low-productivity carbonate reservoir development on example of Turnaisian stage of Tuymazinskoye field. *Oil Industry* **12** 106–110.

[3] Moid F, Rodoplu R, Nutaifi A M, and Kayumov R 2020 Acid Stimulation Improvement with the Use of New Particulate Base Diverter to Improve Zonal Coverage in HPHT Carbonate Reservoirs. *International Petroleum Technology Conference.* DOI: 10.2523/IPTC-20154-ABSTRACT

[4] Mukhametshin V G, Dubinskiy G S, Andreev V E, Mukhametshin V V, and Kuleshova L S 2021 Geological, technological and technical justification for choosing a design solution for drilling wells under different geological conditions. *IOP Conference Series: Materials Science and Engineering* **1064** 012061 DOI: 10.1088/1757-899X/1064/1/012061

[5] Burgos-Cara A, Ruiz-Agudo E, and Rodriguez-Navarro C 2017 Effectiveness of oxalic acid treatments for the protection of marble surfaces *Materials & Design* **115** 82-92 DOI: 10.1016/j.matdes.2016.11.037

[6] Gabzalilova A Kh, Batalov D A, Mukhametshin V Sh, and Andreev V E 2021 Geological and technological justification of the parameters of acid-clay treatment of wells. *IOP Conference Series: Materials Science and Engineering* **1064** 012058 DOI: 10.1088/1757-899X/1064/1/012058

[7] Hall-Thompson B, Ernesto A R, Abdulrahman N, and Alsuhaimi A 2020 Acid Stimulation-Best Practices for Design, Selection and Testing of Acid Recipes in Low Permeability Carbonate Reservoirs. *International Petroleum Technology Conference.* DOI: 10.2523/IPTC-19690-MS

[8] Mukhametshin V V, and Kuleshova L S 2021 Using the method of canonical discriminant functions for a qualitative assessment of the response degree of producing wells to water injection during the development of carbonate deposits. *IOP Conference Series: Materials Science and Engineering* **1064** 012069 DOI: 10.1088/1757-899X/1064/1/012069

[9] Al-Rekabi M A, Aktebanee A, Al-Ghaffari A S, and Saleem T 2020 Carbonate Matrix Acidizing Efficiency from Acidizing Induced Skin Point of View: Case Study in Majnoon Oilfield. *International Petroleum Technology Conference.* DOI: 10.2523/IPTC-20006-MS

[10] Mukhametshin V Sh, Kotenev Yu A, and Sultanov Sh Kh 2018 Assessment of the Need to Stimulate the Development of Hard-to-Recover Reserves in Carbonate Reservoirs. *IOP Conference Series: Earth and Environmental Science* **194(8)** 082027 DOI: 10.1088/1755-1315/194/8/082027

[11] Khuzin R R, Mukhametshin V Sh, Salikhov D A, Andreev V E, Pepelyaev D V, Stefanovich Yu N 2021 Improving the efficiency of horizontal wells at multilayer facilities. *IOP Conference Series: Materials Science and Engineering* **1064** 012066 DOI: 10.1088/1757-899X/1064/1/012066

[12] Batalov D A, Andreev V E, Mukhametshin V V, and Kuleshova L S 2021 Development regulation of oil and gas reservoirs based on effective geological and geophysical information. *IOP Conference Series: Materials Science and Engineering* **1064** 012055 DOI: 10.1088/1757-899X/1064/1/012055

[13] Jennings A R, and Jr P E 2003 Application of hydraulic fracturing (Enhanced Well Stimulation, Inc.)

[14] Zdolnik S E, Nekipelov Yu V, Gaponov M A, and Folomeev A E 2016. Introduction of innovative hydrofracturing technologies on carbonate reservoirs of Bashneft PJSOC *Oil Industry* **7** 92-95

[15] Mukhametshin V V and Kuleshova L S 2019 Prediction of production well flow rates using survey data. *IOP Conference Series: Earth and Environmental Science* **378** 012116 DOI: 10.1088/1755-1315/378/1/012116

IOP Conf. Series: Earth and Environmental Science 1021 (2022) 012018 doi:10.1088/1755-1315/1021/1/012018

Analysis of the state of exploitation of flooded Mesozoic reservoirs of the Grozny oil-bearing region exposed to anthropogenic impact

A Sh Khaladov[1], N D Bulchaev[1], M Z Duskaev[1], M-A E Bakaev[1], Yu Kh Taramov[1], D Kh Khasiev[1], Z Kh Gazabieva[1], A A Yamaletdinova[2]

[1] Institute of Oil and Gas, Grozny State Oil Technical University named after Academician M.D. Millionshchikov,14/53 Avtorkhanov Str., Grozny, 364001, Russia
[2] FSBEI HE USPTU, Republic of Bashkortostan, Ufa, 1 Kosmonavtov Str., 450064, Russia

E-mail: haladov_a_sh@mail.ru

Abstract. One of the main features of the exploitation of Mesozoic reservoirs in the Grozny oil-bearing region is an extremely short life of flooded production wells, which lasts not more than 10% of productive life from water appearance in the well production to its complete flooding.

1. Introduction

Technological calculations are based on the mechanism of piston-like displacement of oil by water; however, the water-oil mixture zone in the reservoir voids is not described. This is mainly due to the fact that oil and water are mutually insoluble, immiscible agents.

Displacement of oil by miscible agents (solvents, compressed gases) leads to the formation of a transition zone between oil and the displacing agent. This transition zone in the section perpendicular to the flow is homogeneous and characterized by a certain composition, which changes from reservoir oil to a set of components of the displacing agent. Its length depends on the length of the section where the process takes place and on the viscosity ratio of filtering agents.

2. Analysis of completed studies

Study of the causes of the considered phenomenon reveals the impact of the type of reservoirs and the mechanism of oil recovery on reservoir flooding. As shown in [1–3], modeling of a fractured reservoir exhibits the qualitative features similar to those typical of computational models of a terrigenous reservoir, but the quantitative features may differ significantly. For example, oil recovery in terrigenous reservoirs changes insignificantly with an increase in the thickness of the separated conventionally homogeneous interlayer of more than 5 m [4,5], in fractured reservoirs, according to the results of field studies, a similar situation can be observed when the thickness of interlayer is about 50 m. Analytically, this can be explained by the impact of the reservoir properties on the process of oil displacement by water. However, the fundamental difference in the duration of the dry and water periods of reservoir exploitation may depend on the oil recovery mechanism, the structure of the void space, and other reasons.

When oil is displaced by water, the transition zone is not represented by a sequence of homogeneous mixtures, but in each section perpendicular to the flow, it is heterogeneous with a gradual increase in water saturation of the void in the direction opposite to the direction of displacement.

Both types of displacement show that oil can remain behind the displacement front, and its amount depends on the displacing and washing properties of the injection agent.

Differences in the nature of the transition zone, which is formed during the displacement of oil by miscible agents, and the water-oil mixture zone are obvious. However, when studying the size of such zones, their unexpected subordination to the general rule was revealed. For example, the results of laboratory experiments on linear models of a reservoir with a porous medium during the displacement of oil by hydrocarbon gas under high pressure are given in [6]. It was found that the length of the transition zone, depending on the length of the section where the process takes place, changes as follows:

$$\frac{1-\eta_1}{1-\eta_2} = \sqrt[3]{\frac{L_1}{L_2}}, \tag{1}$$

where η_1, η_2 are the efficiency of oil displacement before the injection agent appears in the flow for the models with length L_1 and L_2, respectively.

Then the ratio of the lengths of the transition zones for various reservoir models with a porous medium

$$n = \frac{l_1}{l_2} = \sqrt[3]{\frac{L_1}{L_2}}, \tag{2}$$

where l_1,l_2 the size of the transition zone in the models with length, L_1 and L_2, respectively.

It is noted that when oil is displaced by water of similar viscosity, the size of the oil-water mixture zone changes according to the same rule. This phenomenon can be explained as follows. The mixing of mixing agents moving in a porous medium has been described by a number of researchers using hydrodynamic, diffusion and statistical flow models. A lot of acceptable solutions for describing this process were obtained. With regard to the aforementioned generality of some features in oil displacement by miscible agents and water, the hydrodynamic flow model may be the closest to the nature of the mixing of agents filtering in a porous medium, and the most common diffusion model of mixing fluids will be less substantiated.

Taking into account the foregoing, let us analyze the flooding of Mesozoic reservoirs, where oil displacement by water occurs from the bottom up, after which the fluid flow in the horizontal plane is directed to the production wells along the exposed part of the section. This flow schematization is simplified, but the exploitation of reservoirs with low watering of production wells makes it acceptable.

It should be noted that the gravitational separation of filtering agents significantly affects the quantitative indicators of their mixing when moving in the vertical plane. For miscible agents, the dependence of the length of the transition zone on the increase in the inclination angle from 0° to 30° is expressed as follows:

$$l_{inc} = l_h(1 - tg\alpha), \tag{3}$$

where l_{inc} and l_h are the length of the transition zone in the inclined and horizontal reservoir models, m; α is the inclination angle of the reservoir model to the horizon, deg.

Previous studies have shown that the inclination angle of the reservoir to the horizon that exceeds 30 ° practically does not affect the size of the transition zone.

The viscosity ratio of filtering agents significantly affects the length of the transition zone. The viscosity of oil and water in Mesozoic reservoirs of the Chechen Republic are close in absolute value, and with a small error the ratio of their viscosities in reservoir conditions can be taken equal to 1.

To estimate the size of the water-oil mixture zone in Mesozoic reservoirs, the results of experimental studies obtained by SEVKAVNIPINEFT were used [4, 6]. The analysis of displacement of oil by water with equal viscosity from horizontally located models of porous media with a length of 9.5 m to 19.7 m revealed an empirical dependence of the change in the length of the transition zone (3). The size of the water-oil mixture zone for the reservoir model with a length of 19.7 m was 12 m, and for the reservoir model with a length of 9.5 m it attained 9.4 m. With regard to the fact that the saturation in the transition zone (its main part) varied linearly, the volumetric contents of filtering agents are equal. Since oil displacement ceases after passing through the transition zone, the overall displacement efficiency is the sum of the oil displacement efficiency before flooding and that in the transition zone. In the experiments, up to 70% of the initial hydrocarbon reserves were extracted during the dry displacement period, and up to 80% including the transition zone (limit under the conditions of the study). The coverage of the porous medium of the reservoir model by the process is complete, and the resulting displacement efficiency corresponds to the final oil recovery factor. In real conditions, the efficiency of oil displacement by water will correspond to the experimental one, but the total oil recovery due to the void heterogeneity should be lower as a result of incomplete coverage of the reservoir by the process.

Taking as the basis of the hydrodynamic model a drainage section of the reservoir with a thickness of 50 m, the size of the water-oil mixture zone, the amount of water recovered, and the level of flooding during recovery of oil with water can be estimated and compared with the actual data. To do this, let us calculate the thickness of the transition zone by the height of the reservoir section. If the displacement of oil by water was carried out in the horizontal plane, the water-oil mixture zone with the length of the filtration flow of 50 m would occupy 16 m. Since the displacement is carried out in the vertical plane, the water-oil mixture zone in the total flow will take less than 7 m. In calculation, $\alpha=30°$ since further increase in the inclination angle of the models to the horizon does not affect the displacement efficiency before water flooding and the size of the water-oil mixture zone.

In the combined flow zone, the water content is about 50% of the moving fluid volume. Based on the results of the exploitation of all wells draining the selected stratum in the formation section, the calculated total average flooding is close to 7% if the exploitation yields oil recovery from the section. The total thickness of Mesozoic reservoirs significantly exceeds that adopted in the model for calculations, oil displacement by water is continuous, and flooding of part of the section in production wells only partially disrupts the flow mechanism. Therefore, even a complete flooding of production wells does not lead to a complete disruption of the displacement mechanism. The subsequent involvement of the overlying reservoir intervals increases the existing and the newly formed water-oil mixture zone.

Flow analysis using the obtained dependencies shows that early termination of fluid recovery from wells that drain the watering interval not only contributes to a decrease in the total share of flooding, but also increases the exploitation efficiency through lengthening the period of dry oil production. In real conditions, the size of the water-oil mixture zone can be close to stable due to water recovery at the completion of exploitation of each open drainage interval. If production wells in intermediate intervals are completely flooded, the total average flooding over the entire period of reservoir exploitation can be about 2% with a reservoir thickness of 300 m.

Some Mesozoic reservoirs of the Grozny oil-bearing region have been depleted almost completely using efficient water flooding. For example, in the Braguny field, the average total flooding was about 2.3% with a rigid water pressure regime. At the Eldarovo field, the average total flooding was about 1% during exploitation of the Upper Cretaceous reservoir with a natural elastic-water pressure regime. The resulting difference between the calculated and actual parameters is due to the fact that the flooding of production wells did not fully correspond to the technological state of the exploited formation zone, that is water was produced to flush the reservoir zone affected by displacement. This is a natural phenomenon and contributes to the recovery of part of oil remaining in zones with a slow advance of the displacement front, and to an increase in the degree of its recovery from the volume of the reservoir involved in the process [7].

The total actual average flooding ultimately has a tinge of a technical indicator, since it depends on the extremely profitable oil production rate during the period of water flooding, especially at the late and final stages of reservoir exploitation. A high degree of oil recovery from the affected reservoir volumes is evidenced by exploitation and can be explained by the fact that only a small amount of mobile oil can be found behind the lower edge of the oil-water mixture zone. This does not mean that the maximum possible oil recovery of the reservoir is achieved, since in the course of displacement by water, only the oil that can move in this direction is washed out. The oil-saturated voids, which structure peculiarities do not allow oil movement in the direction of the flow, remain unrecovered from the rock. As noted in [9, 10], the changed direction of the flow and properties of the displacing agent can become a technology for the recovery of oil remaining in Mesozoic reservoirs after water flooding.

The rates of annual oil recovery from fractured reservoirs during water flooding were sometimes close to 10% of the initial reserves, but in most cases during the main period of field exploitation the rates were fairly high and attained 3–5%.

The oil recovery mechanism will change completely. When reservoir pressure will decrease by about 10% of the saturation pressure, oil flows from non-flowing voids into flowing ones and through water will be delivered to the bottom of the production wells with practically unchanged initial gas content.

Degasification of residual oil occurs at high pressures ($\geq$ 20 MPa), which ensures the efficiency of oil displacement by its associated gas, which is close to that achieved with the miscible displacement of oil by a hydrocarbon gas of a selected composition compressed to high pressure. The proposed technology for oil recovery at the final period of field exploitation is not the last link in the overall chain – the consistent use of the technology in the reservoirs will allow the produced water to flood the reservoirs developed in this way, and this process will be the last in the field life [9, 10]. This fully meets the requirements of rational use of natural resources of hydrocarbon raw materials and environmental protection.

Drainage of the reservoir domed zones in terms of nature, efficiency and technological features due to their horizontal location will differ significantly from the previous drainage of steeply dipping zones of the reservoirs [11].

The change in the flow of a displacing agent from vertical to horizontal does not contribute to the efficiency of the process and can have a positive effect on production indicators by the increasing impact of the formation strata. The upper boundary of Mesozoic reservoirs of the Grozny oil-bearing region is not a smooth surface, it has multiple local depressions and outcrops along the entire height of the reservoir. In the case of vertical flow, in the inclined part of the formation the near-domed zones of the reservoir are involved in the displacement process to the same extent as the main part of the reservoir. Vertical displacement in the horizontal part of the reservoir will lead to the fact that oil in the near-domed relief zones of the reservoir will be blocked by water and remain unrecovered [11].

3. Conclusions and proposals for restoration of Mesozoic reservoirs exploitation:

The main features of the proposed technology for exploitation of Mesozoic reservoirs in the Grozny oil-bearing region, restoration and increase of their efficiency are as follows [11]:

- opening of the productive parts of the section in the process of drilling wells without the use of weighted clay solutions at the equilibrium of reservoir and bottomhole pressures
- measurement of the volume of drilling fluid absorbed by the formation when opening the productive interval of the formation
- preparatory operations for replacing the clay solution in the well with water and flushing the well with water according to the conventional method
- choosing of an acidic reagent for well treatment
- determination of the required amount of an acidic reagent by its reaction with a clay solution in laboratory conditions with regard to the volume of the solution remaining in the formation
- if large amounts of drill mud is absorbed, the volume of the acid reagent for the first operation is calculated by isolating the volume of the clay solution remaining in the cylinder of the bottomhole zone of the well with a height equal to the thickness of the productive part of the section and a diameter

of 1–2 m. The results obtained reveal the need for repeated operation, which is carried out with the increased diameter of impact on the bottomhole zone. The impact diameter is adjusted with regard to the reservoir void so that a single treatment of the formation is technologically provided with field equipment and acid reagents

- since acidic solutions are prepared on a water basis, further inflow from the well is carried out using conventional means and techniques
- if the productive interval of the formation is drilled with the use of weighted drill mud, the well may have a lower production index than the natural one. If the remainder becomes hydrophilic in the fractures, after the cycles of treatment and inflow stimulation it will be obtained in the well with varying degrees of recovery and brought to the surface. If the residue becomes hydrophobized in the fractures, its removal from the formation can be facilitated by the increased oil pressure during the well exploitation.
- the use of weighted solutions requires acid emulsions on a hydrocarbon or other basis for well treatment operations. An increase in the viscosity of the acid reagent will increase the efficiency of a single treatment or cycle due to a more complete displacement of the solution from the fractures by the reagent and complete use of the acid reagent.
- cyclic treatment of wells with an acid reagent in accordance with the proposed technology can be successfully used in the exploitation of wells drilled with weighted solutions.
- theoretically, the proposed treatment technology makes it possible to abandon the use of inhibitors for the reaction of acid reagents with formation rocks, since the reaction surface of the dispersed mixture of the reagent and the clay solution exceeds manyfold the flat surface of the fractures.
- since the efficiency of displacement in the fractures of the clay solution by the acid reagent to a small extent depends on the created pressure drop, the optimal volumetric injection rate of the reagent will be close to or less than the efficiency of the wells.

The main task of all types of studies is to obtain, if possible, complete information on all parameters of the reservoir, wells and their technical condition for the successful completion of exploitation upon depletion of oil reserves. Control methods should be systematic and reflect the current state of affairs during the exploitation of wells.

4. Conclusion

The displacement of oil by water from the bottom up increases the overall efficiency of the process by increasing the reservoir coverage and reducing the period of exploitation of oil with water in the entire period of the reservoir exploitation. The forces of gravity would be also effective when changing the direction of their action, that is displacement of oil from top to bottom by compressed gas. During exploitation with a fractured reservoir, the consecutive application of both mechanisms is one of the most rational techniques to increase hydrocarbon recovery.

In general, the transition to horizontal displacement of oil by water in the reservoir dome is preferable for geological reasons at a late stage of exploitation, since it enables additional oil recovery in the near-dome zones and active use of the existing field facilities while minimizing additional capital investments.

A new insight into the mechanism of developing and increasing the efficiency of production wells of Mesozoic reservoirs in the Grozny oil-bearing region is especially relevant at the current late stage of their exploitation.

IOP Conf. Series: Earth and Environmental Science 1021 (2022) 012018 doi:10.1088/1755-1315/1021/1/012018

References

[1] Degtyarev N M, Kadyrov L Kh 2007 Analysis of the mechanism of oil recovery of a complex type reservoir in Mesozoic deposits. *Proceedings of the XXXVI scientific and technical conference on the results of the work of the teaching staff of SevKavGTU in 2006*, Vol. 1, pp. 47–48

[2] Muslimov R Kh 2005 *Modern methods of increasing oil recovery: design, optimization and efficiency assessment* (Kazan: Feng)

[3] Sokolovsky E V 1983 Field studies of enhanced oil recovery processes in porous and fractured formations *Univ. Proceed. SevKavNIPIneft,* **38** 128–133.

[4] Maydebor V N 1980 *Features of the development of oil fields with fractured reservoirs* (Moscow: Nedra).

[5] Tokarev M A 1990 *Integrated geological field control over the current oil recovery when oil is displaced by water* (Moscow: Nedra)

[6] Degtyarev N M 1961 Results of laboratory studies of oil displacement by high pressure gas. News of oil and gas technology *Oilfield business* **10** 31–32

[7] Khaladov A Sh 2002 *Improving the efficiency of removing asphalt-resinous and paraffin deposits during oil production with large temperature drops in the fountain lift* Dissertation for a candidate of technical sciences, Ufa: USPTU

[8] Komisarov A I, Yarovoy V A, Lemeshko N N 1983 Intensification of fluid inflows from deep-seated terrigenous deposits of the Chechen-Ingush ASSR *Oilfield business* **9** 18–19

[9] Kadyrov L Kh 2008 Effect of the rate of oil withdrawal on oil recovery of flooded fractured formations of the Chechen Republic *Actual problems of modern science* **2** 302–303

[10] Kadyrov L Kh 2008 Enhanced oil recovery technology of flooded reservoirs of Mesozoic deposits in the Chechen Republic *Natural and technical sciences* **1** 196–198

[11] Khaladov A Sh 2014 Intensification of oil recovery from Mesozoic deposits of fields at a late stage of exploitation in the Chechen Republic. *Innovations, science-intensive technologies in education and economics: Proceedings of the I international (X All-Russian scientific and methodological conference)*, April 29–30, 2014, Ufa: RITsBashGU

IOP Conf. Series: Earth and Environmental Science 1021 (2022) 012019 doi:10.1088/1755-1315/1021/1/012019

Substantiation of geological exploration and modeling of oil deposits for multistage fracturing in a horizontal well

E V Voronova, L V Petrova and A Yu Gutorov

Ufa State Petroleum Technological University, Branch of the University in the City of Oktyabrsky, 54a, Devonskaya St., Oktyabrsky, Republic of Bashkortostan, 452607, Russian Federation

E-mail: razboinica77@mail.ru, info@of.ugntu.ru

Abstract. To date, an increase in the oil recovery factor by commissioning horizontal wells is of relevance for oil fields at the late exploitation stage. In most cases, horizontal wells are developed using multistage fracturing; therefore, it is crucial to determine the criteria affecting its successful implementation. The study results are of use for a more reasonable approach to the choice of the location of horizontal wellbores and increase in the efficiency of geological exploration and multistage hydraulic fracturing.

1. Introduction
To date, maintained rate of oil production and optimal performance, as well as increased oil recovery factor, are crucial for fields at the late exploitation stage [1–3]. The oil production area can be expanded by commisionning horizontal wells. Due to a high geological layer-by-layer heterogeneity of the reservoirs and the number of permeable intervals equal to 5.6–12, as well as low porosity and permeability, most of the wells are developed by multistage hydraulic fracturing to ensure high productivity.

2. Materials and methods
Commisionning of horizontal wells after carrying out the relevant geological exploration and creating a geological model of the reservoir requires the assessment of such parameters as the distribution of density of residual reserves, the flow capacity factor, the distribution of the current net thickness, and their resulting parameter – the Euler-Venn map. In order to determine profitability, it is necessary to assess the amount of cumulative oil production over 7–20 years. To assess the effectiveness of the measures taken, it is necessary to consider a large number of geological and field factors [3,6].

Geological factors include: current net thickness of the reservoir, current oil and water saturation, clay and sand content, coefficient of the layer-by-layer heterogeneity – number of reservoir beds, bound water, distance to the water-oil contact (which affects the initial water cut), porosity, permeability and flow capacitty, the current density of recoverable reserves per well. In the presence of complicating structural tectonic elements such as shears, faults, fractures, it is necessary to know the stress of the main stress axes and the geomechanical parameters that will affect the productivity after hydraulic fracturing [4, 5].

Geomechanical modeling provides mapping of zones appropriate for drilling, for example, a map of the lost-circulation zone gradient shows areas of lost circulation risks.

Field factors include: reservoir pressure value before hydraulic fracturing, distance to the discharge row with line-drive development system, current and cumulative compensation coefficient, washing coefficient, well grid density, etc. The productivity value after multistage hydraulic fracturing is affected by the number of stages and the developed fracture length.

During exploitation, reversible changes in the void space of collectors can be observed under the impact of technogenic factors [1–2]. In order to distribute collector properties, a petrophysical porosity-permeability relationship is constructed. To analyze the effectiveness of the measures taken, it is necessary to first detect the presence or absence of relationships between the above parameters and an increase in the oil production rate, and cumulative oil production. For this, engineering practice employs statistical methods, for example, correlation analysis [7,8].

The effectiveness of the measures taken can be estimated by assessing the value of cumulative oil production over 5 and 20 years and the NPV value, as well as by creating prospects for residual oil reserves in dead-end low-permeability zones of reservoirs.

Terrigenous deposits of one of the fields in Western Siberia were chosen as the study object. This field is multi-layer and unique in terms of recoverable reserves. The commercial oil-bearing capacity of the field is associated with terrigenous deposits of YuS2 reservoirs confined to the Tyumen series; YuS1 of the Vasyugan series; YuS0 of the Bazhenov series, YuS0-Ach of the anomalous section of the Bazhenov series; BS162, BS163, BS17, BS20, BS18-19, BS21, BS221, BS222 of the Achimov stratum; strata of the clinoform complex as part of BS102-3, BS11, BS12 horizons, and BS100, BS101 shelf layers in the upper part of the Sortymskaya series. The main productive objects are the BS10 2-3, BS12, YUS1 reservoirs.

In terms of the geological structure complexity, the field belongs to complexity group II ("complex" or "very complex"), which is characterized by the presence of zones of lithological substitutions of collectors, pinching-out, inconsistency of thickness and reservoir properties. Reservoirs are characterized by significant lithological-facies heterogeneity, which is expressed both in the number of reservoir beds and in the replacement of permeable layers with impermeable differences. All the reservoirs are characterized by a complex distribution of reservoir rocks, both in area and in section.

BS102-3 reservoir is the main object for exploration and exploitation of the field, which contains 55.4% of the initial geological oil reserves of the field. The effective thickness varies from 0.5 to 52.1 m, averaging 16.7 m, the effective oil saturation varies from 0.3 to 46.6 m, averaging 14.3 m. The sand content averages 0.38 unit fractions and varies from 0.05 to 1 unit fractions, the number of reservoir beds averages 10 units. The average oil saturation factor for wells is 0.624 unit fractions.

BS12 horizon includes two reservoirs – BS121 and BS122. It contains a single oil deposit in a complex structured heterogeneous reservoir. The hydrodynamic connection of BS121 and BS122 reservoirs is due to the presence of reservoir confluence zones.

The deposit of BS121 reservoir is of the structural-lithological type. The effective thickness varies from 0.5 to 33.1 m, averaging 11.5 m; the effective oil pay varies from 0.5 to 31.9 m, averaging 9.8 m; the weighted average for the deposit is 6.3 m. The sand content averages 0.241 unit fractions and varies from 0.051 to 0.741 unit fractions, the number of reservoir beds averages 9.2 units. Average oil saturation factor for wells is 0.496 unit fractions. The deposit is of the structural-lithological type.

YuS1 reservoir. Deposits of the Verkhnevasyugan subseries are developed in the reservoirs in most of the area of the study field.YuS1 reservoir is characterized by zonal development of various types of facies, which changes its morphology and reservoir properties. In the region, there are vast areas of distribution of reservoirs with low filtration properties. The reservoir section is represented by uneven layering of sandstones and siltstones with subordinate clay interlayers. YuS1 reservoir in the area of the study field has a total thickness of 17.4 m (5.3–35.6 m) increasing from north to south and from west to east. Net thicknesses vary from 0.3 to 18.4 m, averaging 6.0 m for the reservoir. The net content of the reservoir is equal to 0.453 unit fractions, (0.056–0.985 unit fractions), the number of reservoir beds varies from 1 to 17, averaging 5.6 units.

3. Results and Discussion

When choosing sites for drilling lateral and horizontal wells, it is necessary to have the following geological and production maps: a map of the distribution of current net thicknesses, a map of the density of current oil reserves, an Euler-Venn map, a map of isobars, and facies maps of body distribution [8–10].

Table 1 shows the efficiency of drilling horizontal wells in the study field according to the exploitation targets.

Table 1. Efficiency of drilling horizontal wells in the study field

Parameter	BS10/2-3 reservoir	BS12/1 reservoir	YuS1 (Yu1) reservoir
Specific additional oil production, thousand tons	21602.38222	4749.98	16220.271
Number of horizontal wells	18	4	10

Table 2. Geological and physical characteristics of the study objects

Parameter	BS_{10}^{2-3} reservoir	BS_{12} reservoir	YuS_1 reservoir
Average depth of the roof, m	-2375	-2360	-2757,4
Oil water contact true vertical depth, m	-2423-2460	-2446-2466	-2679-2920
Reservoirs type		terrigenous, porous	
Average net thickness, m	9.9	6.3	4.3
Average effective water-saturated thickness, m	10.3	6.0	4.1
Porosity coefficient, unit fraction	0.200	0.190	0.166
Oil saturation coefficient, unit fraction	0.620	0.508	0.576
Permeability, μm^2	77.6	12.9	7.44
Sand content, unit fraction	0.340	0.241	0.453
Number of reservoir beds, unit fraction	12	9.2	5.6
Initial reservoir temperature, °C	83	83	88
Initial reservoir pressure, MPa	24.7	25.1	28
Oil viscosity in reservoir conditions, mPa · s	1.484	2.376	1.027
Density of oil in reservoir conditions, g/cm^3	0.790	0.802	0.731
Density of oil in surface conditions, g/cm^3	0.858	0.852	0.822
Volumetric oil coefficient, unit fraction	1.130	1.153	1.252
Oil saturation pressure with gas, MPa	8.89	8.52	10.7
Gas content, m^3/t	51	58	103
Water viscosity in reservoir conditions, mPa · s	0.38	0.38	0.36
Compressibility, 1/MPa · 10^{-4}			
oil	10	11.3	15.4
water	4.6	4.6	4.7
Displacement coefficient (by water), unit fraction	0.560	0.488	0.498
Productivity coefficient, m^3/day · MPa	24.5	12.7	9.8

Table 1 shows that the largest number of wells was drilled in BS10/2-3 reservoir, because it exhibits the greatest reserves. The largest specific cumulative oil production per well was obtained for BS10/2-3 reservoir, since its net thickness is greater (9.2 m versus 6.3 in BS_{12} reservoir and 4.3 m in YuS_1 reservoir) compared to other production objects. YuS_1 reservoir occupies the second position likely due to its lower number of reservoir beds equal to 5.6 compared to other reservoirs – 12 and 9.2. The lowest efficiency of commissioning of horizontal wells was obtained for BS_{12} reservoir. Table 2 shows the geological and physical characteristics of the study objects.

The correlation analysis showed how the initial water cut, effective oil-saturated thickness, depression on the reservoir, the length of the horizontal wellbore, the ratio of the current reservoir pressure to the initial one affect the value of cumulative oil production (Figure 1, 2, 4, 5, 6). In addition, the dependences of the inlet water cut on the distance to the discharge row were plotted (Figure 3) and the dependences illustrating how the developed fracture half-length affects an increase in oil production after hydraulic fracturing (Figure 7).

At increased input water cut (which depends on the amount of bound water, the distance to the oil-water contact, the degree of reservoir depletion and the distance of the horizontal well to the discharge row), of cumulative oil production decreases (Figure 1). Lower initial water cut is found for BS_{10}^{2-3} and BS_{12} reservoirs.

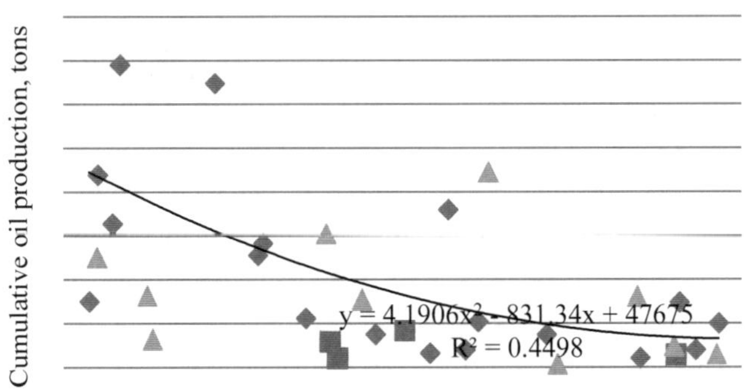

Figure 1. Dependence of cumulative oil production from horizontal wells on water cut values

Figure 2 shows that the increased net oil-saturated thickness exceeding 20 m does not lead to an increase in cumulative oil production, this is likely due to increased number of reservoir beds. The smallest oil-saturated thicknesses of less than 10 m are observed within BS12/1 reservoir, which is due to low cumulative oil production of less than 10 thousand tons.

Figure 5 shows how the horizontal wellbore length affects cumulative oil production. At the same time, the maximum length were drilled into BS10/2-3 reservoir, which is due to a high geological heterogeneity of the object, the number of reservoir beds is 12, while the average oil-saturated thickness is 7.8 m, and the permeability is 77.6 μm^2. For BS12/1 reservoir, the design lengths averaged 170–200 m, and for YuS1 reservoir, the minimum length of horizontal wellbores is 100–150 m (due to the lower number of reservoir beds of 5.6, the average oil-saturated thickness of 19.3 m, permeability of 7.44 μm^2).

When drilling horizontal wells, it is also important to consider the current energy state of the object. For this, the dependence of cumulative oil production on the ratio of the current reservoir pressure to the initial one was plotted (Figure 6). The average value of the ratio for BS10/2-3, BS12/1, YuS1 reservoirs was 0.75, 0.68, 0.83 unit fractions, respectively. The lower this ratio leads to the lower cumulative oil production (BS12/1 reservoir), which means that before drilling the reservoir pressure maintenance system should be improved. It is not recommended to reduce reservoir pressure in the area of planned drilling by more than 0.75 of the initial pressure.

When drilling horizontal wells in conditions of high geological heterogeneity and high number of reservoir beds, it is recommended to carry out development using the multistage hydraulic fracturing technology in order to increase productivity [10–14]. Figure 7 shows how the developed fracture length increases oil production. The maximum increment was obtained for the fracture length of 70 m.

Since the field has an in-line development system, Figure 2 shows both low and high values of water cut at the same distance to the discharge row. The advancement of the injected water front can be considered in the hydrodynamic model using the oil saturation distribution cube, as well as on the maps of current mobile reserves and according to the Euler-Venn diagrams.

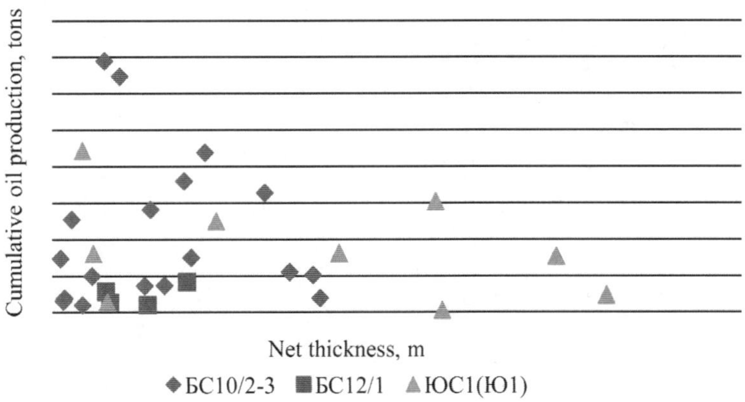

Figure 2. Dependence of cumulative oil production from horizontal wells on the net thickness

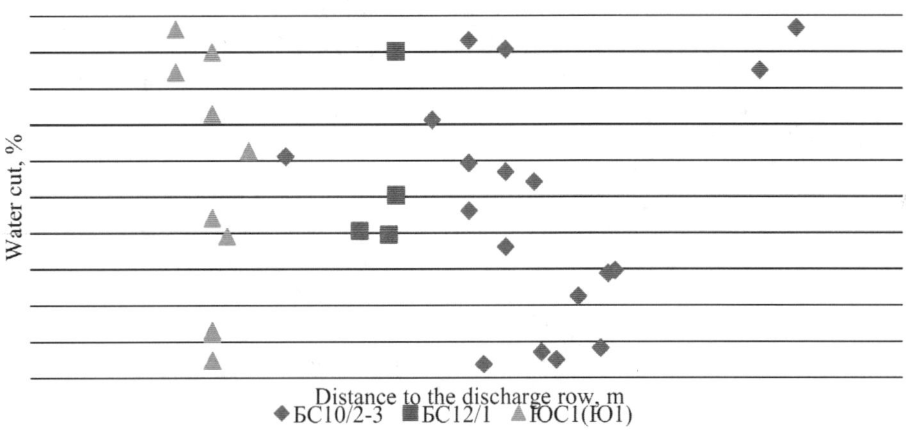

Figure 3. Dependence of the starting water cut on the distance to the discharge row

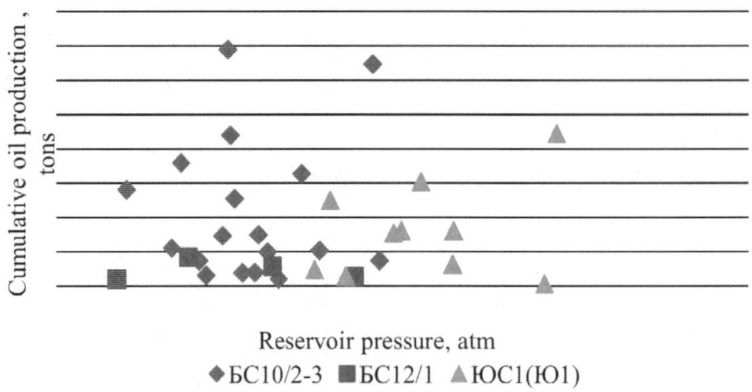

Figure 4. Dependence of cumulative oil production on the reservoir pressure

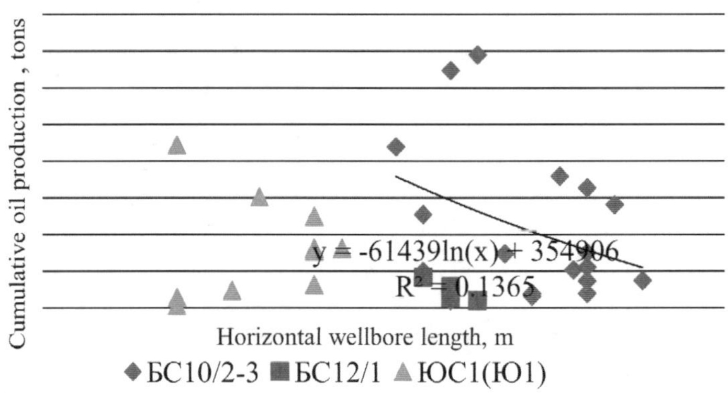

Figure 5. Dependence of the cumulative oil production on the horizontal wellbore length

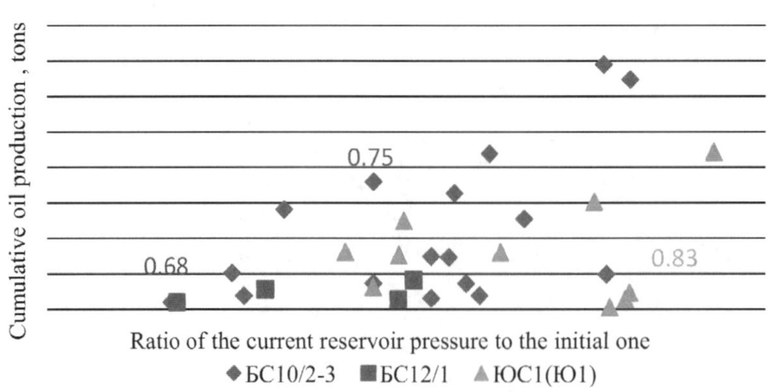

Figure 6. Dependence of cumulative oil production on the ratio of the current reservoir pressure to the initial one

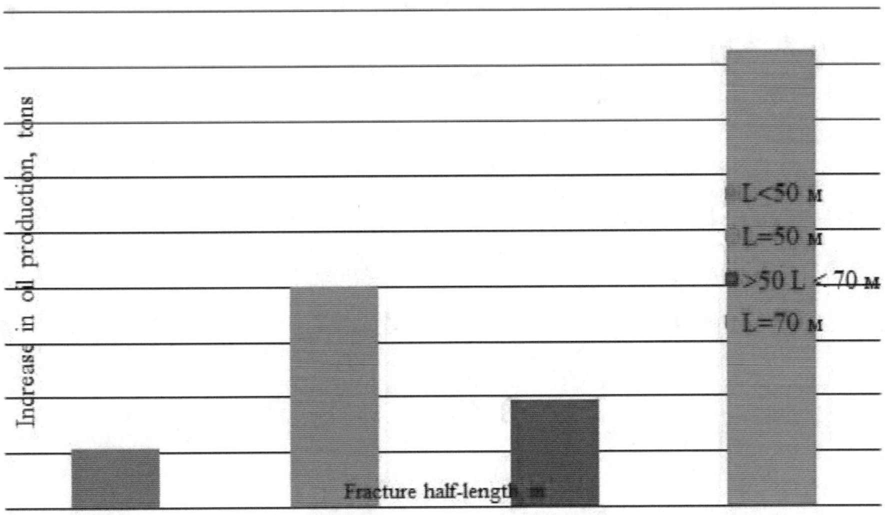

Figure 7. Dependence of the increase in oil production on the fracture half-length

4. Conclusion

In the study of the efficiency of drilling horizontal wells in the study field, the geological and physical characteristics of the field were first investigated in detail, and the distribution of petrophysical properties for the main production object – BS10/2-3, BS12.1, YuS1 reservoirs – was analyzed.

The correlation analysis of the effect of geological and production parameters on the value of cumulative oil production revealed the general trends affecting the efficiency of drilling and the quantitative criteria for its successful implementation [12–14].

Thus, the performed analysis yielded the criteria for a more thorough choice of wells for hydraulic fracturing in the study area in order to achieve higher production rates:

1. At high layer-by-layer geological heterogeneity (exceeding 9), it is recommended to drill wells using a wellbore with the length of at least 250 m to ensure cumulative oil production of more than 20 thousand tons.
2. In this case, the length of the hydraulic fracture should be at least 70 m for each stage of multistage hydraulic fracturing.
3. Reservoir pressure in the area of the proposed drilling of a horizontal well should be no less than 0.75 unit fractions relative to the initial reservoir pressure, otherwise it is necessary to optimize the reservoir pressure maintenance system.
4. The design horizontal wells should not be located in the washed zone or in the zone with the current reserve density of less than 0.15 t/m^2.
5. When organizing an in-line development system, it is necessary to consider the distance to the nearest injection well in order to avoid a breakthrough of the injected water front or possible penetration of a hydraulic fracture into the washed zone.
6. The increased net thickness of more than 20 m does not lead to a significant increase in oil production due to the high number of reservoir beds.

In general, the study field exhibited successful multistage hydraulic fracturing of more than 20 thousand tons. Wells with cumulative oil production of less than 10 thousand tons are located in highly watered, washed areas, or they have been in operation for less than 2 years and have not yet exhibited their potential for profitable production. A hydrodynamic model and analytical characteristics of displacement can predict cumulative oil production for such wells.

References

[1] Zainagalina L Z, Petrova LV and Petrov VA 2019 Analysis of the efficiency of telemetric systems

for drilling wells *IOP Conference Series: Materials Science and Engineering* **560** 012096 doi:10.1088/1757-899X/560/1/012096

[2] Nie Z, Liu Z and Sun X 2018 A Multirelay Cooperation Method for Wireless Transmission of MWD and LWD Signals *IEEE Transactions on Geoscience and Remote Sensing* **56** 1229–37

[3] Almukhametova E M, Fattakhov D I and Zakirov A I 2018 The efficiency analysis of applying hydraulic fracturing of formation at AB1-2 facility of Potochnoye field *IOP C. Ser.: Earth Env.* **194(8)** 082002 DOI:10.1088/1755-1315/194/8/082002

[4] Suleimanov R I, Gabdrakhimov M S, Khabibullin M Y, Zaripova L M and Vasilyeva E R 2018 The study of hydraulic hammer device in drilling tool assembly in hydraulical rotary drilling *Int. J. of Engineering and Technology* **7(2)** 2–30

[5] Akhmetov R T, Mukhametshin V V, Kuleshova L S, Grezina O A, and Malyshev P.M. 2020 The generalized correlating function of capillary curves and the relationship of the filtration-capacitive parameters of reservoirs in Western Siberia with the size distribution of pore channels *Journal of Physics: Conference Series* **1661(1)** 012016 DOI: 10.1088/1742-6596/1661/1/012016

[6] Petrova L V, Buravov V V 2019 Evaluating the technological effect of a horizontal hydraulic fracturing: *Journal of Physics: Conference Series* **1333** 072019 doi:10.1088/1742-6596/1333/7/072019

[7] Nazim N 2008 Dmour Practical Well Test Analysis of a Hydraulically Fractured Low permeability Gas Reservoir: A Case History J.King Saud Univ., *Eng. Sci.* **20(1)** 47–60,

[8] Antonenko D A, Pavlov V A, Surtaev V N, Sevastyanova K K 2008 Integrated Modeling of the Priobskoe Oilfield *SPE* **117413**

[9] Dean L, Mireault R 2008 Production Decline Analysis *Reservoir Engineering for Geologists* **1** 20–22

[10] Zeigman Yu V, Mukhametshin V V, and Kuleshova L S 2020 Management of flooding of low-production deposits according to geological and field data *IOP: Earth and Environmental Science (EES)* **579** 012019 DOI: 10.1088/1755-1315/579/1/012019

[11] Mukhametshin V V and Kuleshova L S 2020 On uncertainty level reduction in managing waterflooding of the deposits with hard to extract reserves *Bulletin of the Tomsk Polytechnic University. Geo Assets Engineering* **331** 140–146 DOI 10.18799/24131830/2020/5/2644

[12] Khabibullin M Ya, Samigullina L Z and Zainagalina L Z 2018 Laboratory studies for the momentum of liquid optimal impact on hydrocarbon formation Advan. in Engineering Res **157** 276–9 DOI:10.2991/aime-18.2018.53

[13] Gabdrkhmanova K F, Izmailova G R , Samigullina L Z 2020 Solution of the Problem of Annular Space Gas Utilization in Wells Operated by Walking-Beam Pumping Unit *IOP Conference Series: Earth and Environmental Science* **459(4)** 042080

[14] Bijjani P, Coscia M, Hammon K, Ashraf M and Ali Hassan H 2015 MWD/LWD challenges and solutions for drilling the longest al hosn gas well in an aggressive subsurface environment Society of Petroleum Engineers - Abu Dhabi Int. Petroleum Exhibition and Conf. (ADIPEC 2015)

Selecting an analytical model for describing the viscosity of an oil-water emulsion in the Aspen HYSYS program

A S Drozdov, S A Leontev

Tyumen Industrial University, 38, Volodarsky street, 38, Tyumen, 625000, Russia

E-mail: Dro3d96@gmail.com

Abstract. When designing technological installations for the preparation of downhole products, it is necessary to know the dynamic viscosity of the oil-water emulsion. It is not always possible to experimentally determine the viscosity and physico-chemical parameters of the oil-water emulsion. In this regard, the formulas used in the Aspen Hysys complex for determining the dynamic viscosity at different temperatures were analyzed. When analyzing the applicability of the analytical dependences, the calculated data were compared with the experimental data obtained by physical modeling of the viscosity of the water-oil emulsion of the Chayandinskoye field. The study was conducted in two stages. The first stage consisted of constructing models of the viscosity of the oil-water emulsion for all dependencies and comparing them with experimental data. This is done to test analytical assumptions and select the most reliable models. The models were constructed for a single temperature T=20 °C. The second stage consists of predicting the viscosity of the emulsion over the entire studied temperature range from -10 to 60 °C and water cut from 0.5 to 95% by weight, followed by verifying the reliability of the results of analytical calculations, as well as identifying existing shortcomings. An alternative formula is proposed, which does not contain empirical constants, describes the entire possible range of water cut, and has a relative forecast error of about 30%. In the future, it is planned to test the formula on water-oil emulsions with oils having similar properties (density, viscosity, amount of paraffins).

1. Introduction

Institutes and scientific organizations that are engaged in the design of technological installations for the preparation of downhole products are far from the projected facilities and the provision of samples of downhole products for research is delayed indefinitely [1]. In this regard, the formulas used in the Aspen HYSYS complex for determining the dynamic viscosity at various temperatures and water levels were analyzed.

The viscosity of oil is the most important physical characteristic, as it directly affects its fluidity (reverse viscosity). Viscous oil is extremely difficult to transport through pipelines and difficult to process, which is difficult for the oil industry. Knowing the viscosity of oil, you can even approximately determine its composition, since this physical characteristic increases with the increase in the molecular weights that make up this mineral fractions. In other words, the heavier the extracted raw material, the more it contains heavy hydrocarbon fractions, which means that its viscosity increases. High-viscosity grades of oil contain a lot of resins and asphaltenes, which significantly complicates the extraction and processing of such raw materials [2].

The Aspen Hysys program was developed in 1996. At the moment, specialists of the oil and gas industries use the complex to create stationary models at the stage of preliminary design of technological systems, to analyze the state of devices and prevent malfunctions, to optimize the modes of operation of the system, business planning and enterprise management [3].

2. Materials and methods

The formulas for describing the viscosity of the oil-water emulsion from the Aspen HYSYS program were taken as the material for the research. The Aspen Hysys computational complex includes several models for describing the viscosity behavior of an oil-water emulsion:

1. Brinkman:

$$\mu_{\text{rel}} = (1 - V_f)^{-2,5};\ V_{\text{inv}} = \frac{1}{1+R_v};\ R_v = (\frac{\mu_{\text{H}}}{\mu_{\text{B}}})^{0,4}; \tag{1}$$

2. Guta-Simcha:

$$\mu_{\text{rel}} = 1 + 2,5V_f + 14,1{V_f}^2; \tag{2}$$

3. Levinton-Leighton:

$$\mu_{\text{rel}} = \exp[2,5A * \left(V_f + V_f^{\frac{5}{3}} + V_f^{\frac{11}{3}}\right); A = \frac{0,4\mu_{\text{H}}+\mu_{\text{B}}}{\mu_{\text{H}}+\mu_{\text{B}}}; \tag{3}$$

4. Barnea-Mizrahi:

$$\mu_{\text{rel}} = \exp\left[\frac{K_1 * V_f}{1-K_2 * V_f}\right]; \tag{4}$$

5. Polynomial:

$$\mu_{\text{rel}} = 1 + K_1V_f + K_2{V_f}^2 + K_3{V_f}^3. \tag{5}$$

where μ_{rel} is the relative viscosity of the oil, equal to the ratio of the viscosity of pure oil to the viscosity of the water on which the emulsion is mixed. V_f is the fraction of the water phase in the oil-water emulsion. V_{inv} is the fraction of water at which the phase inversion occurs. μ_{H} and μ_{B} are the viscosity of pure oil and water. K_1, K_2, K_3 are constants that are determined experimentally.

Analyzing these formulas, an attempt was made to classify them according to various characteristics, highlighting their disadvantages and advantages. The following classification is proposed:

1. By mathematical function. All formulas can be divided into three categories: power, exponential, and polynomial. In this classification, the polynomial functions stand out very strongly. They have an extremum point that can be mapped to the phase inversion point. Based on this, they are not monotonic and give more accurate values when building models based on them. In contrast, power and exponential functions are monotonically increasing, which, depending on their type, will give a large deviation at low or high water cut values [4].

2. By application range. When using formulas under numbers (1), (3), (4) they should have a clear disadvantage: they are limited to use up to the point of inversion. Accordingly, the other formulas have a clear advantage over them.

3. Adaptability to real oil. Since the Aspen Hysys program is used at the field development stage, and laboratory studies of oil are available, a very significant advantage is the ability to adjust the viscosity of the oil-water emulsion to real values, to obtain more accurate values of hydraulic resistances in pipes, various treatment plants, etc. This advantage is enjoyed by forums numbered (4), (5). Also, the formula (5) can be reduced to a simpler quadratic function, which simplifies the task of forecasting.

Based on all of the above, it was decided to conduct the study in two stages. The first stage consisted of constructing models of the viscosity of the oil-water emulsion for all dependencies and comparing them with experimental data. This is done to test analytical assumptions and select the most reliable models. The models were constructed for a single temperature T=20 °C. The second stage consists of predicting the viscosity of the emulsion over the entire temperature range under study (from -10 to 60 °C), checking its reliability, as well as identifying existing shortcomings [5].

3. Analysis of experimental and calculated viscosity values

Figure 1 shows a comparison of the experimental and calculated values of the oil viscosity of the Chayandinskoye field in the Aspen HYSYS program at a temperature of 20 °C.

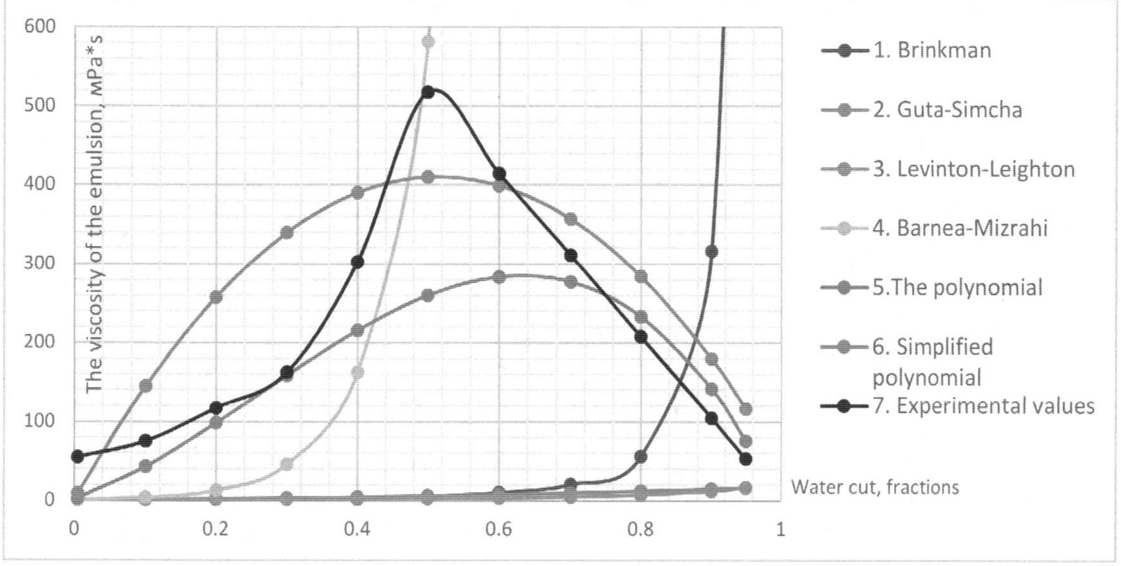

Figure 1. Oil viscosity of the Chayandinskoye field in the Aspen HYSYS program

Analyzing the data obtained, we can draw conclusions about the possibility of using some models. So in models (1), (2), (3), the viscosity is not reliably predicted over the entire circumference interval. The model (4) has a small confidence interval (from 0.4 to 0.5). Thus, only the model number (5) and its simplified form reliably predicts the viscosity. Further research will be conducted using them. The more complex form of the polynomial is the most interesting one, since it describes the model more precisely.

We will calculate the viscosity of the oil-water emulsion according to the formula (5), adapting it to each temperature value. Table 1 shows the calculated values, as well as the arithmetic mean of the deviations from the experimental data.

Table 1. Calculated data on the dependence of the dynamic viscosity of the water-oil emulsion of the Chayandinsky field on the water content and temperature according to the formula (5).

Water cut, fractions, %	The viscosity of the emulsion, MPa*s at a temperature of, °C								
	-10	-5	0	10	20	30	40	50	60
0,5	53.6	27.4	7.9	4.0	2.6	2.0	1.8	1.6	1.5
10	1014.6	546.3	182.6	89.5	42.8	31.6	23.0	17.3	13.1
20	1915.9	1094.7	429.7	222.5	98.2	77.0	54.3	38.8	28.5
30	2657.5	1596.7	700.7	374.2	158.6	128.5	89.4	62.0	44.9
40	3192.1	2002.8	953.7	518.7	215.4	177.8	122.6	83.4	59.8
50	3472.4	2263.4	1147.2	630.3	259.9	216.2	148.3	99.4	71.0
60	3450.9	2329.2	1239.5	683.2	283.4	235.2	160.9	106.5	76.0
70	3080.2	2150.6	1189.0	651.5	277.3	226.2	154.8	101.2	72.6
80	2313.1	1678.0	954.0	509.4	232.9	180.7	124.3	79.9	58.2
90	1102.0	862.1	492.8	231.2	141.5	90.2	63.9	39.1	30.6
95	315.2	309.9	164.3	32.9	75.6	25.4	20.7	10.2	11.1

Based on the obtained data, a new dependence of the viscosity on the water content was created. The polynomial dependence is taken as a basis, but with the exception of the unknown constants K:

$$\mu_{rel} = 1 + 5955{,}7V_f e^{-0{,}047T} + 4622{,}8V_f{}^2 e^{-0{,}047T} + 933{,}45V_f{}^3 e^{-0{,}042T}. \quad (6)$$

This formula is tested in terms of the viscosity of the water-oil emulsion of the Chayandinskoye field, in the future it is planned to test the formula on water-oil emulsions of oils with similar properties.

4. Conclusion

Based on the results obtained, the following conclusions can be drawn:

1) The algorithms embedded in the Aspen Hysys program do not allow us to reliably formulate the viscosity of the oil-water emulsion at low water cut values. This is primarily due to the fact that the program does not have the ability to set a free term in the equations and constants before the exponent. The introduction of this function will significantly increase the accuracy of the viscosity description for emulsions with a water content of less than 20%.

2) For the prediction of high-viscosity oils, most of the embedded formulas are not suitable, their forecast has too large an error. When choosing a prediction model, you should pay attention to the model, since most models do not describe the viscosity over the entire range of possible water cuts.

3) An alternative formula is proposed, which does not contain empirical constants, describes the entire possible range of water cut, and has a relative forecast error of about 30%, which is more accurate than in Aspen HYSYS.

IOP Conf. Series: Earth and Environmental Science 1021 (2022) 012020 doi:10.1088/1755-1315/1021/1/012020

References

[1] Drozdov A 2017 *Research and forecasting of rheological properties of oil-water emulsions* (Tyumen: Tyumen Industrial University)
[2] Galikeev R 2011 *Research of regularities of structure formation of paraffin-containing oils in the system of production and oil collection* (Tyumen: Tyumen Industrial University)
[3] Leontiev S, Galikeev R, Drozdov A, Levitina E 2020 *Prevention and control of asphalt-tar-paraffin deposits in the system of collecting borehole products* (Moscow: RUSAINS)
[4] Zakirov S, Brusilovskiy A, Zakirov E 2000 *Improvement of technologies for the development of oil and gas fields* (Tyumen: Tyumen Industrial University)
[5] Sevastyanov A., Korovin K., Zotova O 2017 Development of fields with hard-to-recover oil reserves (Tyumen: Tyumen Industrial University)

Location of deposits of non-metallic ores in Russia and the technogenic impact of mining on the environment

Kh Kh Kozhiev [1], O Z Gabaraev[1], M S-U Khaliev[3]

North Caucasian Mining and Metallurgical Institute, 44 Nikolaeva St., Vladikavkaz, 362021, Russian Federation
[3]Chechen State University named after A. H. Kadyrov, 17a, Dudaev Boulevard, Groznyy, 364015, Russia

E-mail: hambi@list.ru

Abstract. The article substantiates the need to strengthen the mineral resource base for the production of non-ferrous metals through the involvement in the operation of substandard raw materials, primarily those stored in tailings dumps for the enrichment of metal ores. The results of the analysis of reserves of ore deposits are presented in comparison with the situation in the world. Quantitative data on the state of the mineral resource base of non-ferrous metal ores in Russia and the provision of industry with metal raw materials are given. The role of mineral resources in protecting the geopolitical interests of Russia is shown. It is noted that Russia lags behind technologically developed countries in a number of indicators of production and consumption of mineral resources. It is concluded that the development of technogenic deposits using innovative technologies is a real step towards strengthening the mineral resource base of Russia, therefore, research on the direction of mining mentioned in the article is relevant.

1. Introduction

Modern society needs metals, especially non-ferrous metals. This also applies to industrially developed modern Russia. Data on the reserves of base metal ores are summarized in table.1.

Table 1. Information about the extraction of non-ferrous metal ores in Russia

Metals	Ore reserves, million tons	Production of metals in 2015, thousand tons	Share of global metal production, %
Zinc	59.8	388	2
Lead	8.2	156	4
Nickel	7.3	310	14
Copper	58	741	4
Titan	118	89	1
Tungsten	0.199	3.3	3
Tin	0.776	0.6	0.2
Aluminum	518.4	7	2

Molybdenum	1070	3.2	1

A number of factories are insufficiently supplied with raw materials: Krasnouralsky, Sredneuralsky, Kabashsky, Mednogorsk, Pechenganikel, Severonikel, as well as processing enterprises of the Chita region, Primorye, Khabarovsk and Primorsky Territories.

Mineral resources and products of their processing provide two-thirds of foreign exchange earnings. The possession of strategic mineral resources contributes to the protection of Russia's geopolitical interests [1–3, 14].

2. Materials and methods

Prospects for optimizing the use of mineral resources are assessed by analyzing the reserves of ore deposits [15-17].

The directions of solving the problems of the mineral resource base include:

- development of science and stimulation of scientists;
- updating the theory of prospecting and evaluation of field resources;
- development of a new mineral resource policy taking into account the interests of the regions;
- improvement of state regulation of subsoil use;
- diversification of mining and processing enterprises.

3. Results and discussion

Manganese. The largest deposit of poor hard-to-enrich ores – Usinskoye in the Kemerovo region – is classified as a reserve.

Chromites: there are practically no proven reserves of metallurgical grades [18, 19].

Bauxite: the vast majority is mined in the Urals. The second most important type is nepheline ores.

Copper: the need can be met at the expense of the Udokan deposit.

Nickel: the first place in the world in terms of proven reserves and production.

Lead and zinc: the exploitation of reserves of the flanks and deep horizons of the developed and the identification of new deposits remains relevant.

Tin: the demand exceeds its production by a third.

Tungsten and molybdenum: the availability of reserves is 8–40 years.

Tantalum, niobium, strontium, rare earth metals: tagging behind industrially developed countries in terms of consumption in the presence of a large raw material base.

The availability of proven reserves is unsatisfactory, and the strategic reserve of raw materials at existing fields is insufficient. At the same time, strategically important raw materials are exported [20, 21].

In the tailings of ore enrichment, up to half or more metals are found (Figure1).

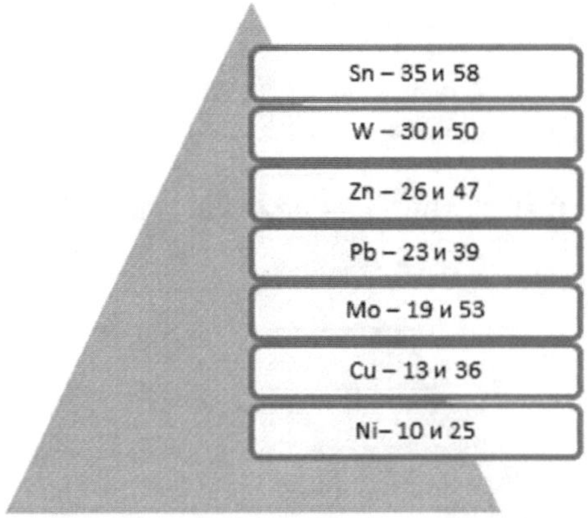

Figure1. The proportion of non-extracted metals, average and maximum values, %.

Preservation of the volume of metal production can be achieved by increasing the volume of processing of reserves of technogenic deposits with leaching technologies (Table 2) [4–6].

Table 2. Leaching technogenic deposits

Metals						
Copper	Nickel-Cobalt	Zinc	Tin	Tungsten	Molybdenum with Uranium	Titanium
Geography of deposits						
Sverdlovsk Region	Murmansk Region, Krasnoyarsk Region	Sverdlovsk Region	Sverdlovsk Region	Buryatia, Trans-Baikal Region	Chita region, Trans-Baikal Region	Trans-Baikal Region
Deposits						
Gumeshevskoye, subterrestrial	Allarechenskoye khvostokhranilishche No. 1	Slag dump	Khvostokhranilishche	Barun-Narynskoye, Calm	Streltsovskoye, khvostokhranilishche	Kruchininskoye

The table shows that the majority of enterprises engaged in leaching metals from man-made raw materials are located in conditions that are not comfortable for year-round operation, so new technologies should ensure the extraction of metals at low temperatures all year round.

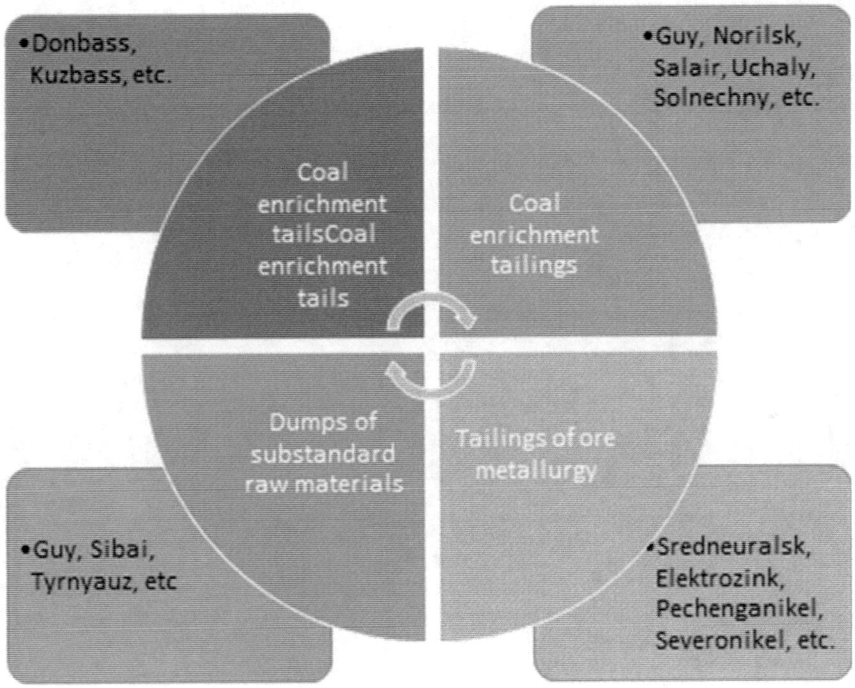

Figure 2. Technogenic deposits of non-ferrous metals.

Technogenic deposits of non-ferrous metals are storages of ore tailings, storages of metallurgy tailings, storages of coal tailings and non-conventional ores (Figure 2).

It is generally known that the total value of metals not extracted during enrichment is comparable to the value of commercial ore processing products.

The reserves of technogenic deposits, for example, the Ural in the North Ossetia-Alania region, are sufficient for their industrial development for many years (Figure 3).

Figure 3. Unal tailing dump.

4. Conclusion

To increase the completeness of the use of natural resources, it is advisable to determine the economic efficiency of the extracted mineral product minus the resources lost during extraction and processing.

Climatic conditions allow them to be developed year-round, compensating for the withdrawal from the number of existing deposits.

One of the reasons for the weak pace of development of the mineral resource base is the increasing gap between the possibilities of ore extraction and processing [6–8].

The research data presented in this article are consistent with the data of foreign specialists [9–13].

References

[1] Valiev N G, Propp V D and Vandyshev A M 2020 The 100th anniversary of the department of mining engineering of UrSMU *News of Higher Educational Institutions. Mining Magazine* **8** 130–143

[2] Dushin A V, Valiev N G, Lagunova Yu A and Shorin A G 2018 Ural Mining University and Moscow Mining University: interaction of higher education institutions *Mining Journal* **4** 4–10

[3] Khasheva Z M and Golik V I 2015 The ways of recovery in economy of the depressed mining enterprises of the Russian Caucasus *International Business Management* **9 (6)** 1210–1216

[4] Komashchenko V I, Golik V I, Belin V A and Gaponenko A L 2014 Improving the efficiency of explosive rebounding on the basis of new methods for initiating borehole charges in quarries *Mining Informational and Analytical Bulletin* **9** 293–304

[5] Golik V I, Gabaraev O Z, Maslennikov S A, Khasheva Z M and Shulgatyi L P 2016 The provision of development conversion perspectives into undeground one for Russian iron ore deposits development *Journal of Social Sciences* **11 (18)** 4348–4351

[6] Shchurov N I, Myatezh S V, Malozyomov B V, Shtang A A, Martyushev N V, Klyuev R V and Dedov S I 2021 Determination of inactive powers in a single-phase ac network *Energies* **14 (16)** 4814

[7] Kongar-Syuryun Ch B, Faradzhov V V, Tyulyaeva Yu S and Khairutdinov A M 2021 Effect of

activating treatment of halite flotation waste in backfill mixture preparation *Mining Information and Analytical Bulletin* **1** 43–57

[8] Shchurov N I, Dedov S I, Malozyomov B V, Shtang A A, Martyushev N V, Klyuev R V and Andriashin S N 2021 Degradation of lithium-ion batteries in an electric transport complex *Energies* **14 (23)** 8072

[9] Rybak J M, Kongar-Suryun Ch, Tyulyaeva Yu, Khairutdinov A M and Akinshin I 2021 Geomechanical substantiation of parameters of technology for mining salt deposits with a backfill *Mining Science* **28** 19–32

[10] Espinosa R D and Rojo J 2017 Towards sustainable mining (Part I): Valuing investment opportunities in the mining sector *Resource Policy* **52** 7–18

[11] Klyuev R V, Bosikov I I, Gavrina O A and Lyashenko V I 2021 Assessment of operational reliability of power supply to developing ore mining areas at a high-altitude mine *Mining Science and Technology (Russia)* **6 (3)** 211–220

[12] Dominguez M S K, Baptista A L F and Diogo M T 2017 Engineering complex systems applied to risk management in the mining industry *International Journal of Mining. Science and Technology* **27** 611–616

[13] Zhukovskiy Y L, Batueva D E, Buldysko A D, Gil B and Starshaia V V 2021 Fossil energy in the framework of sustainable development: analysis of prospects and development of forecast scenarios *Energies* **14 (17)** 5268

[14] Shakhnazarov K Yu, Prykhin E I 2020 Detection of Intermediate Phases in Metal Alloys. *Key Engineering Materials* **854** 23 – 29 Doi:10.4028/www.scientific.net /KEM.854.23

[15] Konyuhov V Y, Gladkih A M, Zott R S 2020 Accelerator as an effective replacement of a business incubator in the Irkutsk region. *Journal of Physics: Conference Series* **1582** 012045

[16] Kondrat'ev V V, Ershov V A, Shakhrai S G, Ivanov N A, Karlina A I 2016 Formation and Utilization of Nanostructures Based on Carbon During Primary Aluminum Production. *Metallurgist* **60(7-8)** 877–882

[17] Nemarov A A, Lebedev N V, Kondrat'ev V V, Kornyakov M V, Karlina A I 2016 Theoretical and experimental research of parameters of pneumatic aerators and elementary cycle flotation. *International Journal of Applied Engineering Research* **11(20)** 10222–10226

[18] Konyuhov V Yu, Gladkih A M, Chemezov A V 2019 Evaluation of the economic feasibility of the introduction of plasma hardening technologies in the Far North enterprises. *IOP Conference Series: Materials Science and Engineering* **560** 012147

[19] Gabov V V, Babyr N V, Zadkov D A 2021. Mathematical modelling of operation of the hydraulic support system of the powered support sections with impulse-free continuous regulation of its resistance to the roof rock lowering. *IOP Conference Series: Materials Science and Engineering* **1064** 012045 doi:10.1088/1757-899X/1064/1/012045

[20] Gladkih A M, Konyuhov V Yu, Galyautdinov I I, Shchadova E I 2019 Green building as a tool of energy saving. *IOP Conference Series: Earth and Environmental Science* **350** 012032

[21] Konyuhov V Yu, Gladkih A M, Galyautdinov I I, Shchadova E I 2020 Calculations of efficiency in implementing progressive mold forming methods. *IOP Conference Series: Materials Science and Engineering* **760** 012027

IOP Conf. Series: Earth and Environmental Science 1021 (2022) 012022 doi:10.1088/1755-1315/1021/1/012022

Justification of the use of a supersonic separator at the Yu. Kuvykin gas condensate field

D V Kirsanov, N N Leticheskaya

Astrakhan State Technical University Astrakhan, 414056, Russia

E-mail: danyakirsanoff@yandex.ru

Abstract. The main trends in the advancement of technology for the development and operation of offshore fields can be called underwater production complexes, which are a system of technological blocks that allow the production and treatment of hydrocarbons on the seabed. This article covers the technology of gas condensate separation using a supersonic separator. This 3S separator is able to perform deep purification of gas condensate with the extraction of CO_2 and H_2S and can be installed on the shelf as part of an underwater production complex. We consider it possible to apply this technology at the Yu. Kuvykin gas condensate field, which has a positive impact on the realization of the field's resource potentials.

1. Introduction

To date, the main increase in the volume of hydrocarbon raw materials produced in Russia is carried out through the introduction of offshore fields into development. And increasingly, development is carried out using technologies for the extraction and treatment of hydrocarbons at the bottom of the sea. An underwater production complex (MPC) is a set of technological units located on the seabed and responsible for certain operations during the operation of offshore wells with an underwater mouth.

Recently, this technology of developing offshore fields has gained a significant scale around the world. In Russia, the only field with subsea production to date is the Kirinskoye gas condensate field, located in the Sea of Okhotsk on the north-eastern shelf of Sakhalin Island. Production at this field is carried out without the use of platforms and other surface structures. The main link of the MPC is the manifold, where all extracted hydrocarbons are collected and transported to the shore via an offshore pipeline [1].

2. Advantages of field development using MPC

Due to the development of the field with the help of MPC, it is possible to make alternate commissioning of drilled wells. As a result, this allows obtaining hydrocarbon raw materials early. The process of installing basic underwater equipment and drilling wells from a drilling vessel is much faster than installing a stationary ice-resistant platform and drilling a cluster of directional wells from it. Among other things, alternate drilling of wells reveals the features of the geological and physical characteristics of the field a little early. Also, one of the advantages of the MPC should be called the protection of all equipment from ice and adverse weather conditions on the water surface, but this arrangement has a downside. This means that the equipment is also difficult to access for specialists, which requires the involvement of experienced divers to install and operate the equipment.

There are two schemes for installing MPC equipment: these are the open and closed (dry) location of the equipment. In the first case, the equipment comes into contact with water and undergoes pressure corresponding to the depth of its location. In the second case, the equipment is placed in special submersible chambers, which retain atmospheric or slightly higher pressure. As a result, the equipment does not come into contact with the water environment and does not experience the pressure that a column of liquid would exert. Systems with an open location of equipment are implemented much more often.

At this stage, all the necessary equipment has been created and is constantly being improved. These are underwater chokes, separators, pumps, heat exchangers, etc.

The project will be implemented with less capital investment. Less resources will also be used, as the development of offshore fields using MPC radically changes the established technological, technical and infrastructure concepts. There is no need to build stationary ice-resistant platforms. All equipment is located directly on the shelf. Well drilling can be performed from specialized vessels or mobile platforms. Staff access to the equipment is limited, but this, on the contrary, is a positive side, since the field will be monitored and managed remotely. Processing of all received information can be carried out using automated systems and artificial intelligence, which will reduce the influence of the human factor and optimize decision-making.

3. 3S separation technology

One of the promising technologies that can become the main link in the MPC for gas or gas condensate treatment at the bottom of the sea is a supersonic separator. The 3S separator is a low-temperature supersonic separator that combines adiabatic cooling of natural gas and cyclonic separation.

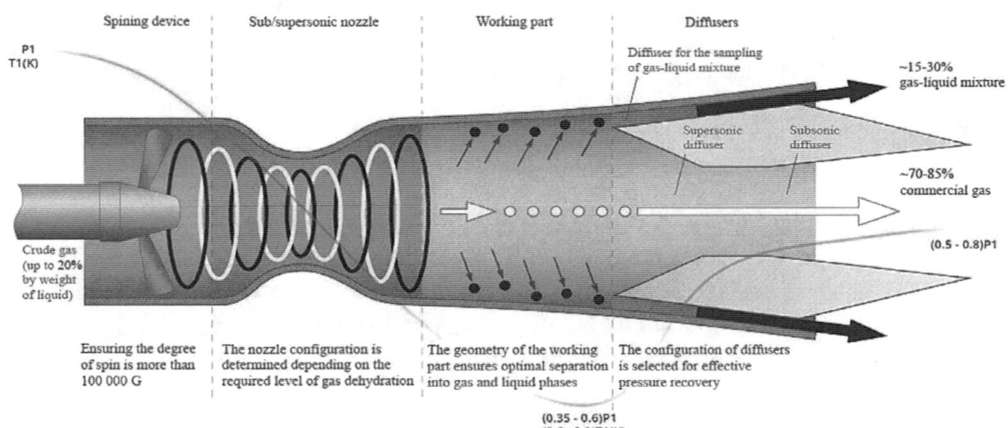

Figure 1. Schematic diagram of the 3S-separator [4]

Figure 1 shows a schematic diagram of the 3Sseparator. At the inlet, the gas is swirled by a swirler, which provides a swirl degree of more than 100,000 G. After that, it enters the Laval nozzle, where it acquires supersonic speed. Since at this stage the kinetic energy of the flow increases and the potential energy decreases, the gas temperature decreases. The flow rate at this stage can be 600-800 m / s, the gas temperature reaches -60 – -100 ° C, and the formation of the liquid phase begins. Then the gas enters the working part, where the components of the gas mixture form a two-phase boundary layer. At such low temperature, everything that is heavier than ethane condenses, including the part of the methane with ethane. This indicates a high degree of recovery ensuring a high degree of gas purification. All condensed liquid due to centrifugal forces in the turbulent flow and acceleration up to 10 m/s ^{2} is thrown to the walls of the nozzle and enters the diffuser for selecting the gas-liquid mixture. The prepared gas exits through the central part of the diffuser, where due to the expansion of the channel

IOP Conf. Series: Earth and Environmental Science 1021 (2022) 012022 doi:10.1088/1755-1315/1021/1/012022

geometry, the flow is slowed down, part of the kinetic energy of the flow passes into the potential, which leads to an increase in the gas temperature and pressure [2].

4. Main advantages of the 3S separator

Below are the main significant technological features of the 3S separator:

1. The most effective use of reservoir gas energy due to the restoration of the pressure of the prepared gas at the exit of the supersonic separator.
2. Small dimensions and low metal consumption, which simplifies its installation.
3. Advanced recovery of C_3 and higher hydrocarbons.
4. Deep purification of acidic natural gases by extracting CO_2 and H2S.
5. There is no need to use chemical reagents.
6. There is no need for constant maintenance, which makes it possible to install this equipment on the shelf.
7. No moving parts, which increases the reliability and environmental safety of the device [3].

Thus, the company "ENGO Engineering", which has supersonic gas separation technology, successfully implements it in Russia at the Yurkharovsky, Yarudeysky and other fields [4].

Based on the studies of industrial tests of the 3S separator technology at the polar NGKM installation, we can conclude that this technology has been successfully applied. Thus, the replacement of the throttle valve with a 3S separator in the composition of the motor fuel processing unit (UPMT) allowed one to:

- Lower the dew point by 10 °;
- Reduce the concentration of target components from 5 and higher, depending on the operating mode, to 56% [5].

When comparing the 3S separator with other, more common low temperature separation technologies such as turbo expanders and JT valves, the following should be noted:

- In accordance with the graph shown in Figure 2, it can be seen that the efficiency of the extraction of heavy components by the C_3+ 3S separator, at a gas temperature of -20 °C, is 9% higher compared to the turboexpander and by 25% compared to the JT valve;
- The turboexpander is, in principle, not designed to operate, provided there is liquid in the incoming gas stream;
- At the moment, there is no turboexpander in the version that allows it to be installed under water;
- The 3S separator allows the extraction of heavy C3 + components at pressures exceeding 100 bar, which is not possible with the use of turbo expanders and JT valves;
- 3S separators are easier to maintain operate and design.

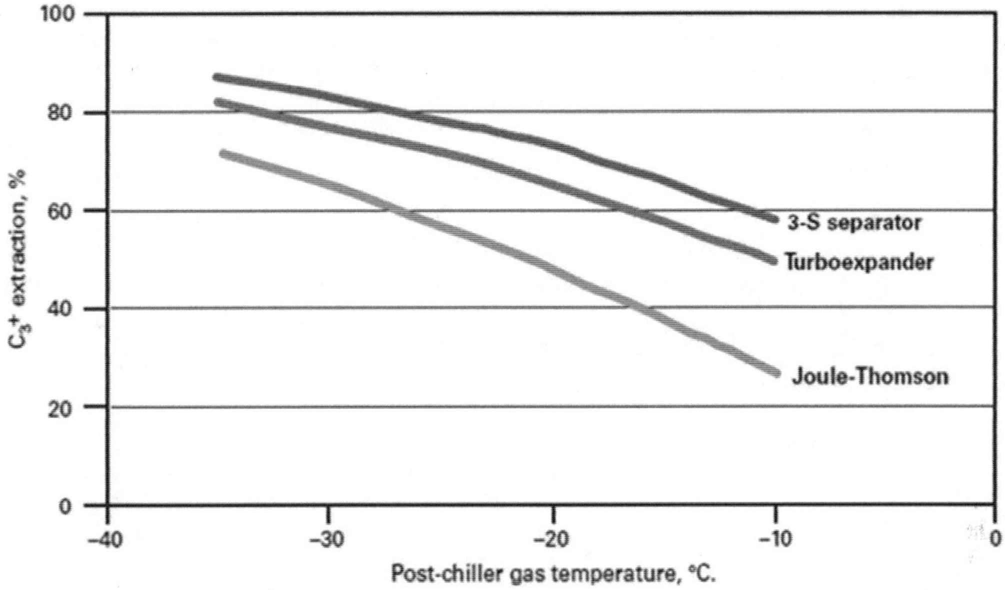

Figure 2. Efficiency of extraction of heavy components C_3+ [6]

5. Prospects for implementation on the Caspian shelf

From an economic point of view, it is necessary to find the optimum based on the state of the gas and the required degree of separation. In some cases, the gas is too wet for continuous transport; in others, the concentration of the liquid phase is so high that the pump cannot operate stably. Although separation technologies have a positive impact on the transportation of hydrocarbons, we must not forget about the infrastructure that is necessary for the transport of separated mixtures.

We consider the 3S separation technology very promising and would like to talk about possible prospects for its use on the shelf of the Caspian Sea. We assume that the Yu. Kuvykin gas condensate field is suitable for the application of this technology. Today, the oil reserves at the field are estimated at 50 million tons, and gas-20 billion m^3. The field is scheduled to be launched between 2022-2023 [7].

1. Installation of technological equipment near production wells, directly on the sea floor, will allow for more efficient field development.
2. The introduction of wells into explanation will be faster.
3. The depth of the sea on the site is 5-8 meters, which will facilitate the installation of underwater equipment, as well as allow you to conduct pilot tests for training personnel, checking equipment, etc., close to real ones.
4. The deposit has a high content of hydrogen sulfide, which is not a problem for a supersonic separator. And the prepared gas will be safe for subsequent elements
5. There is no need to use substances to prevent hydrate formation after 3S separation, since the flow temperature is restored. And in the separator, itself, the formation of hydrates is impossible, due to the flow rate, which is in the nozzle of the order of 10^{-4} s, which is too small for hydrate formation.
6. The existing infrastructure will facilitate the implementation of the project.
7. The 3S-separator, after gas purification, restores the pressure of the prepared gas at the level of ~ 100 atm., which will allow it to be transported over considerable distances.

6. Conclusion

Taking into account the trends in the development of the resource base of offshore fields in accordance with the" Energy Strategy of Russia until 2030", the implementation of such solutions is necessary. This will provide an opportunity for testing and developing these technologies. Their subsequent modernization and gradual introduction as advanced projects will have a positive impact on the realization of the resource potentials of the deposits.

References

[1] *Sakhalin-3* Available at: https://www.gazprom.ru/projects/sakhalin3/
[2] Imaev S Z, Safyannikov M I 2016 Regulation of supersonic separators *Teoriya neftegaza* **9**, 98-103
[3] Kozhukhar E D, Lyamina N F 2014 Application of a separator for underwater preparation of hydrocarbons from gas condensate fields *Geology, geography and global energy* **3** 54
[4] *ENGO Engineering Projects* Available at: http://engo3s.ru/projects
[5] Korytnikov R V et al. 2012 Industrial tests of supersonic separation technology at the NTS UPMT UKGP-1C installation of the Zapolyarny NGKM *Oil field business* **6** 34-40
[6] Alfyorov V. et al. 2005 Supersonic nozzle efficiently separates natural gas components *Oil & Gas Journal,* **103(20)**, 53-58.
[7] *Kuvykina (Sarmatian) gas condensate field* Available at: https://neftegaz.ru/tech-library/mestorozhdeniya/142049-kuvykina-sarmatskoe-gazokondensatnoe-mestorozhdenie-gkm/

Influence of formation drilling conditions on the choice of well stimulation method

A Kh Gabzalilova and Z A Garifullina

Ufa State Petroleum Technological University, Branch of the University in the City of Oktyabrsky, 54a, Devonskaya St., Oktyabrsky, Republic of Bashkortostan, 452607, Russian Federation

E-mail: alfiragabzalilova@mail.ru, info@of.ugntu.ru

Abstract. The paper considers the influence of crank radius change on pumping unit kinematics during the dynamic fluid level control between tubing and casing in order to reduce the string fault rate. To improve the quality of prediction of these factors, the calculation of the movement of the polished rod over time is carried out using the Fourier variable separation method: the product of the balance bar head of a pumping unit defined as the movement of the slider of the pumping mechanism (in a straight line – like the chords of a circle), and the difference of such a chord with the length of the circle section equivalent to the movement of the articulated joint of the balance bar with the connecting rod. As a result, the accuracy of calculations is improved and the solution of the problem is simplified.

1. Introduction

In recent decades, due to the continuous increase in the specific gravity of productive reservoirs with low permeability, the rational choice of the method and modes of well stimulation has become ever more relevant. As a result, well stimulation volumes are expanded by swabbing, with reservoir permeability up to 0.1 μm^2. The deposits with permeability less than 0.03 μm^2 are being developed. Let us consider the problem in more detail.

As is known, according to studies conducted at IFNTUOG, BashNIPIneft and USPTU, the traditional method of formation drilling implies that the downhole zone of the productive formation (PFZ) is contaminated to a depth of 3-7 m at reduced axial loads (Fig. 1). At round-trip operations (RTO) – up to 20-24 m.

These depths are clearly confirmed by the volumes of infiltrates [1] extracted from the formations during well “testing” by formation testers.

2. Materials and methods

To improve the quality of prediction of these factors, the calculation of the movement of the polished rod over time is carried out using the Fourier variable separation method: the product of the balance bar head of a pumping unit defined as the movement of the slider of the pumping mechanism (in a straight line – like the chords of a circle), and the difference of such a chord with the length of the circle section equivalent to the movement of the articulated joint of the balance bar with the connecting rod. As a result, the accuracy of calculations is improved and the solution of the problem is simplified.

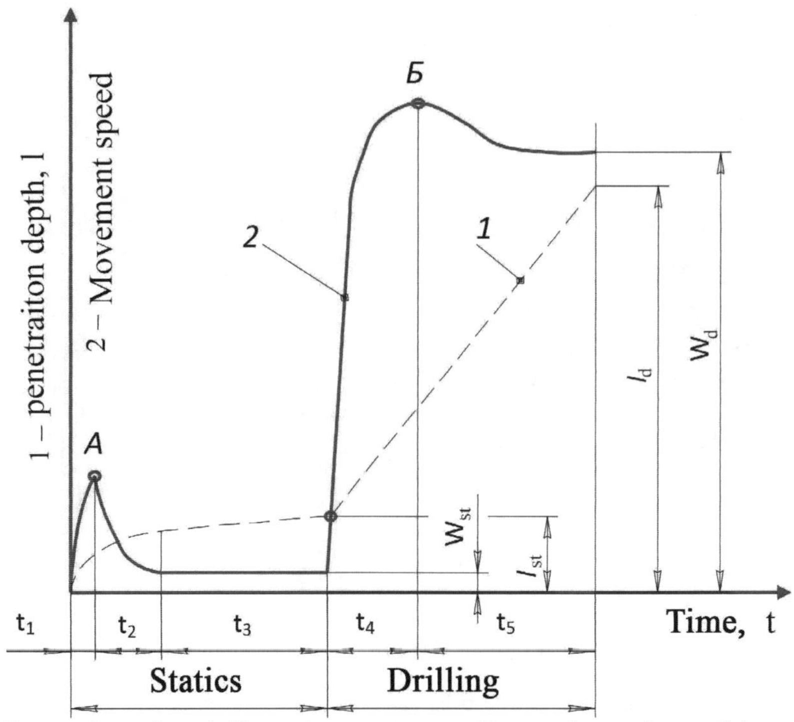

Figure 1. Penetration of mud filtrate into porous medium under static conditions and during drilling

3. Results and Discussion

Infiltrate penetration makes it difficult to restore the natural permeability of PFZ, which in turn leads to a decrease in the final reservoir recovery rate. During drilling the dominant reason for the deep penetration of the washing fluid infiltrate into the PFZ is dynamic rather than static disturbances, which is noted by many researchers [2-4]. Hence, let us consider the intensity and depth of PFZ contamination under the influence of static and dynamic loads during drilling and RTO.

When the washing fluid flows out of the drilling bit jet nozzles, due to the occurrence of accelerations it decomposes into solid (for example, clay) particles and filtrate. Due to a static pressure drop $\Delta P_{ст}$ in the "bottomhole-PFZ" system, the latter penetrates into a porous medium in a zone with undeveloped clay crust. According to Dupuis formula (for flow rate Q definition), the total amount of infiltrate penetrated into the formation with thickness *h* during time *t* is equal:

$$Qt = \frac{2\pi kh\Delta P_{ст}}{\eta_{и} \ln(R_{к}/R_{с}}) t = \pi h(R_{к}^{2} - R_{с}^{2})m_{0}, \quad (1)$$

where Q – flow rate of infiltrate penetrating into the formation; $v = Q(2\pi r)^{-1}$ – flow rate of infiltrate at a distance r; k and m_o – permeability and porosity of the reservoir; h – formation thickness; $\Delta P_{ст}$ – static or hydrodynamic pressure drop between the well radius R_c and at the outer boundary of infiltrate penetration with the contour radius $R_к$; $\eta_и$ – infiltrate viscosity.

Due to the action of hydrodynamic loads, which most often arise from periodic approach of the bit to the bumpy bottom of a well [3], the friction forces of the infiltrate against the walls of the pore channels decrease and, accordingly, its intensity and depth of penetration into the PFZ increase. The difference between penetration intensities of washing fluid infiltrate into PFZ is clearly shown in Fig. 1 [5-7].

Fig. 1 shows that the pattern of infiltration into the formation due to static pressure drop and turbulence of the fluid flow to the bottomhole (quasi-static infiltration) cannot explain the infiltration depth observed in the fields (by the volumes of fluid collected by the reservoir testers) – up to several meters in relatively low-permeability (up to 0.02...0.10 μm^2) reservoirs. Such infiltration is satisfactorily explained only by the hydrodynamics of the "string-well-well bore zone" system.

For the general case of spatial symmetric propagation of hydrodynamic waves in a homogeneous porous medium in the direction of radius r, taking into account the damping coefficient n of the fluid vibration energy by friction forces against the walls of the pore channels, the initial wave equation has the form:

$$\frac{\partial^2(r\Phi)}{\partial t^2 \partial t \partial r^2} + n\frac{\partial(r\Phi)}{\partial t^2 \partial t \partial r^2} = c^2\frac{\partial^2(r\Phi)}{\partial t^2 \partial t \partial r^2}, \tag{2}$$

where Φ – velocity potential; $u=\partial\Phi/\partial r$ – movement of infiltrate particles; $k=md^2/2$, and m – respectively, permeability and porosity of rocks with pore diameters d; $n=u\lambda/(2d)=\nu m^{1.5}/(dk^{0.5})$; λ – hydraulic resistance of porous medium; ν – kinematic viscosity of infiltrate.

The solution (2) for harmonic waves with the sequential use of the Fourier variable separation method and Laplace transformations has the form

$$r\,\Phi = U_{R}e^{i\varphi}\,[sh^2(\alpha\, r) + \cos^2(\beta\, r)]^{0,5}, \tag{3}$$

where $\alpha = \sigma_1$;$\beta = \sigma_2$; $\sigma_{1,\,2} = (\omega/c)\ [0.5\ (1 + n^2\omega^{-2})^{0,5} \mp 1]^{0,5}$;φ = arc tg [tg (α r) tg (β r)] ;ψ = arc tg [cth (α r) tg (β r)].

In a particular case, for flat waves, the solution of equation (2), taking into account the analysis of hydrodynamic waves given in work [5], is characterized by the depth of penetration of the washing fluid infiltrate into the PFZ during drilling and the RTO shown in Fig. 2 and Fig. 3.

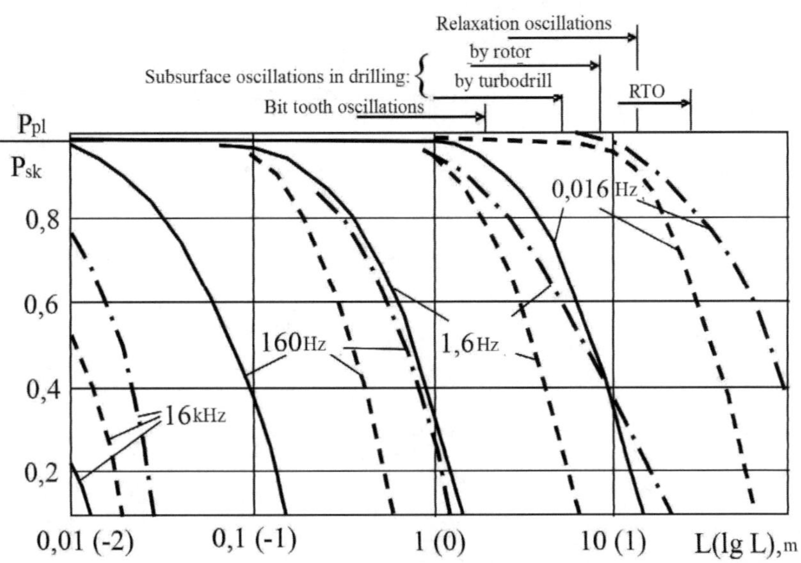

Figure 2. Maximum penetration depth of flat hydrodynamic waves into the formation with 20% porosity: ▬▬ – pore radius; r=1 μm; – – – – 5 μm; ▪▪▬▪▪ – 10 μm

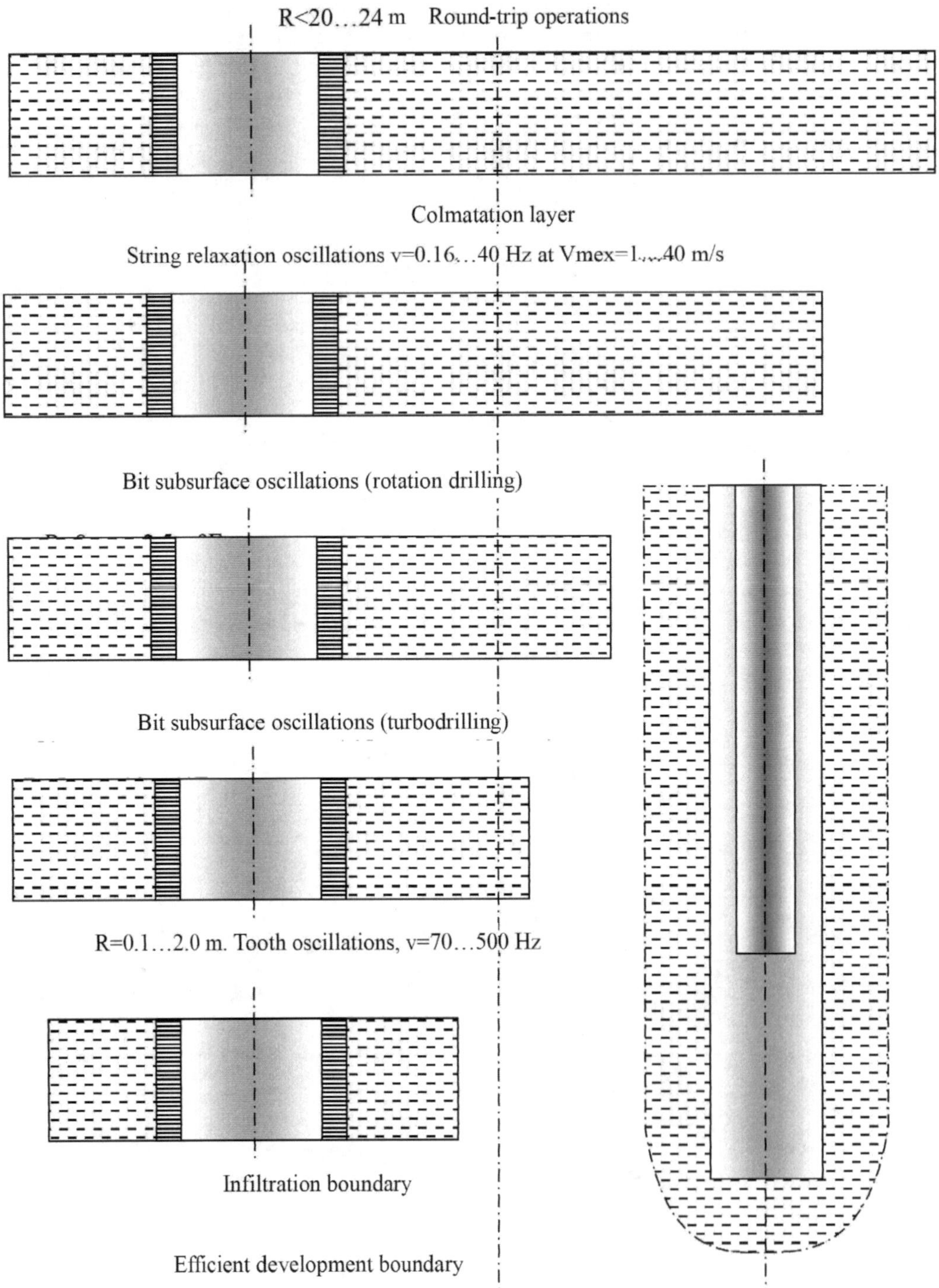

Figure 3. Influence of process operations during formation drilling on contamination of formation downhole zone

Fig. 2 and Fig. 3 show that the intensity of PFZ contamination with washing fluid infiltrate should be taken into account when choosing the method and modes of well development. Otherwise, a difference may be observed between the yields of inclined wells located on the same well pad under similar characteristics and operating conditions of productive reservoirs. The prevention of intensive contamination of PFZ during RTO is ensured by the need, if possible, to penetrate the formation by drilling per one run of a tool into the well.

But in low-permeability reservoirs with pore channel diameters up to 10...20 μm [8, 9], due to friction forces of their formation fluids against the walls, due to the static pressure drop in the "formation-well" system, it is difficult to restore the natural permeability of PFZ at an infiltrate penetration depth of more than 1...2 m. Therefore, in recent decades, domestic fields have increasingly expanded well development by means of swabbing [10]. But at the same time, the problem of cleaning PFZ from the washing fluid infiltrate is only partially solved.

To intensify the purification of PFZ from infiltrate, it is necessary to forcibly reduce the fluid friction forces against the walls of the pore channels using hydrodynamic waves of liquid pressure (HDW) propagating through the fluid, or acoustic waves – along the rock skeleton.

To restore the natural permeability of PFZ in low-permeability reservoirs it is advisable to use swabbing with additional use of downhole hydraulic generators combined with swabs. A relatively simple and efficient method of vibration swabbing is the method developed at VNIITneft (one of the authors of this work) for generating HDW using the GDG-14 hydrogenator consisting of two elements (swabs [11, 12, 13, 14] or identical plugs for cementing casing strings), between which a spring is additionally installed. The prototype of the method is the known effect of pulsed loads during hard braking of railway trains. The imitation of this effect ensured the jump-like movement of piles or plugs in oilfield columns and surface pipelines with their accompanying friction forces against the cylinder walls (as an option – inertia forces) with the HDW amplitude.

The basic tests of the HDW excitation method based on the GDG-14 hydrogenerator (subinfralow-frequency) were first conducted at Demsky UBR ANK Bashneft. When testing the full-scale sample of the hydrogenerator, the main GDG-14 parameters (Fig. 4) were recorded: amplitude of hydraulic shock waves – 0.8...1.2 MPa; generation frequency – 0.02...0.11 Hz.

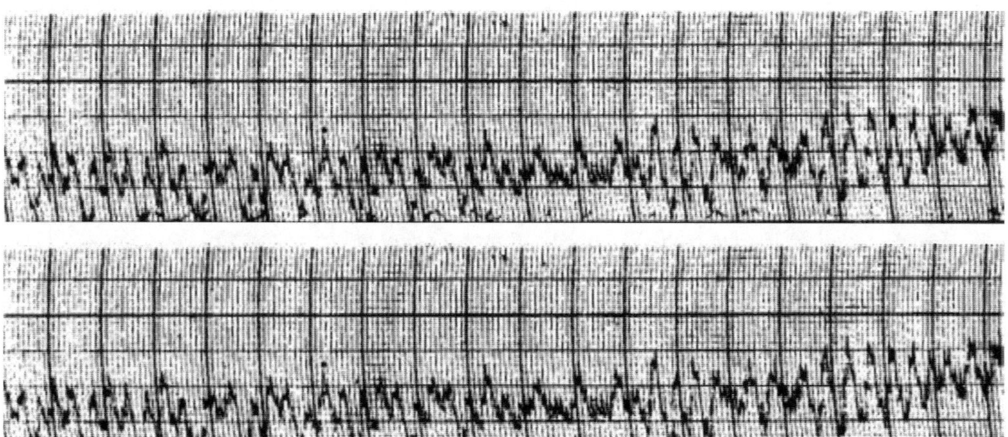

Figure 4. Pressure amplitude ΔP generated by hydrodynamic waves of the GDG-14 hydrogenerator in stationary fluid flow (on a stand) at a flow rate of 0.015 m^3/s (speed of tape traction mechanism – 10 divisions/s; hydrodynamic pressure recording scale ΔP: 1 division – 0.5 MPa)

The results of performed tests with GDG-14 in well No. 797 of Arlanskoye field confirmed the acceptability of the proposed method and the GDG-14 hydrogenerator to excite hydrodynamic waves of fluid pressure in distributed systems (in tubing, casing and for cleaning ground pipelines).

4. Conclusion

In all cases, when choosing the main GDG parameters and well development modes by vibration swabbing, it should be borne in mind that in order to restore the natural PFZ permeability, the frequencies of the infra and subinfrasonic frequencies emitted by the GDG should be slightly lower than the frequencies of the ground vibrations of the multi-rotary bits (from a bumpy bottomhole).

In case of PFZ contamination and during RTO, in particular in horizontal wells, their development must be carried out differentiated according to individual sections of the bore interval located in the developed zone of the productive reservoir. The annular space behind the casing string at specified interval along boundaries of these sections is additionally isolated by means of annulus packers. The use of such technology allows a number of foreign companies increasing the flow rates of horizontal wells up to 5...10 times compared to vertical ones (with the technologies used in domestic fisheries, the flow rates usually increase up to 1.5...2.5 times).

References

[1] Akhmetov R T, MukhametshinVV, KuleshovaLS, GrezinaOA, and MalyshevP.M. 2020 The generalized correlating function of capillary curves and the relationship of the filtration-capacitive parameters of reservoirs in Western Siberia with the size distribution of pore channels*Journal of Physics: Conference Series* **1661** 012016 DOI: 10.1088/1742-6596/1661/1/012016

[2] Zeigman Yu V, Mukhametshin V V, and Kuleshova L S 2020 Management of flooding of low-production deposits according to geological and field data *IOP: Earth and Environmental Science* **579** 012019 DOI: 10.1088/1755-1315/579/1/012019

[3] Korn G A and Korn T M 1984 Mathematical Handbook forScientists and Engineers: Definitions, Theorems, and Formulas for Reference and Review (Moscow: Nauka)

[4] Quan Hui et al 2017 Mathematical modeling for the evolution of the large-and meso-scale vortex in the screw centrifugal pumpwith the buoyancy effect *Adv. Mech. Eng.* **9**

[5] Yan Di et al 2017 Numerical modelling of twin-screw pumps based on computational fluid dynamics P*roc. Of the Institution of Mechanical Engineers Part C-J. Of Mechanical Engineering Sci.* **231** 4617-34 DOI: 10.1177/0954406216670684

[6] Dennis G Z 2013 A First Course in Differential Equations (Los Angeles: Ricard Stratton)

[7] Yuksel S 2014 Differential Equations for Engineering Science (Canada: Queen's University

[8] Gabzalilova A Kh, Batalov D A, Mukhametshin V Sh, and Andreev V E 2021 Geological and technological justification of the parameters of acid-clay treatment of wells *IOP Conference Series: Materials Science and Engineering* **1064** 012058 DOI:10.1088/1757-899X/1064/1/012058

[9] Kuleshova L S, Gabzalilova A Kh, Akhmetshina D I 2021 Stoke's theory in problem solution of cementing slurry sedimentation in horizontal sections *Journal of Physics: Conference Series* **1753** 012082 DOI:10.1088/1742-6596/1753/1/012082

[10] Suleimanov R I, Gabdrakhimov M S, Khabibullin M Y, Zaripova L M and Vasilyeva E R 2018 The study of hydraulic hammer device in drilling tool assembly in hydraulical rotary drilling *Int. J. of Engineer. and Technol.* **7** 28-30 DOI: 10.14419/ijet.v7i2.23.11877

[11] Grezina O A 2018 Assessment of the results of acid implosion stimulation of the near-wellbore area based on statistical data *IOP Conference Series: Earth and Environmental Science***194** 082016 DOI:10.1088/1755-1315/194/8/082016

[12] Habibullin Ya M and Sidorkin D I 2016 Determination of tubing string vibration parameters under pulsed injection of fluids into the well *SOCAR Proc.* **3** 27-32 DOI: 10.5510/OGP2016030028

[13] Almukhametova E M, Fattakhov D I, Zakirov A I and Safiullina A R 2018 The analysis of the hydraulic fracturing efficiency at the Potochnoe field facility AV1-2 *IOP Conf. ser. Earth Env.* **194** 082004 DOI:10.1088/1755-1315/194/8/082004

[14] Almukhametova E M, Zakirov A I, Fattakhov D I, Faizullin A A and Safiullina A R 2018 Analysis of the technology of intensifying oil production through the bottomhole formation zone treatment in the Potochnoe field *IOP Conf. ser. Earth Env.* **194** 082005 DOI:10.1088/1755-1315/194/8/082005

IOP Conf. Series: Earth and Environmental Science 1021 (2022) 012024 doi:10.1088/1755-1315/1021/1/012024

Environmental and natural leaching processes in underground ore mining

S V Galachieva [1], G V Stas[2], I I Izotova[3]

[1]North Caucasian Mining and Metallurgical Institute, 44 Nikolaeva St., Vladikavkaz, 362021, Russia
[2]Tula State University, 92 Lenina Avenue, Tula, 300012, Russia
[3] Baikal State University, 11, Lenin str., Irkutsk, 664003, Russia

E-mail: svetagalachieva@list.ru

Abstract. The minimization of the negative impact of the products of natural leaching of metal ores on the ecosystems of the environment is justified. It is shown that the reason for the formation of ore leaching solutions is the use of development technologies without filling voids. The results of observations are systematized and the concept of environmental management by combining the potentials of enterprises in the region is substantiated. The efficiency of using dolomite waste to reduce the leaching front by filling technological voids with dolomite-based mixtures is shown. The results of the study of the intensity and danger of natural leaching, taking into account the zoning of mineralization within the orefield, are presented. Measures to reduce the level of natural leaching are recommended. A leaching model is proposed that describes physicochemical processes as stages of reactions occurring in the ore mass. The algorithm of substantiation of environmental technologies is given. The ecological and economic criterion of the effectiveness of the implementation of environmental measures is proposed. The processes of natural leaching of metals are adequate, which makes it possible to control their development. Natural leaching is a consequence of the insufficient level of applied development technologies and can be minimized.

1. Introduction

With an increase in the volume of mined ores, the amount of waste increases, and the risk of chemicalization of ecosystems by products of natural leaching of ores increases [1–2].

Minimizing the effects of natural leaching by controlling technological processes contributes to solving environmental and social problems.

The concept of integrated environmental management is to combine the potentials of enterprises in the mining region for the disposal of aggressive waste of their activities.

Many exploited deposits are landfills for natural leaching processes, therefore, the search for ways to minimize their consequences is an urgent problem of mining production [9; 11; 12–14].

The peculiarity of ore mining in the mountains is that water flows and gorges increase the impact of minerals on the natural environment, and mountain ranges prevent the dispersion of products of physicochemical processes.

2. Materials and methods

The possibility and expediency of controlled use of natural leaching solutions is provided by solving a number of tasks, including review, critical analysis, experiment, modeling, and engineering forecasting.

Enrichment tailings are dangerous during storage, but when optimizing the extraction operations of metals from them, they can be raw materials for cost-effective production [3-4, 15].

During leaching in the disintegrator, simultaneously with the destruction of crystals, the working solution is pressed into the cracks formed and dissolves metals [16-20].

In the working organ of the disintegrator, most of the metals are extracted to obtain the desired products [21-23].

3. Results.

In ores separated from the massif and not exposed to the surface, natural leaching processes develop, the parameters of which depend on the impurities of accelerators and process retarders.

Leaching products are installed during the processing of mine effluents in an installation with electrochemical softening, electrodialysis desalination, and concentration, where the volume of condensed brine is about 5% of the volume of the initial volume.

The danger of natural leaching is reduced by implementing environmental measures:

- minimizing the volume of untreated voids;
- underground leaching of ores;
- minimization of voidness due to underground leaching tails.

In matters of environmental protection, the integration of technological capabilities of enterprises into a single system has not only economic but also environmental value. Thus, providing the mine with raw materials for the preparation of filling mixtures increases the volume and improves the quality of ores [5-6].

In the course of mining operations within the orefield, zones of chemicalization of environmental ecosystems are differentiated (Figure 1).

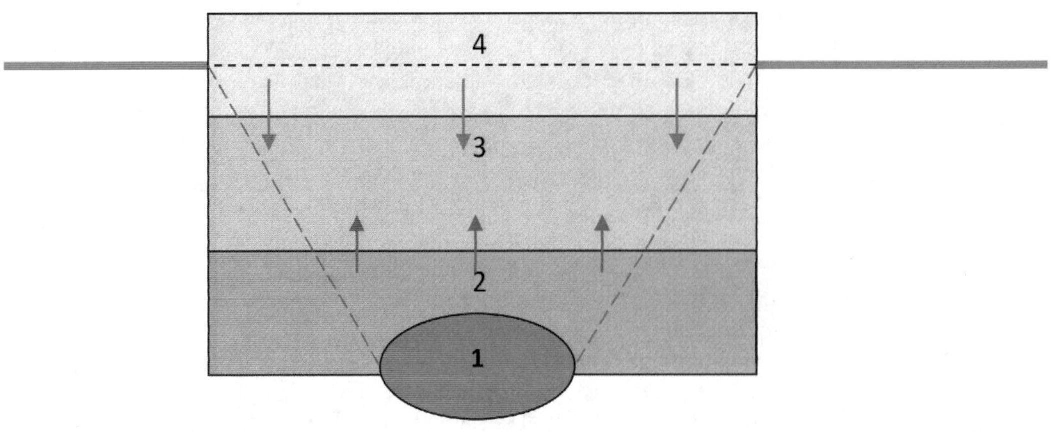

Figure 1. Zones influence of leaching products on the state of the environment: 1 – activation of processes by mining operations; 2 – transportation of chemical agents; 3 – active chemicalization; 4 – effects on living matter.

The physico-chemical processes occurring during leaching are described by the model:

$$M = (P_a + P_n + P_m).(c_n - c_\kappa)K_n$$

where M is the sum of natural leaching products;
P_a – the amount of solvents of atmospheric origin;
P_n – the amount of solvents of natural origin;
P_m – the amount of solvents of technogenic origin;
c_n – the initial concentration of metals;
$c_к$ – the final concentration of metals;
K_n – the coefficient of unaccounted-for chemicalization factors.

The rationale for combining technologies on the basis of environmental conservation includes:

- evaluation of the effectiveness of the association;
- assessment of the costs of compensation for damage caused by uncontrolled natural leaching and when filling the worked-out space with hardening mixtures;
- measures to modernize the technologies of the resources used;
- detailing options for combining technologies.

Filling the technological voids of the mine with mixtures made on the basis of recyclable dolomite tailings improves the quality of ore by reducing dilution by delamination of rocks:

$$E_к = \left[\frac{Z_p}{A_p} K_a K_m + \left(\frac{\Delta E_e + \Delta E_K}{A_к} \right) - \frac{Z_к}{A_к} \right] A_p K_p$$

where $E_к$ is the economic effect of technology coupling, RUB;
Z_p – the cost of ore extraction, RUB/t;
$Z_к$ – the cost for the extraction of dolomite, RUB/m^3;
K_a – the coefficient of dynamism of the volume of production;
K_m – the coefficient of dynamism of time;
ΔE_e – the reduction of operating costs of basic and optimized variants of technologies, RUB/unit.;
$\Delta E_к$ – the reduction of capital expenditures of basic and optimized variants of technologies, RUB/unit.;
A_p – the annual volume of ore production, t;
$A_к$ – the annual volume of dolomite production, t;
K_p – the risk coefficient of market operations.

Filling the worked-out space with monolithic hardening mixtures reduces the contact area of ore pieces with water and air, which reduces the volume of aggressive solutions.

Minimization of environmental pollution by natural leaching products includes the following measures:

- minimization of ore losses during mining;
- laying voids with hardening mixtures;
- leaching of metals in underground blocks;
- demineralization of industrial effluents.

During the underground extraction of metal ores by traditional technologies, the processes of leaching metals from lost ores by natural reagents with the removal of aggressive metal-containing solutions into the environment at all stages of ore extraction and processing are developing in underground workings (Figure 2).

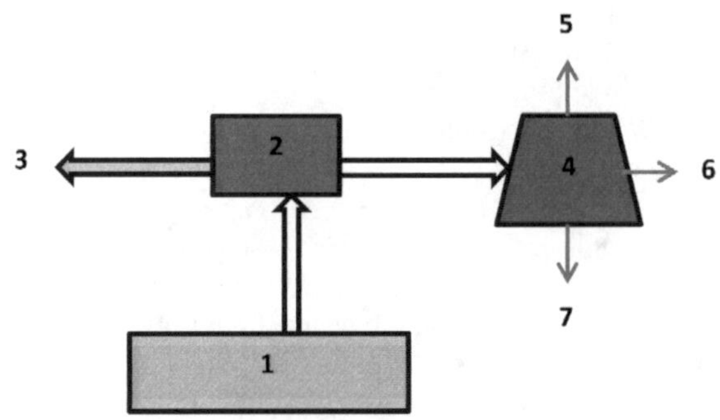

Figure 2. Movement of natural leaching solutions in the traditional development of deposits: 1 – ore deposit; 2 – concentrator; 3 – concentrate; 4 – storage of processing tailings; 5 – water into the atmosphere; 6 – water into the hydrosphere; 7 – water into the lithosphere.

The concept of environmental protection of subsurface use provides for the replacement of uncontrolled natural leaching with technological leaching in a controlled workspace (Figure 3).

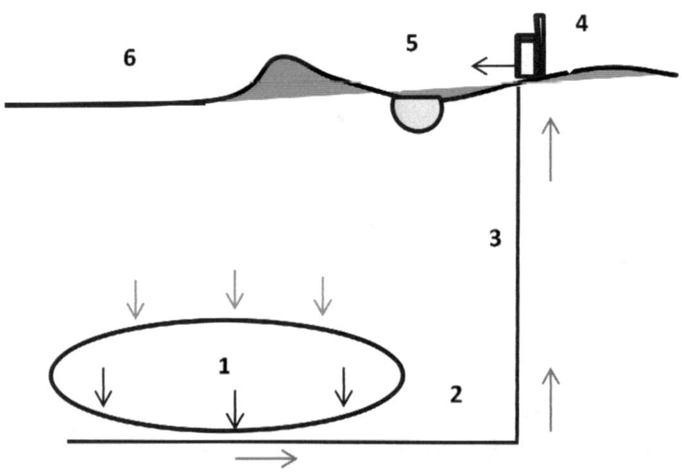

Figure 3. Flow diagram of water flows in the process of traditional field development: 1 – deposit; 2 – mining; 3 – outstanding production; 4 – treatment facility; 5 – reservoir; 6 – relief.

4. Conclusion

The quality of metals extracted by leaching technologies exceeds that of traditional technology, which is important for the use of extracted metals in the production of goods from it [7–11].

Similar studies by foreign experts confirm the materials of the present article [12–13].

References

[1] Golik V I, Dmitrak Yu V, Gabaraev O Z and Kozhiev Kh Kh 2018 Minimizing the impact of mining on the environment *Ecology and industry of Russia* **6** 26–29

[2] Komashchenko V I, Golik V I, Belin V A and Gaponenko A L 2014 Improving the efficiency of explosive rebounding on the basis of new methods for initiating borehole charges in quarries *Mining Informational and Analytical Bulletin* **9** 293–304

[3] Bosikov I I, Klyuev R V, Aimbetova I O and Makhosheva S A 2021 Assessment and analysis of aerodynamic parameters of air flows for effective selection of air supply schemes in coal mines

Sustainable Development of Mountain Territories **13 (3)** 397–405
[4] Golik V I, Gabaraev O Z, Maslennikov S A, Khasheva Z M and Shulgatyi L P 2016 The provision of development conversion perspectives into undeground one for Russian iron ore deposits development *Journal of Social Sciences* **11 (18)** 4348–4351
[5] Dushin A V, Valiev N G, Lagunova Yu A and Shorin A G 2018 Ural Mining University and Moscow Mining University: interaction of higher education institutions *Mining Journal* **4** 4–10
[6] Klyuev R V, Bosikov I I, Gavrina O A and Lyashenko V I 2021 Assessment of operational reliability of power supply to developing ore mining areas at a high-altitude mine *Mining Science and Technology (Russia)* **6 (3)** 211–220
[7] Petrov Yu S, Khadzaragova E A, Sokolov A A, Sharipzyanova G Kh and Taskin A V 2020 Acquisition, transmission and storage of information on production-induced cycle in mining and metallurgy: outlines *Mining Informational and Analytical Bulletin* **11–1** 178–188
[8] Panov V S, Eremeeva Zh V, Sharipzyanova G H, Skorikov R A, Mikheev G V and Ageev E V 2014 The effect of the nature of nanoscale particles and method of mixing on the tribological properties of powder steel 70P *Proceedings of the Southwest State University* **6 (57)** 8–14
[9] Shchurov N I, Myatezh S V, Malozyomov B V, Shtang A A, Martyushev N V, Klyuev R V and Dedov S I 2021 Determination of inactive powers in a single-phase ac network *Energies* **14 (16)** 4814
[10] Shchurov N I, Dedov S I, Malozyomov B V, Shtang A A, Martyushev N V, Klyuev R V and Andriashin S N 2021 Degradation of lithium-ion batteries in an electric transport complex *Energies* **14 (23)** 8072
[11] Zhukovskiy Y L, Batueva D E, Buldysko A D, Gil B and Starshaia V V 2021 Fossil energy in the framework of sustainable development: analysis of prospects and development of forecast scenarios *Energies* **14 (17)** 5268
[12] Doifode S K and Matani A G 2015 Effective industrial waste utilization technologies towards cleaner environment *International Journal of Chemical and Physical Sciences* **4** 536–540
[13] Anderson S G 2016 Alkaline sulfide gold leaching kinetics *Mineral Engineering* **92** 248–256
[14] Sokolov A A, Raus E V and Petrov Y S 2019 A mathematical model for assessing technological damage from the functioning of mining enterprises *Sustainable Development of Mountain Territories* **11 (4) 42)** 554–559
[15] Konyuhov V Y, Gladkih A M, Zott R S 2020 Accelerator as an effective replacement of a business incubator in the Irkutsk region. *Journal of Physics: Conference Series* **1582** 012045
[16] Shakhnazarov K Yu, Prykhin E I, Pomeranets I B 2017 New version of Kurnakov's law about relationship between properties of binary alloys and phase diagrams balance. *IOP Conf. Series: Earth and Environmental Science* **87** 092023 Doi:10.1088/1755-1315/87/9/092023.
[17] Konyuhov V Yu, Gladkih A M, Chemezov A V 2019 Evaluation of the economic feasibility of the introduction of plasma hardening technologies in the Far North enterprises. *IOP Conference Series: Materials Science and Engineering* **560** 012147
[18] Balanovsky A E, Shtayger M G, Kondrat'Ev V V, Nebogin S A, Karlina A I 2018 Complex metallographic researches of 110G13L steel after heat treatment. *IOP Conference Series: Materials Science and Engineering* **411(1)** 012014
[19] Kondrat'ev V V, Ershov V A, Shakhrai S G, Ivanov N A, Karlina A I 2016 Formation and Utilization of Nanostructures Based on Carbon During Primary Aluminum Production. *Metallurgist* **60(7-8)** 877–882
[20] Konyuhov V Yu, Gladkih A M, Galyautdinov I I, Shchadova E I 2020 Calculations of efficiency in implementing progressive mold forming methods. *IOP Conference Series: Materials Science and Engineering* **760** 012027
[21] Nemarov A A, Lebedev N V, Kondrat'ev V V, Kornyakov M V, Karlina A I 2016 Theoretical and experimental research of parameters of pneumatic aerators and elementary cycle flotation. *International Journal of Applied Engineering Research* **11(20)** 10222–10226
[22] Gerasimov D O, Suslov K V 2021 Numerical simulation of multi-energy objects for power supply

of remote areas. *Journal of Physics: Conference Series* **2032** 012103
[23] Konyuhov V Yu, Gladkih A M, Galyautdinov I I, Kiseleva T Yu 2019 Ecological architecture: The green roofs. *IOP Conference Series: Earth and Environmental Science* **350** 012035

IOP Conf. Series: Earth and Environmental Science 1021 (2022) 012025 doi:10.1088/1755-1315/1021/1/012025

Ecology and utilization of ore dressing tailings

O Z Gabaraev[1], T G Konovalova[2], R R Turluev[3]

[1] North Caucasian Mining and Metallurgical Institute, 44 Nikolaeva St., Vladikavkaz, 362021, Russian Federation
[2] Irkutsk National Research Technical University, 83, Lermontova str., Irkutsk, 664074, Russia
[3] Chechen State University named after A. H. Kadyrov, 17a, Dudaev Boulevard, Groznyy, 364015, Russia

E-mail: gabaraev59@mail.ru

Abstract. Based on the study of ways to increase the efficiency of metal production through the use of technologies for leaching substandard ores in terms of metal content, a data bank was formed and an algorithm for controlling the processes of leaching substandard non-ferrous ores in terms of metal content was proposed. The concept of the development of technogenic deposits by leaching metals is proposed as an important direction of rational use of subsurface resources. The development of an algorithm for controlling the process of leaching non-ferrous metals from substandard raw materials is a tool for compensating losses of valuable and scarce raw materials allowed in the technological chain. Aspects of the mechanochemical technology of extracting metals to obtain an ecological and economic effect and radical hardening of the mineral and raw material base of mining have been studied.

1. Introduction

In non-ferrous metallurgy, there is a tendency to weaken the mineral resource base of metallurgical plants.

The urgency of obtaining metals from substandard ores in terms of metal content [25, 26], including leaching, is increasing [1–2]. In this way, 13 deposits of Russia have already been exploited in the past year.

Technogenic deposits are tailings of enrichment, slags of metallurgical plants, and substandard ores. The potential value of non-extracted metals in the waste of processing is often comparable to the value of extracted metals.

Leaching uses the phenomenon of metal conversion into soluble compounds when creating conditions for this.

The speed of the process is determined by the rate of movement of the leaching solution through the leached material.

2. Materials and methods

The possibility of using enrichment tailings as raw materials is proved in the course of full-factor studies.

The enrichment tailings samples from the storage are processed according to the Venken-Box plan with the interpretation of the results in the form of interpolation.

The factors influencing the extraction of metals into the production solution vary at three levels – minimum, zero, and maximum.

The regression analysis algorithm is presented in the form of a computer program with a coefficient of determination for a specific dependence.

The criterion for evaluating options is considered to be the extraction of metals, established by the study of samples taken regularly.

Variable factors regulated during experimentation: the content of reagents, the ratio of liquid and solid phases, leaching time, the speed of rotation of the disintegrator rotors, the number of processing cycles.

3. Results and discussion

Acid leaching is characterized by greater extraction of metals from large raw materials and increased productivity with a shorter duration of processes.

The scheme of underground leaching of metals from ores is shown in Figure 1:

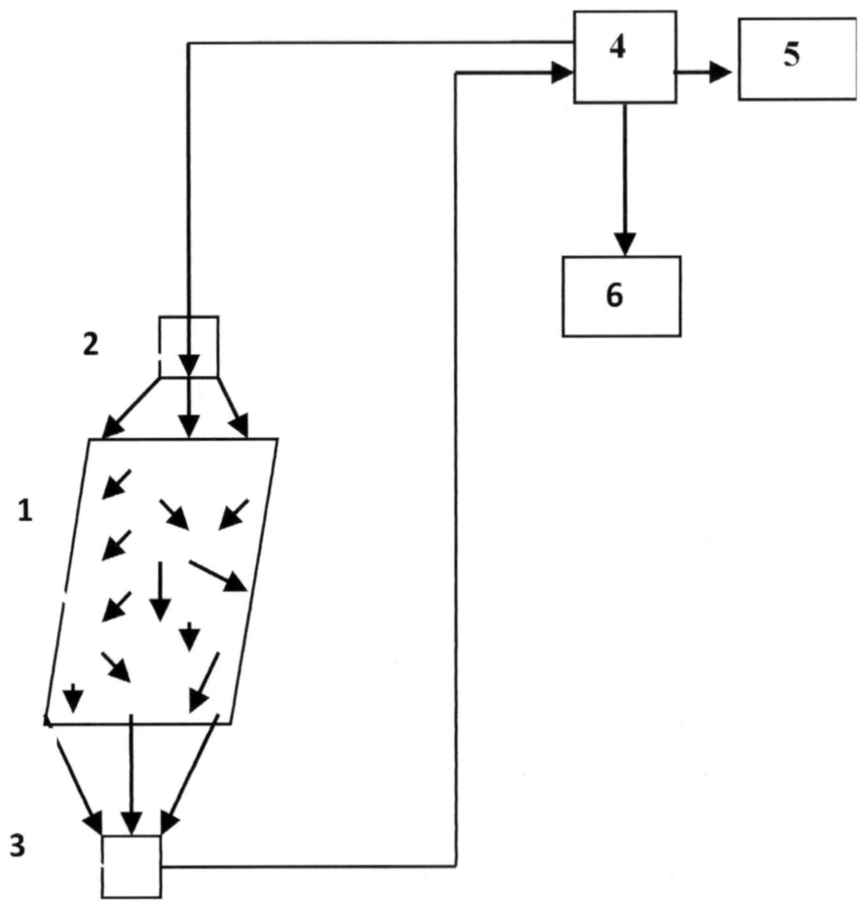

Figure 1. Scheme of underground leaching of metals: 1 – leaching unit; 2 – solution supply horizon; 3 – solution acceptance horizon; 4 – solution processing shop; 5 – metals; 6 – solution processing tails.

The amount of solution leaked through the ore is related to its surface area. The resistance of ore pieces depends on the specific surface of the particles, the pore area, the ratio of the gaps between the particles, as well as the presence of ore and rock fines that prevent the movement of solutions.

IOP Conf. Series: Earth and Environmental Science 1021 (2022) 012025 doi:10.1088/1755-1315/1021/1/012025

Colmatation reduces the filtration rate of the solution, especially when the ore size is 0.1–0.2 mm. To extract metals from such ores, the pulp is mixed.

The efficiency of leaching of metal-containing raw materials is increased by mechanical activation in disintegrators, mills, etc. installations [3–4].

Underground leaching is carried out within the framework of a single production complex, the parameters of which are determined according to the scheme (Figure 2).

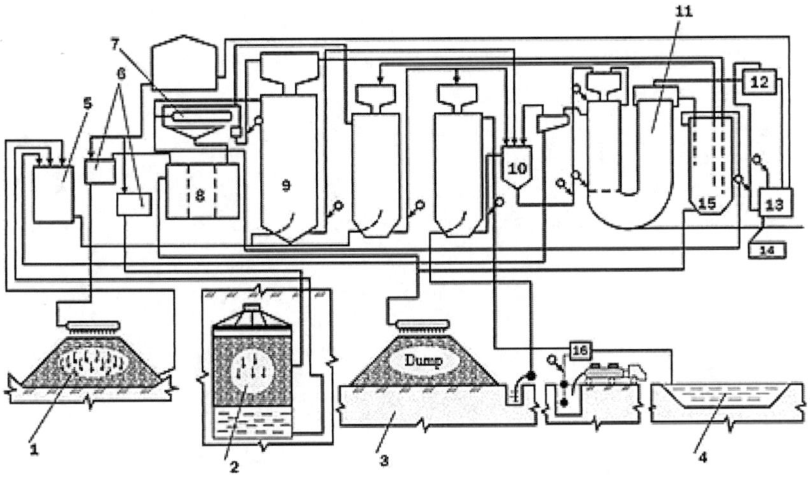

Figure 2. Integration of metal leaching technologies: 1 – heap; 2 – block; 3 – dump; 4 – pond; 5, 6, 7, 8 – tanks; 9, 10 – technological equipment; 11 – sorption and desorption column; 12–16 – auxiliary equipment.

The leaching of the components of complex ores has the following features:

- copper minerals increase the consumption of cyanide;
- antimony reduces the oxygen content;
- carbonaceous components increase the loss of metals.

Methods of intensification of leaching processes include exposure to oxidants, surfactants, bacteria, electromagnetic fields, pressure, vacuum, temperature, etc. [5–6].

The parameters of the underground leaching development system are determined by compliance with the regime of film-drip movement of the solution and the creation of favorable conditions for the oxidation of ores.

Duration of block mining:

$$t = g\,J_{Mi}\;q^{0}{}_{Mi}AV,$$

$$t = J_{Mi}AmS_{б}\,C^{k}{}_{Mi}$$

where A – is the specific surface area of the ore, m^2/mZ;

C^kMi – is the final concentration of metal i in solution, g/m^3.

The selection of controls for leaching parameters is carried out according to the scheme.

The amount of metal transferred to the productive solution:

$$Q = g_1 + g_2 + g_3 + \ldots + g_n$$

where $g_1, g_2, g_3 .. g_n$ – is the amount of metal by process stages.

Net leaching time:

$$t_ч = t_1 + t_2 + t_3 + \ldots + t_n$$

where t_1 , t_2 ...t_n is the duration of the process stages.

The rate of transition of metal into solution per unit of time:

$$g_ч = \frac{Q}{t_ч}$$

where Q – is the number of metals.

The amount of metals transferred to the solution:

$$Q_0 = T_0 \cdot g_в$$

where T_0 – is the contact time of the reagent with the ore.

The number of simultaneously leached blocks:

$$N = \frac{Q_0}{Q} = \frac{T_0 \cdot g_в}{Q}$$

Performance of one leaching block:

$$Q_м = S \cdot V \cdot l_1 \cdot l_2 \cdot c \cdot \gamma,$$

where S – is the irrigated area of the block;

V – is the rate of the leaching front, m/s.;

λ_1 – the coefficient of metal extraction from ore into solution;

λ_2 – is the coefficient of metal extraction from solution; c is the metal content in ore, %;

γ – is the volume weight of ore, kg /m^3.

The results of research on leaching technologies for ores with substandard metal content and tailings of their processing allow us to form a data bank for the design and concept of rational use of subsurface resources [7–12].

Confirmation of the correctness of the concept is the increase in the volume of extraction of metals by leaching. 13 deposits of Russia are already being developed in this way [17, 22-24].

The prospects of the method of underground leaching of rock ores are currently beyond doubt. Having the advantages of a technical and environmental plan, underground leaching is able to occupy the main positions in the extraction of many minerals in the near future [18-21].

4. Conclusion

The success of the introduction of underground leaching largely depends on the quality of ore preparation in the PI blocks. Examples of experimental and industrial work show the unsuitability of traditional development systems for these purposes. At the same time, a number of promising developments have not yet been tested in production conditions.

The study of the mining aspects of leaching and the development of scientific foundations and guaranteed methods of its implementation seem to be one of the most important tasks of mining technology, the solution of which depends on the creation of high-performance flow technology schemes for extracting minerals from the subsoil [13–16].

References

[1] Brigida V S, Kozhiev Kh Kh, Saryan A A and Dzhioeva A K 2020 Time-space problems in geoecology: an inter-disciplinary approach *Mining Information and Analytical Bulletin* **4** 20–32

[2] Golik V I, Dmitrak Yu V, Gabaraev O Z and Kozhiev Kh Kh 2018 Minimizing the impact of mining on the environment *Ecology and Industry of Russia* **6** 26–29

[3] Bosikov I I, Klyuev R V, Aimbetova I O and Makhosheva S A 2021 Assessment and analysis of

aerodynamic parameters of air flows for effective selection of air supply schemes in coal mines *Sustainable Development of Mountain Territories* **13 (3)** 397–405

[4] Kongar-Syuryun Ch B, Faradzhov V V, Tyulyaeva Yu S and Khairutdinov A M Effect of activating treatment of halite flotation waste in backfill mixture preparation 2021 *Mining Information and Analytical Bulletin* **1** 43–57

[5] Valiev N G, Propp V D and Vandyshev A M 2020 The 100th anniversary of the department of mining engineering of UrSMU *News of Higher Educational Institutions. Mining Magazine* **8** 130–143

[6] Petrov Yu S, Khadzaragova E A, Sokolov A A, Sharipzyanova G Kh and Taskin A V 2020 Acquisition, transmission and storage of information on production-induced cycle in mining and metallurgy: outlines *Mining Informational and Analytical Bulletin* **11–1** 178–188

[7] Komashchenko V I, Golik V I, Belin V A and Gaponenko A L 2014 Improving the efficiency of explosive rebounding on the basis of new methods for initiating borehole charges in quarries *Mining Informational and Analytical Bulletin* **9** 293–304

[8] Panov V S, Eremeeva Zh V, Sharipzyanova G H, Skorikov R A, Mikheev G V and Ageev E V 2014 The effect of the nature of nanoscale particles and method of mixing on the tribological properties of powder steel 70P *Proceedings of the Southwest State University* **6 (57)** 8–14

[9] Shchurov N I, Myatezh S V, Malozyomov B V, Shtang A A, Martyushev N V, Klyuev R V and Dedov S I 2021 Determination of inactive powers in a single-phase ac network *Energies* **14 (16)** 4814

[10] Shchurov N I, Dedov S I, Malozyomov B V, Shtang A A, Martyushev N V, Klyuev R V and Andriashin S N 2021 Degradation of lithium-ion batteries in an electric transport complex *Energies* **14 (23)** 8072

[11] Zhukovskiy Y L, Batueva D E, Buldysko A D, Gil B and Starshaia V V 2021 Fossil energy in the framework of sustainable development: analysis of prospects and development of forecast scenarios *Energies* **14 (17)** 5268

[12] Golik V I, Lukyanov V G and Khasheva Z M 2015 The concept of modernization of underground ore mining technologies *Proceedings of Tomsk Polytechnic University. Georesource Engineering* **5** 6–14

[13] Doifode S K and Matani A G 2015 Effective industrial waste utilization technologies towards cleaner environment *International Journal of Chemical and Physical Sciences* **4** 536–540

[14] Vrancken C, Longhurst P J and Wagland S T 2017 Critical review of real-time methods for solid waste characterisation: Informing material recovery and fuel production *Waste Management* **61** 40–57

[15] Konyuhov V Y, Gladkih A M, Zott R S 2020 Accelerator as an effective replacement of a business incubator in the Irkutsk region. *Journal of Physics: Conference Series* **1582** 012045

[16] Balanovsky A E, Shtayger M G, Kondrat'Ev V V, Nebogin S A, Karlina A I 2018 Complex metallographic researches of 110G13L steel after heat treatment. *IOP Conference Series: Materials Science and Engineering* **411(1)** 012014

[17] Salamanova M, Murtazaev S-A, Bataev D, Nakhaev M, Sarsenbayev B 2021 Recipes and properties of multicomponent systems “Aspiration dust - mineral powder - Na2Sio3. *Materials Science Forum* **1017** 143–152

[18] Gorbunov R N, Gorbunova Z V, Gorbunova L N, Kolchin V S, Karlina A I 2021 Evaluation of the efficiency of the road network functioning. *IOP Conference Series: Materials Science and Engineering* **1061(1)** 012018

[19] Konyuhov V Yu, Gladkih A M, Chemezov A V 2019 Evaluation of the economic feasibility of the introduction of plasma hardening technologies in the Far North enterprises. *IOP Conference Series: Materials Science and Engineering* **560** 012147

[20] Nemarov A A, Lebedev N V, Kondrat’ev V V, Kornyakov M V, Karlina A I 2016 Theoretical and experimental research of parameters of pneumatic aerators and elementary cycle flotation. *International Journal of Applied Engineering Research* **11(20)** 10222–10226

[21] Konyuhov V Yu, Gladkih A M, Galyautdinov I I, Shchadova E I 2020 Calculations of efficiency in implementing progressive mold forming methods. *IOP Conference Series: Materials Science and Engineering* **760** 012027
[22] Voropai N, Ukolova E, Gerasimov D, Suslov K, Lombardi P, Komarnicki P 2019 *Simulation approach to integrated energy systems study based on energy hub concept.* 2019 IEEE Milan PowerTech, PowerTech 2019 8810666
[23] Gerasimov D O, Suslov K V 2021 Numerical simulation of multi-energy objects for power supply of remote areas. *Journal of Physics: Conference Series* **2032** 012103
[24] Konyuhov V Yu, Gladkih A M, Galyautdinov I I, Kiseleva T Yu 2019 Ecological architecture: The green roofs. *IOP Conference Series: Earth and Environmental Science* **350** 012035
[25] Perelygin D N, Boichuk I P, Gorlov A S 2019 Rational design of low-tonnage technological complexes for fine and ultrafine grinding of materials. *IOP Conference Series: Materials Science and Engineering* **560 (1)** 012118
[26] Sevostyanov V S, Uralsky V I, Boichuk I P, Ermilov R A, Lukyanova E A 2019 Analysis of separator operating conditions in a technological module with vibration-centrifugal closed-cycle grinding unit for obtaining high-dispersity materials. *IOP Conference Series: Materials Science and Engineering* **552 (1)** 012040

Analysis of the technology for increasing the recovery and intensification of hydrocarbon production

I I Bosikov[1], E V Egorova[2], L A Rapatskaya[3]

[1]North Caucasian Institute of mining and metallurgy (State Technological University), 44 Nikolaeva St., Vladikavkaz, 362021, Russian Federation
[2]Astrakhan State Technical University, Institute of Oil and Gas, 20a Tatishcheva St., Astrakhan 414056, Russian Federation
[3] Irkutsk National Research Technical University, 83, Lermontova str., Irkutsk, 664074, Russia

E-mail: igor.boss.777@mail.ru

Abstract. The analysis and improvement of the method of additional stimulation of horizontal wells – carrying out multi-zone hydraulic fracturing, which allows to significantly increase the productivity of the well, increase the drainage area, which is especially important in formations with low permeability. In this paper, the oil recovery methods used at the Yuzhnoye field, the method of modeling hydraulic fracturing were considered, and hydraulic fracturing of a highly permeable formation was simulated for two wells 1 and 3 (double-wellbore) in this field. The optimization of the hydraulic fracture design is aimed at creating a fracture of maximum conductivity and minimum length. Since neocom is a highly permeable formation, the TSO (tipscreenout) technology is proposed for hydraulic fracturing – this is end screening. This TSO technology is successfully applied at the Prudhoe Bay field (USA), in the Gulf of Mexico, Indonesia, the North Sea, Canada, Brazil, Venezuela, Vietnam, Saudi Arabia. In Russia, an example of successful design and execution is the hydraulic fracturing operation with TSO in well 4370 of the Muravlenkovskoye field, Noyabrskneftegaz company. The hydraulic fracturing method will increase the vertical permeability and unite the disparate parts of the reservoir, which will make it possible to more efficiently develop the reserves of the Neocomian deposits in the field, system for electricity consumption for the purpose of short-term forecasting of electricity consumption.

1. Introduction

Modern Russian and international experience in increasing the extraction and intensification of hydrocarbon production is based on the high technological and economic efficiency of a fairly large number of technologies developed and mastered on an industrial scale [11]. Drilling and operation of horizontal wells and their various modifications, as well as hydraulic fracturing, should be recognized as a priority for the application of offshore field development.

The development of new technologies and equipment for drilling horizontal wells makes it possible to widely use such wells in the development of low-productivity, thin and offshore oil and gas fields. The development of fields located on the shelf, having a small thickness and low filtration properties, with oil rims, using vertical wells, in most cases becomes unprofitable due to the high cost of oil and gas production, low well productivity. The design features of horizontal wells, as opposed to vertical ones, allow obtaining higher production rates.

IOP Conf. Series: Earth and Environmental Science 1021 (2022) 012026 doi:10.1088/1755-1315/1021/1/012026

2. Materials and methods

An effective method of additional stimulation of horizontal wells is the implementation of multi-zone hydraulic fracturing (HF), which allows to significantly increase the productivity of the well, increase the drainage area, which is especially important in formations with low permeability. Over the past decade, there have been great changes in the field of hydraulic fracturing technologies, both in the world and in domestic practice. The use of modern equipment, new technologies, and software products for modeling and analyzing the effectiveness of hydraulic fracturing technologies makes it possible to carry out efficient operations, both in vertical and in deviated and horizontal wells.

While hydraulic fracturing of vertical wells has become a fairly widespread method for increasing production and injection in the oil and gas industry, hydraulic fracturing of horizontal wells is rarely used, mainly in the United States of America and the North Sea [1, 2].

Most of the hydraulic fracturing treatments in vertical wells can be carried out in the same way as already done hydraulic fracturing, and the expected effect can be obtained. For the successful execution of multiple hydraulic fractures in a horizontal well, even before drilling it, serious preparatory work must be carried out to study the following aspects: determining the direction of the minimum and maximum horizontal stress; determination of the spatial orientation of the horizontal wellbore of the planned well; selection of wellbore design; determination of the need for cementing the horizontal section; determination of stimulation intervals; the choice between the creation of transverse or longitudinal hydraulic fractures relative to the horizontal wellbore; horizontal wellbore perforation and related issues - orientation, phasing, type and type of perforators used; selection of the type of hydraulic fracturing, that is, hydraulic fracturing with consolidation of the created hydraulic fracture, acid fracturing, or a combination of these types; the method of isolating each fracture; the optimal distance between cracks, their height and width, as well as the characteristics of the technology used (for example, TSO end shielding); Optimal use of calibration tests to optimize fracture and grit fluid volumes; ways to control the removal of the fixing material (proppant); methods for assessing the productivity of each fracture without testing them separately [3, 4].

To determine the geomechanical properties of the formation, such as the direction of the minimum and maximum horizontal stresses, seismic studies were used, the results of which are data on natural fracturing of the formation, tectonic faults, as well as core studies and MiniFrak tests for wells located in this area and which are characteristic. If there is a mismatch between fracture azimuth and wellbore orientation, hydraulic fracturing becomes risky and difficult [5-9].

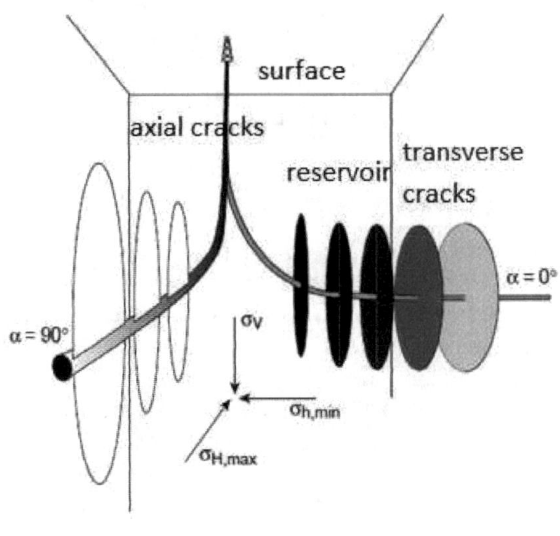

Figure 1. Schematic representation of fracture orientation.

According to the theory of hydraulic fracturing, the fracture propagates along with the maximum stress or perpendicular to the minimum. As shown by Economides, Joshi et al. Authors, it is possible to control the direction of fracture propagation along or across the axis of a horizontal wellbore [10–14]. To orient the cracks, it is necessary to draw the borehole in the direction of maximum or minimum stress, depending on the tasks to be solved (Figure 1).

3. Analysis and evaluation of the reservoir pressure maintenance system

It is planned to operate the western reservoir of the Neocomian superstage of the field with the maintenance of reservoir pressure (RPM) by means of waterflooding. Treated stratal and sea water is used as water supply sources for the reservoir pressure maintenance system. The dynamics of the amount of seawater injection depends on the dynamics of formation water production [17, 18].

When forming the reservoir pressure maintenance system, the following fundamental decisions were taken: injection is performed by one pumping station; the maximum pressure at the injection wellheads is 6 MPa; the quality of the injected water complies with OST 39-225-88 "Water for waterflooding of oil reservoirs".

Requirements for "quality" have the following indicators: the content of suspended solids (vv) – up to 50 mg/l; oil products – up to 50 mg / l; the size of particles of mechanical impurities and emulsified oil in a volume of up to 90% should not be larger than 5 microns; dissolved oxygen in water up to 0.50 mg/l; sulfate-reducing bacteria (SRB) are absent.

The seawater intake for the reservoir pressure maintenance system, the seawater and produced water treatment facilities and the pumping station are located at the central heating station.

Seawater and produced water are treated, respectively, in separate seawater and produced water treatment plants [19-21].

Manifolds are equipped with shut-off valves, metering devices for the amount of injected water, pressure control.

Accounting and regulation of the total flow rate of injected water are carried out at the outlet of the high-pressure pipeline from the pumping station [22, 23].

A corrosion inhibitor is provided in front of the pumps and in front of the injection wells. In addition, it is envisaged to inhibit the deposition of salts, carbonates, and sulfates by introducing reagents before the pumps.

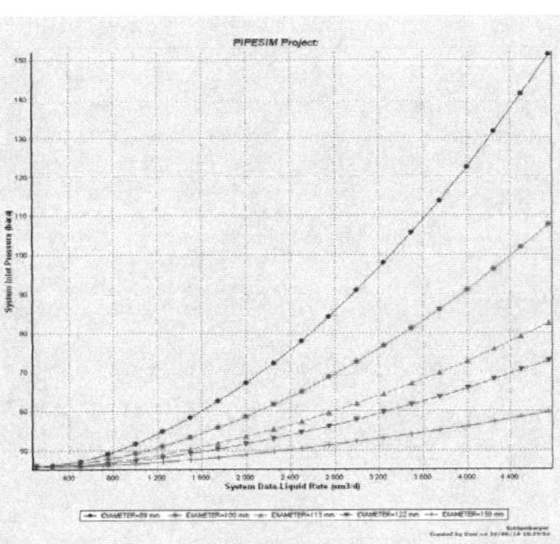

Figure 2. Bottom hole pressure versus wellhead pressure for an injection well of the Neocomian superstage.

Water treatment of seawater for the pressure maintenance system is deep mechanical with the use of chlorine-containing reagents. Before being pumped into the reservoir, seawater passes through a deaeration block. To suppress the vital activity of SRB and prevent internal corrosion, a bactericidal inhibitor is introduced into the purified seawater in front of the pumping station.

For injection wells of the western reservoir of the Neocomian stage, calculations were made to justify the diameter of the tubing.

To select the diameter of the production tubing in the injection well, a graph of the dependence of bottomhole pressure (BHP) on wellhead pressure (THP) was built at the maximum injection volume. The calculation results are shown in Figure 2.From the presented graph it can be seen that to ensure the injection of such a volume of water, at a maximum bottomhole pressure of 19.0 MPa, and taking into account the unification of the equipment, it is recommended to use pipes with a nominal outer diameter of 140 mm (inner diameter 122 mm).

Injection wells are equipped with a tubing string that is run into the injection zone and an isolation packer installed to prevent overpressure of the injected water from affecting the production string.

The fundamental requirements for the reservoir pressure maintenance system for the Aptian reservoir coincide with the requirements for the reservoir pressure maintenance system for the western Neocomian reservoir, given in the previous subsection.

The difference is that the development of the apt deposit is provided with the maintenance of reservoir pressure, in addition to water injection, and gas injection into injection wells. It is planned to commission 4 gas injection wells and 12 water injection wells.

The entire system for the preparation and injection of gas to maintain reservoir pressure has the same composition as the complex for compressor gas lift, which is planned to be used for wells in the Neocomian reservoir when switching to artificial lift.

The injection pressure at the mouth of gas injection wells is regulated by the characteristics of compressor equipment for external transport and is 15.2 MPa.

By technological solutions, it was decided to train working agents of the reservoir pressure maintenance system on a single central technological platform of the central heating station.

Gas preparation consists in drying it from moisture (based on glycol) in order to exclude hydrate formation.

Requirements for water quality have the following indicators: the content of suspended solids (wv) – up to 50 mg/l; oil products – up to 50 mg/l; the size of particles of mechanical impurities and emulsified oil in a volume of up to 90% should not be larger than 5 microns; dissolved oxygen in water up to 0.50 mg/l; sulfate-reducing bacteria (SRB) are absent.

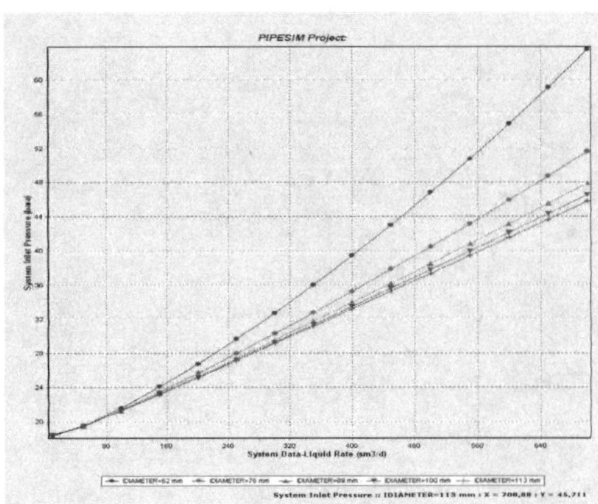

Figure 3. Wellhead pressure versus fluid flow rate, typical water injection water well of the Aptian stage.

The preparation of seawater is carried out at the seawater treatment plant (UPMV), where water, previously purified at the water intake level, is additionally purified on coarse and fine filters, deaeration on a deaeration vacuum column, biocidal treatment (as well as other chemical reagents).

Water treatment is carried out at the same formation water treatment unit as for the Neocomian reservoir.

To select the optimal pipe diameter, calculations were made for a typical water injection well. The calculations were carried out for the maximum design indicator of water injectivity (Figure 3).

As can be seen from the graphs, when calculating with columns with internal diameters of 62, 76, 89, 100, and 113 mm, the difference in pressure losses is significant only for tubing with a diameter of 73 mm. When injecting water into the reservoir into water injection wells, it is recommended to use tubing with a nominal outer diameter of 89 mm, since this ensures the injection of the required volumes, with a pressure at the wellhead at the level of 5.2 MPa.

To select the tubing diameter in a typical gas injection well, a BHP versus THP plot is plotted at maximum gas injectivity. The calculation results are shown in Figure 4.

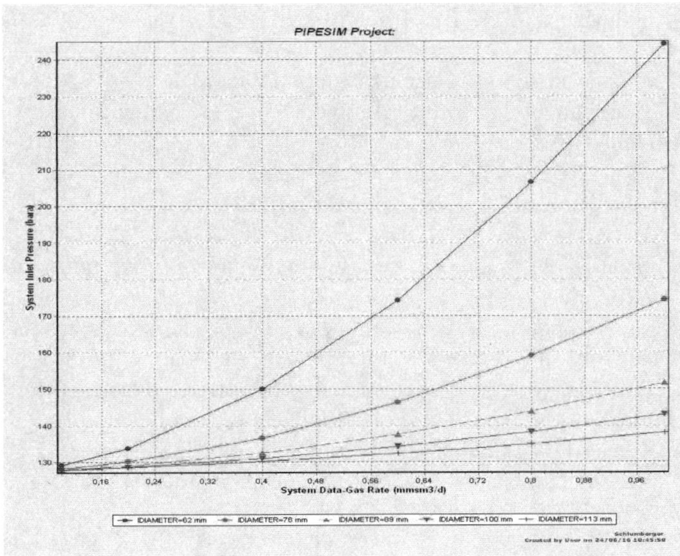

Figure 4. Bottom hole pressure versus wellhead pressure typical Aptian gas injection well.

It can be seen from the above graph that, at the maximum gas injection in the reservoir pressure maintenance system, the difference in pressure loss with the inner tubing diameters of 89, 100, and 113 mm is 1.0 MPa. For gas injection wells, it is recommended to use tubing with an internal diameter of 113 mm (nominal external diameter of 127 mm), which allows the injection of the required volumes of gas at a pressure at the wellhead below the regulated value of 13.8 MPa. In order to unify the range of pipes, it is allowed to use pipes with a nominal outer diameter of 140 mm.

With a decrease in injection volumes below 300 thousand m^3 / day, it is recommended to change the tubing to pipes with a smaller inner diameter of 76 mm (nominal outer diameter of 89 mm).

In order to protect the casing against high pressure during gas injection, gas injection wells are equipped with insulating packers installed 20-30 meters above the tubing shoe.

At the Yuzhnoye field, 100% utilization of the produced gas is envisaged for the entire project period, the utilization factor is 1. Gas is planned to be used for own needs, in the gas lift system, to maintain the reservoir pressure of the Aptian stage and deliver it to the shore for sale.

4. Conclusion

In the research work, the analysis of oil recovery methods used in the Yuzhnoye field, hydraulic fracturing modeling techniques was carried out, and hydraulic fracturing of a highly permeable formation was simulated for two wells 1 and 3 (double-wellbore) in this field. Hydraulic fracturing design optimization is aimed at creating a fracture of maximum conductivity and minimum length.

Since neocom is a highly permeable formation, the TSO (tipscreenout) technology is proposed for hydraulic fracturing – this is end screening. This TSO technology is successfully applied at the Prudhoe Bay field (USA), in the Gulf of Mexico, Indonesia, the North Sea, Canada, Brazil, Venezuela, Vietnam, Saudi Arabia. In Russia, an example of successful design and execution is the hydraulic fracturing operation with TSO in well 4370 of the Muravlenkovskoye field, Noyabrskneftegaz company.

The hydraulic fracturing method will increase the vertical permeability and unite the disparate parts of the reservoir, which will make it possible to more efficiently develop the reserves of the Neocomian deposits in the field.

References

[1] Doifode S K and Matani A G 2015 Effective industrial waste utilization technologies towards cleaner environment *International Journal of Chemical and Physical Sciences* **4** 536–540

[2] Melnikov N N, Kalashnik A I, Kalashnik N A and Zaporozhets D V 2018 Integrated multi-level geomonitoring of natural-and-technical objects in the mining industry *Journal of Mining Science* **54 (4)** 535–540

[3] Litvinenko V S 2020 Digital economy as a factor in the technological development of the mineral sector *Natural Resources Research* **29 (3)** 1521–1541

[4] Makridin E V, Tyulenev M A, Markov S O, Lesin Y V and Murko E V 2020 Overburden management towards higher safety in coal mining regions *Mining Informational and Analytical Bulletin* **12** 89–102

[5] Petrov Yu S and Sokolov A A 2016 Increase of effective management of technological processes of the mountain enterprise on the basis of the analysis of information on technogenic cycles *2nd International Conference on Industrial Engineering, Applications and Manufacturing* ICIEAM 2016 19–20 May (Chelyabinsk) pp 1716–1721

[6] Bosikov I I, Klyuev R V, Khetagurov V N and Azhmukhamedov I M 2021 Development of methods and man-agement tools aerogasdynamics processes at mining sites*Sustainable Development of Mountain Territories* **1** 77–83

[7] Klyuev R V, Bosikov I I, Mayer A V and Gavrina O A 2020 Comprehensive analysis of the effective technologies application to increase sustainable development of the natural-technical system *Sustainable Development of Mountain Territories* **2** 283–290

[8] Shchurov N I, Dedov S I, Malozyomov B V, Shtang A A, Martyushev N V, Klyuev R V and Andriashin S N 2021 Degradation of lithium-ion batteries in an electric transport complex *Energies* **14 (23)** 8072

[9] Rylnikova M V, Vladimirov D Y, Pytalev I A and Popova T M 2017 Robotic geotechnologies as way of improving effi ciency and ecologization of mineral resource management *Physical and Technical Problems of Mining* **1** 92–101

[10] Sinclair L and Thompson J 2015 In situ leaching of copper: challenges and future prospects *Hydrometallurgy* **157** 306–324

[11] Katsubin A V, Khoreshok A A, Tyulenev M A and Markov S O 2020 Technology of advance cutting of sloping and steeply pitching coal seams using hydraulic backhoe excavators *Mining Informational and Analytical Bulletin* **11** 27–36

[12] Zhukovskiy Yu L, Korolev N A, Babanova I S and Boikov A V 2017 The prediction of the residual life of electromechanical equipment based on the artificial neural network *IOP Conference Series-Earth and Environmental Science* **87** 032055

[13] Vrancken C, Longhurst P J and Wagland S T 2017 Critical review of real-time methods for solid waste characterisation: Informing material recovery and fuel production *Waste Management* **61** 40–57

[14] Zhukovskiy Y L and Koteleva N I 2018 Diagnostics and evaluation of the residual life of an induction motor according to energy parameters *Journal of Physics: Conference Series* **1050 (1)** 012106

[15] Martynenko Y V, Bolobov V I, Voronov V A 2021 Use of liquid-gas ejector in liquefied natural gas (LNG) sampling system. *E3S Web of Conferences* **266** 01006
[16] Shutaleva A, Nikonova Z, Savchenko I, Martyushev N 2020 Environmental education for sustainable development in Russia. Sustainability (Switzerland), **12(18),** 7742
[17] Karlina Yu I, Ershov V A, Olentsevich V A, Karlina A I 2021 Justification of the choice of a thermal deburring machine for deburring small parts of microwave electronics. *Journal of Physics: Conference Series* **2032(1)** 012067
[18] Konyuhov V Yu, Gladkih A M, Galyautdinov I I, Kiseleva T Yu 2019 Ecological architecture: The green roofs. *IOP Conference Series: Earth and Environmental Science* **350** 012035
[19] Bataev D K S, Khadisov V Kh, Abukhanov A Z, Salamanova M Sh 2019 Application of man-induced raw material resources as a way to solve ecological problems. *IOP Conference Series: Materials Science and Engineering* **663(1)** 012047
[20] Balanovskiy A E, Astafyeva N A, Kondratyev V V, Karlina A I 2021 Study of mechanical properties of C-Mn-Si composition metal after wire-arc additive manufacturing (WAAM). *CIS Iron and Steel Review* **22** 66–71
[21] Bazhenov Yu M, Murtazaev S-A Yu, Bataev D K-S, Alaskhanov A H, Murtazaeva T S A, Saydumov M S 2021 High-strength concretes based on anthropogenic raw materials for earthquake resistant high-rise construction. *Engineering Solid Mechanics* **9(3)** 335–346
[22] Konyuhov V Y, Gladkih A M, Semenov V V 2019 Measures to the improvement of efficiency of a repair enterprise. *Journal of Physics: Conference Series* **1353** 012046
[23] Gladkih A M, Konyuhov V Yu, Galyautdinov I I, Shchadova E I 2019 Green building as a tool of energy saving. *IOP Conference Series: Earth and Environmental Science* **350** 012032

IOP Conf. Series: Earth and Environmental Science 1021 (2022) 012027 doi:10.1088/1755-1315/1021/1/012027

The combining different-scale studies in a reservoir simulation model of a deposit with a fractured-cavernous type of carbonate reservoir

D A Beltiukov[1], A A Kochnev [1,2], S V Galkin[2]

[1] Branch of LLC «LUKOIL-Engineering» «PermNIPIneft» in Perm, 3a, Permskaya st., Perm, 614015, Russia
[2] Perm National Research Polytechnic University, 29, Komsomolsky prospect, Perm, 614990, Russia

E-mail: doc_galkin@mail.ru

Abstract. When constructing reservoir simulation models of complex reservoirs, it is necessary to take into account the zonal heterogeneity of the filtration and reservoir properties and the influence of secondary processes on the filtration characteristics of the deposit. The underestimation of complex mining and geological conditions negatively affects the forecast for technological and economic indicators of the field development. The paper presents an approach to creating a reservoir simulation model of a complex reservoir based on the integration of different-scale research that allow us to judge the alternation of different types of reservoirs along the lateral and vertical lines. As a result, a refined reservoir permeability cube is constructed which allows us to take into account the influence of the fracturing and cavernous reservoir on the technological indicators of development.

1. Introduction

Currently, the reservoir simulation model is the main tool for assessing the state of the deposit, making design decisions and determining the effectiveness of the technologies used. The reliability of the model directly depends on taking into account all the features of the modeling object and integrating all the available geological and field information. It is especially important to take a detailed approach to constructing a model of a hard-to-recover reserves deposit. The largest deposit in the Timan-Pechora Petroleum Province is characterized by an extremely complex structure of the carbonate reservoir and contains abnormal oil viscocity.

The development of deposits is complicated by the significant spread of fracturing and stylolitization along the section. High permeability and significant inflows are associated with the porous-cavernous intervals of the section, while fracturing ensures communication within the reservoir and the flow of oil into the main filtration channels. Such a collector is classified as micro-homogeneous. The bulk of oil-saturated rocks is characterized by variable porosity and mainly low vapor permeability, but rather high widespread fracturing. In [1], using the example of the analysis of the development of deposits in this region, it is shown that for complex carbonate reservoirs, it is especially important to analyze the maximum range of changes in the filtration and reservoir properties of reservoirs. It is necessary to differentiate petrophysical connections depending on the structure of the void space and the lithological type of rocks when analyzing core researches [1]. As modern experience shows, the method of X-ray tomography of the core can be considered as a method for a detailed assessment of the structure of the void space of fractured-cavernous reservoirs [2, 3]. Taking into account the complex analysis of this information and seismic data, the lithological types can be distinguished and a three-dimensional facies model of the deposit can be constructed [4, 5].

The permeability of pore-type reservoirs, which is estimated from logging, is one or two orders of magnitude lower than the data from hydrodynamic research of wells. This discrepancy is due to the presence of fracture permeability in the formation, as well as partly to the low outflow during drilling with the core of cavernous rocks. The standard approach to modeling a deposit with a similar structure and high-viscosity oil will not allow us to reliably predict the technological indicators of development. Taking into account the fractured permeability of reservoirs in reservoir simulation models can most reliably be implemented by combining hydrodynamic research of wells using data from seismic facies analysis [6-8]. At the same time, it is necessary to develop a methodological approach that solves the problem of adapting the permeability in the reservoir simulation model, taking into account the influence of the crack-cavernous component in the automatic mode [9-11].

2. Materials and methods

In this paper, we consider the approach of combining different-scale studies to determine the zonal heterogeneity of the filtration-capacitance properties. The aim of the study is to localize and account for zones with increased permeability in the reservoir simulation model. Indirect consideration of fracturing, cavernosity, dolomitization, leaching and other secondary reservoir modification processes will improve the quality of displacement simulation in a hydrodynamic simulator.

It is proposed to combine the initial data when constructing the permeability array of the permocarbon deposit of the Usinskoe oil field. Various approaches for taking into account the results of studies of various scales for the analysis of the complex structure of carbonate reservoirs are described in [12, 13].

3. Results and discussion

At the first stage, the zones of probable fracturing are localized based on the fracture intensity based on the results of 3D seismic studies. The discretization allowed us to obtain a distribution in the volume of the parameter of the probable presence of cracks.

At the next stage, the results of field studies of wells and data on the identification of zones of absorption of drilling mud are analyzed. Based on the available research, zones of increased permeability and hydroconductivity are identified. Since the results of wells tests characterize the permeability in the area of the well (bottom-hole and remote reservoir zones), it becomes possible to identify zones of increased permeability in the volume of the deposit on the basis of 2D maps.

Further, the results of multi-scale research were combined, and a total array of zones of increased permeability was created. Figure 1 shows geological and statistical sections of discrete high-permeability arrays for each of the studies and for the total discrete high-permeability array based on seismic data, field research, and drilling mud absorption zones.

When comparing geological and statistical sections (Fig. 1), it can be seen that zones of increased permeability occur in the lower part of bed 1 and the upper part of bed 2. There is also an increase in the frequency of intervals in the lower part of the section.

A map of the probability of the presence of zones of increased permeability over the entire area of the deposit is constructed based on the total account of the discrete zoning of increased permeability in the reservoir volume.

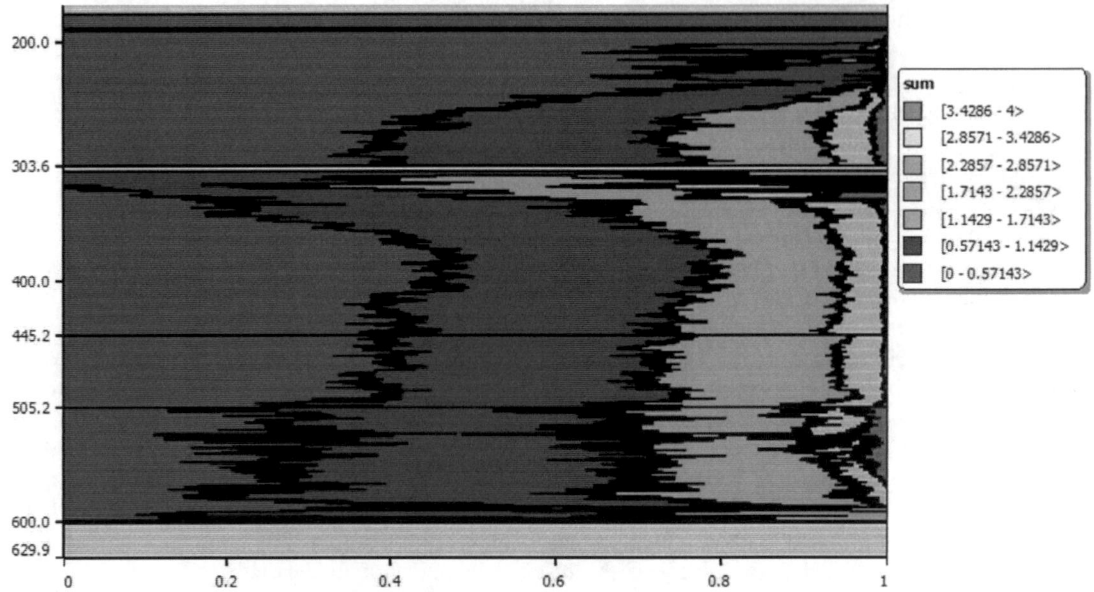

Figure 1. Geological and statistical section of the total discrete array of increased permeability based on seismic data, field research and drilling mud absorption zones

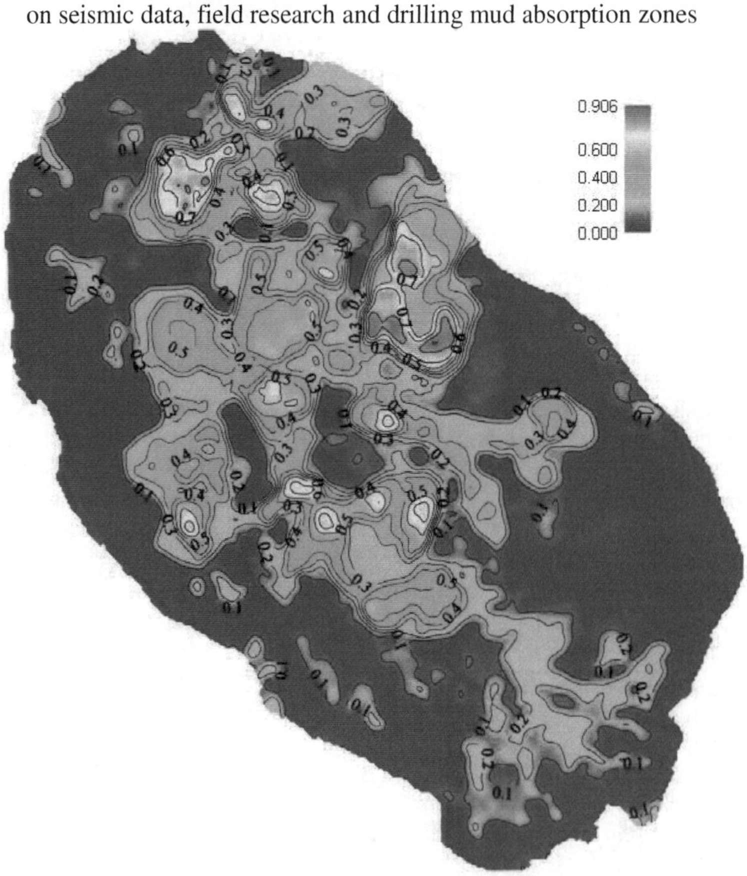

Figure 2. Map of the probabilistic distribution of zones with increased filtration-capacitance properties

Data from hydrodynamic research were used to construct the permeability array. The distribution of properties was carried out in the entire volume of the deposit with a boundary value of the probability of increased filtration properties – 0.5. In the rest of the volume of the deposit, the permeability corresponds to the distribution according to the petrophysical dependence of permeability-porosity.

Zones of fracturing development are determined on the basis of integration of different-scale studies. A modified array of the permeability distribution according to the petrophysical dependence is constructed, taking into account the highly permeable intervals.

4. Conclusion

The resulting permeability array served as the basis for creating a permanent reservoir simulation model of the reservoir. Indirect consideration of the propagation of fracturing allowed one to reduce the degree of uncertainty of the filtration parameters when adapting the model to the actual development indicators and to increase the reliability of the forecast calculations.

This technique can be replicated in the construction of models of complex carbonate reservoirs in conditions of limited initial information aimed at studying fracturing.

Acknowledgments

The study was carried out within the framework of the state task of the Ministry of Science and Higher Education of the Russian Federation, FSNM-2020-0027 for the implementation of fundamental scientific research.

References

[1] Popov N A, Putilov I S, Guliaeva A A, Vinokurova E E, Fairuzova Iu V 2020 Influence of the lithogenesis type of rocks on porosity and permeability properties (the case of the Permo-carboniferous deposit of the Usinskoye field) *Perm Journal of Petroleum and Mining Engineering* **20** **(2)** 104-114 DOI: 10.15593/2224-9923/2020.2.1

[2] Efimov A A, Galkin S V, Savitckii Ia V, Galkin V I 2015 Estimation of heterogeneity of oil & gas field carbonate reservoirs by means of computer simulation of core x-ray tomography data *Ecology, Environment and Conservation* **21** 79-85

[3] Kolychev I Yu 2017 Study of zones of wettability distribution based on lateral logging for oil-bearing Visean reservoirs of the Solikamsk depression *Perm Journal of Petroleum and Mining Engineering* **16(4)** 331–341 DOI: 10.15593/2224-9923/2017.4.4

[4] Putilov I S, Galkin V I 2007 The results of statistical analysis for study facies characterization of T-Fm stage of Sibirskoe oilfield *Neftyanoe khozyaystvo - Oil Industry* **9** 112–114

[5] Putilov I S, Vinokurova E E, Guliaeva A A, Yuzhakov A L, Popov N A 2020 Creation of a Conceptual Geological Model based on Lithological-Petrographic Research on the Example of the Permo-Carboniferous Deposit of the Usinskoe Deposit *Perm Journal of Petroleum and Mining Engineering* **20 (3)** 214-222 DOI: 10.15593/2712-8008/2020.3.2

[6] Cherepanov S S, Ponomareva I N, Erofeev A A, Galkin S V 2014 Determination of fractured rock parameters based on a comprehensive analysis of the data core studies, hydrodynamic and geophysical well tests *Neftyanoe Khozyaistvo – Oil Industry* **2** 94-96

[7] Putilov I S, Gurbatova I P, Popov N A, Chizhov D B, Yuryev A V 2019 Improving the reliability of the results of physical and hydrodynamic studies *Bulletin of Perm National Research Polytechnic University. Geology. Oil and gas and mining* **19 (3)** 216–227 DOI:10.15593/2224-9923/2019.3.2

[8] Martyushev D A, Zaytsev R A 2019 Influence of oil field reef carbonate reservoir petrophysical parameters of tournasian-famennian deposits in Upper Kama on well productivity *Bulletin of Tomsk Polytechnic University - Geo Assets Engineering* **330 (11)** 77-85 DOI: 10.18799/24131830/2019/11/2350

[9] Smetkina M A, Melkishev O A, Prisyazhnyuk M A 2020 Refining the Values of Permeability when Adapting the Hydrodynamic Model *Perm Journal of Petroleum and Mining Engineering* **20** № 3 223-230 DOI: 10.15593/2712-8008/2020.3.3

[10] Poplygln V V, Galkin S V 2011 Forecast quick evaluation of the indices of the development of the oil deposits. *Neftyanoe khozyaystvo - Oil Industry* **3** 112–115

[11] Repina V A, Galkin V G, Galkin S V 2018 The use of integrated accounting of petrophysical characteristics in the adaptation of geological and hydrodynamic models (using the example of the Visean reservoir of the Gondyrevskoye oil field) *Journal of Mining Institute* **231** 268-274 DOI: 10.25515 / PMI.2018.3.268

[12] Hongbing Li, Jiajia Zhang 2018 Well log and seismic data analysis for complex pore-structure carbonate reservoir using 3D rock physics templates *Journal of Applied Geophysics* **151** 175-183

[13] Célio Maschio Denis José Schiozer 2019 A new parameterization method for data assimilation and uncertainty assessment for complex carbonate reservoir models based on cumulative distribution function *Journal of Petroleum Science and Engineering* **183** 106400 DOI: 10.1016/j.petrol.2019.106400

IOP Conf. Series: Earth and Environmental Science 1021 (2022) 012028 doi:10.1088/1755-1315/1021/1/012028

Evaluation of ways to reduce sand production from unconsolidated heterogeneous formations when using screens

M Ya Khabibullin

Ufa State Petroleum Technological University, Branch of the University in the City of Oktyabrsky, 54a, Devonskaya St., Oktyabrsky, Republic of Bashkortostan, 452607, Russian Federation

E-mail: m-hab@mail.ru, info@of.ugntu.ru

Abstract. Experimental work showed that as a result of placing screens in a well with a bottomhole, the formation of which is composed of inhomogeneous, weakly cemented or loose sandstones, often with reduced permeability due to poor-quality completion, it is necessary to strive for a maximum decrease of $\zeta_{2фк}$ due to an increase in the wellbore of the screen shell and thereby an increase in the quality of filtration actions for colmatant breach, which contributes to a decrease of ζ_{2n}. The following conclusions were drawn when examining sand phenomena from weakly cemented and loose formations with a heterogeneity coefficient of more than 10. The particles that make up the rock structure should not be removed from the formation due to the use of a screen. In practice, this means that the screen must not conduct 85-75% (by weight) of the larger sand fractions. To maintain the overall stability of the reservoir skeleton, it is possible to allow the removal of not more than 15-25% (in weight) of fine particles.

1. Introduction

When studying the change in the filtration coefficient of blocks as a result of mechanical colmatage, it is proved that when the block screen is in contact with a large homogeneous fraction of the formation simulator, the calculated and actual filtration coefficients are quite high. However, the presence of fine-grained fractions and clay particles in the formation simulator leads to intensive mechanical colmatage and under the considered conditions the use of block-type screens is not recommended [1-3].

2. Materials and methods

Depending on the imperfection in the nature of the formation opening the formation-screen system resistance (ζ_2) includes hydraulic characteristics of rocks in the screen zone in addition to the above-mentioned hydraulic characteristics of screens. In case of a discontinuous change of rock permeability in a formation [4]:

$$\zeta_{2n} = (\frac{K_n}{K_u} - 1) \cdot ln\frac{r_u}{r_c}, \quad (1)$$

where K_n and K_u – respectively, filtration coefficients of productive formation rocks and near a well in the zone with changed permeability; r_u – radius of the zone with changed permeability, m; r_c – screen radius, m.

The graphs of the considered dependency are given in Fig. 1. The graphs show that ζ_{2n} changes more intensely at lower r_u/r_c ratios, and the K_n/K_u ratio to a lesser extent affects the change of $(-\zeta_{2n})$ than (+

ζ_{2n}).

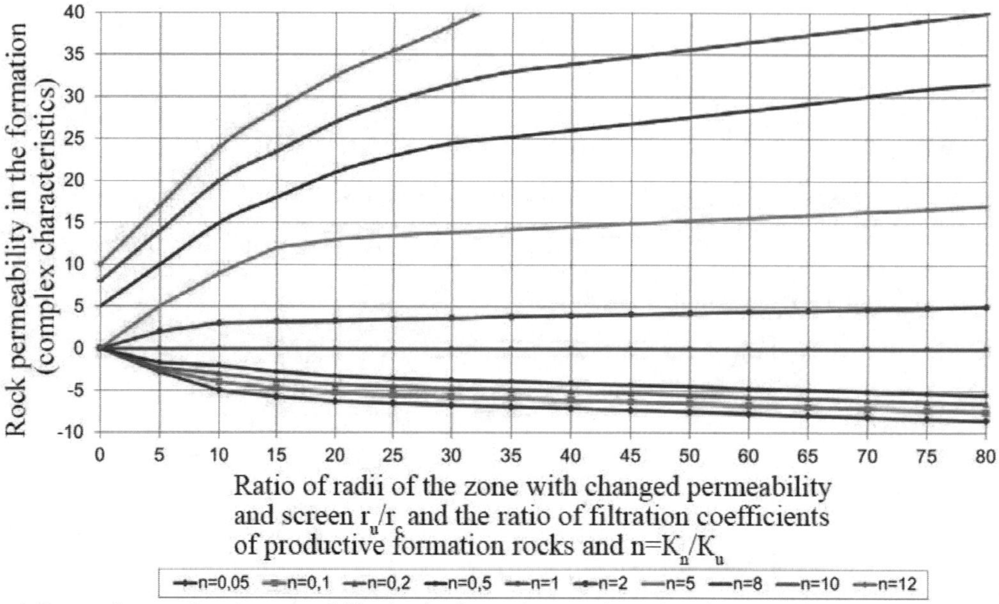

Fig. 1. Dependence of rock permeability in the formation on the ratio of radii and filtration coefficients near a well

As can be seen from the above, the resistance ζ_2 representing a complex characteristic depending on the quality of well completion, hydraulic characteristics of the screen due to the structure, as well as on the decrease of the permeability of the rock in the screen zone during operation, mainly depends on the set of indicators $\zeta_{2фк}$ and ζ_{2n}, and changes thereof in the process of further mineral production [5, 6].

If to imagine the hydraulic formation-screen relation in the form of zones where the filtration modes change, then this will look as follows: *a - b - c - d* (Fig. 2). According to this scheme, let us also give conditional definitions of its components: *a* – formation zone, where the velocity of the produced fluid is less than the critical velocity for the removal of particles that are supposed to be passed through the screen; *b* – zone where the produced fluid velocity is more than $V_{кр}$; *c* – zone of rock contact with a screening shell, which is equal to its thickness; *d* – screen frame which ensures the velocity of removal of particles passed by the screen from the well.

When inserting screens into a well with a bottomhole, the formation of which is composed of inhomogeneous, weakly cemented or loose sandstones, often with reduced permeability due to poor-quality completion, it is necessary to strive for a maximum decrease of $\zeta_{2фк}$ due to an increase in the wellbore of the screen shell and thereby an increase in the quality of the screen of flow paths for passing the colmatant, which contributes to a decrease of ζ_{2n}.

The introduction of a volume of rock with more uniformity and larger size in the screen zone compared to a large fraction of formation sand is an attempt to improve the filtration properties of the screen zone, i.e. a decrease in $\zeta_{2фк}$ and ζ_{2n}. The design of the gravel packing should simultaneously exclude the withdrawal of formation particles into the well.

The analysis of theoretical solutions and experimental data given in numerous literature on the justification of the choice of the ratio of gravel size to the size of formation sand showed that with the development of experimental methods and the accumulation of laboratory and field data, this ratio tends to decrease. Chronologically, this looks as follows.

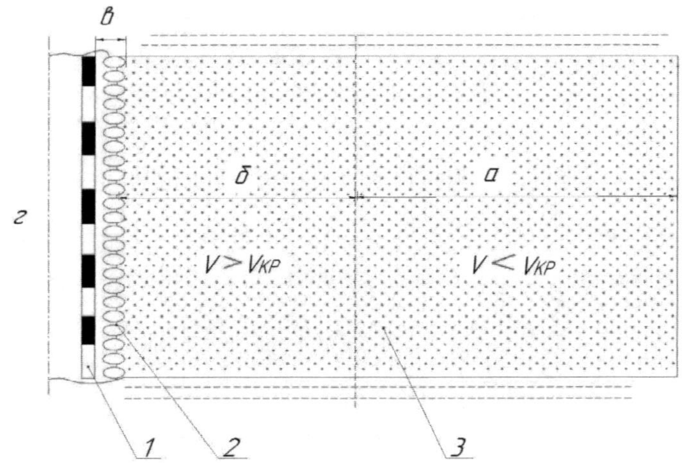

Fig. 2. Formation-screen hydraulic coupling diagram: 1 – screen frame; 2 – screen shell; 3 – formation

The work [7] gives the following ratio at the homogeneous composition of gravel packing and formation sand:

$$D/d = 10...12, \qquad (2)$$

where D – gravel diameter, mm; d – diameter of large sand fractions, mm.

In the works [8], the ratio $D/d = 8...12$, but taken considering arching and grain diameter, which are supposed to be kept from removal, and the requirements for preventing the penetration of sand particles into the screen, i.e. preventing its colmation and excluding arching, are expressed by the ratio D/d = 4…5.

3. Results and Discussion

The studies carried out under conditions close to well conditions, when gravel is mixed with sand in the bottomhole zone or in column when sand settles during backwash, show that this decreases permeability of screen zone and reduces well productivity [9].

The efficiency of gravel packing is also reduced by the quality of its preparation. The study of gravel analyses showed that it is insufficiently cleaned from soil sand when excavated from a quarry, during transportation part of gravel is ground and destroyed and that 30-40% of it is larger or smaller than the nominal size. Besides, gravel and sand are also mixed during well operation if the gravel is large and cannot stop the movement of sand.

The permeability changes of the gravel-sand mixture (Oklahoma, Formation No. 1) are shown in Table 1. These data show that in certain proportions of gravel and sand mixing, the permeability of the mixture may be less than that of the sand. Under downhole conditions (under rock compression) mixing leads to an even greater loss of porosity and permeability. The effect of mixing sand and gravel is significant. The values of permeability of compressed mixtures are given in Table 2, which shows that permeability of sand mixtures with gravel size of 0.25-0.42 is higher than that of others.

The results of experimental work on the evaluation of methods of reducing sand removal from uncemented non-homogeneous formations show that the use of screens in a cased wellbore requires the creation of gravel packing in the annular space between the screen and the column, in the perforated channel and behind the casing. Moreover, the size of gravel should be chosen so as to prevent mixing of gravel packing with formation sand due to migration and intrusion of sand, i.e. $D/d = 4...5$. In this case it is possible to avoid the reduction of well productivity by increasing the size and number of perforations [10].

Table 1. Change in permeability of gravel/sand mixture of different sizes

Gravel size, mm	Gravel content in a mixture, %	Permeability, μm^2	$\frac{K_{mixture}}{K_{sand}}$	Gravel size, mm	Gravel content in a mixture, %	Permeability, μm^2	$\frac{K_{mixture}}{K_{sand}}$
0.25–042	0	11.1	1.0	0.84–2.0	5	11.1	1.00
	15	9.8	0.88		15	12.7	1.02
	25	10.2	0.92		25	16.3	1.47
	40	12.9	1.16		40	12.8	1.15
	50	17.2	1.55		60	28.1	2.53
	75	32.4	2.92		75	67.0	6.04
	100	69.0	6.22		100	652.0	58.7
0.42–0.84	5	10.0	0.90	2.0–2.38	5	12.1	1.09
	15	10.0	0.90		15	0.1	0.91
	25	1.0	0.99		25	14.3	1.29
	40	15.3	1.38		35	17.3	1.56
	75	36.6	3.30		50	14.0	1.26
	100	170.8	15.4		100	1969.9	177.3

K – permeability coefficient

Table 2. Values of permeability of compressed mixtures

Gravel size, mm	Gravel content in a mixture, %	Permeability, μm^2	$\frac{K_{mixture}}{K_{sand}}$	Gravel size, mm	Gravel content in a mixture, %	Permeability, μm^2	$\frac{K_{mixture}}{K_{sand}}$
0.25–042	0	8.74	1.0	0.84–2.0	10	7.3	0.84
	10	7.9	0.9		25	4.5	0.51
	15	7.7	0.88		40	2.9	0.33
	25	6.3	0.72		100	517.4	59.2
	50	7.1	0.81		-	-	-
	75	17.3	1.98		-	-	-
	100	69.7	7.97		-	-	-
0.42–0.84	15	5.5	0.63	2.0–2.38	10	7.4	0.85
	40	3.5	0.40		25	4.7	0.54
	75	5.7	0.65		40	3.0	0.34
	100	120.7	13.81		70	5.5	0.63
	-	-	-		100	2020.0	231.1

When setting screens in an open wellbore, pressure losses at the gravel-formation interface play a significant role, and it should be noted that the greatest productivity and efficiency of the screen in the open wellbore are achieved when there is a packing around it, which can be created by breaking the sand mass of the formation by cyclically changing the flow rate.

When selecting the screen design, at the same time with the possibility of providing them with a reliable hydraulic connection in the formation-screen system, the main task is also solved, i.e. the prevention of sand transfer into the well. It should be noted that preventing the removal of sand means that at optimal steady state well operation, the screen retains the bed skeleton particles but passes fine and clay particles.

During the start-up of the well, the situation is slightly different – part of the skeleton-forming particles is passed until a natural screen of sorted large fractions is formed on the screen shell. The number and size of removed particles is more dependent on the formation of the primer zone, and therefore of $\zeta_{2фк}$ and $\zeta_{2п}$.

The following dependencies are proposed: $D_0 = (3...4.5)\ d_{90}$ – for round holes and $T = (2...3)\ d_{90}$ –

for the width of a rectangular slot; where D_g – hole diameter, mm; t – slot width, mm; d_{90} – diameter of sand particles, less than which 90% of fractions are contained in the rock, mm.

The study of the conditions for the removal of sand particles through flow sections in hole, mesh and slot screens during their operation in both homogeneous and heterogeneous sands made it possible to recommend empirical dependencies for determining the dimensions from the holes given in Table 3 [11].

Table 3. Dimensions of screen holes depending on inhomogeneity coefficient and their average diameter

Screen type	Recommended screen hole sizes: in homogeneous rocks and packaging, $K_н < 2$	Recommended screen hole sizes: in inhomogeneous rocks and packaging, $K_н > 2$
Hole	(2.5–3.0) d_{cp}	(3.0–4.0) d_{cp}
Mesh	(1.5–2.0) d_{cp}	(2.0–2.5) d_{cp}
Slot	(1.25–1.5) d_{cp}	(1.5–2.0) d_{cp}
Wire	1.25 d_{cp}	1.5 d_{cp}

Note: smaller values of coefficients at K_{cp} and K_{50} refer to fine-grained rocks, and larger – to coarse-grained rocks

Johnson wire-wound screens provide a slot width depending on the size distribution of rocks and are designed to pass 50-70% of fine fractions ($\delta = d_{50-70}$) with the assumption that the remaining part will create a reliably working natural screen. In case of fine-grained sands of floating nature above the productive horizon, the carried out fractions should be reduced to 30% ($\delta = d_{30}$).

Experimental data obtained when determining the optimal method of reducing sand removal from uncemented silty sand formation with creation of natural gravel packing from large fractions of sand show that the best results on sand are obtained at size of a smaller slot calculated for initial passage of 95% of fine fractions $\delta = d_{95}$ [12]. It should be noted that these data are given for a three-layer Johnson wire-wound screen in which the size of the slot is reduced from the outer layer to the inner layer in a 3:2:1 ratio. The reduction of sand removal from the weakly cemented formation at Anastasievsko-Troitsky field was achieved using two-layer wire-wound screens of the Thermneft NPO design. The size of the slot was assumed as follows: on the outer layer of the screen shell with an opening of 0.25 mm, equal to or more than the diameter of sand grains at a 95% point according to the integral curve of particle size distribution ($\delta_{нор} = d_{95}$), and on the inner – 0.15 mm. The inner gap is larger than the diameter of sand grains in 50% point, but less than half of grain diameter in 85% point of the integral curve ($d_{50} < \delta_{внутр} < d_{85}$).

4. Conclusion

The following conclusions were drawn when studying sand phenomena in weakly cemented and loose formations with a heterogeneity coefficient of more than 10. The particles that make up the rock structure should not be removed from the formation due to the use of a screen. In practice, this means that the screen must not conduct 85-75% (by weight) of the larger sand fractions. To maintain the overall stability of the reservoir skeleton, it is possible to allow the removal of not more than 15-25% (in weight) of fine particles.

References

[1] Khabibullin M Ya 2019 Managing the processes accompanying fluid motion inside oil field converging-diverging pipes *Journal of Physics: Conference Series* **1333** 042012. DOI: 10.1088/1742-6596/1333/4/042012

[2] Khabibullin M Ya 2019 Managing the reliability of the tubing string in impulse non-slationary flooding *Journal of Physics: Conference Series* **1333** 052012 DOI:10.1088/1742-6596/1333/5/052012

[3] Mukhametshin V Sh 2020 Justification for increasing the performance of hydrochloric acid

treatment in wells of fields with carbonate reservoir *IOP Conf. Ser.: Mater. Sci. Eng.* **905** 1-5 DOI: 10.1088/1757-899X/905/1/012083

[4] Zaynagalina L Z 2020 On the determination of design and geometric parameters of a upper bit slurry grinder *IOP Conference Series: Materials Science and Engineering* **905** 012094 DOI:10.1088/1757-899X/905/1/012094

[5] Zaynagalina L Z 2020 Dynamics of the oscillatory system of the near-bit slurry grinder *IOP Conference Series: Materials Science and Engineering* **905** 012092 DOI:10.1088/1757-899X/905/1/012092

[6] Zaynagalina L Z 2020 On the development of an experimental design and field test of an upper bit tool *IOP Conference Series: Materials Science and Engineering* **905** 012093 DOI:10.1088/1757-899X/905/1/012093

[7] Kadyrov R R, Mukhametshin V V, Galiullina I F, Kuleshova L S, and Safiullina A R 2020 Prospects of applying formation water and heavy brines derived therefrom in oil production and national economy *IOP Conference Series: Materials Science and Engineering* **905** 1-6 DOI: 10.1088/1757-899X/905/1/012081

[8] Haoran Zh, Yongtu L and Xingyuan Zh 2017 Sensitivity analysis and optimal operation control for large-scale waterflooding pipeline network of oilfield *Journal of petroleum science and engineering* **154** 38-48

[9] Sun W and Mun-Hong H 2017 Forecasting and uncertainty quantification for naturally fractured reservoirs using a new data-space inversion procedure *European Assoc Geoscientists & Engineers Computational geosciences 15th Conference on the Mathematics of Oil Recovery (ECMOR) (Amsterdam, Netherlands)* 21(5-6) pp.1443-1458

[10] Mukhametshin V Sh, Andreev V E, and Yaskin S A 2020 Designing measures to increase oil recovery based on the identification and grouping of deposits *IOP: Earth and Environmental Science* **579** 1–11 DOI: 10.1088/1755-1315/579/1/012013

[11] Galimullin M L, Khabibullin M Ya 2020 Experience with sucker-rod plunger pumps and the latest technology for repair of such pumps *Chemical and Petroleum Engineering.* 55 (11-12) 896-901. DOI: 10.1007/s10556-020-00710-1

[12] Stenkin A V, Kotenev Yu A, Sultanov Sh Kh, Mukhametshin V V, Kuleshova L S, and Safiullina A R 2020 Increased production of oil reserves in low production deposits of West Siberian oil and gas province *IOP Conference Series: Materials Science and Engineering* **905** 1-7 DOI: 10.1088/1757-899X/905/1/012088

IOP Conf. Series: Earth and Environmental Science 1021 (2022) 012029 doi:10.1088/1755-1315/1021/1/012029

Possible methods of oil potential prediction for yu2 formation in the North of the Surgut Arch

A O Shadrin

Perm National Research Polytechnic University, Komsomolsky Prospekt 29, 614990, Perm, Russian Federation

E-mail: arseniy.shadrin@yandex.ru

Abstract: This article addresses the geological structure of Yu2 formation in the north of the Surgut arch. The paleo-landscape aspects of the research area are described and the main regularities of paleogeographic conditions at the time of the bed formation are identified. The accumulation conditions of the sediments building up the Yu2 oil reservoir of the Upper-Tyumen sub-suite are defined. The target of research, Yu2 formation, was assessed for sedimentation environments. The configuration and structure of oil deposits across the study area were evaluated. Key features of deposits, which complicate geometrization and allocation of prospecting sites for reserves portfolio expansion, have been identified. A model has been developed for predicting oil potential based on geological and geophysical reservoir characteristics subject to the lithofacies structure.

1. Introduction

The target of research in this work is Yu2 formation of Tyumen suite within the north of Surgut arch, West Siberian oil and gas province. The study area features high exploration maturity, which increases the relevance of the research aimed at further oil potential identification. The oil potential search problem in high exploration maturity conditions is addressed in a number of works, specifically in [1, 2, 3, 4, 5].

Stratigraphically, the Yu2 formation is confined to sediments of Tyumen suite of the middle division of the Jurassic system. The Yu2 formation sediments in the study area feature continental-marine sedimentation environments and a complex geological structure [6, 7, 8, 9].

The Yu2 reservoir is represented over the entire study area and has a proven oil potential over extensive areas, which makes it one of the priorities for the operators' reserves portfolio replacement.

The Yu2 formation is confined to the Upper Tyumen sediments that feature a complex geology in the study area. The Tyumen suite is mainly composed of continental sediments. The end of the Bathonian, identified with the Tyumen suite sedimentation period end, is defined by the marine transgression and the formation of passage and marine sediments of the Yu2 formation upper part.

In the study area, the Yu2 formation is defined by a variety of facies sedimentary environments. The Yu2 formation sediments mainly feature lacustrine-alluvial, channel, deltaic and coastal-marine genetic types of sedimentation. Sub-meridional river systems were the main factors of landscape formation. Rivers acted as a transport system for solid flows, and point bars were the main accumulators of sandy

IOP Conf. Series: Earth and Environmental Science 1021 (2022) 012029 doi:10.1088/1755-1315/1021/1/012029

material. Flowing in backlands, the rivers were fringed by floodplains often connected with marshes and lakes.

The inter-bed areas are also composed of alluvial-lacustrine sediments formed as a result of lateral migration of river beds over time. The humid climate contributed to the development of extensive floodplain parts of river systems, lakes and marshes.

The first signs of the transgression onset are associated with the mid-Bathonian. The sea advance from the north resulted in the gradual formation of successive delta systems. The deltas of the major lowland rivers are complex structures composed of manifold delta complexes that cannot be distinguished.

The sea transgression peaked at the end of the Bathonian. Slopes and most of the Surgut arch became the areas of coastal-marine sedimentation. Coastal-marine facies is identified in the upper part of the formation in the cores sampled from a few wells in the area of study. In areas of shallow-water and coastal-marine sedimentation, along with fine-pelitic sediments, sandy submarine shafts were formed, and sandy material also accumulated on shoals in the coastal part of the sea.

The variable sedimentation environments of the Yu2 formation created a variety of facies and induced complex geology of the area. The reservoir's multi-facies structure resulted in separate reservoirs, hydrodynamic and lithofacies screens that impeded connectivity and free fluids migration in the formation.

The Yu2 geology aspects substantially hamper application of standard methods and approaches to address the task of oil potential prediction and identification of prospects. To solve the oil potential prediction problem for the Yu2 formation in the study area, a model of oil potential prediction shall be developed using a categorical variable that characterises the facies environment of wells areas.

2. Methods

The Yu2 formation's oil potential over the study area has been proven with commercial oil flows from a number of wells, the flow rates having a significant spread of values and reaching 150 m^3/day. No commercial oil flows were obtained in the most of the study area; however, there are numerous indications of oil potential, attributed to oil-saturated interlayers in core samples and non-commercial oil flows.

The difficulty in predicting the Yu2 formation's oil potential is primarily attributed to the complex structure and geometry of the discovered oil deposits. Due to high heterogeneity of their lithofacies structure, the oil deposits have multiple hydrodynamic screens impeding hydrocarbon migration in the formation and oil accumulation in structural traps.

Fluid contact in the oil deposits is poorly dependent on the structural factor. A graph of dependency of the oil/water contact hypsometric position on the Yu2 reservoir roof is shown in Figure 1.

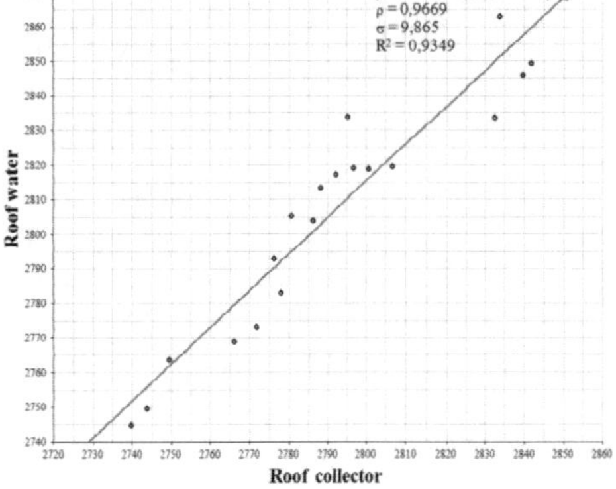

Figure 1. A graph of the fluid contact dependency on the Yu2 reservoir roof

Figure 1 displays poor dependence of the fluid contact on the Yu2 reservoir roof (r=0.967; std.dev=9.865). For comparison, a more homogeneous formation, Yu1, has a correlation coefficient of 0.996 and a standard deviation of 3.563 for a similar correlation.

Complex fluid contact surfaces and deposit configurations hamper the application of prediction methods based on the analysis of structural and geodynamical characteristics of the study areas [10, 11].

Additional complications are derived from the poor continuity of reservoir properties within the genetic types of sedimentation. Table 1 shows statistical characteristics of reservoir properties of the rocks attributed to different sedimentation environments based on the core material and well log data analysis.

Table 1. Reservoir properties of sedimentation environments from core data

Sedimentation environment	Porosity mean min-max	Permeability mean ($1*10^{-3}$ mcm^2) min-max
Riverbed	12.7 4-20.3	1.5 0.001-46
Floodplain	11.1 3.3-17.4	1 0.001-16.52
Meandering riverbed	10.7 3.4-18.3	2.13 0.01-75
Lagoon shore	11.60 2.1-19.8	1.1 0.001-22.12
Coastal slope	11.7 1.4-20.3	7.45 0.01-69

The heterogeneity of reservoir properties, reservoir compartmentalisation, capillary forces and other factors related to lithofacies reservoir properties, along with hydrodynamic screens, complicate the fluid flow structure in the formation. The formation's complex flow structure has a strong effect on the oil accumulation. High formation heterogeneity results in sudden discontinuity of oil/water contact, zones devoid of reservoirs, low-permeability and high-water-cut areas within oil deposits.

Given the complex reservoir structure of the Yu2 formation, the oil potential prediction is only possible using lithofacies models. For oil potential prediction modelling, a training sample set was sampled from the wells that entered the Yu2 formation with flow test results. The sample set was defined by continuous variables: net pay thickness (H); arenosity (Ar); Alpha PS (APS); compartmentalisation (Comp); Yu2 formation reservoir roof (AbsRoof), a categorical variable defining facies.

The facies was determined from well log data and core analysis. To prepare the most representative sampling, the facies set was approximated to the facies that are most common in the area and noticeably differ from each other: riverbed; meandering riverbed; floodplain and lake; other continental facies.

Next, a linear discriminant analysis was used to build a model for oil potential prediction from the lithofacies characteristics of the Yu2 formation.

3. Results

Canonical equations of prediction models for various facies are given in Table 2; a graph of oil potential probability dependency on a canonical score is shown in Figure 2.

Table 2. Canonical equations of linear discriminant models

Sedimentation environment	Canonical equation
Riverbed	Z=Comp*0,0405-Ar*1,7637+1,0633*APS+0,0271*AbsRoof-0,0077*H-73,711
Meandering riverbed	Z=Comp*0,0405-Ar*1,7637+1,0633*APS+0,0271*AbsRoof-0,0077*H-75,733
Floodplain and lake	Z=Comp*0,0405-Ar*1,7637+1,0633*APS+0,0271*AbsRoof-0,0077*H-74,878
Other continental facies	Z=Comp*0,0405-Ar*1,7637+1,0633*APS+0,0271*AbsRoof-0,0077*H-74,462

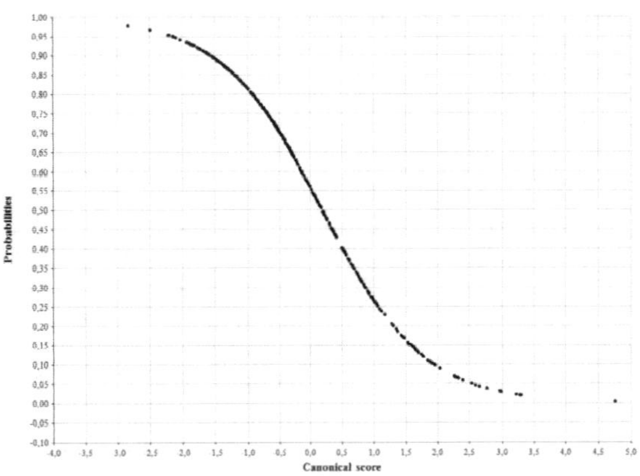

Figure 2. A graph of oil potential probability dependency on a canonical score

The prediction model has an efficiency of 78%, which is a high index for a production site with complex geology.

The analysis of the sedimentation environments and their geophysical characteristics has indicated that the target of research is defined by a complex geological structure. The variability of lithofacies and the heterogeneity of reservoir properties laterally and vertically significantly complicate geometrization of deposits and identification of promising sites. To predict the oil potential of the production site with such a complex geology, a linear discriminant analysis was conducted and oil potential prediction models were built subject to the facies characteristic. The developed prediction model produces a fairly high result and can be applied to identify prospective areas for exploration planning.

References

[1] Voevodkin V L, Galkin V I, Krivoshchekov S N 2012 Investigation of the effect of oil-content and research criteria in the Perm region on the hydrocarbon deposits distribution *Oil Industry* **6** 30-34

[2] Krivoshchekov S N 2011 Establishment of regional and zonal criteria of Perm Region's oil forecast probabilistic and statistical methods *Oil Industry* **10** 10-14

[3] Nosov M A, Galkin V I, Krivoshchekov S N, Melkishev O A, Kochnev A A 2013 On substantiation of license area economic-geological evaluation by hydrocarbon resources and reserves *Oil Industry* **13** 14-17

[4] Putilov I S, Galkin V I, Pyatunina E V 2018 On the need to use new criteria when predicting the oil and gas potential of small-sized structures (by the example of the territory of the Perm Region) *Oil industry* **5** 40-43
[5] Putilov I S, Galkin V I 2016 Features of the use of new criteria in predicting the oil and gas potential of structures of various tectonic elements of the Perm Region *Geology, geophysics and development of oil and gas fields* **2** 4-7
[6] Sevastyanov A A et al. 2017 Geological features and assessment of the mining potential of the deposits of the Tyumen formation *Bulletin of Perm University. Geology* **16 (1)** 61-67.
[7] Vakulenko L G, Yan P A 2010 Sedimentogenesis of the Yu2 oil-bearing horizon and its significance in the search and exploration of hydrocarbon deposits *Proceedings of Tomsk Polytechnic University. Georesource engineering* **316 (1)** 61-66
[8] Popov A Yu, Kazanenkov V A 2010 Features of the Bath sedimentogenesis in the north-east of the Latitudinal Ob region *Proceedings of Tomsk Polytechnic University. Georesource engineering* **316 (1)** 67-71.
[9] Mkrtchyan O M, Filina S I 1985 Features of the structure of the U2 formation in Western Siberia and the placement of oil and gas deposits in it *Geology of oil and gas* **3** 48-53.
[10] Krivoshchekov S N, Kozlova I A 2012 Geodynamic characteristics of the conditions of French-Famennian strata dispersed organic matter immersion and catagenesis in Perm Region *Oil Industry* **7** 82-85
[11] Shadrin A O and Krivoshchekov S N 2021 *IOP Conf. Ser.: Earth Environ. Sci.* **666** 052056

The structure and bearing capacity of rock structures during underground mining

V B Zaalishvili[1], O G Burdzieva[1], M S-U Khaliev[3]

[1]Vladikavkaz Scientific Centre of the Russian Academy of Sciences, 93a Markova St., Vladikavkaz, 362002, Russia
[3]Chechen State University named after A. H. Kadyrov, 17a, Dudaev Boulevard, Groznyy, 364015, Russia

E-mail: vzaal@mail.ru

Abstract. The processes of controlling the ore-containing mass are refined and detailed, and the mechanism of interaction of stresses and deformations during underground mining of ore deposits is detailed. The properties of the massif and the role of faults and macrofractures in the development of deposits of complex structure are systematized. A methodology for studying stresses and deformations of rocks is given. Regularities of array maintenance at the interface site are established. Quantitative values of changes in physical and mechanical properties and stability of rocks in the interface zone are given. The array is differentiated by the magnitude of the acting stresses in various phases of the connection. The calculation of stresses for specific conditions with the determination of the reliability of the mining site was performed. Patterns of stress and strain changes over time are noted. An explanation of the phenomenon is given from the standpoint of the interaction of structural rock blocks in a discrete field of the array.

1. Introduction

The exploited deposit is an underground structure with an internal structure and a system of elements. The calculation of the parameters of the structure is carried out by methods of structural mechanics, which include the principles and methods of calculating structures for strength, rigidity, and stability. For the calculation of underground structures, general laws of mechanics and ratios are used, taking into account the physical and mechanical properties of the rock mass, the conditions of the interaction of elements, parts, and the basis of the structure [1–3].

The safety and efficiency of mining of the deposit are ensured by preventing critical stresses in the elements of the ore-rock-earth surface system arising from the interaction of man-made and natural factors. An increase in the depth of field development and an increase in the volume of mineral extraction increase the impact on ore-containing massifs, the stability of which is the result of the interaction of the emerging force fields [4–5].

The result of insufficient consideration of the state of ore-containing arrays is a decrease in the quality of mined ores, which complicates the use of metals in related industries [6–8, 11].

The purpose of research in this area is to predict the development of mountain pressure for timely management decision-making [12-14].

2. Research methods

During the construction of underground workings, geological and surveying tools are used, including:

- measurement of stresses in the support by strain gauges;
- measurement of support deformations by surveying leveling (Figure 1).

The dynamics of stresses in the array with the approach of the mine to the ore body were studied by the strain gauge method, and the deformation of the support was studied by reference points mounted on a metal frame [15-19].

3. Results and discussion

The sensor indicators were generalized and processed by methods of mathematical statistics. Analysis of the readings of geophysical instruments made it possible to differentiate the deformation of the support with the allocation of the phases of the development of deformations of the support are distinguished:

- appearance when the mine approaches the ore body;
- magnification when cut from the hanging side up to 19 mm;
- maximum development up to the damage of the support.

The maximum deformations of the support were recorded near the cleaning face and in the area of the break. The impact of the cleaning works began to be felt at a distance of 15–18 m from the ore body.

Prior to the opening of the ore body by mining, deformations increased monotonously, and after the hanging side was reached, the load increased sharply (Table 1).

Table 1. Dynamics of stresses and strains

Measurements	Deformations, mm Sensor numbers				of Stress, M P a Sensor numbers		
	4	10	3	7	24	44	11
1	0.9	0.15	0.24	0.09	–	–	–
2	3.6	0.55	0.21	0.22	2.3	0.12	0.04
3	–	2.2	0.26	0.24	9.7	2.35	0.85
4	–		0.50	0.26	7.65	5.1	2.65
5	–		0.90	0.62		6.35	3.6
6	–		1.00	0.8		–	3.7

Displacement of weakened rocks of the fault zone sharply reduces the stability of the workings. Other things being equal, the inter-port ore pillars deform more than the pores (Figure 1).

The array is characterized by three sections of operating voltages. The initial section is characterized by a slight increase in stress. With the approach of the front of the cleaning works and the increase in the area of mining of the substage, the ore massif is experiencing an increasing load.

If the structural blocks of rocks do not jam, the roof takes the form of a vault, the addition of empty rocks increases the dilution of ores and metal losses during ore enrichment, therefore, the flat shape of the roof is preserved by using such a technology in which the rocks in the lower layer above the mine were wedged and did not collapse from the rocks created by the mass inside the vault of the natural stress equilibrium (Figure 2).

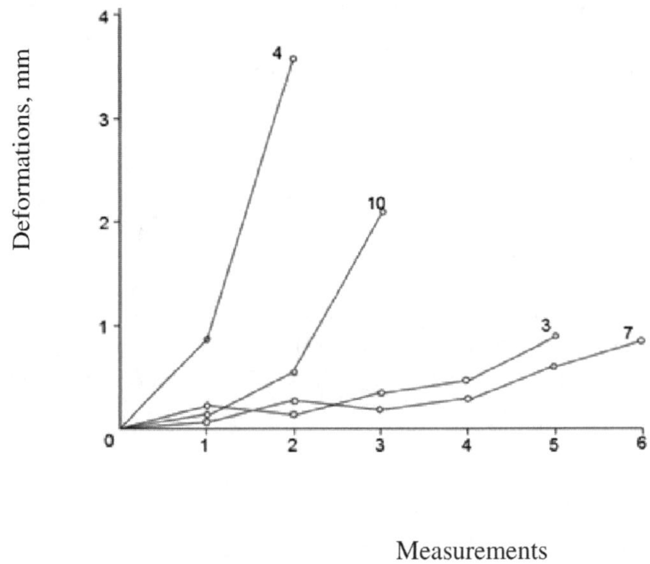

Figure 1. Deforming of the walls of wells over time: 3, 4, 7, 10 – sensor numbers.

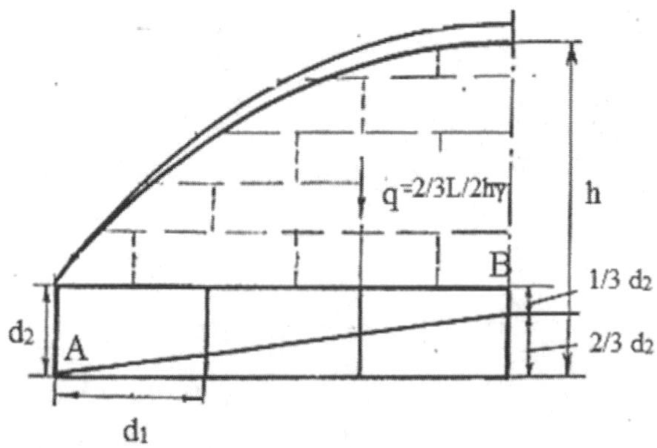

Figure 2. To determine the bearing capacity of wedged roof rocks: *h* – the height of the arch of rocks, m; d_1 – the length of rock blocks, m; d_2 – the height of rock blocks, m; γ – the volume weight of rocks, t/m3; *q* – strength; *A* – the heel of the arch; *B* – the lock of the lower layer of rocks.

In monolithic rocks, a force acts on half of the span:

$$q = \frac{2}{3} \cdot \frac{L}{2} h \cdot \gamma \tag{1}$$

where *q* is the length of the output, m;
L – the capacity of the main roof, m;
h – volume weight of rocks, t/m3;
γ – the moment of force q relative to point A, N*m.
The moment of force T relative to point A:

$$M_q = \frac{L^3 \gamma}{18 V_0} \tag{2}$$

Rasp force:

$$T = \frac{10R''_r}{K_2} \cdot \frac{d_2}{3} \quad (3)$$

where d_2 is the height of the structural block of the rocks of the main roof, m.
From the equality of moments, for

$$R''_r = R'_r = R_r \quad (4)$$

where R''_r – the resistance of rocks to compression in the direction of the strut of the arch of rocks;
R'_r – resistance of rocks to compression in the direction of the action of the mass of rocks;
K_2 – the stock ratio.
In rocks disturbed by small cracks , the force acts on half of the span:

$$q = \frac{L}{2} b \cdot \gamma \quad (5)$$

where b is the power of the immediate roof, m.
The moment of force qH relative to point A:

$$M_q = \frac{L}{2} b \cdot \gamma \cdot \frac{L}{4} = \frac{L^2 b \gamma}{8} \quad (6)$$

Strut force:

$$T = \frac{10R_r}{K_2} \cdot \frac{1}{3} d_{n2} \quad (7)$$

where d_{n2} is the height of the rock block, m;
The moment of the expansion force T relative to point A:

$$M_T = T \frac{5}{6} d_{n2} = \frac{10R_r 5d_{n2}^2}{18K_2} \quad (8)$$

From equality of moments:

$$L = 1.9 d_{n2} \sqrt{\frac{10R_r}{K_2 \gamma b}} \quad (9)$$

The span of the fixed outcrop. The force acting on the half span:

$$q_3 = \frac{L_3}{2} \cdot b\gamma \quad (10)$$

The moment of force q_3 relative to point A:

$$M_q = \frac{L_3}{2} \cdot b \cdot \gamma \cdot \frac{L}{4} = \frac{L_3^2 b \gamma}{8} \quad (11)$$

The strength of the rasp:

$$T = \frac{10R_r}{K_2} \cdot \frac{2}{3} d_{n2} \tag{12}$$

The moment of the expansion force T relative to point A:

$$M_T = T \frac{5}{3} d_{n2} = \frac{10R_r \cdot 10 d_{n2}^2}{K_2 \cdot 9} \tag{13}$$

From equality of moments:

$$L_3 = 2.98 d_{n2} \sqrt{\frac{10R_r}{K_2 b\gamma}} \tag{14}$$

$$L_f < L_3 = 1.49 d_{n2} \cdot n \sqrt{\frac{10R_r}{K_2 b\gamma}} \tag{15}$$

where *n* – is the number of bonded roof layers.

4. Conclusion

The use of rock properties adequately improves the quality of mined ores and improves labor safety.

The research results indicate that the dynamics of stresses and deformations in the mining zone is predicted, which allows you to adjust the parameters of development technologies.

The results of the study are confirmed by the conclusions of foreign experts [9–10].

References

[1] Golik V I 2013 Conceptual approaches to creating low- and non-waste mining based on combination of physical-technical and physical-chemical geotechnologies *Mining Journal* **5** 93–97

[2] Golik V I, Burdzieva O G and Ganin M P 2019 An ecological concept of natural leaching of metal ores *Proceedings of Tula State University. Earth Sciences* **2** 16–26

[3] Valiev N G, Propp V D and Vandyshev A M 2020 The 100th anniversary of the department of mining engineering of UrSMU *News of Higher Educational Institutions. Mining Magazine* **8** 130–143

[4] Golik V I, Dmitrak Yu V, Gabaraev O Z and Kozhiev Kh Kh 2018 Minimizing the impact of mining on the environment *Ecology and Industry of Russia* **6** 26–29

[5] Dushin A V, Valiev N G, Lagunova Yu A and Shorin A G 2018 Ural Mining University and Moscow Mining University: interaction of higher education institutions *Mining Journal* **4** 4–10

[6] Bosikov I I, Klyuev R V, Aimbetova I O and Makhosheva S A 2021 Development of methods and management tools aerogasdynamics processes at mining sites *Sustainable Development of Mountain Territories* **3** 397–405

[7] Shchurov N I, Myatezh S V, Malozyomov B V, Shtang A A, Martyushev N V, Klyuev R V and Dedov S I 2021 Determination of inactive powers in a single-phase ac network *Energies* **14 (16)** 4814

[8] Shchurov N I, Dedov S I, Malozyomov B V, Shtang A A, Martyushev N V, Klyuev R V and Andriashin S N 2021 Degradation of lithium-ion batteries in an electric transport complex *Energies* **14 (23)** 8072

[9] Yang X X, Kulatilake P H, Jing H and Yang S 2015 Numerical simulation of a jointed rock block mechanical behavior adjacent to an underground excavation and comparison with physical model test results *Tunnelling and Underground Space Technology* **50** 129–142

[10] Panov V S, Eremeeva Zh V, Sharipzyanova G H, Skorikov R A, Mikheev G V and Ageev E V

2014 The effect of the nature of nanoscale particles and method of mixing on the tribological properties of powder steel 70P *Proceedings of the Southwest State University* **6 (57)** 8–14

[11] Lavrenko S A, Shishljannikov D I 2021 Performance evaluation of heading-and-winning machines in the conditions of potash mines. *Applied Sciences (Switzerland)* **11(8)** 3444 doi: 10.3390/app11083444

[12] Konyuhov V Yu, Gladkih A M, Galyautdinov I I, Kiseleva T Yu 2019 Ecological architecture: The green roofs. *IOP Conference Series: Earth and Environmental Science* **350** 012035

[13] Balanovsky A E, Shtayger M G, Kondrat'Ev V V, Van Huy V, Karlina A I 2018 Plasma-arc surface modification of metals in a liquid medium. *IOP Conference Series: Materials Science and Engineering* **411(1)** 012013

[14] Konyuhov V Yu, Gladkih A M, Chemezov A V 2019 Evaluation of the economic feasibility of the introduction of plasma hardening technologies in the Far North enterprises. *IOP Conference Series: Materials Science and Engineering*Kondratiev V V, Karlina A I, Guseva E A, Konstantinova M V, Kleshnin A A 2018 Processing and Application of Ultra disperse Wastes of Silicon Production in Construction. *IOP Conference Series: Materials Science and Engineering* **463(3)** 032068

[15] Konyuhov V Y, Gladkih A M, Zott R S 2020 Accelerator as an effective replacement of a business incubator in the Irkutsk region. *Journal of Physics: Conference Series* **1582** 012045

[16] **560** 012147

[17] Gladkih A M, Konyuhov V Yu, Galyautdinov I I, Shchadova E I 2019 Green building as a tool of energy saving. *IOP Conference Series: Earth and Environmental Science* **350** 012032

[18] Konyuhov V Yu, Gladkih A M, Galyautdinov I I, Severina Y D 2019 Economic aspects of green technologies. *IOP Conference Series: Earth and Environmental Science* **350** 012036

[19] Gabov V V, Zadkov D A, Babyr N V, Fangwei X 2021 Nonimpact rock pressure regulation with energy recovery into the hydraulic system of the longwall powered support. *Eurasian Mining* **2021(2)** 55-59. 10.17580/em.2021.02.12

IOP Conf. Series: Earth and Environmental Science 1021 (2022) 012031 doi:10.1088/1755-1315/1021/1/012031

Improving the efficiency of the physical protection system for oil and gas facilities (by the example of trunk pipelines)

G A Tsvetkov and A A Aleshkov

Perm National Research Polytechnic University, 29 Komsomolsky prospekt, Perm, Perm Krai, 614990, Russia

E-mail: zvetkov71043@mail.ru

Abstract. The article discusses the features of ensuring technospheric safety at oil and gas facilities (using the example of trunk pipelines) in the implementation of national interests. To achieve strategic goals, the developed structure of the physical protection system of trunk pipelines can be used to detect and prevent outside interference in their work, based on the use of passive seismic-analytical and active bioradioinformative methods for monitoring the security zone. An innovative approach is considered, which is based on the effect of the dependence of intruders' own electromagnetic radiation on the strength of an external low-intensity electromagnetic field of the radio wave range. Its implementation in the creation of technical means provides high selectivity, initially assuming a response only to living organisms, as well as the possibility of classifying biological objects according to their most informative features in any natural and climatic conditions.

1. Introduction

Trunk pipelines are strategically important facilities that provide the most reliable and efficient transportation of liquid and gaseous resources over long distances. Economic and political stability in many regions of the world depends on their uninterrupted functioning, and the transportation process itself is one of the important elements of industrial, economic, environmental, and national security.

In general, over the past decades, there has been a tendency for a decrease in the number of accidents in pipeline transport due to traditional causes (corrosion and defects) almost all over the world [1, 2], which can be explained by the end of the equipment run-in period, improved technologies for the construction and operation of pipelines, the use of accumulated knowledge of previous accidents. In modern conditions of operation of trunk pipelines, the main threat to the safe transportation of hydrocarbons is acts of unlawful interference in their work, which are the predominant cause of accidents and leaks of hydrocarbons, especially oil and oil products. This gives particular importance to the issues of detecting and predicting (preventing) such activity during the operation of facilities of the pipeline transport system.

The analysis of traditional and advanced safety measures used to prevent and mitigate the consequences of deliberate impacts on pipelines shows their imperfection and the presence of many challenges that still must be solved. The search continues for the most effective, economical, and reliable configuration and structure of the physical protection system, as well as the development of new means and methods for ensuring the physical security of trunk pipelines [1, 2].

2. Combined system of physical protection of trunk pipelines

Applied in practice, technical methods for monitoring the protected zones of pipelines, aimed at detecting external influences, are characterized by several disadvantages [1, 2]:

- long time of consequences (detection of interventions after impact on trunk pipelines);
- lack of "intelligence", which consists in the impossibility of classification and recognition of detected targets;
- unacceptable false alarm levels due to high environmental interference.

Based on this, it is possible to formulate the requirements for the developed system of physical protection of trunk pipelines:

1. early detection of offenders: before the start of a dangerous impact directly on the pipeline or its individual elements;
2. identification of offenders in automatic mode, as well as obtaining additional data on offenders for prompt adoption of adequate response measures: position, route, and speed of movement;
3. decision support (assistance to operators);
4. high sensitivity and, at the same time, low probability of false alarms, high noise immunity - minimal influence of external factors of natural and man-made nature;
5. high economic efficiency, ease of installation, operation and maintenance, the possibility of further modernization and improvement.

To compensate for the shortcomings of various means and methods of ensuring the safety of trunk pipelines and the implementation of the above requirements, it is necessary to use several devices based on different physical principles at the same time: the use of an intelligent multisensor physical protection system.

One of the promising directions for creating security systems for industrial facilities is the development of point seismic systems of passive location (detection), which are most convenient for blocking sections of extended perimeters of objects on rough terrain, that is, they are ideal for direct pipelines and other objects of the pipeline transport system. In general, seismic-analytical geolocation is a remote, non-contact, relatively reliable and economically feasible way to observe dynamic processes in the protected zone of trunk pipelines [3].

However, seismic systems are characterized by insufficient efficiency of existing methods of object recognition and obtaining additional information, as well as low noise immunity (the probability of false alarms is high). This is primarily due to the imperfection of approaches to seismic signal processing. Even though such algorithms have been developed in the world since the early 70s, their efficiency remains insufficiently high [3, 4]. This implies the need to improve the reliability of means of localization and classification of intruders in seismic security systems and to provide information support for the physical protection system using an additional subsystem for identifying intruders and making decisions.

The implementation of this subsystem is possible using modern active sensors with high signaling reliability and efficiency in eliminating false alarms and great potential in the recognition and classification of offenders (more information). However, the use of such devices is characterized by the high cost of implementation and operation (extended barrage zones), the need for powerful energy sources and new methods of transferring and processing large amounts of data [4]. Here we come to the need to deploy a hidden active information subsystem that has both the advantages of active and masked sensors, but with low power consumption and noise level. In particular, this can be information delivered by the own electromagnetic radiation of biological objects – the method based on fixing the intrinsic electromagnetic radiation of biological objects under the influence of an external electromagnetic field with specified parameters using bio-radioinformative technology. It is this method that is proposed for information support of the physical protection system of pipelines and for solving the problem of identifying violators and obtaining additional information about them.

IOP Conf. Series: Earth and Environmental Science 1021 (2022) 012031 doi:10.1088/1755-1315/1021/1/012031

3. Electromagnetic identification of violators of the security zone of trunk pipelines using bioradioinformative technology

At present, it has been theoretically substantiated and experimentally proven that any living organism, including humans, is a source of electromagnetic and acoustic fields, including chaotic ones, of various frequency ranges [5]. Under the influence of external electromagnetic fields, under certain conditions, a reaction of all studied levels of the structural and functional organization of living organisms occurs, which leads to the formation of a response of the biological system in the form of generation of physical fields that determine the information content of the external electromagnetic field [6, 7].

A necessary and sufficient condition for achieving the required information content of the external electromagnetic field is some correspondence between the intensity and modulation-time parameters of the field of the structural organization and the control parameters of the biological object. According to the authors [6-8], such fields can control the release of free energy of a biological object without introducing significant energy from the outside into it. In this case, the criterion of information interaction is the predominance of the energy of the response of the organism over the energy of the controlling external field.

When external fields are exposed to a biological system, the radio wave range is of particular interest. From this point of view, the works [9, 10] deserve attention, in which the experimentally established regularities and features of the biological action of low-intensity electromagnetic fields and the most probable mechanisms of their perception and generation are given.

Information about the change in the magnitude of the biological system's own electromagnetic fields, depending on the values of the external field parameters, on the one hand, and on the characteristics of the biological system itself, on the other, can form the basis for a new principle of identifying (recognizing) and selecting offenders for decision-making subsystems of the physical protection system of trunk pipelines.

The electromagnetic field formed by the body in the range of 0-30 MHz can form electromagnetic radiation outside the body, propagating into the surrounding space at a speed of about 300,000 km/s. This makes it possible to fix the impulse parameters of a specific object at a distance and analyze them using bioradioinformative technology - studying the regularities of changes in the parameters of its own chaotic electromagnetic radiation formed by the body of a biological object in the frequency range of 0-30 MHz under the influence of an external informative low-intensity electromagnetic field of the radio wave range with specified intensity and modulation-time parameters [7, 10, 11].

Accordingly, the essence of the system in which the bio-radioinformative technology is applied consists in the creation by ground means (source) of an external (in relation to the object) informative electromagnetic field, the informative parameters of which determine the generation of electromagnetic radiation by biological objects. Radio reception by ground means of this electromagnetic radiation in the form of broadband chaotic signals of the radio range and their decoding will make it possible to determine the location (coordinates) of the offender, his route and speed of movement, and classify it by type (animal or human).

Nevertheless, the problem of analyzing the intrinsic electromagnetic radiation of biological objects arising under the action of an external field is ambiguous. The data on the aspects of the influence of the electromagnetic field on living organisms are currently rather fragmentary, and sometimes contradictory and do not allow the process to be fully represented. The difficulty lies, first of all, in the need to take into account all the variety of factors that determine the amount of deposited energy, dielectric properties of biological tissues, mass and size characteristics and orientation in space of a biological object, intensity, modulation-time and polarization parameters of the external field.

Currently, there are two basic directions for further research in the analysis of natural electromagnetic radiation in the system of physical protection:

1. Evaluating the structure of chaotic oscillations by mathematically determining their fractal properties associated with the capacitive dimension, where the Hausdorff exponent is used as a criterion, which varies depending on the observation time [12].

2. Evaluating the dynamics of changes in the components of the field strength of the own electromagnetic radiation of biological objects, depending on the intensity and modulation-time parameters of the external electromagnetic field. There is a kind of "saturation" mode, which makes it possible to identify an object at a given value of the external field strength. In this case, the intrinsic electromagnetic radiation of objects also depends on their mass and size parameters at given values of the external field strength [7].

To test the operability of bioradioinformative technology in combined physical protection systems, we present a model of an intruder (biological system), based on the second direction of research (as the most elaborated).

If we imagine a biological object in the form of a conducting sphere of the radius a, surrounded by a dielectric shell b with a permittivity ε and a potential φ, then the dynamics of changes in the components of the field strength of the biological system's own electromagnetic radiation in the near-field zone (Figure 1) can be described by the equation:

$$E_r = -\frac{\partial \varphi}{\partial r} = \frac{E_{ext}}{4\pi\varepsilon_0} \cdot \frac{2p_\omega \cdot cos\theta}{r^3}$$

$$E_\theta = -\frac{\partial \varphi}{r\partial \theta} = \frac{E_{ext}}{4\pi\varepsilon_0} \cdot \frac{2p_\omega \cdot sin\theta}{r^3}$$

where θ is the angle characterizing the direction of scattering; p_ω is the dipole moment.

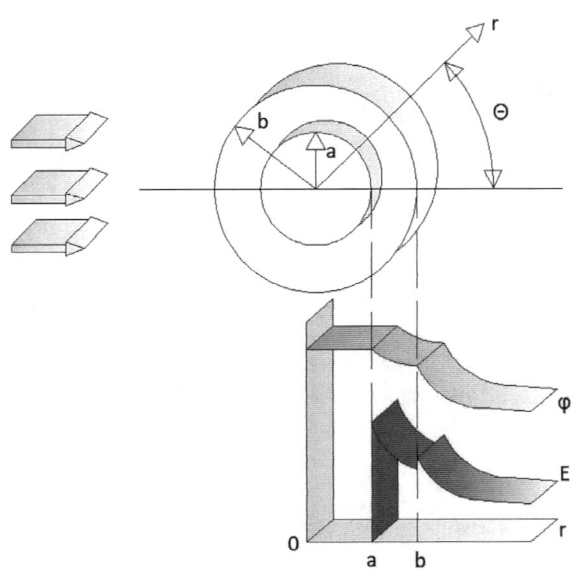

Figure 1. Dynamics of changes in the field strength of the own electromagnetic radiation of biological objects.

To test the adequacy of the model, the influence of weak structured fields on the electromagnetic radiation of biological objects was experimentally measured. The results of the experiment are shown graphically in Figure 2.

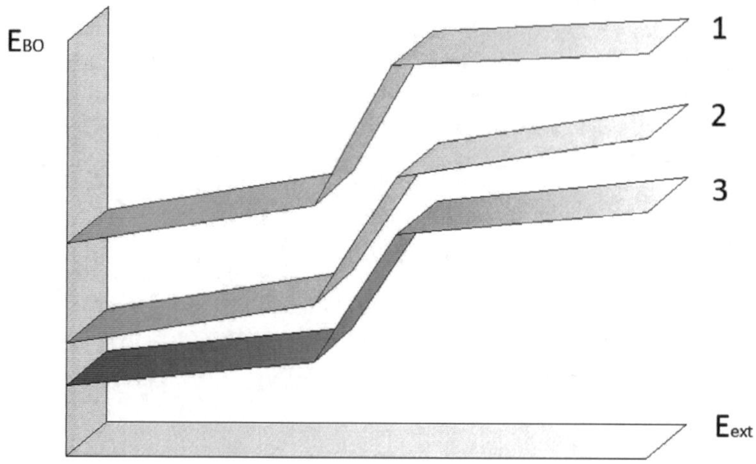

Figure 2. Results of experimental studies.

The characteristic form of the experimentally obtained dependence of the electromagnetic radiation inherent in a biological system on the strength of the external field proves that the response of biological objects to this external influence is essentially nonlinear and this confirms the reliability of the analytical conclusions.

The dependence has three main segments. The first (E_{ext} = 0 - 159 μV / m) and the third (E_{ext} = 810 - 8000 μV / m) segments correspond to the model of a passive environment, when the reaction of a biological object to an external effect is linear. This is mainly caused by multi-pole radiation, capacitance and inductance effects, and an increase in the effective length of the receiving antenna. The second segment (E_{ext} = 159 - 810 μV / m) shows the case when the biological object's own environment becomes active and contributes to the biological object's own radiation. In this case, the response is almost twice the external effect, which can only be explained by resonance phenomena.

A significant excess of the total signal "intruder + external field" over the level of the natural electromagnetic background provides the necessary sensitivity and selectivity for identifying a biological object.

The specific "saturation" mode (segment 3) presented on the graph confirms that identification of the biological system is possible at a given value of the external field strength.

4. Conclusions

1. A combined bio-radioinformative and seismic system of physical protection of objects of the pipeline transport system is proposed and substantiated. The detection system based on seismic sensors is designed to ensure the detection of signs of the presence of intruders and a high probability of their correct finding, including by increasing false alarms. The task of the subsystem interacting with it is to provide additional information for precise localization and classification of offenders in order to make correct decisions to prevent external influences on the protected object from the offenders.

2. Analytical studies of the relationship between the characteristics and the magnitude of the radiation inherent in the biological system and the intensity of the external structured electromagnetic field of low intensity of the radio wave range have been carried out.

The dynamics in space and time of its own low-intensity complex-structured electromagnetic fields when exposed to biological systems of external electromagnetic radiation of the radio wave range carries purely individual information about the structure and functioning of biological objects, which can be used to identify (recognize) and select informative signs of violators.

3. Two most promising areas of analysis of the intrinsic electromagnetic radiation of biological systems arising under the action of an external field are considered:

- based on the assessment of the dynamics of changes in the field strength components of the own electromagnetic radiation of biological systems in the near zone, depending on the intensity and modulation-time parameters of the external electromagnetic field.
- construction of "strange attractors" based on chaos rhythms (the function of changing the Hausdorff exponent), changing under the influence of external electromagnetic radiation on a person.

Based on the first approach, a preliminary mathematical model of the intruder is presented.

4. The adequacy of the developed model and the presence of a relationship between the electromagnetic radiation of a biological object and the intensity and parameters of the temporal modulation of the external field were fully confirmed by experimental studies.

5. Both directions require further research in the direction of their adaptation and effective application in physical protection systems of linear objects. In the future, it is advisable to implement:

- development of a general model of functioning and structure of a combined bio-radioinformative and seismic system for physical protection of trunk pipelines;
- highlighting new informative features of biological objects, compiling an alphabet of classes of objects (phenomena);
- development of a general method and algorithm for analyzing the characteristic properties of processes and phenomena in the protected area, including for difficult natural and climatic conditions (arctic, sea, etc.);
- implementation of special software and algorithms for automated data processing.

References

[1] Aleshkov A A and Tsvetkov G A 2021 Modern problems of ensuring the physical safety of trunk pipelines *Trends in the development of science and education* **70-2** 109-118

[2] Talarico L, Sörensen K, Reniers G and Springael J 2015 Pipeline Security *Securing Transportation Systems* 281-311

[3] Vorobev V F 2009 Modern technologies and prospects for the development of technical means of protection and supervision *Bulletin of the penitentiary system* **11** 5-9

[4] Bondarchuk A S 2011 *The system of protection and defense of important state facilities and the assessment of the effectiveness of its functioning: monograph* (Perm: PMI MI)

[5] Yashin S A 2013 System for registration of own low-intensity electromagnetic fields of the human body *Bulletin of new medical technologies* **3** 157-163

[6] Marchuk A V and Tsvetkov G A 2010 Human and Habitat: Another Latent Security *Threat Problems of ensuring safety in industry, construction and transport* (Perm: PNTU) 84-91

[7] Marchuk A V 1999 Fundamentals of the theory of electromagnetic identification of biological objects *Russian Journal of Biomechanics* **3 (4)** 94-105

[8] Conclusion on the results of experimental measurements No. 1399 (Main Directorate of State Supervision over Communications in the Russian Federation under the State Committee of the Russian Federation for Communications and Informatization: Directorate of State Supervision over Communications in the Perm Region)

[9] Dodd R K, Eilbeck J C, Gibbon J D and Morris H C 1982 *Solitons and Nonlinear Wave Equation* (San Diego: Inc. Harcourt Brace Jovanovich Publishers)

[10] Chubiy A D, Zhukov V O and Tsvetkov G A 2002 Bioradioinformative method for evaluating the effectiveness of protection against technogenic electromagnetic radiation *Problems and Prospects of Rational Arms Operation Management* (Perm: PMI RT)

[11] Swanson G D 1977 Biological Signal Conditioning for System Identification *Proc. IEEE* **65 (5)** 735-740

[12] Chubiy A D and Zhukov V O 2011 *Determination of the possibility of remote personal identification of a person by his own chaotic electromagnetic radiation* (Perm: SNIB Elbrus)

Prospects for waste-free utilization of ore dressing tailings

V I Golik [1,2], R R Turluev[3], A T Alpeisov[4]

[1]North Caucasian Mining and Metallurgical Institute, 44 Nikolaeva St., Vladikavkaz, 362021, Russian Federation
[2]Moscow Polytechnic University, 38b Semenovskaya St., Moscow, 107023, Russian Federation
[3]Chechen State University named after A. H. Kadyrov, 17a, Dudaev Boulevard, Groznyy, 364015, Russia
[4]Satbayev University, 22 Satbaev str., Almaty, Kazakhstan

E-mail: v.i.golik@mail.ru

Abstract. The information about the state of the use of natural resources in Russia is given and the need to search for new sources of raw materials for industry is justified. There is a tendency to reduce the production capacity and the rate of extraction of metal ores, which does not allow the use of the infrastructure of mining enterprises and creates the problem of processing substandard ore deposits and disposal of accumulated waste. The concept of deep processing of substandard metal-containing raw materials and a strategy for the survival of mining enterprises, including the use of metal leaching technologies, is formulated. The advantages of technologies with the leaching of metals from mineral raw materials at the stages of its processing are characterized. The advantages of leaching in disintegrators are shown by comparison with the indicators of the basic variant. It is noted that mechanochemical activation improves the quality of concretes made on the basis of recycled tailings. Managing the properties of recyclable materials allows you to implement the principles of resource conservation with an economic effect. It is concluded that the recommended technology is the real direction of technological modification of production.

1. Introduction

The extracted mineral resources are only partially used, the waste of their processing is accumulated on the earth's surface, polluting the natural environment with the products of disintegration and natural leaching [1–2].

Rising prices for non-ferrous and rare metals stimulate the use of tailings mining and processing of ores [24, 25]. So, in Bohemia, useful components have been extracted from old dumps since the 80s of the last century.

Waste, except for self-igniting, is used to fill voids formed by mining operations.

If at the initial stage of the development of technologies with the laying of hardening mixtures, the needs of mines were provided by reserves of standard construction sands, and cement or granulated blast furnace slags were used as a binder, now the needs for inert materials are met through the disposal of tailings of processing plants.

The tailings of metal ore processing can be used in hardening mixtures after the extraction of metals from them, because natural leaching of them can strengthen artificial arrays.

The problem of strengthening the base of the metallurgical industry on the basis of efficient processing of technogenic raw materials is actualized due to depletion of reserves of deposits, complication of development conditions and increased negative impact of production waste on the environment.

At the same time, the enterprises have accumulated technogenic reserves of processing tailings, the content of which is often close to the content of metals in mined ores. The solution of the problem of extraction of metals from them is hindered by the insufficient development of scientific and technical foundations of metal extraction technologies.

Demographic changes in the structure of the human community force us to look for new sources of resources [3–4].

In the post-reform years, the decline in metal production does not allow the use of the infrastructure of mining enterprises, many of which are city-forming, and aggravates the problem of mining stocks of substandard ores and disposal of accumulated mineral waste, the cost of useful components in which often exceeds the value of the ingredients already extracted from the mined ores.

The reason for the decrease in the rate of extraction of metals, most often, is not the depletion of ore reserves, the use of mining technologies focused on rich ores.

Traditional technologies do not provide full disclosure of minerals due to the use, mainly, of mechanical energy alone, therefore, modernization of the processes of processing metal-containing raw materials should be carried out by involving operations of hydrometallurgical and chemical processing.

Technologies with leaching of metals from ores meet the requirements. The technology of underground leaching of minerals makes it possible to put into operation reserves of substandard minerals, the development of which by traditional methods is not feasible, which radically expands the raw material base of the industry [13-15].

The super-task of technology development includes the development of its scientific foundations, regulating the solution of problems of ore preparation, chemical and technological processes of transferring metals to a mobile state, hydraulics and hydrogeology, extraction of metals from solution, ensuring the safety of associated work.

The use of mining waste is being developed with the development of the technology of mechanochemical activation of mineral leaching processes in high-speed disintegrator mills [5–6].

2. Materials and methods

To substantiate the relevance of the research topic, according to the literature data, the state of the use of waste from the processing of extracted mineral metal raw materials, the reasons for the decrease in the rate of metal extraction and the prospects for improving the situation are analyzed.

Based on the analysis by the method of engineering forecasting, the concept of deep processing of substandard metal-containing raw materials is being developed with the involvement of technologies for the extraction of metals from rock-exposed ores.

A super-task of technology development is formulated, including the development of scientific foundations that regulate the solution of the problem.

Experimental proof of the advantages of mechanochemical activation of mineral leaching processes in high-speed disintegrator mills is being carried out.

The effectiveness of innovative technologies is proved by a complex method, including analysis, experiments and forecasting [23].

The compliance of the compositions of the hardening mixture based on activated processing tailings with the requirements is determined by modeling using statistical data. The working hypothesis of the behavior of artificial arrays is formulated on the basis of experimentally established patterns and is the basis of the methodology for calculating the normative strength of hardening masonry for specific conditions.

3. Results and discussion

The scheme of the plant for processing the tailings of enrichment is shown in Figure1.

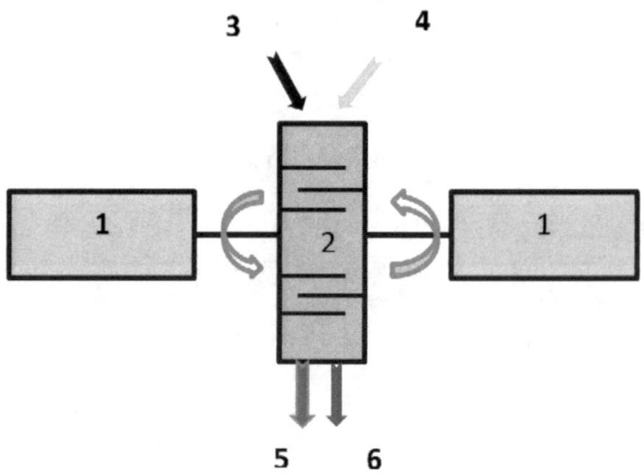

Figure 1. Installation diagram: 1 – electric motors; 2 – working body; 3 – reagent; 4 – enrichment tails; 5 – metal-containing solution; 6 – leached tails.

Mechanochemical activation is characterized by higher rates due to the fact that intermolecular bonds of minerals are weakened in it at a high processing speed, and the solution is pressed into deformation cracks under high pressure.

Extraction of metals from substandard raw materials is carried out in all areas of mining production (Figure 2).

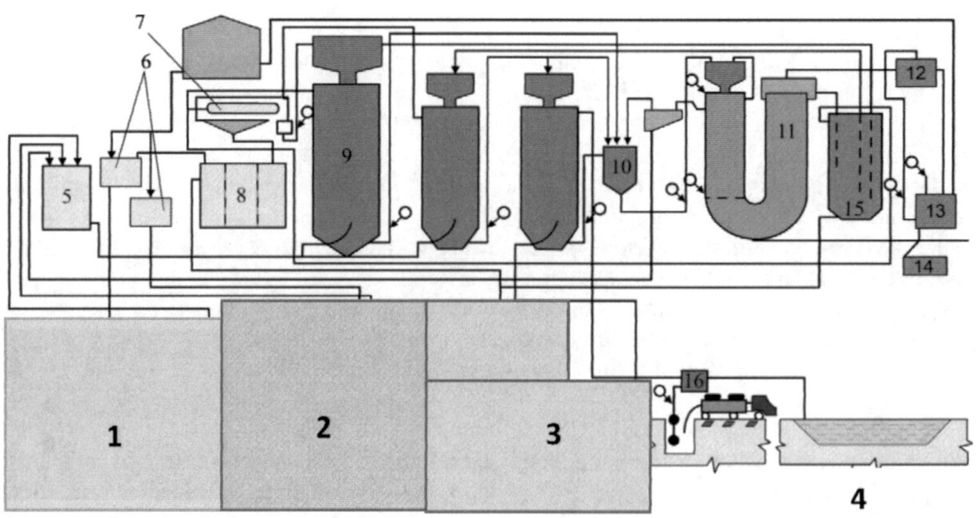

Figure 2. Leaching of metals from substandard raw materials: 1 – stack; 2 – underground leaching unit; 3 – dump; 4 – pond; 5, 6, 7, 8 – preparation group; 9, 10, 11 – sorption-desorption group; 12–16 – auxiliary group.

Mechanochemical activation of the leaching process improves the quality of concrete to a greater extent than in other activators (Table 1).

Table 1. The strength of the mixture based on tails after activation

Components of the mixture, kg/m³				Strength and coefficient of variation Age, p.		
Cement	Binding tails	Inert tails	Water	14	28	90
		Activation in a ball mill (fineness 40%)				
40	400	1200	350	0.40	0.44	0.6
80	360	1200	350	0.49	0.59	0.71
120	320	1200	350	0.70	0.89	1.1
180	260	1200	350	0.97	1.18	1.27
		Activation in the disintegrator (subtlety 40%)				
40	400	1200	350	0.51	0.62	0.78
80	370	1200	350	0.67	0.82	0.99
120	320	1200	350	0.84	0.92	1.12
180	260	1200	350	122	1.34	1.40

It follows from the table that under the same conditions, the strength of products based on activated tails increases.

The phenomenon of component activation is used in the combination of activators in the processes of preparation, transportation, and laying of hardening mixtures (Figure 3).

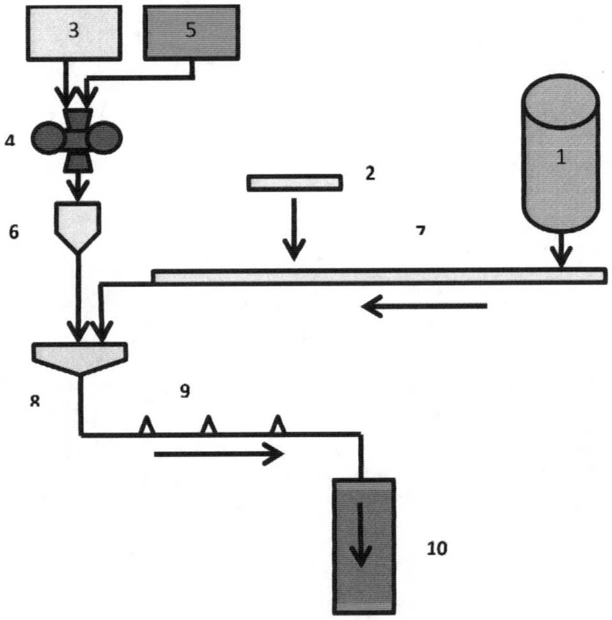

Figure 3. Activation of components of hardening mixtures: 1 – cement; 2 – vibrating screen of aggregates; 3 – slag; 4 – disintegrator; 5 – activated water; 6 – vibrating mill; 7 – conveyor; 8 – laying pipeline; 9 – mixer; 9 – vibrators; 10 – block chamber.

Managing the properties of recyclable materials makes it possible to implement the principles of resource saving with obtaining an economic effect and reducing dependence on suppliers of raw materials for the manufacture of concrete [16-19].

Processing of unclaimed and dangerous substances during storage in the disintegrator produces marketable products, including sand for construction, stone casting and glass production, sludge fraction, etc [20-22].

An important advantage of the disintegrator technology is the possibility of supplying raw materials for the production of hardening mixtures in development systems with the laying of voids. The high cost of binders and inert components reduces the scope of their application. Therefore, the development of technologies for managing the state of arrays with maximum use of the properties of artificial masses is an urgent scientific task.

The purpose of research in this direction is to optimize the rationing of the strength of masonry massifs with an increase in the completeness and quality of ore extraction, including the use of the effect of differentiating the normative strength of hardening mixtures while maintaining the load-bearing capacity of the laying array.

4. Conclusion

The results of the study allow us to summarize:

- tailings processing of technologically opened ores are cheap and affordable raw materials for the production of marketable products with technological diversification of production;
- mechanochemical activation of the leaching process in the disintegrator adequately increases the extraction of metals from the tailings of processing;
- the equipment for the implementation of the leaching process can be manufactured by the enterprises' own forces.

The results of the study are of interest to metal consumer industries, since the quality of metal products often determines the ecological and economic value of goods [7–12].

References

[1] Golik V I, Razorenov Yu I, Puzin V S and Stas G V 2021 Differentiated assessment of stability of exposed rock surfaces in sublevel stoping with backfill *Physico-Technical Problems of Mineral Development* **5** 85–93

[2] Golik V I, Burdzieva O G and Ganin M P 2019 An ecological concept of natural leaching of metal ores *Proceedings of Tula State University. Earth Sciences* **2** 16–26

[3] Klyuev R V, Bosikov I I, Gavrina O A and Lyashenko V I 2021 Assessment of operational reliability of power supply to developing ore mining areas at a high-altitude mine *Mining Science and Technology (Russia)* **6 (3)** 211–220

[4] Valiev N G, Propp V D and Vandyshev A M 2020 The 100th anniversary of the department of mining engineering of UrSMU *News of Higher Educational Institutions. Mining Magazine* **8** 130–143

[5] Shchurov N I, Myatezh S V, Malozyomov B V, Shtang A A, Martyushev N V, Klyuev R V and Dedov S I 2021 Determination of inactive powers in a single-phase ac network *Energies* **14 (16)** 4814

[6] Golik V I 2013 Conceptual approaches to creating low- and non-waste mining based on combination of physical-technical and physical-chemical geotechnologies *Mining Journal* **5** 93–97

[7] Panov V S, Eremeeva Zh V, Sharipzyanova G H, Skorikov R A, Mikheev G V and Ageev E V 2014 The effect of the nature of nanoscale particles and method of mixing on the tribological properties of powder steel 70P *Proceedings of the Southwest State University* **6 (57)** 8–14

[8] Shchurov N I, Dedov S I, Malozyomov B V, Shtang A A, Martyushev N V, Klyuev R V and Andriashin S N 2021 Degradation of lithium-ion batteries in an electric transport complex *Energies* **14 (23)** 8072

[9] Doifode S K and Matani A G 2015 Effective industrial waste utilization technologies towards cleaner environment *International Journal of Chemical and Physical Sciences* **4** 536–540

[10] Dominguez M S K, Baptista A L F and Diogo M T 2017 Engineering complex systems applied to risk management in the mining industry *International Journal of Mining. Science and Technology* **27** 611–616

[11] Tayebi-Chorami M, Edraki M, Corder G and Golev A 2019 Re-thinking mining waste through an integrative approach led by circular economy aspirations *Minerals* **9 (5)** 286

[12] Ben-Avuah E, Richter O, Elkington T and Pourrahimian Yu 2016 Strategic mining options optimization: open pit mining, underground mining or both *International Journal of Mining Science and Technology* **26 (6)** 1065–1071

[13] IVANOV V V, SIDORENKO S A, SIDORENKO A A, KORZHAVYKH P V 2014 The prospects of introduction of surface miners in technology of development of deposits of contiguous seams of the limestone. *Advances in Environmental Biology* **8(13)** 298-300

[14] Filatova I, Nikolaichuk L, Zakaev D, Ilin I 2021 Public-private partnership as a tool of sustainable development in the oil-refining sector: Russian case *Sustainability (Switzerland)* **13(9)** 5153 doi: 10.3390/su13095153

[15] Shakhnazarov K Y, Pryakhin E I, Mikhailov A V 2021 630°C ± 30°C - Nodal (Critical) Temperature of Iron and Carbon Steel. *Materials Science Forum* **1040** 191–199 Doi: https://doi.org/10.4028/www.scientific.net/MSF.1040.

[16] Konyuhov V Yu, Gladkih A M, Galyautdinov I I, Severina Y D 2019 Economic aspects of green technologies. *IOP Conference Series: Earth and Environmental Science* **350** 012036

[17] Balanovsky A E, Shtayger M G, Kondrat'Ev V V, Nebogin S A, Karlina A I 2018 Complex metallographic researches of 110G13L steel after heat treatment. *IOP Conference Series: Materials Science and Engineering* **411(1)** 012014

[18] Kondrat'ev V V, Ershov V A, Shakhrai S G, Ivanov N A, Karlina A I 2016 Formation and Utilization of Nanostructures Based on Carbon During Primary Aluminum Production. *Metallurgist* **60(7-8)** 877–882

[19] Konyuhov V Y, Gladkih A M, Semenov V V 2019 Measures to the improvement of efficiency of a repair enterprise. *Journal of Physics: Conference Series* **1353** 012046

[20] Nemarov A A, Lebedev N V, Kondrat'ev V V, Kornyakov M V, Karlina A I 2016 Theoretical and experimental research of parameters of pneumatic aerators and elementary cycle flotation. *International Journal of Applied Engineering Research* **11(20)** 10222–10226

[21] Gladkih A M, Konyuhov V Yu, Galyautdinov I I, Shchadova E I 2019 Green building as a tool of energy saving. *IOP Conference Series: Earth and Environmental Science* **350** 012032

[22] Konyuhov V Yu, Gladkih A M, Galyautdinov I I, Shchadova E I 2020 Calculations of efficiency in implementing progressive mold forming methods. *IOP Conference Series: Materials Science and Engineering* **760** 012027

[23] Gabov V V, Zadkov D A, Babyr N V, Fangwei X 2021 Nonimpact rock pressure regulation with energy recovery into the hydraulic system of the longwall powered support. *Eurasian Mining* **2021(2)** 55-59. 10.17580/em.2021.02.12

[24] Sevostyanov V S, Uralsky V I, Boichuk I P, Ermilov R A, Lukyanova E A 2019 Analysis of separator operating conditions in a technological module with vibration-centrifugal closed-cycle grinding unit for obtaining high-dispersity materials. *IOP Conference Series: Materials Science and Engineering* **552 (1)** 012040

[25] Perelygin D N, Boichuk I P, Gorlov A S 2019 Rational design of low-tonnage technological complexes for fine and ultrafine grinding of materials. *IOP Conference Series: Materials Science and Engineering* **560 (1)** 012118

Methods for forecast of oil-bearing capacity by the example of the JS2/1 layer within the Surgut arch

A V Turabaeva*, V A Belkina

Tyumen Industrial University, 38 Volodarskogo st., Tyumen, 648000, Russia

E-mail: sashenka_surgut@mail.ru

Abstract. In order to increase the efficiency of geological exploration (GE), adequate geological models (AGM) must be created, enabling the identification of the most prospective oil-bearing areas. It should be pointed out that an extensive base of methods for building AGM exists nowadays. At the same time, there are practically no publications on substantiation of formalised criteria for ranking zones for GE with various geological risks. This paper proposes a model that provides ranking of the area according to the quality of reservoir rocks, based on the model's strict decision rule. Based on the joint analysis of the quality maps of reservoir rocks and paleorelief, an oil potential map has been built. Efficiency of the method is shown by the example of the JS2/1 layer of continental genesis, characterised by the complex geological structure.

1. Introduction

Development of methods for the identification of channel-type areas is required to increase the efficiency of GE during prospecting for hydrocarbons of continental genesis [1]. These zones are characterised by increased net pay (h_{nt}) and reasonably high reservoir properties (RP).

To identify the zones with high oil-bearing prospects in continental deposits, a technique consisting of the following goals was proposed:

1. to conduct a regional correlation of the studied deposits;
2. to build maps of: net oil pay (h_{nop}), porosity ratio (R_{por}), permeability ratio (R_{per}), net-to-gross ratio (R_{ntg}), shaliness ratio (R_{sh}) and initial oil saturation ratio (R_{ios});
3. to select channel facies with sufficiently extensive axial regions of zones and increased h_{nt};
4. to build a paleostructure map at the beginning of the formation of the studied sediments;
5. to create a two-dimensional RP model (maps R_{por} and R_{per}), reflecting the variability of RP on the study area;
6. to detect significant regressional relationship of axial regions of paleochannel areas with increased values of h_{nop} with paleostructure map, R_{por} and R_{per};
7. to create a two-dimensional reserves quality model based on maps, listed in item 2;
8. to verify the quality of the reserves model based on external criteria.

2. Data

The proposed method has been tested on the deposits of the JS2/1 layer for two reasons: first, it is assessed as regionally oil-bearing, and second, it is one of the most promising objects for replenishing the resource base of "Surgutneftegas" [2 - 5]. To solve the problem of substantiating the prospective oil-bearing areas of the JS2/1 layer of the Surgut arch (SA), correlation schemes for the reference network of profiles for 850 exploration and production wells were created. Structural constructions, analysis of time slices, analysis of core material for 15 wells within the Yuzhno-Konitlorsky, Severo-Yuryevsky, Larkinsky subsurface sites were performed. Maps for 25 geological and complex parameters of the JS2/1 layer according to 10764 prospecting, exploration and production wells were built. In addition to well data, the results of seismic prospecting of 85 seismic crews were used [6].

3. Reserves quality model for the JS2/1 layer within central and eastern regions of the Surgut arch

Since the JS2/1 layer only presents commercial interest within central and eastern regions of the SA, the AGM was built for that part of the SA. Cutoff values were identified according to the characteristics of reservoir rocks, which made it possible to distinguish three classes of reservoir rocks.

Critical values of "high" quality reservoir rocks are given in Table 1. It should be pointed out that this area is characterised by the highest values of geologic parameters, excluding R_{sh}, which reaches its highest values in "high" quality areas. The core in the rocks of this zone is represented by fine-, medium-grained, massive sandstone (Figure 1). The largest areas of absence of reservoir rocks of "high" quality are noted in the northern, western and south-western regions and fragmentarily in the rest of the studied territory. The well data coverage of the analysed areas is the highest in comparison with other areas.

Table 1. Geological features of "high" quality reservoir rocks

Reservoir rocks quality	R_{por}, unit fractions	R_{per}, $*10^{-3}\mu m^2$	R_{ios}, unit fractions	R_{ntg}, unit fractions	R_{sh}, unit fractions	Occurrence of areas with increased h_{nt}
High	> 0.13	> 5	> 0.3	> 0.3	< 0.2	+

Areas of "medium" quality reservoir rocks include sections, in which values of most parameters are close to those of "high" quality areas, but one parameter being lower than cutoff value is acceptable. For reservoir rocks of "medium" quality within JS2/1 layer these parameters are, primarily, ratios of porosity (less than 0.13 unit fractions) or permeability (less than $5*10^{-3}\mu m^2$) (Table 2). Lithologically the reservoir rocks are composed of fine-, medium-grained sandstones, interbedded with clay rocks and coals (Figure 1). These areas are marked by lower number of axial regions of paleochannels with increased net oil pay. "Medium" quality reservoir rocks are evenly distributed throughout the study area. The well data coverage of the analysed area is compared to that of areas with "low" quality reservoir rocks.

Table 2. Geological features of "medium" quality reservoir rocks

Reservoir rocks quality	R_{por}, unit fractions	R_{per}, $*10^{-3}\mu m^2$	R_{ios}, unit fractions	R_{ntg}, unit fractions	R_{sh}, unit fractions	Occurrence of areas with increased h_{nt}
Medium	>, < 0.13	> , < 5	> 0.3	> 0.3	< 0.2	+/-

Areas with reservoir rocks of "low" quality are characterised by higher values of R_{sh} (more than 0.2 unit fractions), lower values of R_{por} (less than 0.13 unit fractions), R_{ntg} (less than 0.3 unit fractions), R_{ios} (less than 0.3 unit fractions), R_{per} (less than $5*10^{-3}\mu m^2$) and the absence of axial regions with increased values of h_{nop}. The core contains the maximum amount of highly argillaceous sandstones

IOP Conf. Series: Earth and Environmental Science 1021 (2022) 012033 doi:10.1088/1755-1315/1021/1/012033

(Figure 1). The largest areas of these reservoir rocks are associated with the south-western and northern regions of the studied territory, fragmentarily they are present in the rest of the studied territory. The well data coverage is the lowest, compared to other areas.

Table 3. Geological features of "low" quality reservoir rocks

Reservoir rocks quality	R_{por}, unit fractions	Rper, $*10^{-3}\mu m^2$	R_{ios}, unit fractions	R_{ntg}, unit fractions	R_{sh}, unit fractions	Occurrence of areas with increased h_{nt}
Low	< 0.13	< 5	< 0.3	< 0.3	> 0.2	--

The maps of geological parameters and core data allowed us to systemize reservoir rocks of JS2/1 layer within central and eastern sections by quality (Figure 1).

Oil prospects map was build based on joint analysis of reservoir rocks types map and a map of paleorelief at the beginning of the formation of Tyumen suite deposits (Figure 2). It allows ranking of individual areas by the efficiency of GE.

The highest efficiency of GE is estimated in reservoir rocks of high and medium quality (class 1 zone). They are characterised by increased density of axial regions with higher h_{nt}. Within Novobystrinsky, Yuzhno-Konitlorsky, and at the junction of Severo-Yuryevsky and Russkinsky subsurface sites (SS) class 1 zones are recommended for exploration by prospective drilling. Within Bystrinsky, Vostochno-Surgutsky, Rodnikovy SS, it is most effective to conduct additional exploration by drilling sidetracks and deepening a number of production wells. Within the "Class 1" zone, it is advisable to increase the area of the Novobystrinsky SS in the southern part at the expense of the unallocated subsurface fund.

"Class 2" zones are associated with reservoir rocks of medium quality in which the estimated axial regions of zones with increased net pay may be encountered, which extend from adjacent zones of higher reservoir quality. Within Larkinsky, Bystrinsky, Vostochno-Elovy, Tundrinsky subsurface sites these zones can be explored by prospective drilling. Within lyantorsky, Russkinsky, Savuysky, at the junction of Severo-Yuryevsky and Tonchinsky, Vachimsky, at the junction of Tonchinsky and Yaunlorsky, Fedorovsky, Vostochno-Surgutsky SS these zones may be explored by drilling sidetracks and deepening of production wells.

"3rd class" zones are related to reservoir rocks of low and medium quality. Within Larkinsky, Rodnikovy SS it is advisable to explore "class 3" zones by prospective drilling. The area of Zapadno-Solkinsky SS is recommended for exploration by drilling sidetracks and deepening of production wells.

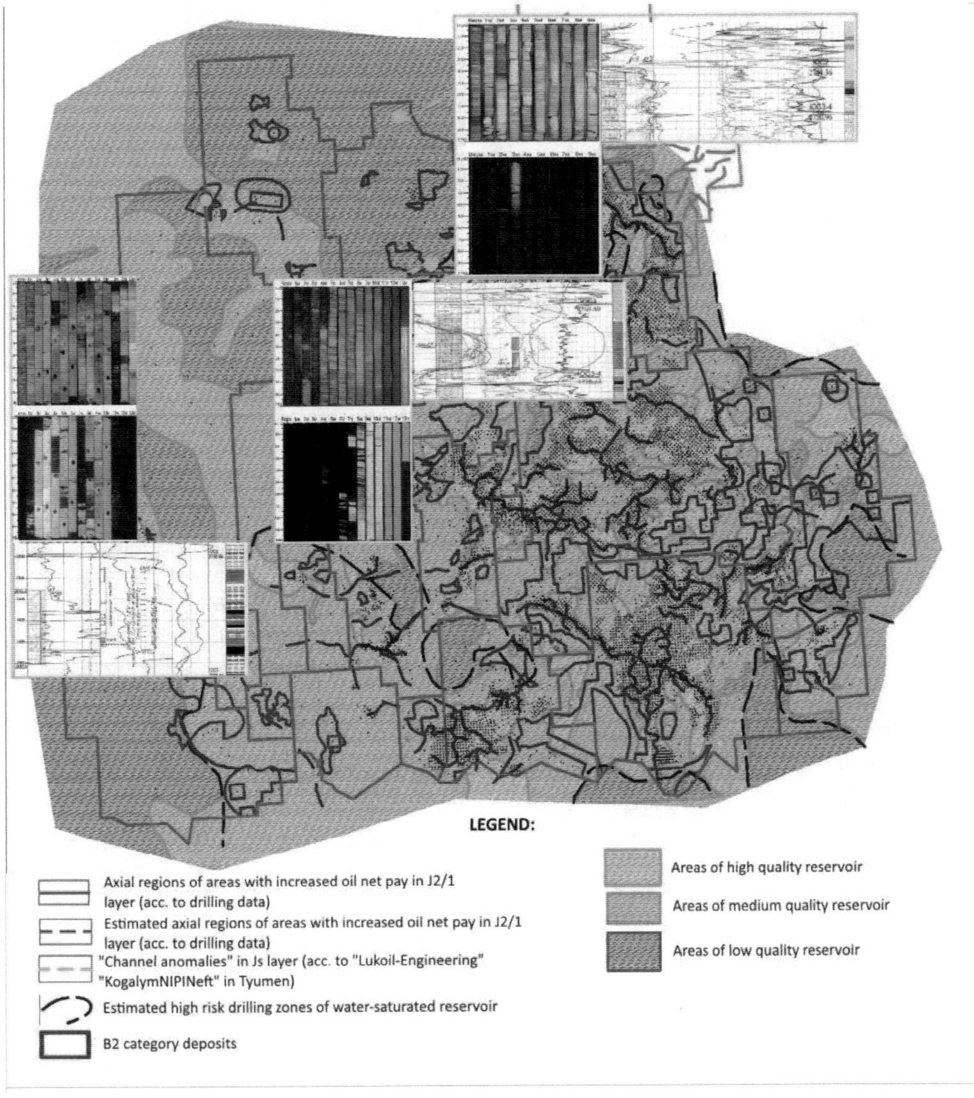

Figure 1. Map of reservoir rocks types based on GIS data and coring of the JS2/1 layer of the Surgut arch.

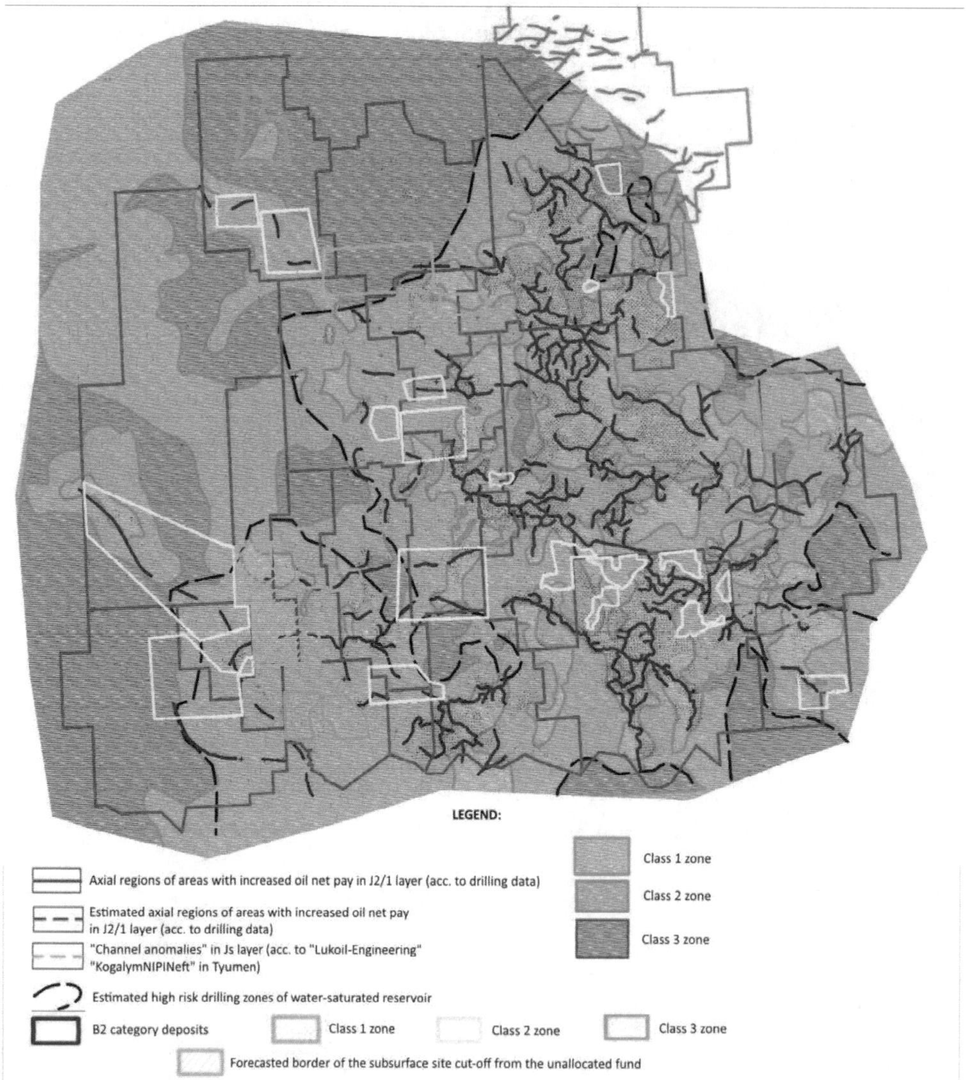

Figure2. Oil potential map with areas of substantiated geological exploration.

4. Conclusion

As a result of the performed research:

1. regional correlation was conducted, updating the stratigraphical borders of JS2/1 layer;

2. maps of reservoir properties and a number of other geological parameters were built, based on which the axial regions of zones with increased net pay were determined;

3. two-dimensional model of paleorelief at the beginning of the formation of tyumen suite deposits was created;

4. comparative analysis of RP and paleorelief maps highlighted a dependence of zones with increased net pay to paleorelief;

5. oil potential map was built, and, according to it, the territory was ranked by efficiency of GE; 6. Based on the model of the JS2/1 layer, specific most effective methods for additional study of select subsurface sites within central and eastern parts of the Surgut arch.

IOP Conf. Series: Earth and Environmental Science 1021 (2022) 012033 doi:10.1088/1755-1315/1021/1/012033

The success ratio of the described method for forecast of oil-bearing capacity is rated based on data, not used in the construction of the model. External verification of the presented model was carried out on a sample of a fairly large volume - 1054 wells. Empirical data confirmed the estimated values in 754 wells. The success rate of 70% is rather high, especially if we take into account the complex geological structure of the continental deposits of the JS2/1 layer.

References

[1] Turubaeva A V, Belkina V A 2021 Geological structure of J2/1 layer of the Surgut arch in order to clarify the prospects of oil potential and justify geological exploration *Oil and Gas Studies.* **1**, 30-39 (in Russian).

[2] Sudakova V V 2005 *Forecasting oil and gas traps based on a comprehensive analysis of seismic survey and drilling materials on the territory of the Surgut arch of hydrocarbons* Cand. Sci. Dissertation, IUT. Tyumen. (in Russian).

[3] Kos I M 2004 Regularities of the formation and placement of lithological traps in the Middle-Upper Jurassic sediments of the Surgut arch in connection with the forecast of oil and gas content Cand. Sci. Dissertation, VNIGRI, Saint-Petersburg

[4] Baturin A Ju, Kultyshov A K, Popov D A 2007 Geological modeling of facialy variable formations by the example of the JS2 layer of the Surgut arch *Oil Industry,* **8,** 105-109. (in Russian).

[5] Svastyanov A A, Korovin K V, Zotova O P, Zubarev D I 2017 Geological Characteristics and Assessment of the Potential Production of the Tyumen Suite Deposits *Bulletin of Perm university* **16** **(1)** 61-67 (in Russian).

[6] Simm R, Bacon M 2014 Seismic amplitude: An Interpreter`s Handbook. (Cambridge Univ. Press.)

Karachay-Cherkessia – international landscape and geoecological landfill

P A Kipkeeva[1], Yu Ya Potapenko[2], N V. Shubaeva[2], A A Abaykhanova[1], F M Borlakova

[1]Karachay-Cherkess State University named after U.D. Aliev, 29 Lenin Str., Karachaevsk, 369202, Russia
2 ICC "Safety of Hydraulic Structures", 178 Baklanovsky Ave., Novocherkassk, Rostov Region, 346400, Russia

E-mail: kipkeeva62@mail.ru

Abstract. Landscape and geoecological study served a response to global climate change and the threat of a global environmental crisis. The scientists of the Teberdinsky State Biosphere Reserve, Stavropol State University and the Department of Natural Geography of Karachay-Cherkess State University made the greatest contribution to the development of this sphere. Modern study on geology, geography, biology, toponymy, geomorphology, geoecology was conducted, an improved methodology for assessing the degree of anthropogenic impact on the natural environment was proposed, GIS technologies were mastered and introduced into the education and research. In 2019, the 1st International Conference was held in Karachaevsk, the main principle of which was the concept of balanced development of mountain territories.

1. Introduction

Russia preserved the largest array of natural ecosystems on the planet, which may be considered a reserve for the geobiosphere stability. In the Caucasus, such ecosystems include, first of all, the Northwest Caucasus, which, in turn, unites a significant part of the territory of the Karachay-Cherkess Republic.

The establishment of the Karachay-Cherkess Republic in 1990 contributed to the transformation of Karachaevsk into a historical, ethnic and social information center. The Department of Department of Natural Geography in Karachaevsk, opened in 1992 became the guardian of historical ties and the connecting link.

The ***purpose of the study*** was to substantiate the suitability of the territory as a training ground for landscape and geoecological research.

The ***object of the study*** at the regional level is the modern earth surface of the republic, which includes a relief from foothill plains to the nival belt inclusive.

The ***subject of the study*** is the structure of the landscape shell and the degree of its anthropogenic change, i.e. modern landscapes that arose on the basis of natural landscapes in the process of culture genesis. The main morphostructures in the republic are inherited and repeat the spatial image of the lithogenic base. Morphostructures (MF) spread from south to north: the Main Ridge, the Front Ridge, the North Jurassic Depression, monoclinal ridges – Rocky and Pasture ridges.

IOP Conf. Series: Earth and Environmental Science 1021 (2022) 012034 doi:10.1088/1755-1315/1021/1/012034

2. Materials and methods

The field of landscape and geoecological problems was explored through interdisciplinary geographical methods, including geoinformation mapping, decryption of aero- and space imagery, landscape, geomorphological, geoecological research thus ensuring territorial and regional nature management.

3. Results

Landscape and geoecological study served a response to global climate change and the threat of a global environmental crisis. The scientists of the Teberdinsky State Biosphere Reserve, Stavropol State University and the Department of Natural Geography of Karachay-Cherkess State University made the greatest contribution to the development of this sphere.

In 1974, the international Franco-Soviet symposium "Alps-Caucasus" was held in the Karachay-Cherkess Autonomous Region (now the Republic) and in Georgia. The main objective of the symposium was to exchange scientific experience and achievements of French and Soviet geographers in the field of integrated study, rational development and protection of mountain areas. The Soviet delegation was represented by large scientific geographers, employees of academic institutes of geography, headed by academician I.P. Gerasimov. The French group of scientists was headed by the director of Grenoble Alpes University prof. P. Weir. Two symposiums were held – one in the French Alps, the second in the Western Caucasus, where several routes were carried out. In the first route Mineralnye Vody – Cherkessk – Karachaevsk – Dombay, two topics were discussed: *Steppe ecosystems of the Precaucasian foothill plain and their transformation by humans* and *Vertical zone of ecosystems of the northern slope of the Western Caucasus*. In the second route, the climb to the Alibek glacier was made and the topic *Medium and high mountain landscape belts and their development* was discussed. In total, 10 routes were completed in Karachay-Cherkessia and then in Kabardino-Balkaria to study all high-altitude belts, nival and avalanche processes, alpine and rock vegetation. When climbing the Elbrus, French colleagues identified the analogues of the alpine moraines of the Fernau and Egessen stages.

In 1992, the Department of Natural Geography (DNG) was opened in Karachaevsk. Since that time, the republic began to train specialists in geography and biology, while the university professors were actively involved in scientific research.

In 1996, a republican branch of the Russian Geographical Society (RGS) was opened at the department, which intensified education and research.

The society included large and active scientists of Karachay-Cherkessia, as well as the representatives of foreign countries – Ukraine, Australia, Germany. The RGS member was a famous writer, journalist Vasily Peskov, who actively participated in the educational process through the weekly section "Window to Nature". The activities of the society intensified scientific research in Karachay-Cherkessia. The educational collection of rocks and minerals, sets of space photos of the Caucasus and mountain countries of different continents were created at the DNG. In collaboration with the State Center "Nature" (Moscow), the latest techniques and the use of GIS technology were introduced into the student training and field research of scientists.

DNG professors carried out a number of studies in various fields of scientific knowledge aimed at ensuring environmentally sound territorial and regional nature management.

DNG staff compiled modern essays on geology, geography, biology, toponymy, geomorphology, geoecology of the territory of the republic, and proposed an improved methodology for assessing the degree of anthropogenic impact on the natural environment.

For almost every region of the republic, the graduates of the department compiled geoecological maps and marked the functional types of anthropogenic load. The dissertation of L.A. Chagarova [1], one of the DNG graduate students, on the landscape structure of the Front Range interested the chairman of the Russian Geographical Society, Professor K.V. Chistyakov. At his invitation, the study was defended in 2011 in St. Petersburg. Other graduate students – L.I. Appoeva [2], P.A. Kipkeeva [3] and N.S. Dega [4] successfully defended their dissertations in Rostov-on-Don The topics included the zoning of comfort, the principles of environmental and geographical education by means of geographic information systems and the dynamics of the main components of the landscape. Based on many years of observations V.V. Onishchenko [5] presented the monograph "Phenology of Dendroflora of the Teberdinsky State Reserve".

S.A. Khapaev developed a toponymic system for the territory of the Karachay-Cherkess Republic. Ornithologist A.A. Karavaev revealed more than 100 new bird species in the republic [6]. Yu.Ya. Potapenko compiled a modern geological essay of the region and textbooks on landscape science and geoecology [7-9].

The Teberdinsky Reserve continued fundamental study of mountain areas with the annual edition of scientific papers. The scientists of Moscow State University (biologists, soil scientists) and environmentalists of Pushchino University actively participated in the study of the nature environment of the reserve. Their work is characterized by high professionalism and use of the latest methods and technologies. A large amount of scientific information was obtained with a thorough study of phytocenoses of the Alpine landscape belt, conducted under the supervision of Professor V.G. Onipchenko of Moscow State University [10].

Since the 1970s, the territory of Karachay-Cherkessia has served as a laboratory for the development of versions of landscape maps of mountain areas having methodological importance for the development of landscape mapping of mountain areas [11]. In 1991, one of the leading landscape scientists of the USSR A.G. Isachenko compiled a landscape map of the Teberda Valley [12]. The legend for the map looks as a matrix with vertical high-altitude belts and horizontal landscapes indicating the age of the lithogenic base. On the map of V.A. Shalnev (1996) [13], the corresponding taxa are distinguished vertically: geobotanical belt, landscape belt and landscape tier. The matrix legend was also used on the landscape map of the republic compiled by Yu.Ya. Potapenko in 1996 [8]. The vertical row includes plant belts, while the horizontal – geological-geomorphological components starting from morphostructures to smaller taxa. It is worth noting the contribution of professor of Stavropol State University V.A. Shalnev to the study of landscapes of the North Caucasus. For half a century, he created a series of monographs covering both general theoretical problems of geography and regional landscape science and the latest methods of large-scale mapping (using the example of the Teberdinsky reserve) using GIS technologies [14, 15].

In 2019, a notable event in the scientific sphere was the 1st International Conference on Balanced Development of Mountain Territories held in Karachaevsk-Teberde-Dombay [16]. Here we should quote a phrase from the report of the Commission by G. Brundtland that humanity is able to make the development balanced and "seeks to take into account the needs and aspirations of the present without jeopardizing the ability to take them into account in the future [17, p. 55].

In subsequent years, Russia, and in particular, the Caucasus, began to use the term "sustainable development" instead of "balanced development" due to the inaccurate translation of the English "sustainable development".

During 1991-2005 Russian scientists, primarily geographers published a large number of works covering the problems of nature management in the transition to a market economy. At the same time, these problems were discussed by the scientific community at conferences of various levels, usually referring to a popular concept of "sustainable development".

In the 60-80s of the 20th century, the USSR carried out environmental planning at the regional level mainly within the framework of Territorial Integrated Nature Conservation Schemes (TINCS). For example, the TINCS for the Caucasus Mineral Waters region was completed in 1990. This document was developed by a scientific team of 18 institutes under the leadership of Giprogor Institute (Moscow).

The main disadvantages of territorial planning were the following: 1) priority of economy, economic development of the territory; 2) reliance on general settlement schemes and urban planning solutions; 3) subordinate role of environmental approach; 4) predominance of medium-scale planning developments (map scale 1:200 000-1:100,000).

By the beginning of the 90s, the principles of the new environmental policy of the Russian Federation were formulated: 1) systematic approach to nature management, 2) inventory of natural resource potential and determination of environmental restrictions on its use, 3) optimization of development conditions by choosing options for sparing nature management, 4) compliance with ecological equilibrium within specific landscapes.

Since then, conferences on sustainable development of mountain areas have been held in the North Caucasus almost annually. However, no practical results were obtained. According to B.I. Kochurov [18], the mechanical transfer of the program of sustainable development to the conditions of Russia was erroneous, since it did not take into account the peculiarities of the country's historical development: focus on the development of large spaces, imperfection of organization and arrangement of territories. The liquidation of the Ministry of Ecology in 2002 sharply weakened the role and capabilities of environmental services.

Based on the above, the term "sustainable development" was not used at the conference in Karachay-Cherkessia. A new relevant and achievable goal had to be identified. The proposed name of the conference "balanced development" is most closely related to the concept of ecological and economic balance (EEB) of the territory developed in 1995-2003 by B. I. Kochurov [18] and successfully tested in a number of regions of Russia. This concept involves new spatial forms of environmental management –environmental and economic structures, determination of the degree of anthropogenic impact and environmental tension, which ensures the reproduction of renewable natural resources and does not cause environmental changes and consequences.

Of the numerous options for the taxonomy of geographical sciences, the authors of the paper believe that the most adequate, besides V.A. Shalnev [19], is the geosystem concept based on the ideas of synergy, integrity of the object and scientific knowledge. The geosystem is understood as a complex space-time formation, including three subsystems – nature, population and economy. Due to historical circumstances, the geosystem of Karachay-Cherkessia turned out to be very studied in detail, which allows referring to this territory as to a training ground for landscape and geoecological research.

4. Conclusions

The territory of Karachay-Cherkessia has all modern cartographic material, which allows conducting high-level training in the basics of landscape science, geomorphology, geoecology, geobotany. A large number of natural monuments are registered in the region. Together with nature reserves and nature sanctuaries, nature monuments expand the significance of the republic as a unique natural object.

References

[1] Chagarova L A 2019 Using medium and large-scale aerial photographs in landscape and geochemical studies of the highlands of the Front Range *Current directions of balanced development of mountain areas in the context of interdisciplinary approach. Proceedings of the I International Scientific Conference,* pp. 88-92

[2] Appoeva L I 2003 *Spatial differentiation of natural and ecological-social comfort of the Karachay-Cherkess Republic.* Dissertation for the degree of the candidate of geographical sciences, Rostov-on-Don

[3] Kipkeeva P A 2013 Optimization of ecological and geographical education in Karachay-Cherkessia *News of higher educational institutions. North Caucasus region. Series: Natural Sciences* **4 (176)** 99-103

[4] Dega N S 2010 *Dynamics of the main components of the landscape of Karachay-Cherkessia in the context of a changing climate and economic activity.* Dissertation for the degree of the candidate of geographical sciences, Southern Federal University. Rostov-on-Don

[5] Onishchenko V V 2005 Fenology of the dendroflora of the Teberdinsky State Reserve. *Works of the Teberdinsky reserve.* **36** (Kislovodsk: MIL)

[6] Karavaev A A, Khubiev A B, Kaziev U Z 2009 Changes in fauna and population of birds of the Karachay-Cherkess Republic *News of higher educational institutions. North Caucasus region. Series: Natural Sciences* **2 (150)** 86-89
[7] Potapenko Yu Ya 2004 *Geology of Karachay-Cherkessia.* (Karachaevsk: KChSU),
[8] Potapenko Yu Ya 2009 *Landscape and geoecological research and mapping of mountain areas.* (Karachaevsk: KChSU)
[9] Kipkeeva P A, Potapenko Yu Ya 2015 Main factors in ensuring sustainable tourism in Karachay-Cherkessia *Bulletin of Moscow University. Series 5: Geography* **5** 76-81
[10] Zernov A S, Onipchenko V G 2011 *Vascular plants of the Karachay-Cherkess Republic: conception of flora.* (Moscow)
[11] Kipkeeva P A, Volosukhin Ya V, Potapenko Yu Ya, Shubaeva N V 2018 Landscape and geoecological analysis of basins of the small rivers in nature protection zones of mountain territories *International symposium "Techniques and sciences about Earth: applied and basic research" (ISEES 2018)*, pp. 26-31 DOI: https://doi.org/10.2991/isees-18.2018.6.
[12] Isachenko A G 1991 *Landscape science and physical and geographical zoning* (Moscow: Higher School)
[13] Shalnev V A 2004 *Landscapes of the North Caucasus: evolution and modernity.* (Stavropol)
[14] Shalnev V A 2000 *History and methodology of general geography.* (Stavropol)
[15] Shalnev V A 2006 *Problems of interaction between society and nature: a geographer's view* (Stavropol)
[16] 2019 *Current directions of balanced development of mountain areas within the interdisciplinary approach: Materials of the I International Scientific Conference* (Karachaevsk)
[17] 1987 *Report of the International Commission on Environment and Development "Our Common Future"*, available at: http://www.un.org/ru/ga/pdf/brundtland.pdf
[18] Kochurov B I 2001 Environmentally safe and balanced development of the region *Izvestia AS. Series: Geography.* **4** 87-92
[19] Shalnev V A, Lyashenko E A, Melnichuk V V 2019 *History, theory and methodology of geography: educational manual (lecture course)* (Ministry of Science and Higher Education of the Russian Federation, North Caucasus Federal University. Stavropol)

Improving the efficiency of exploration and development of oil and gas fields in the conditions of frequently interbedding rocks

T N Minnivaleev, D Sh Nosirov , L S Kuleshova, A N Minnivaleev and A A Gizzatullina

Ufa State Petroleum Technological University, Branch of the University in the City of Oktyabrsky, 54a, Devonskaya St., Oktyabrsky, Republic of Bashkortostan, 452607, Russian Federation

E-mail: timxn@yandex.ru, info@of.ugntu.ru

Abstract Exploration and development of mineral deposits largely affects the dynamics of development of the oil and gas industry in the country. An important stage in the field development is the design of technologies for constructing wells, and the drilling process itself. The appropriate modes and drilling tools create conditions for the maximum oil recovery. In this paper, based on the analysis of existing well drilling technologies and features of the mining and geological conditions of the fields, the use of a new technological solution is substantiated and recommendations are given for its implementation in the Yuzhno-Priobskoye field. The feature of the Yuzhno-Priobskoye field is frequent interbedding, that is, alternation of some rock layers by others. These layers exhibit different mechanical properties, such as hardness, elasticity, and abrasiveness, which arise certain difficulties in the field exploration and development. The proposed technological solution significantly increases the efficiency of mining and geological operations.

1. Introduction

The Yuzhno-Priobskoye field was discovered in 1982. According to experts, the reserves of the field were estimated at 5 billion tons. Over 38 years of exploitation of the Yuzhno-Priobskoye field, 2.4 billion tons of oil have been produced. The products of this field from the Yuzhno-Priobskoye field belong to the class of naphthenic hydrocarbons, which means that the concentration of paraffin in oil is not more than 3 percent and the sulfur content is 1–2%. The oil recovery factor is 0.35, which indicates prospects for further exploitation of this field [1].

More than 150 wells are currently operating in the field. The Tyumenskaya suite with low reservoir porosity and permeability occurs in the thickness of the field. Reservoirs located in the range of 2800–3000 m, group AC1-12, are classified as low permeability reservoirs. The reservoir thickness is 20–40 m.

However, the Yuzhno-Priobskoye field has a specific feature that distinguishes it from others, namely heterogeneity of the rock layers. In simple terms, there is frequent interbedding of one rock with another, that is, argellites with sandstones and siltstones [2]. Due to this feature of the Yuzhno-Priobskoye field, which implies the interbedding of one rock with another, a special approach is required to design the exploration and development of the field. The development of this approach to the design and construction of wells improves technological measures. Thus, a qualitative selection of existing and development of new technologies and methods of drilling and construction of wells in the special mining and geological conditions of the Yuzhno-Priobskoye field and other fields characterized by frequent rock layer interbedding is of relevance [3].

2. Materials and methods

Three different types of drill bits were the object of this study – a roller cone, PDC, and a combined bit that exhibits the elements of the previous two types of bits. The advantages and disadvantages of rock cutting tools are analyzed. The study presents a comparative characteristic of these bits in the conditions of well drilling with frequent interbedding of rocks wit different physical and mechanical characteristics.

3. The content of the method and assessment of its effectiveness

The aim of the study is:

- to study the effective use of various types of bits in the construction of wells in the mining and geological conditions of frequent interbedding of one rock with another, in particular, in the Yuzhno-Priobskoye field;
- on the basis of this study, to identify the type of bits that, in terms of the features of exploration and development of the above field, will be the most feasible.

At present, there is a tendency to build deep wells in the fields of Western Siberia and in the Yuzhno-Priobskoye field, in particular. With regard to the mining and geological conditions of the Yuzhno-Priobskoye field, special emphasis is placed on the choice of a bit and the requirements for it. The cost of a bit typically exceeds the amount intended for the construction of wells, from 1 to 5%. The bit is a basic drilling tool, which determines the duration of the drilling process [4–5]. The duration of the drilling process is affected by:

- drilling rate;
- penetration per bit unit;
- type of drilling.

All the above factors are the requirements for bits.

Carry out a theoretical and experimental analysis of the feasibility of using shear-cutting bits, which include PDC bits and crushing-shearing action, which include roller cone bits.

Roller cone bits appeared far in the last century with the active development of the oil industry [6]. At first, they were used as rock cutting tools for geological exploration.

The efficiency of roller cone bits decreases due to the complexity of the mining and geological conditions of drilling, as evidenced by the experience of using roller cone bits in the Yuzhno-Priobskoye field. This is due to the fact that the bodies of the cutters fail after prolonged mechanical stress. As a rule, the main element of roller cone bits are cones, and when the latter fails, the bit is out of order [7]. Well profile also affects the wear of roller cone bits, in particular, the construction of multilateral directional wells has been developing recently.

It is should be noted that the experts concluded that roller cone bits, namely three-cone bits, are effective for drilling directional wells, due to the presence of 3 supporting movable cones in them. A relatively low price per bit unit is also beneficial [8].

When drilling wells, as described above, the main indicator of the efficiency of the bit used is its mechanical rate of penetration (ROP). In drilling, several types of drilling rates can be distinguished, which characterize the complex indicators of drilling. The following rates characterize the technical and economic indicators of drilling:

- mechanical rate;
- cruising rate;
- technical rate;
- commercial rate;

- cycle rate

It should be noted that the mechanical rate, one way or another, affects all other rates.

As noted above, an indicator of the efficiency of the bit used is its mechanical rate. To assess the efficiency of using a roller cone bit in the construction of wells in the Yuzhno-Priobskoye field, see Figure 1, which shows the dependence of the bit ROP on the well depth.

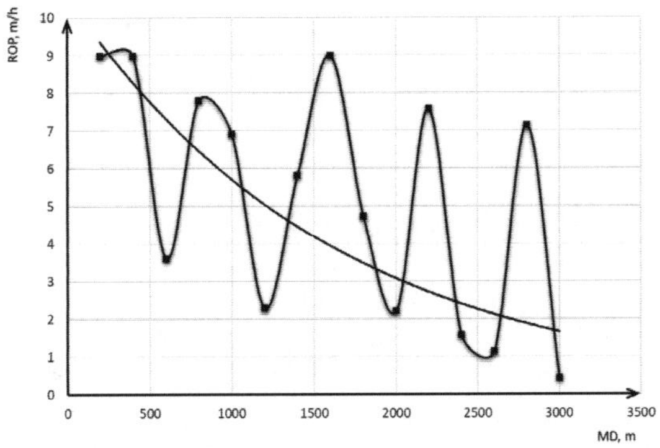

Figure 1. Dependence of the ROP changes on the well depth when drilling with the roller cone bit

Figure 1 shows that during drilling ROP changes within a significant interval. Sharp jumps presented in the form of humps and depressions indicate heterogeneity of the rock layers, that is, their layering and interbedding. Due to the frequent interbedding of the rock layers in the Yuzhno-Priobskoye field, the efficiency of using roller cone bits is reduced.

During the transition from one rock layer to another, significant vibration occurs in the transition zone and affects the sevice life of the cones, which indicates the low reliability of the cone bits [9].

The average mechanical rate during drilling using a roller cone bit was 4.97 m/h.

The next stage in the development of rock cutting tools was the development of a PDC (policristalline diamond compact) bit design. A bit of this type differs from the roller cone in te absence of moving parts. The body of these bits is milled from a steel billet. Self-sharpening cutters are installed on the blades in certain places, with a diamond layer 0.6–1.0 mm thick in the upper part. The use of diamond cutters in the construction of PDC bits is due to the fact that diamond is one of the hardest materials with an extremely high degree of wear resistance [10].

As a rule, a high degree of wear resistance of a bit indicates its high penetration rate of several hundred meters in medium and medium-hard rocks. Due to the fact that the self-sharpening effect is characteristic of diamond cutters, the required axial load on the bit decreases [11].

Since there are no moving parts in the design of PDC bits, this reduces the risk of bit failure. The absence of movable parts of PDC bits shows its poor controllability when drilling directional and multilateral wells. However, the absence of moving parts contributes to the fact that the transmitted torque, in particular, by the rotor is transmitted almost completely (90–95%), compared to the roller cone bit with the torque of the transmitted rotation equal to 70–80%. Thus, in terms of this parameter, PDC bits are more effective [12].

Frequent changes in the mechanical characteristics of the rocks that occur during drilling initiate significant axial vibrations in the drilling tool assembly. Such frequent interbedding is typical, for example, of the Yuzhno-Priobskoye field.

To evaluate the effectiveness of PDC bits in the alternating mining and geological conditions, the ROP of the bit was evaluated throughout the entire well construction stage on the example of the Yuzhno-Priobskoye field, which is shown in Figure 2.

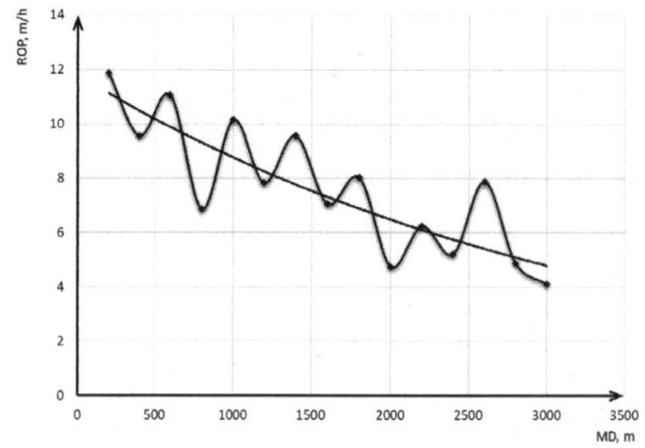

Figure 2. Dependence of the ROP changes on the well depth when drilling with the PDC bit

Figure 2 shows that the range of the amplitude of changes in the mechanical rate is significantly lower in comparison with roller cone bits. At the same time, the provided average ROP in the entire interval of wells drilling was 7.15 m/h. This suggests that the cycle rate increases due to a decrease in the total time for well construction, which is accompanied by an increase in the indicators of well drilling.

Another type of bit to consider is a hybrid bit. This type of bit has appeared recently and combines the advantages of the above two bits.

Hybrid bits are characterized by a high degree of wear resistance due to the combination of characteristics of PDC bits and roller cone bits.

A high degree of wear resistance indicates a significant reduction in longitudinal and transverse vibrations in the transition zones, that is, during the transition from a rock of one hardness to another one. Reduction of the dynamic load on the bit increases its operating characteristics and ROP.

As described above, to assess the effectiveness of the use of hybrid bits, in particular, Kymera bits, the ROP value was investigated in the range of 0–3000 m (Figure 3).

Analysis of the figure shows steady operation of these types of bits, which provides a significant reduction in longitudinal and axial vibrations. It addition, ROP ranges from 12.1 to 14 m/h and the average ROP is 12.8 m/h, which is higher by 61% compared to roller cone bits and by 44% compared to PDC. Based on the above, the Kymera hybrid bit is the most efficient in terms of exploration compared to roller cone and PDC bits. Kymera hybrid bit makes it possible:

- to reduce the cost of drilling wells;
- to reduce the risk of failures due to axial and radial vibrations;
- to ensure good controllability when drilling directional wells;
- to provide high mechanical rate;
- to provide high wear resistance when drilling interbedded rock layers.

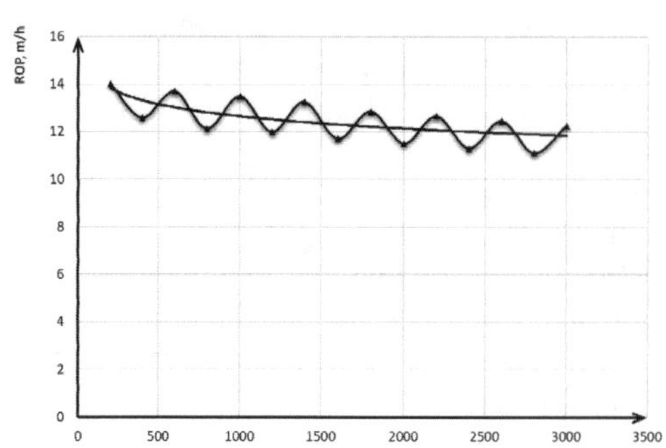

Figure 3. Dependence of the ROP changes on the well depth when drilling with the Kymera hybrid bit

4. Conclusion

For the Yuzhno-Priobskoye field with the mining and geological characteristics represented by frequent interbedding of some rock layers with others, the roller cone bit is less effective due to the presence of moving roller cutters in it. This is accompanied by a high risk of failure and low wear resistance, that is, the penetration rate (bit resource) provided by this type of bit is very low, which increases financial costs for drilling.

PDC bits exhibit a relatively high wear resistance compared to roller cone bits and the absence of moving part in the structure, which ensures their greater reliability. However, controllability of drilling directional wells decreases, and the possibility of significant longitudinal and lateral vibrations of bits in the transition zones increases.

A hybrid version of the Kymera hybrid bits is most effective in difficult geological conditions. Due to the combination of the advantages of the above two types of bits, it provides high wear resistance, significant drill penetration, and a high degree of technical and economic indicators of drilling, which are presented in the above figures.

Based on the above, the use of the Kymera hybrid bit ensures a significant increase in ROP compared to roller cone and PDC bits when drilling wells in the conditions of frequent interbedding of rocks, in particular, in the Yuzhno-Priobskoye field. In addition, the durability of drilling equipment is increased, and the risk of failures is minimized.

References

[1] Okoronkwo A A, Carigali P A and Ibrahim T B 2014 Tackling drilling challenges through engineered BHA and optimized drilling parameters: A case study from the Niger Delta IADC/SPE *Asia Pacific Drilling Technology Conf. 2014: Driving Sustainable Growth Through Technology and Innovation,* Bangkok, Thailand: Society of Petroleum Engineers, 25 -27 August 2014.

[2] Mukhametshin V V 2018 Bottomhole formation zone treatment process modelling with the use geological and geophysical information *IOP Conference Series: Earth and Environmental Science* **194(2)** 022024 DOI: 10.1088/1755-1315/194/2/022024.

[3] Ugbogu I N A and Nagib M B 2013 A new well control model to optimize drilling performance *Kuwait Oil and Gas Show and Conf. (KOGS 2013)*, Mishref, Kuwait: Society of Petroleum Engineers, 7 -10 October 2013, pp. 206–219.

[4] Shagapov V S, Tazetdinova Y A, Gizzatullina A A 2019 On theory of thermal recovery of high-viscosity oil *Journal of Applied Mechanics and Technical Physics* **60(5)** 882–888

[5] Batalov S A, Andreev V E, Lobankov V M and Mukhametshin V Sh 2019 Numerical simulation of oil formation with regulated disturbances. Oil recovery quality simulation *Journal of Physics: Conference Series* **1333** 032006 DOI: 10.1088/1742-6596/1333/3/032006.

[6] Minnivaleev T.N. 2018 Efficiency determination of amplitude decay of pressure fluctuation of circulating fluid with the help of a multi-stage piston-type downhole compensator *Advances in Engineering Research* **157** 397–400 DOI:10.2991/aime-18.2018.76.

[7] Minnivaleev T N, Arslanov I G, Yagafarova K N 2018 Graphene aerogel is modern and promising material *IOP Conference Series: Earth and Environmental Science* **194(4)** DOI: 10.1088/1755-1315/194/4/042014. (Scopus).

[8] Batalov D A, Soloviev N N, Mukhametshin V Sh, Andreev V E, Dubinskiy G S, and Safiullina A R 2020 Forecasting the use of non-stationary waterflooding in the conditions of oil deposits in Western Siberia *IOP: Earth and Environmental Science (EES)* **579(1)** 012020 DOI: 10.1088/1755-1315/579/1/012020.

[9] Mukhametshin V Sh, Andreev V E, Yaskin S A 2020 Designing measures to increase oil recovery based on the identification and grouping of deposits *IOP: Earth and Environmental Science* (EES) **579(1)** 012013 DOI: 10.1088/1755-1315/579/1/012013.

[10] Kadyrov R R, Mukhametshin V V, Galiullina I F, Kuleshova L S, and Safiullina A R 2020 Prospects of applying formation water and heavy brines derived therefrom in oil production and

national economy *IOP Conference Series: Materials Science and Engineering* **905(1)** 012081 DOI: 10.1088/1757-899X/905/1/012081.

[11] Gabdrakhimov M S, Suleymanov R I, Shaislamov V Sh and Minnivaleev T N 2012 RF, Patent No. 2464404 *Flush Water Compensator* (20.10.2012).

[12] Shagapov V S, Tazetdinova Y A, Gizzatullina A A 2019 On the theory of extraction of high-viscosity oil from the stratum under thermal action *Journal of Engineering Physics and Thermophysics* **92(6)** 1415–1422

Description and prospects of oil and gas potential of the Middle Devonian – Lower Frasnian complex of the south-east of the Timan-Pechora province

E A Kuznetsova

Perm State University, 15 Bukireva Str., Perm 614990, Russia.

E-mail: e.lena.kuznetsova@yandex.ru

Abstract. The article is devoted to the assessment of the petroleum potential of the deep Middle Devonian – Lower Frasnian oil and gas complex in the south-east of the Timan-Pechora oil and gas province. Within the Verkhnepechorskaya depression of the Pre-Ural foredeep and in the south of the Pechora-Kolvinsky aulakogen, several wells were drilled with a depth of more than 5 km, some of which opened deposits of the Lower Paleozoic. These strata are difficult to access and poorly studied, and the prospects for their oil and gas potential are unclear. The article describes the composition of the complex, gives geochemical characteristics, describes reservoir properties, and presents the results of 1 and 2D basin modeling. Models of the zoning of catagenesis are presented. The oil and gas complex includes a variety of oil and gas source rocks. It is possible to allocate collectors, as well as seals. In the Middle Devonian – Lower Frasnian sediments, the processes of oil, gas and gas condensate generation took place, which could ensure the formation of deposits both in the deep strata of the Middle Paleozoic, and in the overlying horizons. The generation and accumulation of hydrocarbons in deep-buried sediments occurred at a favorable time for the formation of deposits.

1. Introduction

The south-eastern regions of the Timan-Pechora oil and gas province are characterized by a wide range of industrial oil and gas potential. Hydrocarbon deposits were discovered in the Middle and Upper Devonian, Lower Carboniferous series, as well as in the Permian and Triassic systems, some oil inflows were obtained from the Silurian. Oil and gas occurrences and gas inflows were detected at depths of more than 5 km, and oil destruction products were found at a depth of 7 km. Deep-buried deposits are widespread within the territory under consideration, which are little and very unevenly studied. When studying geological objects with insufficient data, including deep sediments, modern basin modeling technologies are effective.

At great depths of the specified territory, the following oil and gas complexes have been discovered in various volumes: Ordovician – Lower Devonian, Middle Devonian – Lower Frasnian, Domanik – Tournaisian, Lower–Middle Visean. In some wells at a depth of more than 4.5 km, the basal part of the Upper Visean – Cisuralian oil and gas complex is also uncovered. This article discusses the main features and prospects of the oil and gas potential of the deep Middle Devonian – Lower Frasnian oil and gas complex based on basin modeling.

In the south-eastern part of the oil and gas province, deep-buried sediments are confined to the Verkhnepechorskaya depression of the Pre-Ural foredeep and the south of the Pechora-Kolvinsky aulakogen.

The structure of the Verkhnepechorskaya depression is typical of foredeeps. There is a clear asymmetry with the allocation of the western and eastern zones, which have different thickness of the sedimentary cover. The maximum thickness of the sedimentary cover (up to 11 km or more) is observed in the zones of development of thrusts. The following structural zones are distinguished in the structure of the depression: the Pechora-Ilychskaya monocline in the west and the Vuktyl thrust, the Kuryinskaya anticline zone, the Saryudinskaya and Patrakovskaya folded zones in the east. The latter are often attributed to the fold-thrust belt of the Western Urals.

Within the Verkhnepechorskaya depression, there are many wells that have uncovered deep-buried sediments, 13 of them are below 5.5 km [5]. Most of the wells that opened deep strata are confined to the Vuktyl thrust (the deepest well is Vuktylskaya-58, has a bottom of 7026 m), in addition, they were opened by wells of the Saryudinskaya uplift zone and the Pechora-Ilychskaya monocline of the Verkhnepechorskaya depression.

The Pechora-Kolvinsky aulakogen is one of the most complex in structure and development of large negative structures of the Timan-Pechora oil and gas province. Here we considered the Timano-Pechorskaya deep reference well (drilled depth 6903.5 km), which on the lower horizons entered the Pechora-Kozhvinskaya (Podcherem-Kamenskaya) paleodepression of the aulakogen, and on the upper ones is in the arch of the Zapadno-Soplesskaya disjunctive brachianticline located within the Middle Pechora transverse uplift of the Pre-Ural trough.

2. Results and Discussion

For basin modeling, data obtained during drilling and exploration of wells that opened deep subsurface were collected: 1) the thickness of the layers; 2) the lithological composition of each layer and 3) the measured values of pressure, temperature, and porosity.

The strata of the middle and Upper Devonian are exposed by most of the deep wells of the Verkhnepechorskaya depression and the Timan-Pechora deep reference well. The Middle Devonian sediments are represented by the interlayer of sandstones, siltstones, and mudstones with interlayers of clays, gravelites, conglomerates. The sections of the Middle Devonian in the work area are characterized by different completeness and thickness, which varies from 88 m (Zapadno-Vuktylskaya-1 well) to 320 m (Zapadnoye Dutovo-1 well) and up to 2000 m (Timano-Pechorskaya well). The Frasnian, with a capacity from 167 m (Zapadno-Vuktylskaya-1 well) to 952 m (Lebyazhskaya-2 well), is represented by limestones, sandstones, and clays.

Also, in the process of compiling the database for modeling, the data for determining the maturity of the organic matter of sediments – the vitrinite reflectance of and geochemical characteristics reflecting the generation of hydrocarbons, etc. were generalized. The results of Rock-Eval pyrolysis made it possible to determine the content of organic carbon, the genetic type of organic matter, the stages of its transformation and the hydrocarbon-generating potential, as well as to classify deep oil and gas mother rocks and to determine their quality.

The organic matter of the rocks of the Middle Devonian – Lower Frasnian oil and gas complex is mainly represented by a mixed type, but the sapropel type of organic matter is determined in the upper part of the Devonian system. According to the Rock-Eval pyrolysis data, the generation potential and organic carbon content of these complexes are also small, except for the Upper Devonian deposits mentioned earlier. The values of the productivity index indicate that the strata reached the thermobaric conditions characteristic of oil and gas generation, but the scale of hydrocarbon generation was probably not high. The achievement of rigid thermobaric conditions and high catagenesis values is also indicated by the determination of the vitrinite reflectance (gradations of catagenesis MK_4–MK_5, AK_1–AK_3) [1, 5].

According to the data of wells in which organic carbon was determined, the density of the content of organic matter was calculated, which made it possible to identify deep oil and gas mother formations.

In the process of preparing the database for modeling, data on reservoir properties were generalized. The Middle Devonian – Lower Frasnian deposits in the wells Vuktylskaya-58, Zapadno-Vuktylskaya-1, Belaya-1, Lebyazhskaya-2 are characterized by porosity values from 0.1 to 11 % (on average – 1.3 %), the rocks are dense, impermeable. The filtration and capacitance properties of the thickness are low. In this complex, the regional Kynovo-Sargaev seal is distinguished [5].

The results of borehole studies were prepared and uploaded to the PetroMod basin modeling software complexes (modules 1 and 2D) and Genex (SingleWell and MultyWell modules).

After studying the geological structure of the territory, "boundary conditions" were selected and entered the used software complexes – data on the thermal condition of the sedimentary cover of the studied area and paleo water depth. For this purpose, the lithological-paleogeographic and paleotectonic maps of the Timan-Pechora oil and gas province were reviewed, and the main tectonic events and changes in sedimentation conditions were considered. The model of the formation of the territory and the geothermal condition are consistent with the main stages of the development of the Pre-Ural foredeep. The capacities of the eroded deposits and heat flow were corrected when calibrating the 1D models according to the values of the vitrinite reflectance and temperature.

After all the preparatory work was carried out, 1D basin modeling of the previously listed wells and several others, whose depth exceeds 4.5 km, and less deep, but opened the foundation, was performed, and then there was 2D modeling along the line of the regional seismic profile 22-RS (Figure 1).

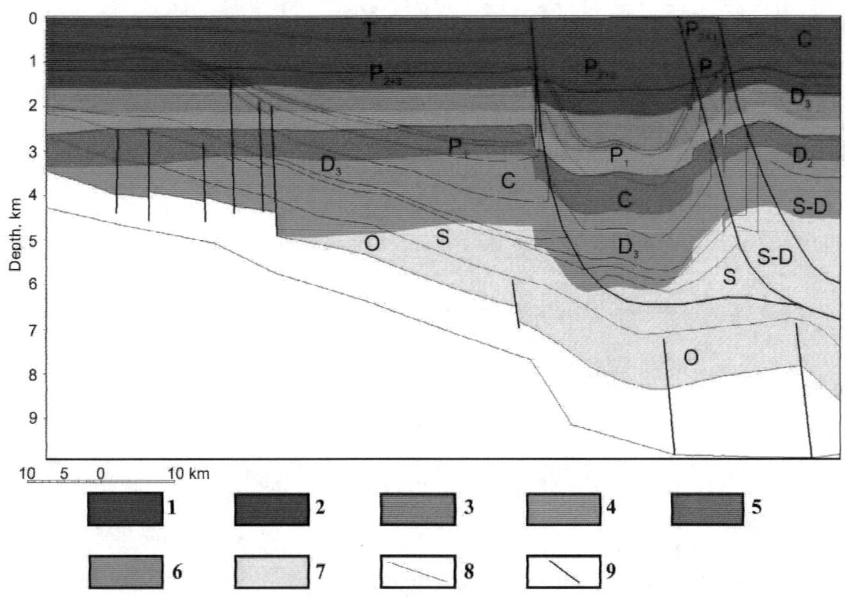

Figure 1. Model of modern zoning of catagenesis in the section of the Verkhnepechorskaya depression: 1–7 – gradations of catagenesis: 1 – PK_1–PK_3, 2 – MK_1, 3 – MK_2, 4 – MK_3, 5 – MK_4–MK_5, 6 – AK_1–AK_3, 7 – AK_4; 8 – boundaries of layers; 9 – faults

The calculation results are presented in the form of numerous graphs that can be combined into three groups. The first is models of the modern structure of the section, showing changes in properties (temperature, porosity, vitrinite reflectance, etc.) depending on the depth – Depth Plots. The second is models of the evolution of the selected strata – changes in properties over geological time – Time Plots. The third is models of immersion – Burial Plots. The quality of the calculated models was evaluated by independent parameters, such as the current measured values of reservoir temperature, vitrinite reflectance, pressure, porosity, and pyrolysis results using the Rock-Eval method, as well as literature data.

The reconstruction of the history of sedimentation and thermal evolution showed that the deposits of the Middle Devonian – Lower Frasnian stage entered the main phase of oil formation starting from the Early Carboniferous time within the north and west of the Verkhnepechorskaya depression and the Vuktyl thrust, and in the center of the Pechora-Ilychskaya monocline – from the Kungur age. The main phase of gas formation in most of the Verkhnepechorskaya depression began at the end of the Permian period, only in the west of the Pechora-Ilychskaya monocline – from the Triassic. And on the Middle Pechora transverse uplift, these processes began even earlier, from the Frasnian and Famennian centuries, respectively.

It is known that the tectonodynamic evolution of the depressions of the foredeeps and the largest buried uplifts occurred actively not only in orogenesis, but also during the pre-orogenic stage of development. The formation of structural traps occurred in three major stages: Ordovician–Devonian, Carboniferous–Cisuralian and Lopingian–Triassic [3]. The calculated critical moment indicates that the processes of generation, migration and accumulation of hydrocarbons occurred after the formation of local uplifts of the ancient foundation in deep strata. At the same time, it is worth noting that the transformation ratio of organic matter for given thicknesses reaches 99–100 %.

The rate of deflection had a great influence on the scale of generation and accumulation of hydrocarbons. The Middle Devonian – Lower Frasnian oil and gas complex in the south-east of the Timan-Pechora oil and gas province is characterized by sedimentation rates favorable for oil generation [4].

According to the literature data [2] and the results of basin modeling, schematic maps of the zoning of catagenesis along the base and top of the deep-submerged Middle Devonian – Lower Frasnian oil and gas complex were constructed (Figure 2).

The maps showed that the same-age sediments of the eastern and western zones of the Verkhnepechorskaya depression are characterized by different stages of catagenesis – the level of maturity of organic matter in the east of the depression is higher than that in the west. Thus, at the base of the complex within the Vuktyl thrust, the Saryudinsky folded zone and the Pechora-Kolvinsky aulakogen, the organic matter corresponds to the catagenesis gradation $\geq AK_4$, and in most of the Verkhnepechorskaya depression corresponds to the main gas formation zone.

In the upper part of the complex, in most of the territory, organic matter also corresponds to the gradations of catagenesis MK_4–MK_5 and AK_1–AK_3. $\geq AK_4$ is released only within the Vuktyl thrust and the Pechora-Kolvinsky aulakogen. Also, in the very west of the Pechora- Ilychskaya monocline, the main zone of oil formation is distinguished (gradation of catagenesis MK_3).

This catagenetic zonality is explained by the immersion of the foundation surface in the north-east direction. It can also be related to the difference in the rate of hydrocarbon generation, which depends on the type of kerogen, the gradient of temperature changes, pressures, lithological composition, sedimentation rate, etc. In addition, it is necessary to consider that the upper part of the section was eroded and the zoning of catagenesis can be attributed to a tectonically disturbed type reduced from above.

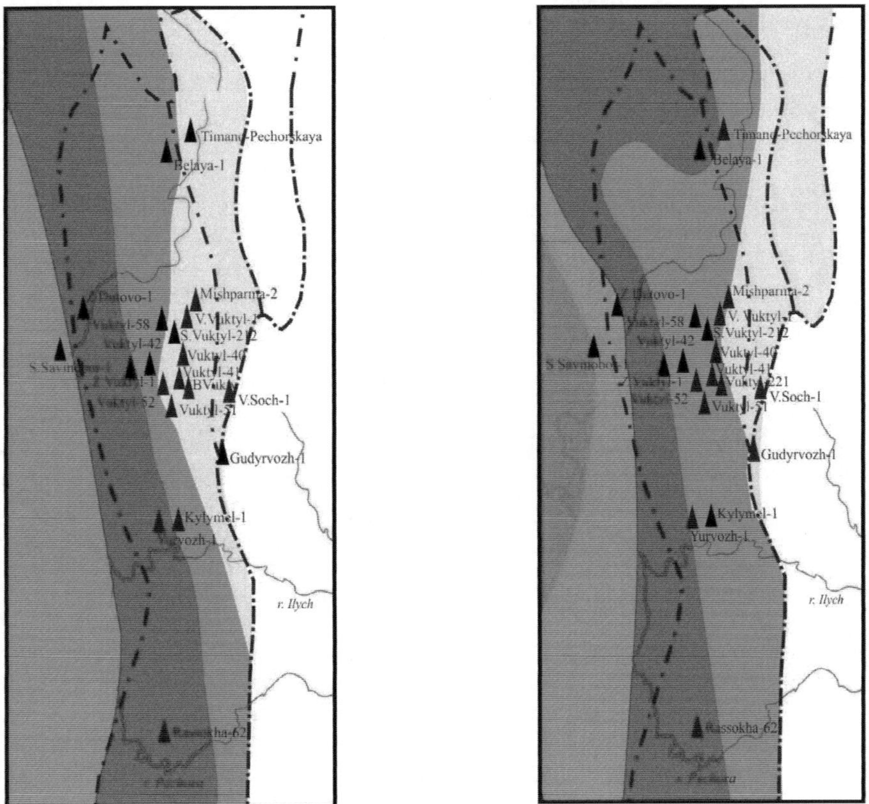

a b

Figure 2. Schematic maps of the zoning of catagenesis along the base (a) and top (b) of the Middle Devonian – Lower Frasnian oil and gas complex. For the symbols, see Figure 1

3. Conclusion

According to the results of basin modeling, it can be concluded that the organic matter of the sediments of the Middle Devonian – Lower Frasnian series reached catagenesis gradations of MK_3, MK_4–MK_5 and higher. That is, at great depths, the process of generating oil, gases and gas condensates occurred, which could ensure the formation of deposits both in the deep-buried strata of the Middle Paleozoic and in the overlying horizons. In addition, the generation and accumulation of hydrocarbons in deep sediments occurred at a favorable time for the formation of deposits.

References

[1] Belyaeva G L 2014 Catagenesis of organic matter of rocks of deep-submerged horizons of the Timan-Pechora NGP and its relation to oil and gas potential *Geology, geophysics and development of oil and gas fields* **7** 36–39

[2] Danilevsky S A, Sklyarova Z P, Trifacheva Yu M 2001 *Geofluid Systems of the Timan-Pechora province* (Ukhta: Regional Press House)

[3] D`yakonov A I, Ovcharova T A, Shelemej S V 2008 *Assessment of oil and gas potential of autochthons and allochthons of the Pre-Ural regional deflection on the evolutionary genetic basis* (Uxta: UGTU)

[4] Kuznetsova E A 2019 Influence of the sedimentation rate on oil and gas generation in the sediments of the south-eastern regions of the Timan-Pechora oil and gas province *New directions of oil and gas geology and geochemistry. Development of geological exploration works* 293–298

[5] Kuznetsova E A 2020 Geology and Petroleum Potential of Deep Deposits of the

Verhnepechorskaya Depression *Bulletin of Perm University. Geology* **19(2)** 175–182

Comprehensive approach to carbonate rocks typing based on core laboratory experiments data

A V Raznitsyn[1,2], I S Putilov[1,2], O A Melkishev[1]

[1] Perm National Research Polytechnic University, 29, Komsomolskiy ave., Perm, 614990, Russia
[2] Branch of OOO «LUKOIL-Engineering» «PermNIPIneft» in Perm, 3a, Permskaya st., Perm, 614015, Russia

E-mail: alexandrraznitsyn@gmail.com

Abstract. The paper presents results of the comprehensive approach to identification of complicated carbonate rocks types. The subject of research is the Mishrif formation core from one of the Mesopotamian oil-and-gas province field wells. At the first stage the 3 lithological petrophysical rocks types were identified based on core nuclear magnetic resonance experiments taking into account structural features, which are determined with description data of thin sections. At the second stage the 5 petrophysical rocks types were identified based on routine and special core analyses taking into account secondary changes influence, which is determined by thin sections description data. The identified petrophysical rocks types were described in detail, linear discriminant functions were deduced. These functions allow predicting the identified petrophysical types with the suggested parameters. Comparison of the lithological petrophysical and petrophysical types has shown that the pore space structure of the studied deposits depended on two factors: the first is structural features and the second is postdepositional alterations.

1. Introduction

Carbonate rocks are characterized by an extremely heterogeneous pore space structure both vertically and laterally, which is caused by some reasons. On the one hand various facies conditions of deposits accumulation within a hydrocarbon field result in formation of a wide range of rock structural textural features and consequently markedly different flow and capacity properties. On the other hand secondary changes (such as solution, fracturing, stylolitisation, calcitization, etc.) alter original petrophysical properties in part or in whole.

Oil and gas reserves calculation and design of a field development in complicated reservoirs are sophisticated problems that require an unconventional approach. One of the approaches to petrophysical modeling is identification of rock types.

According to G E Archie [1] a «type of rock is a formation whose parts have been deposited under similar conditions and have undergone similar processes of later weathering, cementation or re-solution… a specific formation or rock type will have certain effective pore-size distributions which will produce a particular family of capillary pressure curves. The pore-size distribution controls the porosity and is related to the permeability and water saturation».

The purpose of this paper is presentation of the comprehensive approach to rock typing of complicated carbonate productive deposits based on core laboratory analyses data.

2. Subject of research and methods

The subject of research is the Mishrif formation core from one of the Mesopotamian oil-and-gas province field wells. In general terms studied deposits are non-uniformly oil saturated limestones.

The results of routine core analysis (RCAL) (porosity and gas permeability measurements) and special core analysis (SCAL) (capillary pressure measurements with porous plate method (gas-water system), nuclear magnetic resonance (NMR) experiments) of 91 plugs and thin sections description were applied for this study. The methods of mathematical statistics were used for data processing.

3. Identification of carbonate rocks types

As mentioned above, the pore space structure of carbonate rocks is depended on two factors: the first is structural features that are formed during sediment genesis and early diagenesis stages, the second is postdiagenesis (secondary) changes. Hence identification of carbonate rocks types should be performed in two directions.

In [2] lithological petrophysical types (LPRTs) of the studied deposits were identified taking into account their structural characteristics. Rock typing is based on integration of NMR experiments and thin sections description data. Identification of the classes is supported with the mathematical statistics methods (Student's t-test, correlation analysis, linear discriminant analysis, principal components method). The LPRTs are the 1st class (wackestone and wackestone/floatstone rock structures [3]), 2nd class (packstone, grainstone and rudstone rock structures) and 3rd class (framestone rock structure). Figure 1 shows averaged distributions of normalized NMR-intensities over transverse relaxation times T_2 of the identified LPRTs. As it is seen from Figure 1, the LPRTs are well distinguished with regard to NMR experiment results. Contributions to total porosity of different porosity types are presented under T_2 distributions. Also thin sections photographs of the identified LPRTs are shown upon T_2 distributions. NMR experiment results are consistent with thin sections images: the size of pores becomes bigger with increase of a T_2 distribution mode.

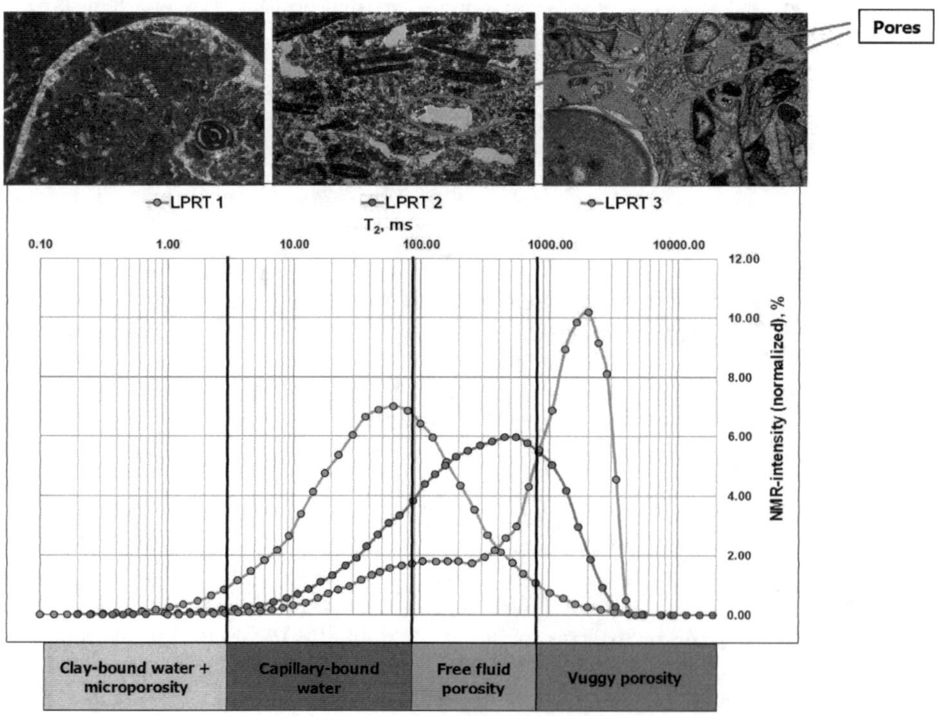

Figure 1. Averaged distributions of normalized NMR-intensities over transverse relaxation times T_2 of identified LPRTs [2].

We suggested usage of the two characteristics to identify petrophysical types (PRTs) based on secondary changes studying:

- The structural parameter (term as N) is determined from RCAL and can be found from [4, 5]:

$$N = \left(\frac{k_g}{\varphi}\right)^{\frac{1}{2}}$$

where N – the structural parameter, $mD^{0.5}$; k_g – gas permeability, mD; φ – porosity, unit fraction. This characteristic reflects the flow potential of the rock and it is sensitive to occurrence of open fractures and stylolites in core plugs.

- The logarithmic mean of transverse relaxation times T_2 is determined from the NMR experiment and can be found from [6]:

$$T_2logmean = 10^{\frac{\sum \varphi_i log(T_{2j})}{\sum \varphi_i}}$$

where $T_2logmean$ – the logarithmic mean of transverse relaxation times T_2, ms; φ_i – i-th porosity, unit fraction, corresponding to the j-th value of T_{2j}, ms. This characteristic reflects the capacity potential of rock and it is sensitive to occurrence of vugs in core plugs.

For futher PRTs identification we introduce the complex characteristic, which equals:

$$P_{comp} = \frac{lg\left(\frac{k_g}{\varphi}\right)^{\frac{1}{2}}}{lgT_2logmean}.$$

Figure 2 shows a cumulative curve of complex characteristic P_{comp}, where straight lines correspond to identified PRTs. It should be noted that boundaries of straight lines are adjusted taking into account thin sections description data. Comprehensive analysis of petrophysical and petrographical investigations results shows that the pore space structure of the studied deposits and consequently flow and capacity properties are most influenced by such secondary changes as fracturing and solution that lead to open fractures and vugs formation. Figure 3 presents a secondary changes distribution diagram for the identified PRTs. The diagram shows the percent of thin sections within the PRT for which one or another postdepositional process is common. As it is seen from figure 3, PRT 1 is characterized by absence of fractures and vugs, open fractures formation is a predominant secondary change for PRT 2, vugs formation is a predominant process for PRT 3, PRT 4 is characterized by both open fractures and vugs, open fracture formation is the only secondary change for PRT 5.

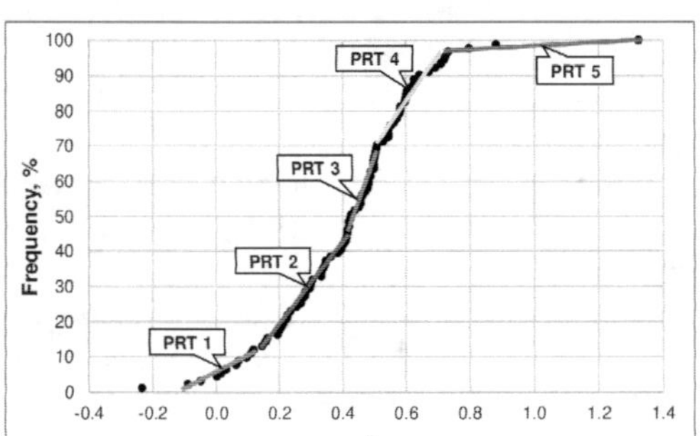

Figure 2. Cumulative curve of complex characteristic P_{comp}.

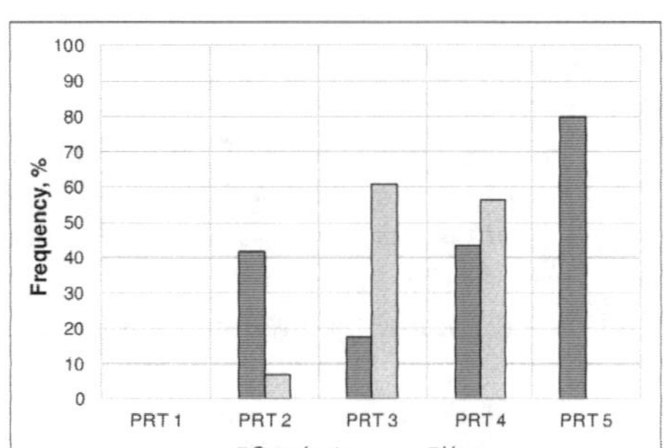

Figure 3. Secondary changes distribution diagram for identified PRTs.

Linear discriminant analysis (LDA) [7] is applied to predict the identified PRTs with suggested characteristics (N and T_2logmean). As a result of LDA application further linear discriminant functions (LDF) are deduced:

$$Z_1 = -6.86597lg\left(\frac{k_g}{\varphi}\right)^{\frac{1}{2}} + 1.34445lgT_2logmean + 0.27803;\ R = 0.809,\ \chi^2 = 39.309,\ p < 0.05;$$

$$Z_2 = -7.78547lg\left(\frac{k_g}{\varphi}\right)^{\frac{1}{2}} + 1.66833lgT_2logmean + 2.42816;\ R = 0.853,\ \chi^2 = 63.676,\ p < 0.05;$$

$$Z_3 = -10.34344lg\left(\frac{k_g}{\varphi}\right)^{\frac{1}{2}} + 5.07592lgT_2logmean + 0.94893;\ R = 0.839,\ \chi^2 = 52.331,\ p < 0.05;$$

$$Z_4 = -6.66023lg\left(\frac{k_g}{\varphi}\right)^{\frac{1}{2}} + 4.11913lgT_2logmean + 0.17892;\ R = 0.748,\ \chi^2 = 20.517,\ p < 0.05.$$

Identification certainty of PRT 1 plugs with LDFs is 100.00 %, PRT 2 – 96.55 %, PRT 3 – 100.00 %, PRT 4 – 95.65 % and PRT 5 – 80.00 %.

Table 1 presents the main statistical characteristics of PRTs petrophysical properties, which quantitatively describe the identified PRTs: the structural parameter (N), the logarithmic mean of transverse relaxation times T_2 (T_2logmean), mean pore throat radius (r), gas permeability (k_g), porosity (φ) and irreducible water saturation (S_{wirr}). A numerator is an arithmetic mean value (a geometric mean value for k_g) ± a standard deviation, a denominator is a value range. As it is seen from Table 1, the PRT 4 has the best reservoir properties and the PRT 1 has the worst ones.

Table 1. The main statistical characteristics of PRTs petrophysical properties

PRTs	N, $mD^{0.5}$	T_2logmean, ms	r, μm	k_g, mD	φ, %	S_{wirr}, %
PRT 1	1.11±0.33	33.976±7.330	0.515±0.174	0.064	6.21±2.23	80.69±8.66
	0.45-1.56	21.700-46.139	0.313-0.939	0.010-0.240	3.00-9.91	65.41-91.74
PRT 2	3.42±1.40	68.797±63.491	1.108±1.164	1.028	11.26±5.04	68.54±18.08
	1.54-7.74	19.019-346.615	0.280-6.638	0.136-14.080	5.11-23.48	27.32-93.35
PRT 3	10.82±4.17	173.387±110.658	2.393±1.488	18.991	19.60±5.98	42.29±20.65
	6.08-19.84	51.646-487.220	0.437-5.930	3.610-115.300	7.20-29.29	10.25-80.41
PRT 4	26.13±16.99	284.774±248.480	4.087±3.420	74.599	16.70±6.01	41.98±27.52
	9.27-79.36	38.055-795.678	0.635-10.266	7.770-1399.960	7.64-31.04	8.27-82.52
PRT 5	27.34±16.23	74.001±12.446	1.368±0.390	28.732	7.62±2.35	75.93±11.44
	15.46-54.73	12.633-42.142	1.012-1.901	18.540-48.860	4.47-10.16	64.19-91.03

Description of the pore space according to the P.W. Choquette and L.C. Pray's classification [8] for the PRTs is presented further.

PRT 1: the pore space consists of moldic, intercrystal and intraparticle pores. Flow and capacity properties predominantly depend on pores (a porous-type reservoir).

PRT 2: the pore space predominantly consists of moldic pores, also intraparticle pores and fractures. Interparticle pores and vugs contribute to total porosity occasionally. Flow and capacity properties depend on pores and fine open fractures (a fractured-porous-type reservoir).

PRT 3: the pore space predominantly consists of moldic pores, vugs and soluted intraparticle and interparticle pores. Fractures and channels contribute to total porosity occasionally. Flow and capacity properties predominantly depend on pores and vugs (a vuggy-porous-type reservoir).

PRT 4: pore space predominantly consists of moldic pores, vugs and intraparticle pores, less often – interparticle pores, fractures and channels, occasionally – intercrystal pores. This PRT comprises plugs with the growth-framework porosity. Generally void is soluted. This PRT can be separated into two groups: the first is a porous-vuggy-type reservoir, the second is a fractured-porous-vuggy-type reservoir.

PRT 5: the pore space consists of open fractures, moldic and intercrystal pores. Flow and capacity properties depend on pores and open fractures (porous-fractured-type reservoir). Fracture opening of

PRTs plugs is bigger in contrast to the PRT 2.

4. Comparison of identified lithological petrophysical (LPRT) and petrophysical (PRT) types of carbonate rocks

Contribution of every PRT to the LPRTs is calculated, Table 2 shows the results.

Table 2. Content (%) of every PRT in LPRTs

LPRTs	PRT 1	PRT 2	PRT 3	PRT 4	PRT 5
LPRT 1	17.46	36.51	20.63	19.05	6.35
LPRT 2	0.00	25.00	41.67	29.17	4.17
LPRT 3	0.00	0.00	0.00	100.00	0.00

As it is seen from Table 2, the LPRT 1 is characterized by presence of all identified PRTs with the PRT 1 predominated. It should be noted that presence of PRTs with vugs (the PRT 3 and PRT 4) in this LPRT is associated with newly formed vugs [9], because LPRT 1 rocks were nonporous originally.

The LPRT 2 is characterized by predominantelly presence of the PRT 3. Vug formation is inherent process for this LPRT, because LPRT 2 rocks were porous originally.

The LPRT 3 is characterized by the only presence of the PRT 4. This is caused by presence of growh-framework porosity of LPRT 3 rocks at a sediment genesis stage and later pore enlargement due to postsedimentational solution.

5. Conclusion

Application of the comprehensive approach allows identifying and describing lithological petrophysical and petrophysical types of studied carbonate rocks.

After conducted comparison of the LPRTs and PRTs, it is possible to conclude that the pore space structure formation of the studied deposits and consequently flow and capacity properties are controled by two factors: the first is structural features, which are formed during sediment genesis and early diagenesis stages (primary factor), the second is postsedimentational processes (secondary, imposed factor).

References

[1] Archie G E 1950 Introduction to Petrophysics of Reservoir Rocks *Bulletin of the American Association of Petroleum Geologists* **34(5)** 943-961

[2] Putilov I S, Raznitsyn A V 2020 Identification of Carbonate Rocks Petrophysical Types with Nuclear Magnetic Resonance Experiments Data Taking into Account Their Lithological Characteristics *Bulletin of Tomsk Polytechnic University. Geo Assets Engineering* **331(9)** 186-198

[3] Dunham R J 1962 Classification of Carbonate Rocks According to Depositional Texture *American Association of Petroleum Geologists Memoir* ed Ham W E (Denver: the Association and the Society of Economic Paleontologists and Mineralogists) vol. 1 pp 108-121

[4] Tul'bovich B I 1990 *Petrophysical Support of Effective Hydrocarbons Recovery* (Moscow: Nedra)

[5] Wibowo A S, Permadi P 2013 A Type Curve for Carbonates Rock Typing *International Petroleum Technology Conference* (Beijing: International Petroleum Technology Conference) pp 1-12

[6] Coates G R, Xiao L, Prammer M G 1999 *NMR Logging. Principles and Applications* (Houston: Halliburton Energy Services)

[7] Davis J C 1990 *Statistica and Data Analysis in Geology* (Moscow: Nedra)

[8] Choquette Ph W, Pray Ll C 1970 Geologic Nomenclature and Classification of Porosity in Sedimentary Carbonates *The American Association of Petroleum Geologists Bulletin* **54(2)** 207-250

[9] Bagrintseva K I 1999 *Formation Conditions and Properties of Oil and Gas Carbonate reservoirs* (Moscow: RGGU)

IOP Conf. Series: Earth and Environmental Science 1021 (2022) 012038 doi:10.1088/1755-1315/1021/1/012038

Investigation and analysis of cement slurries

O N Shemelina[1], V P Ovchinnikov[1], S S Shemelin[1], U V Dolgova[1], Yu N Pavelyeva[2], Yu S Minchenko[3]

[1] Tyumen Industrial University, 72, Melnikaite st., Tyumen, 625000, Russia
[2] FSBI VSEGEI, Sredny pr., 74, St. Petersburg, 199106, Russia
[3] North Caucasus Federal University, Stavropol, Kulakova str., 16/1, 355035, Russia

E-mail: onshemelina@mail.ru

Abstract. Laboratory studies of grouting slurries and cement compositions corresponding to specific mining and geological conditions are an important link in the design and construction of wells. In this work, a study of the rheological and physical-mechanical properties of cement grouting mixture and grouting cements was carried out. In the research for laboratory tests, samples of ground-bed mortars "GranTSEM - 7 BG" and "CemFrost", "CemFrost Lite - 5" were used. For the qualitative process of cementing, the analysis of grouting slurries was carried out with the study of changes in the consistencies of grouting slurries over time. Thus, the acceleration of the process of drilling observation and production wells depends on the correct selected cement slurry. As a result, of the studies carried out, it was revealed that the introduction of additives into the cement slurry that bind the barrier water and the drilling fluid helps to reduce fluid loss and increases the specific surface area, which speeds up the drilling process.

1. Introduction

The construction of observation and production wells is carried out in accordance with the approved design documentation [1].

The designs of observation and production wells are presented in Tables 1 and 2, respectively.

Table 1. The construction of injection wells

Name	Ø, mm	Descent interval, m	Height of cement lift, m	Test pressure, MPa	WOC, h
Surface casing	244,5	0-550	to the well head	9,0	36
Production string	168,3	0-1260	to the well head	11,5	48

Table 2. The construction of production wells

Name	Ø, mm	Descent interval, m	Height of cement lift, m	Test pressure, MPa	WOC, h
Surface casing	244,5	0-550	to the well head	9,0	36
Production string	168,3	0-1465	to the well head	11,5	48
Jetting stringer	114,3	1435-1616	-	-	-

Reservoir pressure in the productive reservoir PK_1 is 30 kgf/cm^2. Drilling under the 168.3 mm casing section is carried out on inhibited polymer-clay drilling mud with a density of 1,24 g/cm^3.

Drilling under the liner section is carried out using oil-based mud with a density of 1.07 g/cm^3.

An obligatory stage of well construction is casing, which includes lowering the casing strings and cementing them, which allows avoiding the collapse of the walls of the wells and preserving the reservoir properties of the productive formations [2].

Currently, an urgent task is the use of cement slurries with improved physical and mechanical characteristics. The most promising direction for preventing the contamination of productive formations during their fastening is to reduce the fluid loss of plugging materials [3].

2. Materials and methods

Laboratory studies of grouting slurries and cement compositions corresponding to specific mining and geological conditions are an important link in the design and construction of wells.

For laboratory tests, samples of ground-bed mortars "GranTSEM - 7 BG" and "CemFrost", "CemFrost Lite - 5" were used.

Complex-formulated multicomponent oil well cements are unique in the variety of their properties and the complexity of their compositions. The cement grouting mixture is intended for the preparation of grouting slurries used for cementing casing strings of oil, gas and gas condensate wells at low and moderate temperatures. The success of works to combat lost circulation is largely determined by the quality of the used grouting mixtures.

The mixing of the cement grouting mixture and grouting cements was carried out in accordance with GOST 26798.1-96 on a laboratory mixer LMR-1 at a rotation speed of 1500 rpm for three minutes.To determine the density of the grouting slurry, an ABR-1M hydrometer was used.

Determination of the flowability of the cement slurry was carried out using the cone shape. Determination of water separation of the cement slurry was carried out using a measuring cylinder with a volume of 250 ml at a temperature of 20°C.

For the cement grouting mixture "GranTSEM - 7 BG", the determination of water yield was carried out on a 7120 filter press (at a temperature of 20° C and a pressure drop of 6.8 MPa after conditioning for 30 minutes).

Determination of the thickening time was carried out on a sealed consistometer 7025A10 (HPHT) at a temperature of 20° C and a pressure of 14.9 MPa.

The setting time was determined using an IGLA-VIKA device at a temperature of 26°C. Determination of the compressive strength by ultrasonic analysis was carried out on an analyzer (HPNT) at a temperature of 26° C and a pressure of 20.6 MPa.

For backfill cements "CemFrost Lite - 5" and "CemFrost", fluid loss was determined using a 7120 filter press (at a temperature of 15°C and a pressure drop of 6.8 MPa after conditioning for 30 minutes). Determination of the thickening time was carried out on a sealed consistometer 7025A10 (HPHT) at a temperature of 15° C and a pressure of 7.3 MPa.

The setting time was determined using an IGLA-VIKA device at a temperature of 4°C. The determination of the compressive strength by ultrasonic pressure was carried out on an analyzer (HPHT) at a temperature of 4° C and a pressure of 20.6 MPa, after conditioning for 30 minutes at a temperature of 6°C.

Tables 3 - 6 show the test results of the cement grouting mixture and grouting cements.

Table 3. Technological properties of cement slurries

Cementing slurry	Density of cement slurry, kg/m^3	Water separation of the cement slurry, ml	Water separation of cement slurry, ml	Water loss of cement slurry, cm^3/30 min
«GranTSEM - 7 BG»	1710	220	0,2	95,0
«CemFrost»	1800	277	0,0	122,7
«CemFrost Lite - 5»	1520	260	0,6	140,5

Table 4. Technological properties of cement slurries

Cementing slurry	Static shear stress		Twisting angle at rpm						
	10 sec	10 min	300	200	100	60	30	6	3
«GranTSEM - 7 BG»	15	27	72	54	32	23,5	18	17	14
«CemFrost»	10	15	42,5	35,5	21	16,5	12,5	10	10
«CemFrost Lite - 5»	7	12	38,5	28	18	13	10	6	6

Table 5. Thickening time and setting time of cement slurries

Cementing slurry	The initial consistency of cement slurry, Vs	Thickening time of the cement slurry, min			Setting time of cement slurry, h-min	
		30 Vs	50 Vs	70 Vs	start	stop
«GranTSEM - 7 BG»	18	44	140	170	7-00	7-35
«CemFrost»	6	185	192	197	5-10	6-05
«CemFrost Lite - 5»	10	237	250	254	7-55	8-35

Table 6. Strength characteristics of plugging stones

Cementing slurry	Ultimate strength of plugging stone under compression according to UZA, MPa							
	50 psi	100 psi	500 psi	8 h	12 h	24 h	36 h	48 h
«GranTSEM - 7 BG»	5 h 31 m	6 h 02 m	8 h 46 m	2,52	7,83	20,33	25,48	27,69
«CemFrost»	4 h 40 m	5 h 10 m	12 h 31 m	2,93	3,36	3,67	3,85	3,99
«CemFrost Lite - 5»	5 h 46 m	6 h 11 m	6 h 59 m	6,91	8,42	8,84	9,04	9,14

3. The results of laboratory tests

The results of laboratory tests showed that the grouting slurry / stone "GranTSEM - 7 BG" is characterized by optimal rheological parameters and accelerated strength gain (after 24 hours, the compressive strength according to UZA was 20.33 MPa).

Figures 1 and 2 show the consistograms of the grouting slurry and the graph of the strength of the cement stone over time according to UZA.

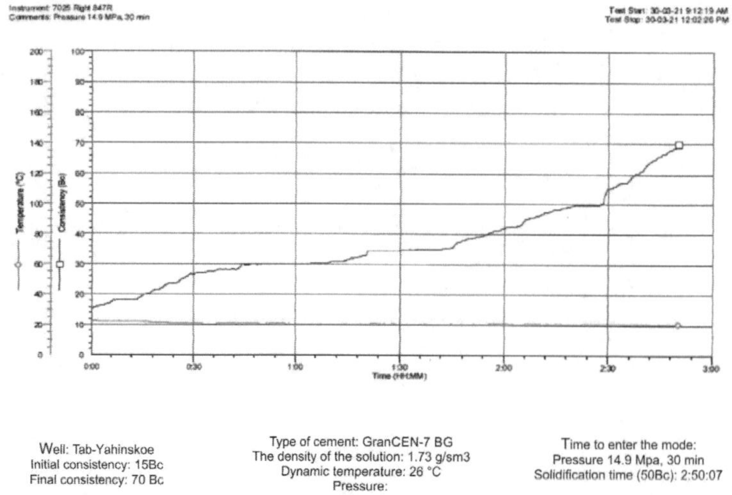

Figure 1. Change in the consistency of the grouting solution GranCEM - 7 BG

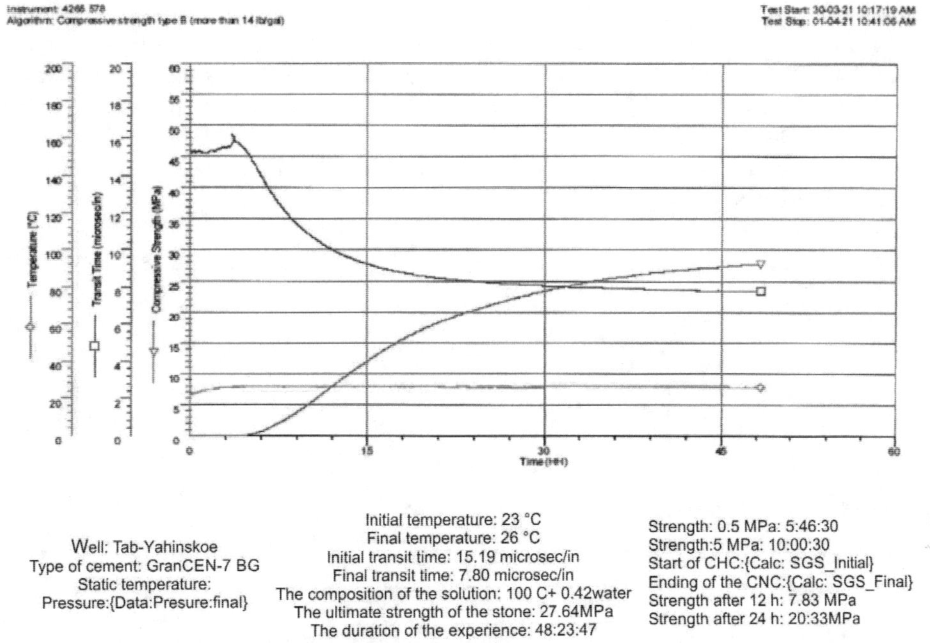

Figure 2. Strength of grouting stone GranTSEM - 7 BG

The results of laboratory tests showed that the cement slurry "CemFrost" has a high flow rate (277 mm), which allows one, if necessary, to increase the density of the cement slurry to a maximum value of 1900 kg/m^3 by reducing the water-cement ratio [4]. It is also noted that the tested backfill cement is characterized by an accelerated strength gain (after 24 hours, the compressive strength according to USA was 3.67 MPa).

Figure 3 shows the change in the consistency of the cement slurry CemFrost.

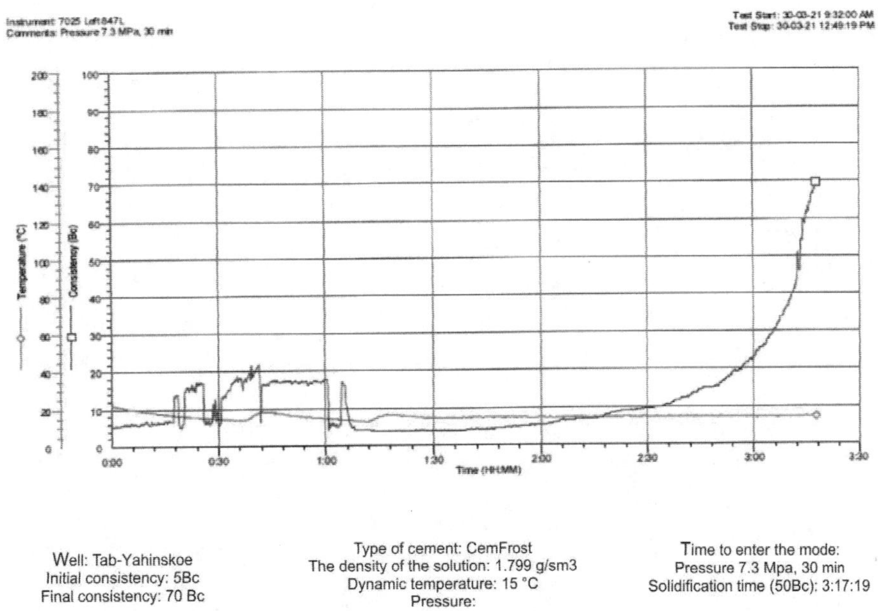

Figure 3. Change in the consistency of the cement slurry CemFrost

The results of laboratory tests have shown that the cement slurry / stone "CemFrost Lite - 5" has a high flow rate (260 mm), which allows one, if necessary, to increase the density of the cement slurry to a maximum value of 1600 kg/m^3 by reducing the water-cement ratio. It was also noted that the tested backfill material is characterized by an accelerated strength gain (after 24 h, the compressive strength according to USA was 8,84 MPa) [5].
Figures 4 and 5 show the consistograms of the grouting slurry and the graph of the strength of the cement stone over time according to UZA.

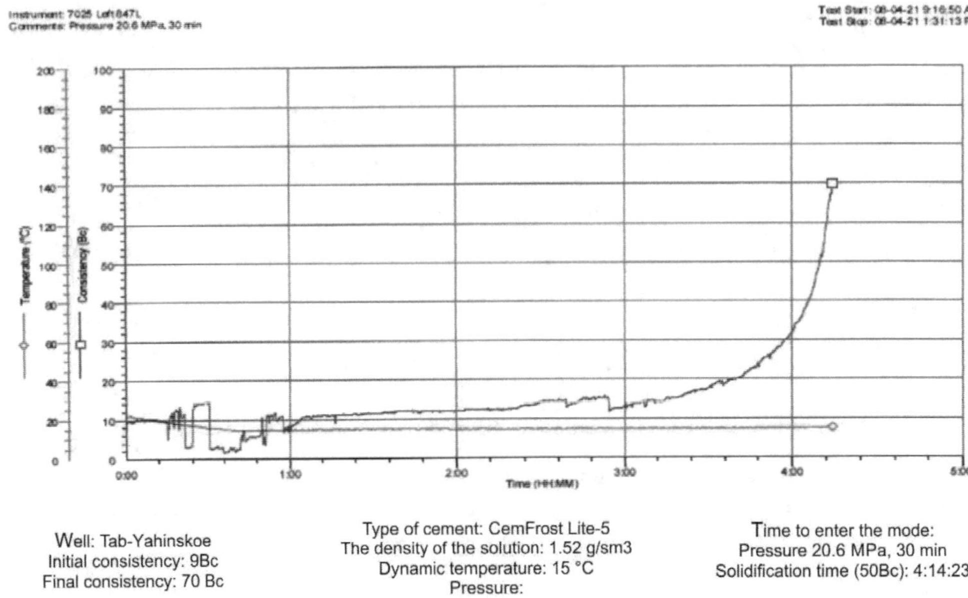

Figure 4. Change in the consistency of the cement slurry CemFrost Lite – 5

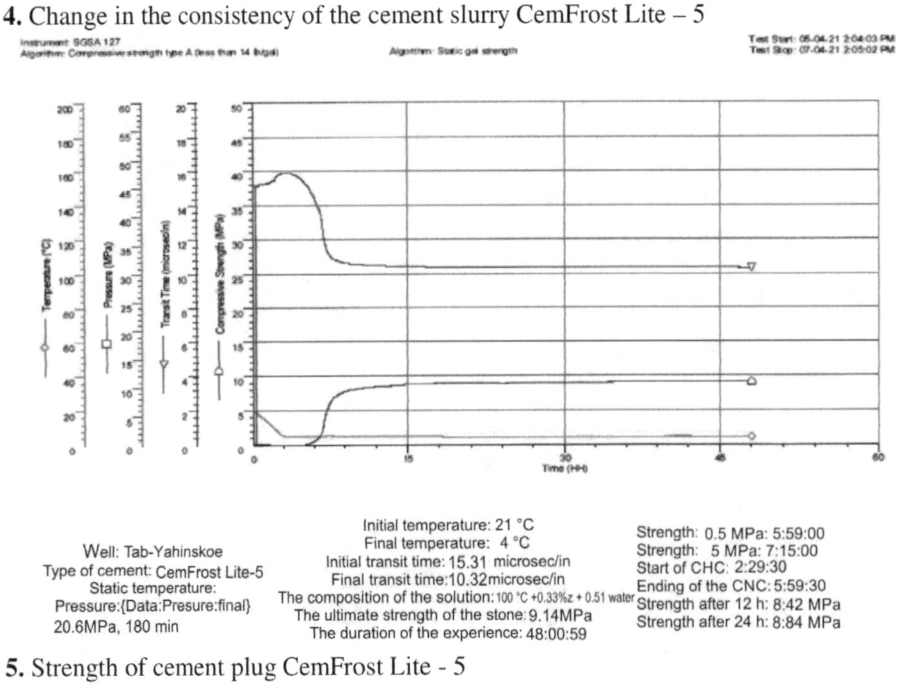

Figure 5. Strength of cement plug CemFrost Lite - 5

4. Conclusion

The introduction of additives into the cement slurry, which bind free mixing water and drilling fluid, reduces fluid loss, increasing the specific surface area. Fluid loss reducers block pore channels, which helps to reduce the volume of filtrate in the near-wellbore area and the degree of well contamination [6]. Thus, the acceleration of the process of drilling observation and production wells depends on the correct selected cement slurry.

5. Acknowledgments

I express my gratitude to the laboratory of Tyumen Industrial University, where the tests were carried out.

References

[1] Dvoinikov M N, Ovchinnikov V P, Budko A A, Gazgireev Yu O 2004 Well cementing at the Vanyeganskoye field *Drilling and oil* **5** 18-20
[2] Pavelyeva O N, Pavelyeva Yu N, Parshukova L A, Ovchinnikov V P 2020 Changes in filtration properties in reservoir rocks during drilling *Subsoil use XXI century* **1 (83)** 64-69
[3] Ovchinnikov V P, Kuznetsov V G, Gazgireev Yu O 2004 Analysis of the properties of plugging materials for cementing wells in the permafrost zone of the group *Tez. regional scientific. tech. conf.* (Tyumen) p 44.
[4] Shemelina O N 2019 Analysis of filtration properties in reservoir rocks during drilling *Collection of articles based on the materials of the VI All-Russian conference of young scientists* (V. 2) (Surgut: Science and innovations of the XXI century) pp 84-89.
[5] Ovchinnikov V P, Podnebesnykh A V, Yakovlev I G, Ovchinnikov P V, Saltykov V V, Rozhkova O V, Gerasimov D S, Kharitonova T A, Pavelyeva O N 2019 Processes of changing the filtering properties of oil and gas collectors during the construction and operation of wells *Textbook* (Tyumen) p 331. - ISBN 978-5-9961-2046-8.
[6] Shemelina O N, Ovchinnikov V P 2020 Development of the drilling mud formulation *Proceedings of the XXIV International Symposium named after Academician M. A. Usov students and young scientists dedicated to the 75th anniversary of Victory in the Great Patriotic War* (V. 2.) (Tomsk: Problems of geology and development of mineral resources) pp 439-440.

The effect of polymer-cement mortars with additives of aluminosilicate microspheres on the reservoir properties of exploration and production wells

R R Kadyrov[1], V V Mukhametshin[2], L S Kuleshova[1], D I Kobishcha[1] and A R Safiullina[1]

[1] Ufa State Petroleum Technological University, Branch of the University in the City of Oktyabrsky, 54a, Devonskaya St., Oktyabrsky, Republic of Bashkortostan, 452607, Russian Federation
[2]Ufa State Petroleum Technological University, 1, Kosmonavtov St., Ufa, Republic of Bashkortostan, 450064, Russian Federation

E-mail: markl212@mail.ru, info@of.ugntu.ru

Abstract. The article shows that the use of lightweight grouting solution with hollow aluminosilicate microspheres can reduce the time and the cost of materials spent on securing exploration wells, ensure the rise of solution to the design height. Improved strength and fracturing of the stone ensures resistance to fracture during perforation, and a sufficiently strong bond can ensure tightness of the annulus in exploration and production wells.
Keywords: polymer cement, acetone-formaldehyde resin, Portland cement, aluminosilicate microspheres, cementing of exploration wells.

1. Introduction
Provided that all environmental protection requirements are met, oil deposit development depends on the state of wells, the continuity of their operation, which determines the technical and economic indicators [1–8]. In this case, the quality of well casing is important.

Research in this area was aimed at solving two main problems: to lift the cement slurry to the wellhead, or at least to overlap the casing shoe and to separate the reservoir of the productive section.

2. Materials and methods
At the fields of the Ural-Volga region, the first problem was solved by cementing wells in two stages or by complicating the structure by running and cementing an intermediate string. The simple and economically viable single-stage cementing is excluded due to the absorption of cement slurry after the fracture of hydrodynamically unstable deposits located in three different intervals of the section. When eliminating this, for the upper part of the wellbore (1000 m), a lightweight grouting slurry is used based on alumina cement slurry (clay powder from local clays - calcium bentonite).

Cementing performance of production casing has improved. However, pressure gradients, reaching an average of 0.187 kPa / m, and insufficient resistance to hydraulic fracturing of rocks in the Artin and Upper Carboniferous (upper complex), Namyur and Serpukhovsko-Oka (middle complex), Lower Famennian and Upper Famennian (lower complex) sediments cause the absorption of cement slurry.

and the need for re-cementing [9].

Therefore, in order to lift the grouting slurry to the wellhead and form a continuous low-permeability cement stone in the annulus, it is possible to use lightweight grouting compositions with a density required for specific conditions; their hardening is not accompanied by shrinkage phenomena.

Lightweight grouting compositions are prepared by combining a binder with additives that have a significantly lower density compared to a binder or reagents that increase the water content of grouting solutions.

The additives used to obtain lightweight cement slurries are divided into the following classes [10]:

- additives with a low density;
- air-entraining additives, aerating cement slurries;
- highly dispersed additives that reduce the cement slurry density by increasing the amount of mixing water.

Traditional lightweight fillers are natural mineral substances of sedimentary (flask, diatomite, clay) or volcanic (tripoli, perlite, pumice) origin. However, these additives have a high water demand and are not technologically efficient.

Air-entraining additives reduce density of the grouting slurry, but increase porosity of the stone. This can destriy the material due to tensile stresses in closed pores under the action of water. Clay powders with a density of 2300-2600 kg / m3, which are usually added in an amount of up to 20% by weight of the solid phase, are widely used as lightweight additives.

Gel-cement solutions prepared from them are a mixture of Portland cement and clay powder. However, gel cements have serious disadvantages (insufficiently low density of gel cement mortar - 1650 kg / m3, high fragility of the stone during long-term storage). Clay powder weakens the cement stone, since an increased water content is required to ensure the flowability, which is the reason for the low rate of structure formation, low strength of the cement stone and its high permeability. This material does not ensure the necessary contact with the rock and metal of the casing string. This is the reason for the leakage of the annular space, low insulating characteristics of the lining and rapid water cut in the well production.

An analysis of literary sources and results of experimental studies made it possible to substantiate the requirements for a lightweight cement slurry and lightening additives. The most important are participation in the formation of the cement stone, low density, low water demand, manufacturability [11, 12]. In terms of cement stone structure formation and grouting slurry lightening, finely dispersed silica-containing materials are most effective.

The TatNIPIneft Institute developed a lightweight grouting slurry with a density of 1300-1600 kg / m^3.

Hollow aluminosilicate microspheres are mass-produced in Russia and represent a light free-flowing gray powder. They are obtained from the water flotation fraction of ash from coal combustion at thermal power plants. Mineralogical composition is represented by SiO2 - 50-60%; Al2O3 - 25-35%. Fracture strength is up to 28 MPa.

Hollow aluminosilicate microspheres are used as a lightening additive due to the presence of silicate and aluminate phases, which contributes to the participation of microspheres in the formation of the cement stone structure; in addition, they are crystallization centers in the cement slurry, which reduce the energy barrier for the hydration reactions.

Experimental studies have shown that the disadvantage of hollow aluminosilicate microspheres is their ability to bind large amounts of mixing water. The introduction of acetone-formaldehyde resin into the mixing liquid makes it possible to eliminate this drawback and improve the structure and properties of the grouting material (Table). A 6% aqueous solution of calcium chloride was used as the mixing liquid. The mixing liquid covers the surface of microspheres and cement grains with a thinnest film. The result of this is good plasticization, the fragility of the stone decreases and the resistance to alternating loads increases.

The basic properties of lightweight polymer-cement mortar and stone were determined at 20± 2°C

IOP Conf. Series: Earth and Environmental Science 1021 (2022) 012039 doi:10.1088/1755-1315/1021/1/012039

and atmospheric pressure in accordance with GOST 26798.1-96 "Backfill cements. Test methods". Fluid loss was determined using the VM-6 device. The brittleness of the resulting stone was determined as the ratio of ultimate compressive strength to ultimate bending strength.

Table 1. Composition and technological parameters of lightweight grouting mortar and stone

The composition of the lightweight polymer-cement mortar,% wt.					Density, kg / m3
portland cement	alumino-silicate microspheres	acetone-formaldehyde resin	6% calcium chloride solution	fresh water	
51.00	10.20	0.10	38.70	-	1550
51.65	10.33	0.40	37.62	-	1530
51.85	10.37	0.60	37.18	-	1520
52.35	10.47	0.80	36.38	-	1420
52.70	10.54	1.00	35.76	-	1350
50.95	10.19	0.1	38.76	-	1350
66.6	-	-	-	33.3	1830

Table 1. Continuation

Spread, cm	Setting time, h-min		Water drainage, cm3 / 30 min	Tensile strength of stone, MPa		Fragility
	beginning	end		bending	compression	
22.5	7-45	8-20	30.0	2.25	8.55	3.80
23.0	7-20	7-55	28.0	2.30	8.76	3.81
23.5	6-45	7-10	22.0	2.43	9.21	3.79
24.0	5-50	6-40	19.0	2.45	9.25	3.76
24.5	5-00	6-15	18.0	2.50	9.27	3.71
24.5	7-15	8-30	19.0	1.29	5.45	4.22
210	6-50	9-05	63.2	2.88	13.61	4.73

3. Results and Discussion

The table shows that a decrease in the mass content in the Portland cement solution below 51% deteriorates strength characteristics of the plugging stone and increases fragility.

Lightweight solution was prepared by the standard technology. A dry mixture consisting of oil well cement and microspheres was prepared in advance. Depending on the required amount of dry mixture, it can be prepared by two schemes:

- preparation of a dry mixture in an amount of more than 20 tons on a stationary unit;
- preparation of dry mix less than 20 tons using two cement carriers.

The scheme for preparing a dry mixture through two cement trucks was implemented in the following sequence:

- a cement truck with a compressor picks up the required amount of microspheres and cement;
- the mixture is mixed by repackaging from one cement truck (3AS-30) to another;
- the mixture is repackaged into a cement mixer (US 6-30) or a separate container intended for storing this mixture.

The mixing fluid is prepared by cementing units TsA-320M. The cement was mixed with microspheres in a traditional way using mixing machines and cementing units.

The technology has successfully passed industrial tests at the fields of PJSC TATNEFT. The efficiency of the field tests was 90%.

Some difficulties were caused by the impossibility of determining the height of the backfill slurry behind the casing, assessing the quality of stone adhesion based on the results of ASC in wells cemented with lightweight slurries with additives of aluminosilicate microspheres, due to their partial damping of acoustic vibrations.

To determine the lift height and the quality of adhesion to the column of lightweight cement slurry with sound-absorbing additives, geophysical studies are required.

The economic effect of using lightweight grouting slurries can be achcieved by reducing the waiting time for cement hardening (WPC), reducing the time and materials for cementing, improving the quality of cementing, which increases the turnaround time of wells and their durability.

The economic effect from lightweight grouting slurry was 350 thousand rubles per well.

4. Conclusion

Thus, lightweight grouting solution with hollow aluminosilicate microspheres can reduce the time and material costs for casing wells, and ensure that the solution is raised to the design height. The increased strength and fracture toughness of the stone makes it possible to exclude re-cementing when perforating the casing, and a sufficiently strong bond will ensure the tightness of the annulus.

Field tests were obtained in 9 wells; in 7 of them positive results were obtained.

References

[1] Kuleshova L S, Kadyrov R R, Mukhametshin V V, and Safiullina A R 2019 Design changes of injection and supply wellhead fittings operating in winter conditions *IOP Conference Series: Materials Science and Engineering* **560** 012072 DOI: 10.1088/1757-899X/560/1/012072

[2] Soloviev N N, Mukhametshin V Sh and Safiullina A R 2020 Developing the efficiency of low-productivity oil deposits via internal flooding *IOP Conf. Ser.: Mater. Sci. Eng.* **952** 012064 DOI: 10.1088/1757-899X/952/1/012064

[3] Mukhametshin V Sh 2020 Rationale for the production of hard-to-recover deposits in carbonate reservoirs *IOP Conf. Ser.: Earth Env.* **579** 012012 DOI: 10.1088/1755-1315/579/1/012012

[4] Kuleshova L S, Kadyrov R R, Mukhametshin V V, and Akhmetov R T 2019 Auxiliary equipment for downhole fittings of injection wells and water supply lines used to improve their performance in winter *IOP Conference Series: Materials Science and Engineering* **560** 012071 DOI: 10.1088/1757-899X/560/1/012071

[5] Mukhametshin V Sh, Andreev V E, and Yaskin S A 2020 Designing measures to increase oil recovery based on the identification and grouping of deposits *IOP: Earth and Environmental Science (EES)* **579** 012013 DOI: 10.1088/1755-1315/579/1/012013

[6] Malyarenko A M, Kotenev Yu A, Mukhametshin V Sh, Bogdan V A, Kotenev A Yu, Blinov S A, and Umetbaev V G 2020 The influence of the rocks material composition on the reservoir properties of the reservoir *Journal of Physics: Conference Series* **1582** 012061 DOI: 10.1088/1742-6596/1582/1/012061

[7] Mukhametshin V Sh 2020 Justification for increasing the performance of hydrochloric acid treatment in wells of fields with carbonate reservoir *IOP Conf. Ser.: Mater. Sci. Eng.* **905** 012083 DOI: 10.1088/1757-899X/905/1/012083

[8] Kadyrov R R, Mukhametshin V V, Galiullina I F, Kuleshova L S, and Safiullina A R 2020 Prospects of applying formation water and heavy brines derived therefrom in oil production and national economy *IOP Conference Series: Materials Science and Engineering* **905** 012081 DOI: 10.1088/1757-899X/905/1/012081

[9] Muslimov R Kh, Shavaliev A M, Khisamov R B and Yusupov I G 1995 *Geology, development and exploitation of the Romashkinskoye field* (Moscow: VNIIOENG Publ)

[10] Savenok N B and Mariampolskiy N A 1992 Methods for improving lightweight grouting systems *Oil Industry* **12** 8-9

[11] Yakevich V F 2002 Application of lightweight grouting solutions for cementing productive layers *Construction of oil and gas wells on land and sea* **7-8** 40-43

[12] Oreshkin D V 2002 Lightweight grouting materials *Construction of oil and gas wells on land and sea* **11** 21-23

A study of the impact of oil recovery on the well grid compaction in the oil-containing layers

E M Almukhametova and I A Fayzullin

Ufa State Petroleum Technological University, Branch of the University in the City of Oktyabrsky, 54a, Devonskaya St., Oktyabrsky, Republic of Bashkortostan, 452607, Russian Federation

E-mail: elikaza@mail.ru, info@of.ugntu.ru

Abstract. The paper examines applications of the technology of compacting the well grid by drilling small-diameter wells in oil-containing layer N and identifies its influence on the oil recovery factor. The effectiveness of the results of grid drilling using small-diameter wells in oil-recoverable reservoir N and an increase in the efficiency of developing the oil-saturated layer by compaction of the existing development grid by drilling small-diameter wells are analyzed. Research methods used are the methods of systems analysis, mathematical modeling; mathematical programming, mathematical statistics, cluster analysis, numerical methods, computational experiments; database design theory.

1. Introduction

The relevance of the study is due to the fact that most oil fields are complex and heterogeneous, the reservoir properties of the productive layers are variable and have extensive water-oil zones.

As the fields are developed, there is a change in the structure of oil reserves; the share of hard-to-recover reserves begins to grow [1].

The efficiency of oil production decreases, which requires adjustments to the operating modes of the wells. It is possible to implement geological and technical measures for injection and production wells. Methods for increasing the volume of oil recovery, depending on the conditions of a particular field should be used. Therefore, it is an urgent task to develop innovative geological, technical and organizational and technological solutions aimed at increasing the efficiency of complex fields under he financial and economic conditions in Tatarstan [1]. The development of hard-to-recover oil reserves and an increase in the efficiency of the existing wells are largely constrained by the fact that there is a very limited number of proven methods. The problems include the choice of optimal density of the well grid and layer opening. It is necessary to determine the most efficient well operation modes and select the optimal value of the grid density. At TATNEFT company, construction of wells on a compacted grid implies drilling of small-diameter wells. The production casing diameter of these wells varies from 102 mm to 114 mm. It is also worth noting that drilling of small-diameter wells can reduce material costs by 58% [2].

2. Materials and methods

The experience of designing systems for the development of low-productive deposits and fields of the Middle and Lower Carboniferous shows that their development using traditional methods is not efficient and sometimes unprofitable. The main reason is the high cost of traditional wells and the low oil recovery factor (ORF). The recovery factor can be increased by compacting the grid by drilling inexpensive small diameter wells (SDW).

Pilot-industrial infill drilling of SDWs in the Republic of Tatarstan increased the rate of decline in oil production in the old well stock. At the same time, the rate of decline in production for SDWs is higher than for the old ones.

Oil reserves involved in development differ tenfold and amount to 1.6 and 17.4 thousand tons per field, respectively. Consequently, during the development of a deposit in a natural mode, wells drilled with the aim of compacting the well grid involve reserves of fractured zones, wells interact, and the economic indicators of development deteriorate [12].

Infill injection wells during the development of the waterflooding system involve three times more reserves compared to the old producing wells.

Consequently, in the context of modern market relations and the wide-scale use of modern technologies in the oil industry, as well as taking into account the results of the analysis of the development of pilot sites for drilling out carbonate wells, multiple mesh compaction with small-diameter wells is advisable.

For field N, over the last two years the total volume of injected water is noticeably less than the accumulated fluid production for the same period (compensation is less than 100%).

The consequence of insufficient compensation of production by injection and a decrease in bottomhole pressure is a decrease in reservoir pressure [8].

The complexity of carbonate deposits, and poor knowledge of reservoir properties of rocks lead to the fact that the deposit and oil and gas reserve parameters are justified with low reliability. Errors are large when calculating recoverable reserves. However, to determine the technological indicators of the development of such deposits and to plan production, a more reliable estimate of reserves is required. When analyzing the development of many carbonate deposits in the Ural-Volga region, new patterns were revealed [1, 2.7].

Recoverable oil reserves are usually determined by multiplying the value of balance reserves by the oil recovery factor expected by the time the reservoir is depleted. The latter is justified and accepted according to the data on the operating mode of the deposits. Since the balance reserves and the oil recovery factor are obtained with large errors, the product of these two inaccurate values significantly reduces the accuracy of calculation of recoverable reserves [4].

3. Results and discussion

The method for determining recoverable reserves using direct depletion is more accurate. Using the method developed by Kopytov A, it is possible to eliminate regular errors in determining some reserve parameters and estimate the real value of the oil recovery factor based on field oil production data.

Table 1 presents data for assessing the dependence of the oil recovery factor, estimated by the Kopytov's displacement characteristic on the well grid density (WGD, S) for the experimental areas of the Bashkir deposits in the Republic of Tatarstan [1].

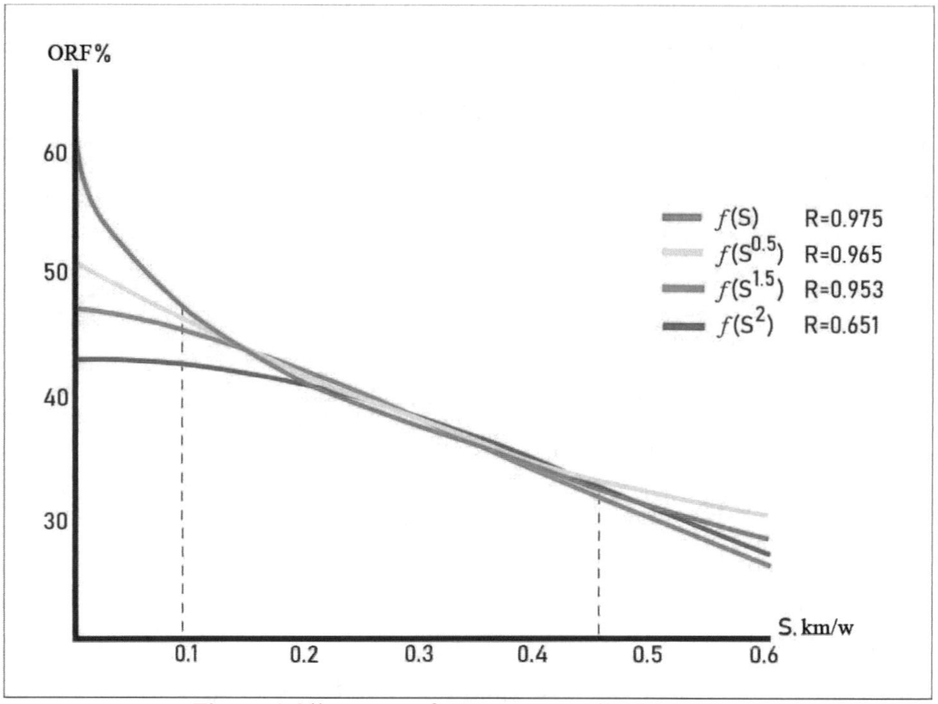

Figure 1 Oil recovery factor versus well grid density

The areas of applicability of this formula were established depending on the WGD and the value of exponent n. The study used the method of grouping objects by the same geological and physical parameters (in relation to the specific balance reserves to the amplitude production rate of wells). Three groups were identified. The actual data with a minimum error were approximated by a generalized dependence at ϖ =1 and 1.5, respectively, in the areas of sparse (0.1 ÷ 0.5 km^2 / well) and dense (0-0.25 km^2 / well) well grids.

Table 1. Summary data of the assessment of the dependence of oil recovery on oil recovery factor for experimental areas

No	Well	OGE	Field	Experimental site	PSS, ha / well	ORF, un.	Ln ORF*	$S^{1,5,}$ $(km^2/w)^{1.5}$	S, km^2/w
1	Bashkirsky	YaN	Yamashinskoe	2	4.5	0.332	-1.103	0.0095	0.0450
2		YaN	Yamashinskoe		8.8	0.193	-1.645	0.0261	0.0880
3		YaN	Yamashinskoe	1	7.2	0.226	-1.487	0.0193	0.0720
4		LN	Romashkinskoe	1	5.5	0.367	-1.002	0.0129	0.0550
5		LN	Romashkinskoe	2	3.9	0.342	-1.073	0.0077	0.0390
6		LN	Romashkinskoe	3	4.8	0.256	-1.363	0.0105	0.0480
7		LN	Romashkinskoe	4	8.1	0.298	-1.211	0.0231	0.0810
8		NN	Aksubaevo-Mokshinskoe	Balandinsky	38	0.015	-4.200	0.2342	0.3800

9	NN	Aksubaevo-Mokshinskoe	Mokshinsky	35	0.045	-3.101	0.2071	0.3500
10	NN	Nurlatskoe	-	37.0	0.032	-3.442	0.2251	0.3700
11	NN	Bureykinskoe	-	41.0	0.018	-4.017	0.2625	0.4100

In a narrow interval of the WGD study (0.2 ÷ 0.35 km² / well), ϖ can take any values.

Figure 2 shows the dependence of the oil recovery factor on the WGD for the experimental areas of the Bashkir deposits [2].

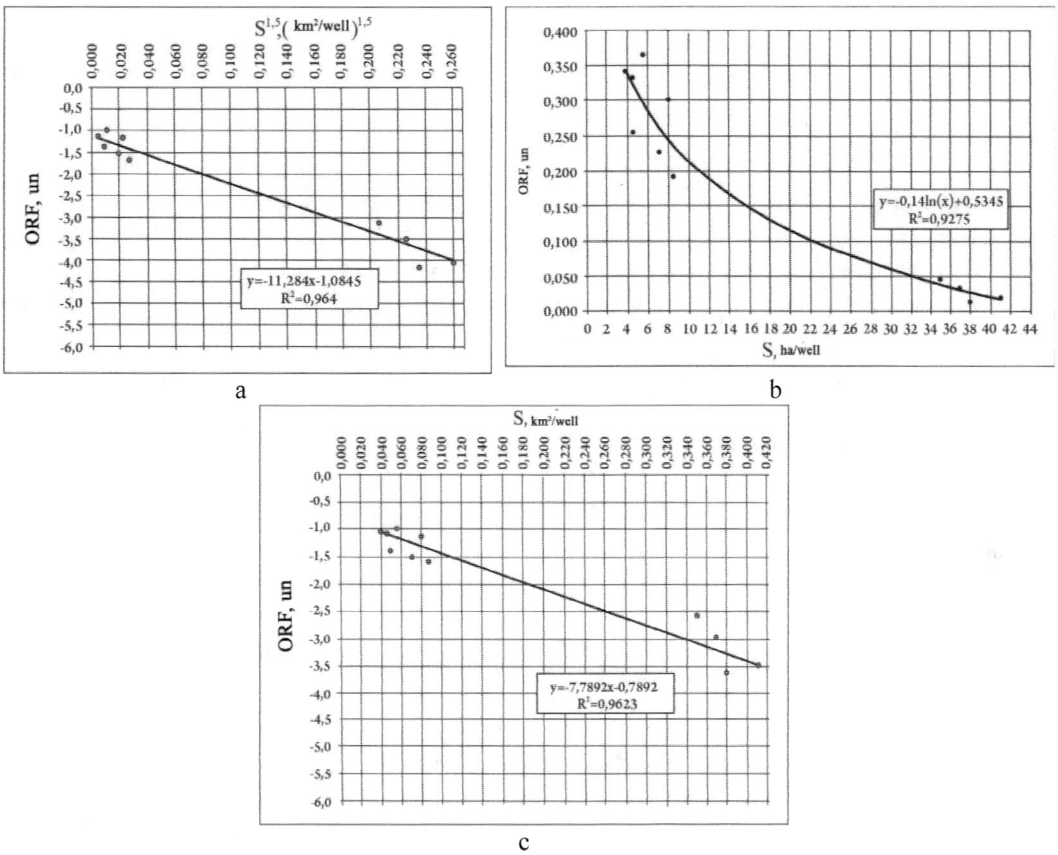

Figure 2. Dependence of the oil recovery factor on well grid density for the experimental sections of the Bashkir deposits

Extrapolation of these dependences in the area of dense grids S <0.010 km² / well leads to a significant deviation of the curves from each other. The highest correlation ratio in the processing of actual data in the dense well grids S = 0.01 ÷ 0.25 km² / well has a dependence at ϖ=1,5.

The degree of influence of the WGD on oil recovery in carbonate reservoirs was estimated by processing the final oil recovery coefficients depending on the well grid density using formula (2). The processing of initial data on this dependence allows determining their parameters using 2-3 points (values).

Figure 2 shows the dependence of the recovery factor on the WGD in semi-logarithmic coordinates for the experimental areas of the Bashkir deposits at ϖ=1and ϖ=1,5, respectively.

Linear trends for the ORF - WGD dependence were obtained and linear equations were developed. The results are shown in Table 2.

Table 2 Dependences of the oil recovery factor on the density of the well pattern for carbonate Bashkir sediments

Well	FGD interval km2 / well	ORF interval	Equation type n=1 - 1,5	Square correlation coefficient
Bashkir	0.039-0.41	0.015-0.367	η=1, $\eta = 0.454 \cdot e^{-7.72 \cdot S}$	0.9623
			η=1.5, $\eta = 0,338 \cdot e^{-11,3 \cdot S^{1,5}}$	0.964

As can be seen from Table 2, at ϖ=1, overestimated values of coefficient A were obtained, which confirms the need for denser well grids (most of the points through which the trend line is drawn are just points with denser well grids) to use the dependence at ϖ=1.5.

Table 3 shows values of the recovery factor obtained from the equations in Table 2 at ϖ=1.5 and using common WGD values.

Table 3. KIN at various FGD values

FGD			ORF.		
m×m	ha/w	km²/w	Bashkir	Turney	Verey
		Triangular well grid			
100×100	0.87	0.0087	0.335	0.450	0.386
150×150	1.95	0.0195	0.328	0.440	0.381
200×200	3.46	0.0346	0.314	0.421	0.371
250×250	5.41	0.0541	0.293	0.392	0.355
300×300	7.79	0.0779	0.264	0.352	0.332
350×350	10.61	0.1061	0.229	0.303	0.304
400×400	13.86	0.1386	0.189	0.248	0.269
		Square well grid			
100×100	1.00	0.0100	0.334	0.449	0.385
150×150	2.25	0.0225	0.325	0.436	0.379
200×200	4.00	0.0400	0.309	0.413	0.367
250×250	6.25	0.0625	0.283	0.378	0.347
300×300	9.00	0.0900	0.249	0.331	0.320
350×350	12.25	0.1225	0.208	0.275	0.286
400×400	16.00	0.1600	0.164	0.215	0.246

Thus, studies on the influence of well grid density on the oil recovery factor revealed the following dependencies for the three carbonate deposits of the Tatarstan fields:

- for the Bashkirian layer:

$$\eta = 0.338 \cdot e^{-11.3 \cdot S^{1.5}} \quad (1)$$

- for the Turneystage:

$$\eta = 0.454 \cdot e^{-11.7 \cdot S^{1.5}} \quad (2)$$

- for the Verey horizon:

$$\eta = 0.338 \cdot e^{-7.1 \cdot S^{1.5}} \quad (3)$$

IOP Conf. Series: Earth and Environmental Science 1021 (2022) 012040 doi:10.1088/1755-1315/1021/1/012040

4. Conclusion

In the context of falling oil prices, each development option can be characterized by three main indicators: the volume of oil production, the economic indicators of hydrocarbon field development and the oil recovery factor. These indicators characterize the efficiency of field development, often contradict each other; the subsoil user and the government are interested in achieving the maximum value of various indicators [8].

It is necessary to search for a compromise between these criteria of rationality, to find the best balance of interests between the user of the subsoil, the government and the requirements for the protection of subsoil and the environment [6].

Taking into account interests of the subsoil user, the development option or individual geological and technical measures should bring maximum income to the subsoil user. According to this principle, the optimal grid density was selected by the first option, in which the oil recovery factor is 0.381 (with the approved one of 0.294) and the NPV is 3.4 billion rubles [2].

References

[1] Almukhametova E M 2018 Modelling and assessing the effectiveness of developing the Shumovskoye field *IOP Conference Series: Earth and Environmental Science* **194** 062003

[2] Akhmetov R T, Mukhametshin V V, and Andreev A V 2017 A quantitative assessment method of the productive formation wettability indicator according to the data of geophysical surveys *SPE Russian Petroleum Technology Conference (16-18 October 2017, Moscow)* DOI: 10.2118/187907-MS

[3] Almukhametova E M 2006 Optimization of FPM system in Barsukovskoye deposit with hydrodynamic modeling and analysis of inter-well interaction *J. Phys.: Conf. Ser.* **1015** 032006

[4] Gabdrakhmanov N Kh, Davydova O V and Valeyev M D 2018 Studying gas separation at the inlet of the oil well bottom pump *IOP Conference Series: Earth and Environmental Science* **194** 082011 DOI:10.1088/1755-1315/194/8/082011

[5] Galiullina I F and Kadyrov R R Technical and economic background for siting production of well-killing liquid at oil fields *IOP Conference Series: Earth and Environmental Science* **194(8)** 082013 DOI:10.1088/1755-1315/194/8/082013

[6] Hoefner M L, Fogler H S 1988 Pore evolution and channel formation during flow and reaction in porous media *AIChE Journal* **34 (1)** 45-54

[7] Mukhametshin V V 2018 Rationale for trends in increasing oil reserves depletion in Western Siberia cretaceous deposits based on targets identification *Bulletin of the Tomsk Polytechnic University. Geo Assets Engineering* **329(5)** 117–124

[8] Nierode D E, Williams B B 1971 Characteristics of acid reaction in limestone formations *SPE J* **11**(4) 406-418

[9] Schechter R S, Gidley J L 1969 The change in pore size distribution from surface reactions in porous media *AIChE J.* **15(3)** 339-350

[10] Shakurova Al F, Shakurova Ay F 2018 The influence of hydraulic fracturing on the estimated ultimate recovery *IOP Conference Series: Earth and Environmental Science* **194** 082039 DOI:10.1088/1755-1315/194/8/082039

[11] Valko P, Norman L, Daneshy A 1998 Petroleum Well Construction Chapter 17 Well Stimulation Wiley

[12] Wang Y, Hill A D, Schechter R S 1993 *The optimum injection rate for matrix acidizing of carbonate formations* SPE Annual Technical Conference and Exhibition

IOP Conf. Series: Earth and Environmental Science 1021 (2022) 012041 doi:10.1088/1755-1315/1021/1/012041

Assessment of the impact of geological and production parameters on the effectiveness of acid fracturing in the fields with carbonate reservoirs

E V Voronova, L V Petrova and A Yu Gutorov

Ufa State Petroleum Technological University, Branch of the University in the City of Oktyabrsky, 54a, Devonskaya St., Oktyabrsky, Republic of Bashkortostan, 452607, Russian Federation

E-mail: razboinica77@mail.ru, info@of.ugntu.ru

Abstract. A number of different geological and production parameters affecting successful acid fracturing are considered in the conditions of carbonate reservoirs of oil fields in Bashkortostan. The study results are of use for assessing the effectiveness of geological and technical measures and for a more reasonable choice of candidate wells for acid fracturing.

1. Introduction

To date, the main share of active oil reserves in Bashkortostan fields is confined to the carbonate reservoirs of the Kashiro-Podolsk deposits, and Kashirskiy, Vereiskiy, and Bashkirian stages, as well as the Kizelovskiy horizon of the Tournaisian stage. At the same time, reservoirs are distinguished by a complex geological structure, significant variability of the section and lateral, and in most cases they exhibit low reservoir properties, such as permeability, which can vary from 7 to 40 mD, and in some cases, increased clay content [1–3].

Development of oil reserves from such reservoirs requires various geological and technical measures to increase the natural permeability of the reservoirs during well development. In carbonate reservoirs, bottomhole treatments with various chemical reagents, hydrochloric acid treatments, thermal gas and chemical treatment and acid fracturing, as well as multistage hydraulic fracturing for horizontal wells, significantly contribute to productivity increase.

2. Materials and methods

To assess the effectiveness of the measures taken, it is necessary to consider a large number of geological and field factors. Geological factors include: current net thickness of the reservoir, current oil and water saturation, distance to the water-oil contact (which affects the initial water cut), porosity, cement composition, the presence of natural fracturing, permeability and conductivity, coefficients of lithological heterogeneity in the section, such as the number of reservoir beds and the number of interlayers, and in the lateral, such as the discontinuity coefficient, current recoverable reserves per well, and geomechanical parameters that will affect directly during hydraulic fracturing [3,6].

Geomechanical modeling provides mapping of zones appropriate for drilling, for example, a map of the lost-circulation zone gradient with clusters of lost circulation risks, and enables estimation of the critical pore pressure. In the geomechanical modeling, the tasks of fracture modeling include: estimation of the fracture geometry, determination of the optimal technological parameters of injection, analysis of the fracture capacity sensitivity to the input parameters, determination of the ultimate strength of rocks, assessment of the stress state of rocks during hydraulic fracturing and the height of the development of fractures and their direction, which is very important if a pressure maintenance system is implemented within the reservoir. In this case, it is important that the developing fracture does not penetrate into the watered zone [8].

Capabilities of geomechanics for modeling fracturing include forecasting the development of spatial orientations of fluid-conducting fractures, and probabilistic assessment of effective properties, such as fracture porosity, aspect ratio, cohesion, quantitative assessment of the proportion of fluid-conducting fractures [2–5]. In addition, in the production drilling, geomechanical modeling provides mapping of various lost circulation risks for drilling fluid, which enables taking measures to reduce risks in zones with medium and high absorption to adjust the well trajectory and, if necessary, geomechanically accompany drilling of the horizontal section of wells.

Field factors include: reservoir pressure before hydraulic fracturing, distance to the injection well, injection capacity, current and accumulated compensation ratio, washing coefficient, well grid density, etc. In addition, the choice of acids for acid fracturing and their ratio, and the holding time, which ultimately affect productivity after the above operations.

To analyze the effectiveness of geological and technical measures, such indicators as the increase in oil production after the impact, the multiplicity of increase in oil production, the value of cumulative oil production over 7–20 years, profitability, the duration of the effect, the NPV value, as well as operation expenses and capital expenditures (OPEX and CAPEX).

For carbonate rocks, especially with natural fracturing and caverns, a petrophysical porosity-permeability relationship is created for imposing data from hydrodynamic studies. In this case, fractured properties are characterized by higher permeability at the same porosity value. As noted in [1–2], during the field development, the natural properties of layers can undergo changes associated with the stress-strain state of rocks (for example, compaction) under the impact of anthropogenic factors (changes in the field development system).

The presence or absence of relationships between the above parameters and an increase in oil production can be identified using the well-known correlation and regression analyzes, which highlight the most significant parameters that have the greatest impact and establish the presence or absence of the relationship.

The criteria for the effectiveness of the measures taken can be as follows: the profitability of oil production over a certain period of time (typically 5–20 years), the increase in oil production rate, payback [8–9].

A number of deposits of the Bashkorstotan fields that includes carbonate reservoirs of the Middle Carboniferous (Kashirian C_2ks, Vereiskiy C_2vr horizons and Bashkirian C_2b stage) were chosen as the study object.

The oil content of the Kashirian horizon is associated with organogenic detrital and dolomite limestones [3], and dolomites. The reservoir porosity varies from 9 to 20%, averaging 13%; the permeability is 12 mD; the average net thickness is 2 m; the initial oil saturation is 68–70%. The reservoirs of the Vereya horizon are organogenic and organogenic-clot-lumpy limestones, with porosity of 11–14%, permeability of 3–16 mD and initial oil saturation of 69–74% (for C_2vr_1 and C_2vr_3 benches). The initial oil saturation of C_2vr_4 is very low, which, together with thin layering (less than 0.8 m) and clay content, creates the conditions for the non-participation of the reservoirs of this bench in the recovery process.

In the considered deposit, C_2vr_4 bench is water-bearing. The oil-bearing capacity of the bench of the Bashkirian stage is associated with the upper part of its section. Oil deposits throughout the entire field (including the studied deposit) are underlain by bottom water. The reservoirs of the bench are developed in the form of three or more interlayers of porous-permeable organogenic-lumpy and organogenic-clotted varieties among dense organogenic-detrital, organogenic-detrital and crystalline limestones [10–12]. Secondary processes are widely developed. The share of reservoirs is 0.29, the coefficient of heterogeneity is 7.4. According to the data of hydrodynamic studies of wells, permeability is an order of magnitude higher (it passes along the upper boundary of the cloud of petrophysical dependence), which indicates the developed fracturing of the reservoirs.

In this case, part of the wells in the reservoir were in the joint exploitation of the Kashirsky and Vereisky horizons, the Bashkirian stage was developed by its own system. The main measures taken were: 5 acid fracturing, 2 treatments of the bottomhole of the reservoir zone, 1 hydrochloric acid treatment, 1 transition to the higher horizon, and 1 reperforation.

3. Results

The oil flow from carbonate reservoirs with low permeability can be initiated as follows: hydrochloric acid treatment and acid fracturing, which is especially important for horizontal wells, the profitability of which depends on the amount of cumulative oil production over a certain period. The number of stages required for multistage fracturing is determined with regard to the stress-strain state of rocks, the length of the horizontal wellbore, and the reservoir properties [1–2, 13–14].

In this case, the analysis of all the geological and technical measures taken yielded the following: oil production exhibits 6.5-fold increase after hydraulic fracturing, 2-fold increase after reperforation, 1.03-fold increase after hydrochloric acid treatments, including 1.2-fold increase in the well injection capacity. Thus, the implementation of acid fracturing in this reservoir can be considered successful.

It was revealed how porosity, distance to the reservoir contour, permeability, distance to the nearest injection well, and current reservoir pressure affect the multiplicity of increase in oil production rate.

Thus, distance to the injection well increased from 200 to 600 m increased the multiplicity of increase in oil production from 1 to 3 times. The reservoir pressure in the sampling zone increased from 30 to 70 atm, the multiplicity of increase in oil production also increased from 1 to 3 times. The energy state in the zone of the proposed production immediately before hydraulic fracturing is of great importance. There are two problems arising in carbonate reservoirs: the need to ensure sufficient compensation for production by injection in order to prevent a decrease in reservoir pressure by less than 25% of the initial one and premature water breakthrough from the front of the injected water and formation of flooding tongues. In addition, the presence of natural fractures can lead to inappropriate injection of fluid due to multiple excess of injection pressure over the reservoir or to unwanted breakthroughs of the water front from injection to production wells due to the formation of dagger water breakthroughs through channels of pronounced anisotropy associated with fractures.

For small deposits with an active contour, the distance to the contour is of importance. In the analyzed example, an increase from 400 to 700 m increased the multiplicity of increase in oil production from 2 to 3 times, the most insignificant increase in oil production was characteristic of wells with a distance of 200–300 m. Since permeability is petrophysically related to porosity, it was found that increased porosity leads to a decrease in oil production rate. This is due to the fact that wells with good permeability were located in the peripheral part of the reservoir, close to the contour zone, as a result of which, due to the good connectivity of this area, there was a faster breakthrough of water along the hydraulic fracture from the contour and an increase in water cut. At the same time, the largest increase was observed in wells with porosity of 10–14%. In this case, the multiplicity of increase in oil production was 1.5–3 times, and with porosity increased from 14 to 16%, the multiplicity of increase in oil production was 1 time.

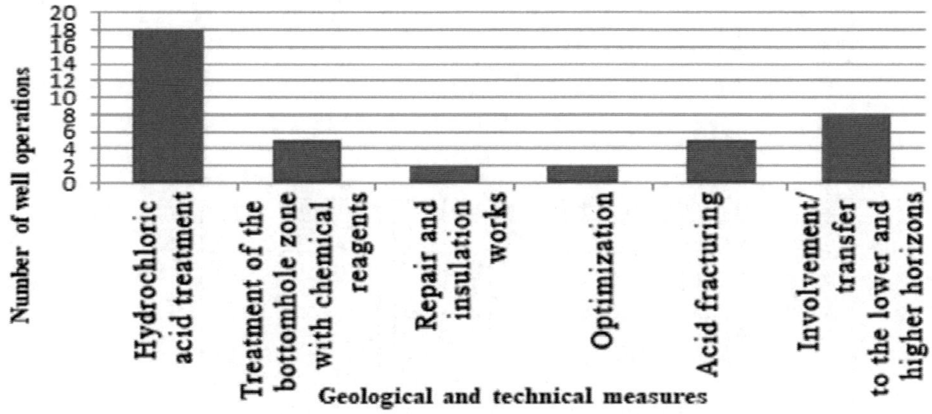

Figure 1. Statistics of the geological and technical measures taken for the studied deposit

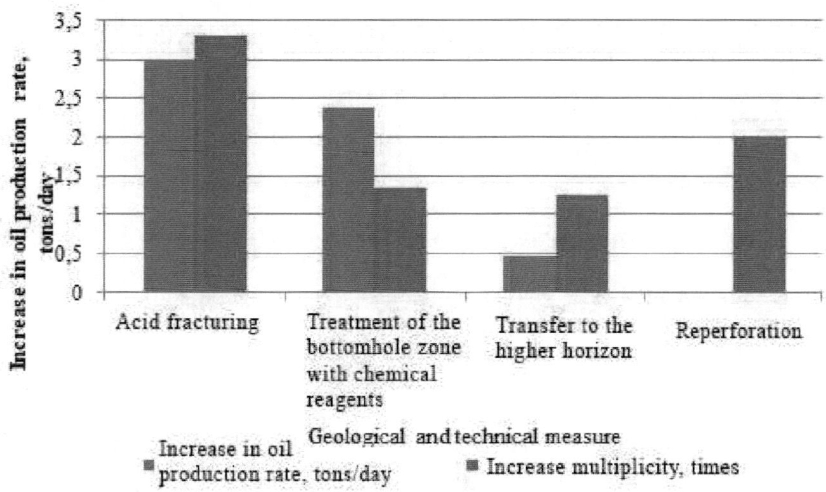

Figure 2. The effectiveness of geological and technical measures taken for the studied deposit

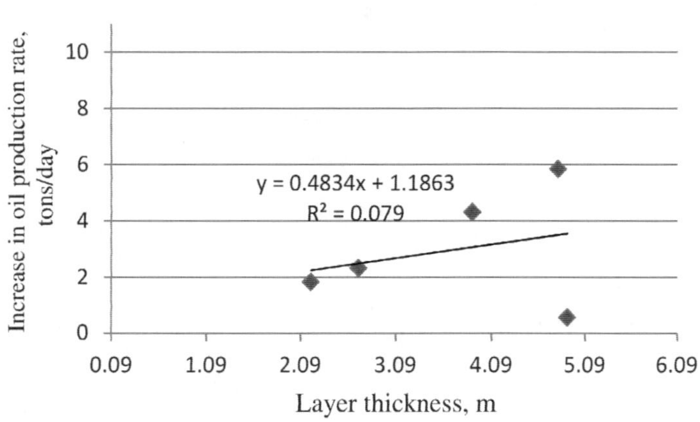

Figure 3. Dependence of the increase in oil production rate after acid fracturing on the current net thickness of the reservoir

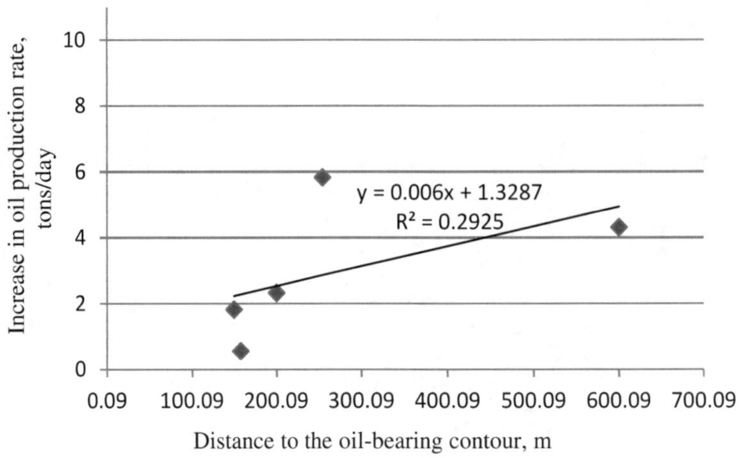

Figure 4. Dependence of the increase in oil production rate after acid fracturing on the distance to the oil-bearing contour

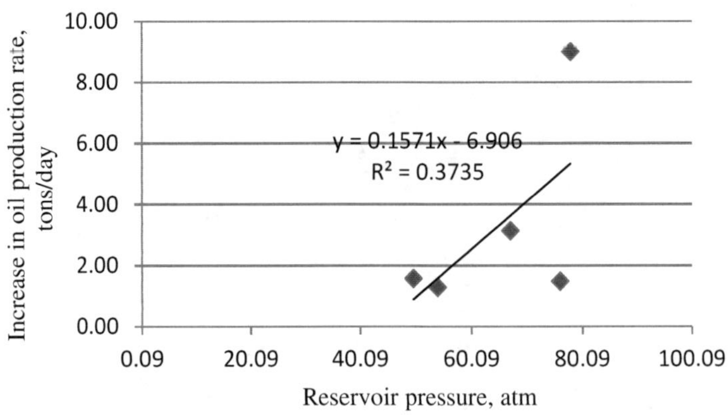

Figure 5. Dependence of the increase in oil production rate after acid fracturing on the reservoir pressure before hydraulic fracturing

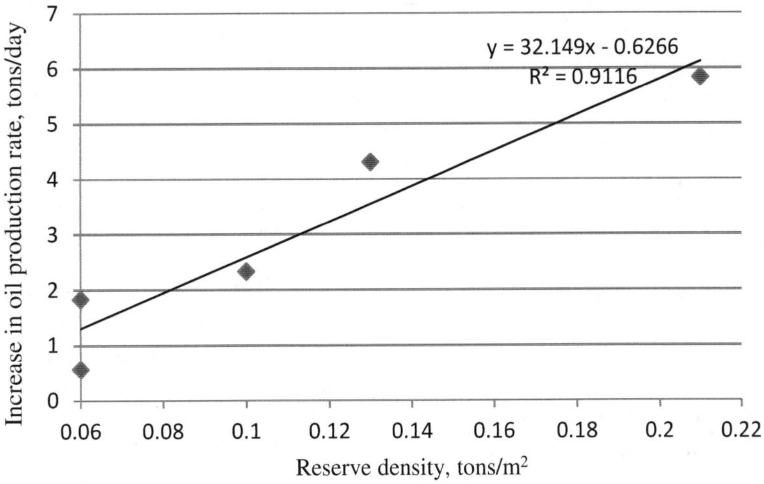

Figure 6. Dependence of the increase in oil production rate after acid fracturing on the density of residual reserves in the well drainage zone

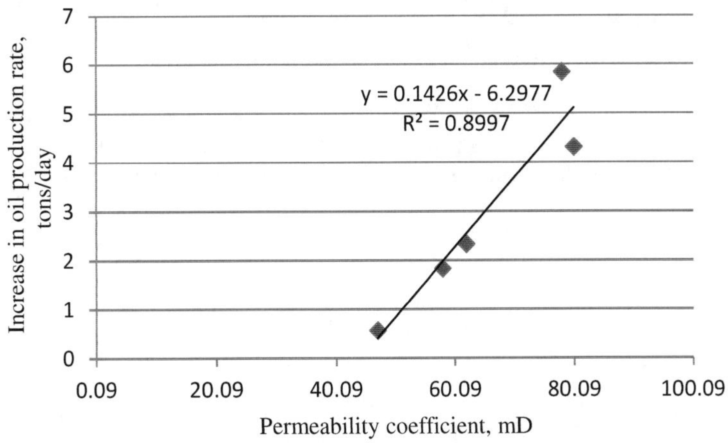

Figure 7. Dependence of the increase in oil production after acid fracturing on the permeability of the well

4. Conclusion

Thus, the study of the relationship between the above geological and technical factors based on the correlation analysis enables a reasonable approach to the choice of candidate wells for hydraulic fracturing.

The analysis yields the criteria for a more thorough choice of candidate wells for hydraulic fracturing in the study area in order to achieve higher production rates:

1. Permeability is at least 45 mD.
2. The distance to the oil-water contact is at least 160–200 m, otherwise there is a risk of the contact coning.
3. Net thickness is at least 2.5–3 m.
4. Reservoir pressure in the sampling zone is at least 30 m (decreased reservoir pressure by no more than 25% of the initial 78 atm is allowed).

5. The density of residual reserves is at least 0.08 tons/m^2.

In general, acid and multistage fracturing is more efficient compared to hydrochloric acid treatments and reperforations, which indicates the need for its further implementation, especially for newly commissioned horizontal wells. In this case, multistage fracturing technologies are more appropriate compared to "blind" acid fracturing due to their greater efficiency.

References

[1] Zainagalina L Z, Petrova LV and Petrov VA 2019 Analysis of the efficiency of telemetric systems for drilling wells *IOP Conference Series: Materials Science and Engineering* **560** 012096 doi:10.1088/1757-899X/560/1/012096

[2] Nie Z, Liu Z and Sun X 2018 A Multirelay Cooperation Method for Wireless Transmission of MWD and LWD Signals *IEEE Transactions on Geoscience and Remote Sensing* **56** 1229–37

[3] Almukhametova E M, Fattakhov D I and Zakirov A I 2018 The efficiency analysis of applying hydraulic fracturing of formation at AB1-2 facility of Potochnoye field *IOP C. Ser.: Earth Env.* **194** 082002 DOI:10.1088/1755-1315/194/8/082002

[4] Suleimanov R I, Gabdrakhimov M S, Khabibullin M Y, Zaripova L M and Vasilyeva E R 2018 The study of hydraulic hammer device in drilling tool assembly in hydraulical rotary drilling *Int. J. of Engineering and Technology* **7(2)** 28–30

[5] Mukhametshin V V and Kuleshova L S 2020 On uncertainty level reduction in managing waterflooding of the deposits with hard to extract reserves *Bulletin of the Tomsk Polytechnic University. Geo Assets Engineering* **331(5)** 140–146 DOI 10.18799/24131830/2020/5/2644

[6] Petrova L V, and Buravov V V 2019 Evaluating the technological effect of a horizontal hydraulic fracturing *Journal of Physics: Conference Series* **1333** 072019 DOI:10.1088/1742-6596/1333/7/072019

[7] Nazim N 20008 Dmour Practical Well Test Analysis of a Hydraulically Fractured Low permeability Gas Reservoir: A Case History J.King Saud Univ., Eng. Sci. **20(1)** 47–60

[8] Antonenko D A, Pavlov V A, Surtaev V N, Sevastyanova K K 2008 Integrated Modeling of the Priobskoe Oilfield *SPE* **117413**

[9] Dean L., Mireault R 2008 Production Decline Analysis Reservoir E*ngineering for Geologists* **1** 20–22

[10] Malyarenko A M, Kotenev Yu A, Mukhametshin V Sh, Bogdan V A, Kotenev A Yu, Blinov S A, and Umetbaev V G 2020 The influence of the rocks material composition on the reservoir properties of the reservoir *Journal of Physics: Conference Series* **1582(1)** 012061 1–11 DOI: 10.1088/1742-6596/1582/1/012061

[11] Akhmetov R T, Mukhametshin V V, and Kuleshova L S 2019 Simulation of the absolute permeability based on the capillary pressure curves using the dumbbell model *Journal of Physics: Conference Series* **1333(3)** 032001 1-8 DOI: 10.1088/1742-6596/1333/3/032001

[12] Khabibullin M Ya, Samigullina L Z and Zainagalina L Z 2018 Laboratory studies for the momentum of liquid optimal impact on hydrocarbon formation *Advan. in Engineering Res.* **157** 276–9 DOI:10.2991/aime-18.2018.53

[13] Gabdrkhmanova K F, Izmailova G R , Samigullina L Z 2020 Solution of the Problem of Annular Space Gas Utilization in Wells Operated by Walking-Beam Pumping Unit *IOP Conference Series: Earth and Environmental Science* **459(4)** 042080

[14] Bijjani P, Coscia M, Hammon K, Ashraf M and Ali Hassan H 2015 MWD/LWD challenges and solutions for drilling the longest al hosn gas well in an aggressive subsurface environment Society of Petroleum Engineers - Abu Dhabi Int. Petroleum Exhibition and Conf. (ADIPEC 2015

Reassessment of the potential of oil reserves in thin-layered "hazel grouse" type reservoirs

E R Agishev[1], V E Andreev[2], L S Kuleshova[3], V V Mukhametshin[2], Sh G Mingulov[3] and D I Akhmetshina[3]

[1]LLC "RN-Bashnipineft", 3/1, Bekhtereva St., Ufa, Republic of Bashkortostan, 450103, Russian Federation
[2]Ufa State Petroleum Technological University, 1, Kosmonavtov St., Ufa, Republic of Bashkortostan, 450064, Russian Federation
[3]Ufa State Petroleum Technological University, Branch of the University in the City of Oktyabrsky, 54a, Devonskaya St., Oktyabrsky, Republic of Bashkortostan, 452607, Russian Federation

E-mail: agishev@yandex.ru, info@of.ugntu.ru

Abstract. The article describes an integrated approach to the development of solutions for the effective involvement of oil in the development of a thin-layered "hazel grouse" type reservoir. The principles of justification based on facies, geological-hydrodynamic and geomechanical modeling are analyzed. In the course of the work, new core data and a petrophysical model were obtained, which made it possible to substantiate the TRIZ status of the object. A conceptual geological model was built and an assessment of geological reserves was carried out to determine the priority zones for putting wells into operation.

1. Introduction

The modern structure of the resource base of oil producing enterprises in the West Siberian oil and gas province has forced oil engineers to extract hard-to-recover oil reserves associated with low-permeability reservoirs [1–6]. Involvement of unconventional reservoirs creates the need for adaptaing available technologies to specific cases and searching for economically optimal solutions [7–9]. The formation AV1 (1-2) is a striking example of an anisotropic, thin-layered highly-clayey reservoir, composed of sandstone of the "hazel-grouse" type, mixed-grained siltstone and clay interlayers. The use of hydraulic fracturing is limited by the risks of breakthrough into the underlying highly productive water-saturated reservoir, separated by an unstable clay dam with a thickness of tens of centimeters to 13 m.

2. Materials and methods

The AV1 (1-2) reservoir is the upper object in the oil-bearing level of fields in the territory of the producing company. The AV1 (1-2) reservoir is a single reservoir, almost entirely located in the pure oil zone. The formation was formed in marine and shallow-sea conditions of sedimentation in the setting of a gentle slope of a large river delta under the strong impact of various marine processes [10–12]. East of the area, there is a proximal part of the delta with thick monolithic reservoirs and high permeability values of 10^{-7}-10^{-6} m^2 (Fig. 1), further westward movement, and a distal part of the delta and reservoir properties with permeability up to 10^{-8} m^2. In the same direction (from east to west), there is a decrease in the effective thicknesses, porosity and permeability properties (FES) and an increase in dissection (Fig. 2).

The formation was formed in the conditions of the prodeltaic slope, and the reservoir properties vary within 0.5–5.0 10-9 m^2.

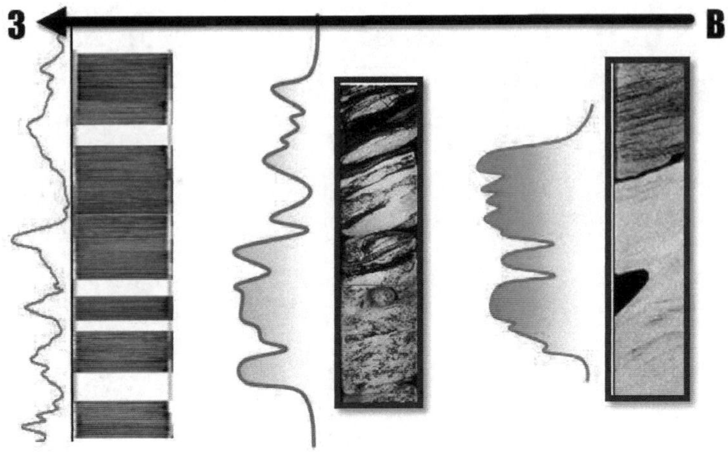

a) b) c)

Fig. 1. Changes in the facies pattern from east to west based on core samples and characteristic curves of gamma-ray logging: a) proximal; b) the distal part of the delta; c) pro-delta slope

3. Results and Discussion

For calculating reserves, a single petrophysical model was adopted for the entire group of formations AB (AV1-AV8), based on logging of the intrinsic polarization potential (PS), gamma-ray logging (GK) and data from core studies taken in the most productive parts of the reservoir. Samples from more than 700 wells were taken, the average density of samples is less than three samples per 1 meter of depth. Based on the results of the first stage, the samples were additionally taken from 10 wells to clarify and confirm the method.

The core is typical of formations AV2-8, which were formed in the continental conditions of sedimentation and have the best reservoir properties. They ghave high critical values of the double difference potential of the PS. This petrophysical model does not allow for correct lithological dissection of the section and determination of reservoir properties for highly dissected, weakly permeable, highly mudded formations. The use of well-known techniques (Doll; deconvolution method; thin-layered analysis) to determine the properties of such reservoirs [13, 14] is not applicable due to the limited well logging complex. In the wells, surveys were carried out using a standard logging system; for the main part of the well stock, the PS, IK, BK, GK and NK methods were used.

An in-house petrophysical model was developed and substantiated based on the curves of the radioactive logging methods (GK and NK). Basing the model on the RK curves justified an increase in the average effective thickness by 31.5% and an increase in the production potential. To identify reservoirs using the extended logging complex, the boundary values of the double difference parameter of HA ($\Delta I\gamma$) and hydrogen content were determined. Dependences of porosity and permeability on $\Delta I\gamma$ were determined.

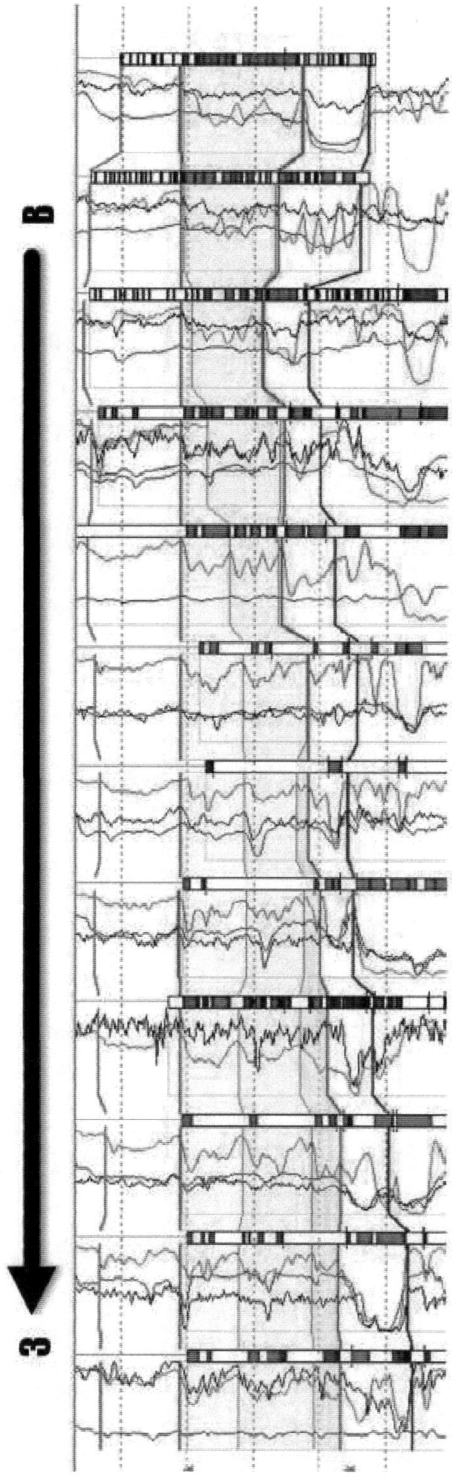

Fig. 2 The Regional Correlation Scheme

Comparison of effective thicknesses on core samples and by the proposed method based on ΔIγ, showed a high convergence; the discrepancy did not exceed 5%. Based on the results of the revision of the petrophysical model, the object was categorized as TRIZ and a privilege for oil production in 4 license areas was gained.

To identify the most promising zones in accordance with the new petrophysical model, it is necessary to revise the interpretation of the log curves of more than 7000 wells. This work requires significant labor costs. At the stage of building a conceptual model, for the subsequent selection of the most promising areas, there is no need for absolute accuracy and you can use automatic interpretation with a random check of the results. The data from the HK and Tubing curves for all wells were standardized and normalized. Then, according to the dependencies obtained at the previous stage, permeable intervals were identified and reservoir properties were calculated.

Classical methods of geological modeling do not always allow reconstructing the distribution of an anisotropic reservoir with significant changes in the FES along the vertical and lateral. Therefore, a spectral modeling method was used to obtain three-dimensional cubes of geophysical properties based on the well data. Using the obtained boundary values and dependencies, it becomes possible to quickly recalculate the cubes of geophysical properties into cubes of lithology and FES. By varying the boundary values, it is possible to give a probabilistic assessment. Based on the simulation results, a variable estimate of reserves was obtained; in the scenario P50, an increase in initial geological reserves of more than 50% was obtained relative to the ones which were on the balance sheet of the company. An increase in reserves is associated with an increase in the oil-saturated thickness and the oil-bearing area due to the exclusion of previously existing clay zones. The factor analysis of changes in parameters for estimating reserves is shown in Fig. 3.

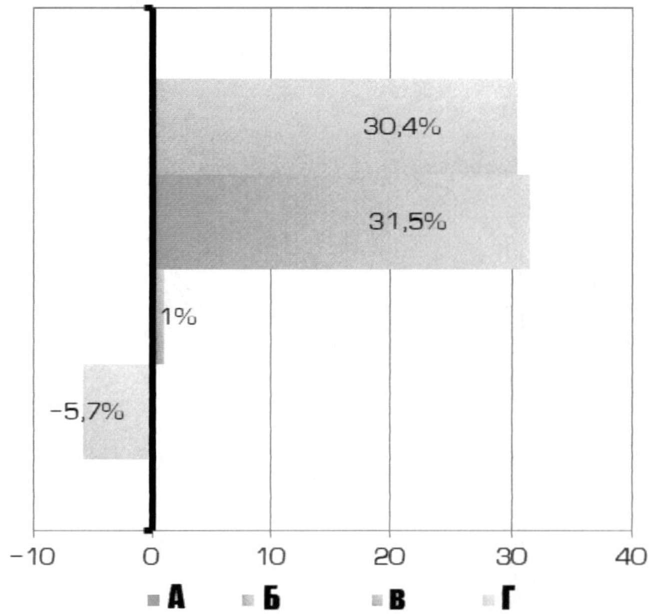

Fig. 3. The factor analysis of changes in parameters for assessing reserves: a) oil-bearing area; b) average oil-saturated thickness; c) porosity coefficient d) oil saturation coefficient

4. Conclusion

The new core data obtained and the petrophysical model made it possible to substantiate the TRIZ status of the object. The oil production incentive significantly increases the interest in the involvement of reserves in development. A conceptual geological model was built and geological reserves were estimated to identify priority zones for wells commissioning. To develop a further strategy for the effective involvement of the reservoir in development, it is planned to build hydrodynamic models to justify the development system, select well pads for priority implementation, including within pilot works, to select and study core samples for petroelastic properties in order to determine the optimal design of hydraulic fracturing.

References

[1] Alvarado V, Stirpe M, La Roque C, Ponce R, Farias M 2002 Streamline Simulation for Enhanced-Oil Recovery: Review and Laboratory Tests *Proceedings of INGEPET, EXPL-4-VA-84. Lima*

[2] Soloviev N N, Mukhametshin V Sh and Safiullina A R 2020 Developing the efficiency of low-productivity oil deposits via internal flooding *IOP Conf. Ser.: Mater. Sci. Eng.* **952** 012064 1–5 DOI: 10.1088/1757-899X/952/1/012064

[3] Akhmetov R T, Mukhametshin V V, Kuleshova L S, Grezina O A, and Malyshev P.M. 2020 The generalized correlating function of capillary curves and the relationship of the filtration-capacitive parameters of reservoirs in Western Siberia with the size distribution of pore channels *Journal of Physics: Conference Series (International Conference on Information Technology in Business and Industry (ITBI 2020))* **1661** 012016 DOI: 10.1088/1742-6596/1661/1/012016

[4] Mukhametshin V Sh 2020 Rationale for the production of hard-to-recover deposits in carbonate reservoirs *IOP Conf. Ser.: Earth Env.* **579** 012012 DOI: 10.1088/1755-1315/579/1/012012

[5] Zeigman Yu V, Mukhametshin V V, and Kuleshova L S 2020 Management of flooding of low-production deposits according to geological and field data *IOP: Earth and Environmental Science (EES)* **579** 012019 DOI: 10.1088/1755-1315/579/1/012019

[6] Mukhametshin V Sh, Andreev V E, and Yaskin S A 2020 Designing measures to increase oil recovery based on the identification and grouping of deposits *IOP: Earth and Environmental Science (EES) (International Symposium «Earth sciences: history, contemporary issues and prospects»)* **579** 012013 DOI: 10.1088/1755-1315/579/1/012013

[7] Kuleshova L S, and Mukhametshin V V 2020 Justification for the selection of distances between wells to increase the efficiency of flooding by deposits in terrigenous reservoirs of Western Siberia *Journal of Physics: Conference Series* **1582** 012057 DOI: 10.1088/1742-6596/1582/1/012057

[8] Mukhametshin V Sh 2020 Justification for increasing the performance of hydrochloric acid treatment in wells of fields with carbonate reservoir *IOP Conf. Ser.: Mater. Sci. Eng.* **905** 012083 DOI: 10.1088/1757-899X/905/1/012083

[9] Malyarenko A M, Kotenev Yu A, Mukhametshin V Sh, Bogdan V A, Kotenev A Yu, Blinov S A, and Umetbaev V G 2020 The influence of the rocks material composition on the reservoir properties of the reservoir *Journal of Physics: Conference Series* **1582** 012061 DOI: 10.1088/1742-6596/1582/1/012061

[10] Brilliant L S, Sharifullin F A, Barakin V A and Alexandrov V M 2000 Clarification of the sedimentation model of the AB 11-2 object of the Samotlorskoye field using lithofacial analysis *Oil Industry* **9** 17-22

[11] Zakrevsky K E, Aksarin V A, Anokhina M S, Buyakina I V, Gritchenko M A, Zverev K V, Zunde D A, Kudamanov A I, Kuznetsov A G, Malshakov A V, Nassonova N V, Khabarov A V and Chernovets L V 2017 Geological modelling of the coastal-marine sediments (on the example of reservoir AV1 (AB11+2+AB13) Samotlor oil field) (Tyumen: Vector Book)

[12] Allen J R L 1982 Sedimentary structures: Their character and physical basis *Developments in Sedimentology* **30A(1)**

[13] Chanh Cao Minh, Jean Clavaud, Padmanabhan Sundararaman, Serge Froment, Emmanuel Caroli, Olivier Billon, Graham Davis, Richard Fairbairn 2007 Graphical Analysis of Laminated Sand-Shale Formations in the Presence of Anisotropic Shales *SPWLA 48th Annual Logging Symposium* **5** 1-11

[14] Klein J D and Martin P R 1997 The Petrophysics of Electrically Anisotropic Reservoirs *The Log Analyst* **38(3)**

IOP Conf. Series: Earth and Environmental Science 1021 (2022) 012043 doi:10.1088/1755-1315/1021/1/012043

Oil Potential Prospects of Domanik sediments in the North-East of the Volga-Urals

S N Krivoshchekov[1]

[1] Petroleum Geology Department, Perm National Research Polytechnic University, 29, Komsomolsky Prospekt, Perm, 614990, Russian Federation

E-mail: krivoshchekov@pstu.ru

Abstract: Due to the depletion of available mineral and raw material resources of conventional sediments there is a need to search for new promising areas of oil and gas exploration in Perm Krai. This article addresses Domanik sediments as a new source of hydrocarbons. Stratigraphically, these sediments belong to the Sargayev, Domanik and Mendym horizons and lithologically they are mainly represented by tight carbonate sediments and shales. These sediments are the main oil-source strata in the Volga-Ural province. The paper analyses the geological structure and distribution of these sediments in the study area. Thereupon, exploration recommendations have been formulated for the next few years. These mainly include subject-matter research, core sampling in the Domanik Formation sediments, special studies of reservoir properties, reactivation and study of discovered oil deposits in the Domanik strata. The main activities should be concentrated in the south and south-west of Perm Krai: in the area of the Bashkir Arch, in the south of the Upper-Kama Depression. The Solikamsk Depression is also of interest in this respect.

1. Introduction

The increasing complexity of the mining and geological environment for exploration and prospecting of new oil and gas accumulations in Russia is currently affecting the exploration efficiency to a significant degree and will undoubtedly intensify in the near future; it is largely determined by the increasing role of unconventional hydrocarbon deposits as prospecting targets.

A rather high degree of development of initial potential oil and gas resources in Perm Krai and in the Ural-Volga region (about 73%, in general) urges to search for new exploration directions to develop an exploration strategy and set up regional geological surveys [1, 2].

2. Problem Statement

Unconventional hydrocarbons, specifically gas, were first used in the late 20th century in North America. The gas was applied for local needs in small volumes. No unconventional hydrocarbon project was profitable. Starting from the early 1980s, with the enactment of the Natural Gas Policy Act of 1978, exploration led to the discovery of significant reserves of conventional gas and associated unconventional gas and oil. Due to the complexity and high cost of unconventional hydrocarbon production technologies, cost-effective large-scale projects were carried out at a later stage.

Unconventional hydrocarbon reserves can be divided into the following categories [3, 4]:

- Tight Gas, Tight Oil - gas and oil in very tight sandstones and low-permeability reservoirs.
- Coalbed Methane (CBM) - methane in coal beds.

IOP Conf. Series: Earth and Environmental Science 1021 (2022) 012043 doi:10.1088/1755-1315/1021/1/012043

- Shale Gas, Shale Oil - gas and oil in shale formations.
- Methane Hydrates - gas hydrates, gas in solid form in a formation.

Unconventional hydrocarbons are found in tight formations that are often oil and gas source rocks or sediments of an intermediate complex between a basement and a sedimentary cover (Fig. 1). Low or virtually no permeability in reservoirs with unconventional hydrocarbon accumulations hampers the development of such HC deposits using conventional methods to ensure cost-effective projects.

The key to effective development of unconventional hydrocarbon reserves from low-permeability reservoirs is innovative technologies, including drilling of long horizontal-section wells (1,000-1,500 m or more) and inflow stimulation by multi-stage interval hydraulic fracturing to develop multiple fractures or natural-fracture system connectivity.

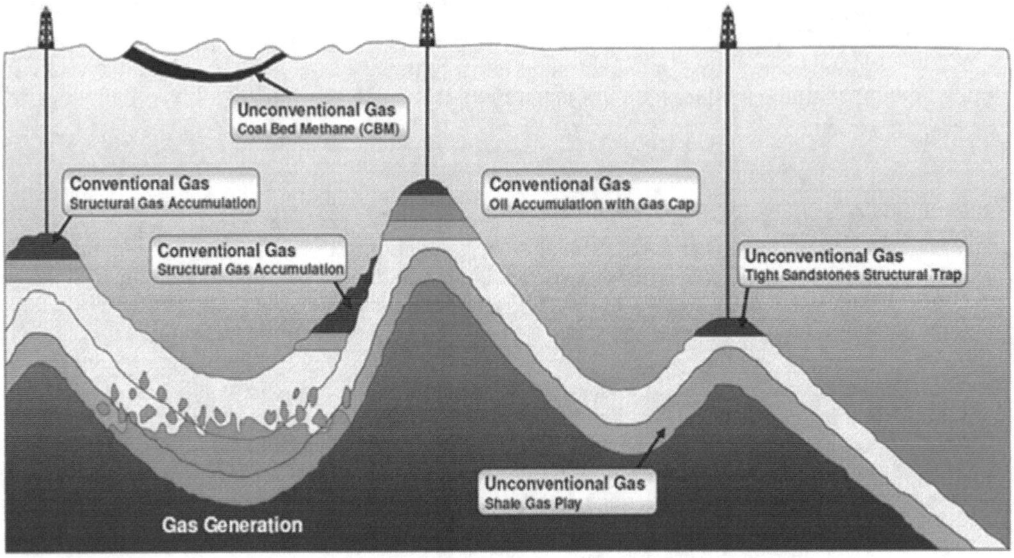

Figure 1. Conventional and unconventional hydrocarbon accumulations

The North America region presently has the largest proven reserves and HC production from unconventional low-permeability reservoirs due to state-of-the-art technologies. Production of unconventional hydrocarbons in other parts of the world has a great potential as well. Projects for shale gas production in Europe (France, Poland, Hungary, Germany, the Netherlands and Sweden) are being actively studied and developed; a common information centre has been established to study its production potential in Europe.

In the north-east of the Volga-Urals, such unconventional HC deposits include, in particular, accumulations in Domanik sediments. Domanik sediments are highly bituminous siliceous-clayey-carbonate strata of predominantly biogenic autochthonous marine genesis. These sediments are the main oil-source formations in practically all the world's petroleum basins and excellent regional seals for hydrocarbon deposits. In recent years, these sediments are increasingly considered as oil and gas bearing and even gas-bearing. Domanikites are a very specific and very individual geological formation that corresponds to the maximum transgression epochs and suggests the existence of special conditions in sedimentary basins of the past. It differs from many sedimentary formations by its specific geochemistry, a special composition of biota that inhabited paleobasins and specific set and relationships of rocks lithology.

Oil and gas in such sediments exist in free state in natural fractures. Hydrocarbon molecules are trapped in the formation by adsorption on kerogen, the shale's bituminous components. Most commonly, such deposits are attributed to a generative rock rather than to an exploratory prospect.

The Domanik sediments are characterised by:

- interbedded oil shale, bituminous, sometimes chert-bearing limestones and marls, and siliceous shales;
- wide development of fine horizontal bedding and absence of textural features indicative of shallow water (shrinkage cracks, etc.);
- subordinate benthonic fauna, its remains only found in some interlayers;
- enrichment in sapropel matter.

Over 30 deposits related to domanikites were described in the USA. The Bazhenov Formation in the West Siberia is a bright example of domanikites in Russia. High oil flow rates (up to 100 t/s) were obtained in the Fore-Caucasus region from the clayey Batalpashin Formation of the Oligocene in the Zhuravskaya area.

3. Materials and Methods

Such strata in the Volga-Ural petroleum and gas province include sediments of Mendym, Domanik (Semiluki) and Sargayev horizons developed in axial and edge parts of the Kama-Kinel trough system [5]. These sediments were developed along with drilling for terrigenous Devonian datum beds, especially in the territory of Tartary, Bashkiria and Samara region. At the same time, intensive oil shows up to commercial inflows with high flow rates were obtained from domanikites in a number of wells. Established has been a trend in location of hydrocarbon deposits of Upper Devonian-Tournaisian carbonate sequence with a density of tectonic lineaments, indicating that the maximum fracturing zones development is most likely in the highest tectonic stress areas [6].

In the north-east of the Volga-Urals, the Mendym and Domanik-Sargayev sediments were studied by drilling 940 exploration wells, including 215 wells where core was sampled from. Cased-hole tests were conducted in 11 wells, and 3 wells produced commercial inflows: well #276-Etyshskaya that entered the Rakinskoye field (oil inflow of 4.2 t/day), well #14-Iskorskaya that entered the Isanevskoye field (oil inflow of 12.5 t/day) and well #155-Shurtanskaya that entered the Tyushevskoye field (oil inflow of 5.6 t/day). No special geological studies were carried out for commercial development of the deposits.

The Domanik sediments are represented throughout almost the entire territory of the Volga-Ural petroleum and gas province and are characterised by a variety in lithology. In the north-west of Perm Krai, sediments were destroyed by pre-Visean and later erosion. The erosion boundary runs along the north-western margin of the Solikamsk Depression, the Visimskaya monocline and the north of the Upper-Kama Depression.

In the extreme north-west of the Kama Arch, the Domanik sediments are completely eroded away. In the area of well 18-Ust-Chernaya, the Serpukhov sediments overlap transgressively on the Timan horizon sediments. No sediments of the complex are present in the Ksenofontovsky section of the Timan, where Kashira horizon sediments overlie Proterozoic rocks.

In the Mendym (Bureg) horizon, the area of depression sediments is reduced as compared to the Semiluki horizon and is less promising as there are hardly any oil shows.

Domanikoid depression formation of Voronezh and Evlanovo-Livny horizons of the Upper Frasnian sub-stage, Famennian stage and Zavolga horizon of the Tournaisian stage is developed in the axial parts of troughs of the Kama-Kinel system of depressions and occupies a much lesser area compared to the Domanik-Mendym age.

Facial varieties of adjacent shallow-shelf formations sediments – edge and marginal depression zones – are more favourable in terms of oil potential.

Carbonate rocks are represented by lighter, less bituminous and siliceous, but often more clayey varieties of limestones and dolomites. The faunal changes consisted in the smaller number of

Coniconchia and pelecypods of the Domanik appearance, near-extinction of goniatids and appearance of a large number of brachiopodas.

The Domanik sequence is still poorly studied despite the facies conditions diversity. For exploratory drilling in the areas of its distribution, comprehensive core sampling and setup of specialised lithology studies are needed.

The Domanik Formations feature possible presence of oil deposits in any part of the exploration target. Commercial oil potential is related to lithology 'windows': decompacted fractured zones. Generally, reservoirs are layered varieties of paleo-arch limestones and shale clays of the axial part of the Kama-Kinel system of depressions. Fracturing and cavernosity produce a significant influence on reservoir properties of the rocks. Along with pores and caverns, the development of voids in carbonate sedimentary strata involves lithological or primary and tectonic or secondary macro- and microfractures. Primary rock fracturing is more typical of the platform areas, while secondary fracturing is more specific to the fore deep and folds of the Urals. The presence of areas of good reservoirs is proved by high oil flow rates at a number of fields in the Volga-Ural province.

In the centre and west of Perm Krai, permeable zones are identified and partially correlated in the lower, middle and upper parts of the Devonian carbonate strata. They are heterogeneous, net-to-gross ratio varying from 30 to 80%.

In the the north-east of the Volga-Urals, the effective porosity of reservoirs in sediments of the Semiluki horizon represented by interbedding of typical rocks of Domanik facies with subbituminous and non-bituminous limestones and subsiliceous dolomites is as much as 11-12 %. The reservoir properties across the range of development of clayey-siliceous facies degrade sharply. The effective porosity decreases down to 3-5%. Within the paleo-depressions corresponding to the Yuryuzan-Sylva and Solikamsk Depressions, virtually no reservoirs have been discovered. In total, out of 388 core samples, reservoir rocks were identified in 30 samples. In the Semiluki horizon rocks the processes negatively affecting reservoir properties (bituminisation, claying, silicification, etc.) have maximum development, whereas the processes of positive impact (leaching, recrystallization, etc.) prevail in the Domanikoid Formation horizons of the Bureg-Tournaisian complex.

The reservoir potential has been confirmed by manifold oil shows and 3 commercial oil deposits: Rakinskoye, Isanevskoye and Tyushevskoye. These are characterised by poor area and section variability of reservoir properties resulting in mosaic-like distribution of permeable zones. The thickness of these zones ranges from several to a few dozen meters.

The commercial oil potential of domanikoids is directly proved by the results of their testing in wells. Out of 1,267 tested wells, no inflow was obtained from 717 wells, water was found in 327 wells, and in 223 wells found were oil inflows with rates ranging from 1 to 100 t/day and more. Therefore, the success ratio is approximating 20%; besides the tests were generally carried out ad hoc, as an associated target during the terrigenous Devonian strata exploration.

Most oil deposits in the Domanik sediments are associated with clinoform bodies located on the structures slopes or with lithologically screened traps. The possible oil deposit sites in the Upper Devonian-Tournaisian carbonate complex may include axial and marginal parts of the Kama-Kinel trough system, as well as zones with intensive growth of fractured bioherm bodies of the Bashkir Arch and Solikamsk Depression. Oil accumulations normally open at the generation zones margins, where a certain geochemical barrier is created through the contact of oil-source and covering rocks. The former are characterised by microaccumulation areas, while the latter by significant accumulations of hydrocarbons.

The Volga-Ural province's Domanik sediments are lithologically are classified into screened, fault-screened (associated with graben-like troughs) and lenticular.

The analysis of structural and thickness maps of the Domanik and Sargayev horizons suggests three main zones: 1 - a zone of intense subsidience (subsea depth of -2,300 to -2,900, east of the platform, the sediments thickness ranges from 0 to 20 m); 2 - a zone of moderate depths (subsea depth of -1,900 to -2,300, the sediments thickness ranges from 10 to 30 m), 3 – a zone of basement uplift and graben-like highs associated with local minima (subsea depth from -1,300 to -1,900, the Domanik sediments thickness ranges from 30 to 100 m, the Sargayev sediments thickness varies from 50 to 300 m).

The following figures indicate of domanikoids' share in hydrocarbon resources of the Volga-Ural petroleum and gas province: the initial total reserves of this formation account for 29% of the Upper Devonian-Tournaisian complex resources and for 9% of resources of the province's whole Paleozoic carbonate complex [5]. Therefore, domanikoids present a real interest as one of the possible sources of hydrocarbon reserves growth in Perm Krai.

4. Results

Based on the analysis of the above, research and development areas for oil and gas potential study and subsequent exploration can be identified and recommendations can be proposed for the study of Domanik sediments in Perm Krai:

- subject-matter regional research at the identified sites, including geodynamic and catagenetic simulation.
- core sampling in the Domanik Formation sediments at drilling of prospecting, appraisal and production wells for Devonian terrigenous sediments (Kustovskoye, Malo-Usinskoye, Andreyevskoye and Garyushkinskoye deposits);
- well-deepening and testing of Domanik sediments at drilling of prospecting and exploratory wells on prospects of the Bashkir Arch central part;
- development of a well survey programme (re-entry, hydraulic fracturing and testing) with oil deposits in the Domanik strata at the Rakinskoye, Tyushevskoye and Isanevskoye fields;
- drilling of long horizontal-section wells and multihole hydraulic fracturing.

5. Conclusions

Due to the depletion of available mineral and raw material resources of conventional sediments there is a need to search for new promising areas of oil and gas exploration in the north-east of the Volga-Urals. Siliceous-carbonate varieties of the Sargayev, Domanik and Mendym horizons represent a promising target to expand the hydrocarbon reserves portfolio. Exploration for these sediments should be conducted in the south and southwest of Perm Krai, this part of the study area constituting the most oil and gas potential.

References

[1] Krivoshchekov S N, Kozlova I A 2012 Geodynamic characteristics of the conditions of french-famennian strata dispersed organic matter immersion and catagenesis in perm region *Neftyanoe Khozyaystvo - Oil Industry* **7** 82–85
[2] Galkin V I, Kozlova I A, Nosov M A, Krivoshchekov S N 2015 Solutions to regional problems of forecasting oil bearing according to geological and geochemical analysis of dispersed organic matter of Domanic type rocks *Neftyanoe Khozyaystvo - Oil Industry* **1** 21–23
[3] Kuusik R, Uibu M, Kirsimäe K 2005 Characterization of oil shale ashes formed at industrial-scale boilers *Oil Shale* **22(4)** 407–420
[4] Anthony E J, Jia L, Wu Y 2005 CFBC ash hydration studies *Fuel* **84(11)** 1393–1397
[5] Nosov M A, Galkin V I, Krivoshchekov S N, Melkishev O A, Kochnev A A 2013 On substantiation of license area economic-geological evaluation by hydrocarbon resources and reserves *Neftyanoe Khozyaystvo - Oil Industry* **3** 14–17
[6] Galkin S V, Kochnev A A, Zotikov V I 2019 Estimate of radial drilling technology efficiency for the Bashkir operational oilfields objects of Perm Krai *Journal of Mining Institute* **238** 410–414

Possibility of using machine learning algorithms to assess the effectiveness of acid hydraulic fracturing

A A Kochnev[1], D V Torsunova[1], V Y Boyarov[1], S N Krivoshchekov[1] and E V Silaev[2]

[1] Perm National Research Polytechnic University (PNRPU), 29, Komsomolsky prospect, Perm, 614990, Russia
[2] OOO "PITC Geofizika" 3, Electronic passage, Perm Region, Dobryanka, Polazna, 618706, Russia

E-mail: sashakoch93@gmail.com

Abstract. At the fields of the Perm Territory, there is a significant application of hydraulic fracturing technology to stimulate the production of carbonate reservoirs. The analysis of the efficiency of the acid hydraulic fracturing technology by the example of the fields of the Perm Territory is presented in the article. The effectiveness of the technology is largely determined by the geological and physical conditions of the fields. Using the methods of mathematical statistics, the main geological and physical parameters which most significantly affect the effectiveness of measures for acid hydraulic fracturing have been identified. A multiple regression model was built to find out the complex influence of the parameters on the increase in oil production after the measures. Various machine learning algorithms have been used to predict the increase in oil and fluid input rates after acid fracturing. Different algorithms showed different results. For the problems of forecasting oil production after the application of acid fracturing, the most successful one was the support vector method.

1. Introduction

A significant proportion of the deposits in Russia and the Perm Region is at the late stages of development. At such fields, various geological and technical measures are widely used to intensify oil production and increase oil recovery. The efficiency of measures varies depending on geological and technological conditions.

One of the most common methods of oil production intensification in the Perm Region is hydraulic fracturing [1], including its modification - acid hydraulic fracturing. This technology is carried out on carbonate reservoirs, which are characterized by heterogeneity of reservoir properties and low productivity coefficients [2-4]. During acid hydraulic fracturing, a highly viscous liquid is injected into the formation under high pressure to form a crack and fill highly permeable channels, as well as a hydrochloric acid solution, when interacting with which the crack surface becomes rough. The injection of proppant with this technology is usually not performed [5]. As a result, there is an increase the permeability of the bottom-hole zone, which contributes to increasing the productivity of the well. Carrying out the technology is not always economically viable. There is also a risk of water breaking through to the producing well [6-8] and not achieving the planned success indicators. To increase the

effectiveness of measures and reduce the associated risks, it is necessary to select candidate wells taking into account the complex influence of geological and technological parameters [9].
The existing forecasting methods and approaches to assessing the effectiveness of geological and technical measures do not allow us to reliably predict the potential of candidate wells, so there is a need to improve them.

2. Materials and methods

During the analyzed period, 392 acid hydraulic fracturings in 42 fields of the Perm Region were implemented. The events were held at the sites of deposits of Famen, Tournaisian, Bashkir and Moscow age.
At the first stage, the analysis of the effectiveness of measures in various geological and physical conditions was carried out. To assess the impact of geological, physical and technological parameters on the efficiency of the acid hydraulic fracturing, a consolidated database was created with data on wells at the fields of the Perm Region where this technology was used. Using the Student and Pearson criteria, geological and technological parameters affecting the efficiency of the acid hydraulic fracturing were identified.
According to the results of calculations of the Student's t-test, the most statistically significant parameters were: oil flow rate before the event, product water content, separation coefficient, hydroconductivity of the remote formation zone, skin factor, average oil-saturated thickness, porosity, average oil saturation, permeability coefficient of the bottom-hole and remote formation zone, sandiness and separation coefficients, average oil viscosity in reservoir conditions, hydroconductivity of the bottom-hole formation zone and remote formation zone, piezo conductivity.
Further, multiple regression was used to identify the complex influence of parameters on the indicator of the increase in the input increase in the oil flow rate.
Significant parameters that comprehensively affect the increase in the oil flow rate were: oil flow rate up to acid hydraulic fracturing (q_0), water content (W), average oil-saturated thickness (H_o), dissection coefficients (K_d), porosity (K_{por}), permeability (K_p), skin effect (sk) and reservoir pressure (P_{res}). The multiple regression equation (1) is obtained:

$$Q_o = 1{,}27q_0 - 0{,}17W + 0{,}03H_o + 0{,}24K_d + 0{,}3K_{por} + 11K_p + 0{,}54sk + 0{,}06P_{res}. \quad (1)$$

R=0.73.

The statistical characteristics of the model are presented in table 1.
So a set of parameters that have the greatest impact on the increase in oil flow rate is obtained. The model allows us to accurately predict the potential increase in oil production (R=0.73).
At the next stage, various machine-learning methods are used to select the optimal method for predicting oil production based on a set of geological and technological parameters, which can be effective in developing a comprehensive methodology for predicting the effectiveness of geological and technical measures [10].
The neural network method also allows solving the regression problem [11-14]. Based on the identified set of parameters, networks with different architectures are built. In general, the training showed satisfactory results, but the network MLP-9-4-1 was chosen for further use because of a simpler architecture and correlation coefficients between the fact and the forecast from 0.71-0.77. The results of training networks with different architectures are presented in table 2.

Table 1. Multiple regression results

	B	t-crit.	p-value
Free parameter	5.25	1.34	0.18
Oil flow rate before acid hydraulic fracturing, t/day	1.27	6.68	0.00
Water content, %	-0.17	-3.84	0.00
Average oil-saturated thickness, m	0.03	0.58	0.56
Dissection coefficient, units.	0.24	1.73	0.09
Porosity coefficient, %	0.30	1.28	0.20
Coefficient of permeability, mm^2	11.00	1.55	0.12
Skin effect	0.54	3.09	0.00
Reservoir pressure, MPa	0.06	0.91	0.36

Table 2. Results of training neural network methods

Architecture	Learning	Control	Test	Learning algorithm	Error function	Function of active hidden neurons	Function of active output neurons
MLP 9-4-1	0.73	0.76	0.68	BFGS 3	Sum. sq.	Identical	Logistics
MLP 9-6-1	0.7	0.78	0.72	BFGS 5	Sum. sq.	Identical	Logistics
MLP 9-10-1	0.71	0.77	0.69	BFGS 3	Sum. sq.	Hyperbolic	Exhibit
MLP 9-4-1	0.71	0.77	0.74	BFGS 3	Sum. sq.	Hyperbolic	Exhibit
MLP 9-12-1	0.72	0.78	0.7	BFGS 4	Sum. sq.	Hyperbolic	Exhibit

To compare the forecast results, another machine learning method was used – the support vector machine method, which is also widely used for solving regression and classification problems [8]. Parameters set for the model are as follows. The core type is the radial basis function, gamma-0.11, C-5, epsilon-0.1. The results of the forecasts of the support vector model turned out to be better than the results of using multiple regression and neural networks, since the correlation coefficient between the actual and forecast values of the oil flow rate after acid hydraulic fracturing was from 0.72 to 0.86 on the training and test samples, respectively.

3. Conclusion

According to the results of the study, it can be noted that higher values of the flow rate increase are expected under the following geological and geophysical conditions. They are large values of the oil flow rate before the event (the average value of 4.87 t/day), filtration characteristics (the average value of the hydroconductivity of the remote reservoir zone is 3.67 (mcm^2*cm)/ (MPa*s), piezoconductivity – 202.48 cm^2/s). They also include the dissection coefficient (averaged 9.96), as well as increased

values of the oil saturation pressure with gas (averaged 10.52 MPa), oil saturation (average value – 0.74), and saturation (average – 73.53 m^3/m^3).
This acid hydraulic fracturing is more effective at low values of water production (average of 8.72 %), less viscous oil (average of 11.13 mPa*s), a lower porosity (average value of 12.62 %), low values of the ratio of net-to-gross (average of 0.31) and dirty bottom zone (the average skin factor amounted to -0.94).
It is revealed that the combined use of statistical methods and machine learning can reliably predict the increase in oil production (R=0.71-0.86). The method of support vectors most accurately allows you to predict the increase in oil production after the acid hydraulic fracturing (the correlation coefficients on the training and test samples are 0.72 and 0.86 respectively).
The obtained models can be used in the future to predict the potential for additional oil production from the acid hydraulic fracturing, taking into account geological and technological parameters.

Acknowledgments
This research was funded by the state assignment of the Ministry of Science and Higher Education of the Russian Federation as part of a government assignment, grant number FSNM-2020-0027.

References
[1] Ilyushin P Y, Kolychev I Y, Rakhimzyanov R M and Solovyov D Y 2015 Analysis of geological and technical measures to increase the productivity of producing wells in the oil fields of the Perm Region B*ulletin of Perm National Research Polytechnic University. Geology. Oil and gas and mining* **15** 81-89
[2] Galkin S V, Krivoshchekov S N, Kozyrev N D, Kochnev A A and Mengaliev A G 2020 Accounting of geomechanical layer properties in multi-layer oil field development *Journal of Mining Institute* **244(4)** 408-417
[3] Hutahaean J, Demyanov V and Christie M 2019 Reservoir development optimization under uncertainty for infill well placement in brownfield redevelopment *Journal of Petroleum Science and Engineering* **175** 444-464
[4] Caeiro M H, Demyanov V and Soares A 2015 Optimized history matching with direct sequential image transforming for non-stationary reservoirs *Mathematical Geosciences* **47(8)** 975-994
[5] Cherepanov S S, Chumakov G N and Ponomareva I N 2015 Results of acid hydraulic fracturing of a reservoir with proppant at the Tournaisko-Famenskaya deposit of the Ozernoye field *Bulletin of Perm National Research Polytechnic University. Geology. Oil and gas and mining* **16** 70-76
[6] Mengaliev A G and Martyushev D A 2020 Assessment of the technological and economic efficiency of hydraulic fracturing using a geological and hydrodynamic model that takes into account the structural features of carbonate reservoirs *Izvestiya, Tomsk Polytechnic University. Georesource engineering* **7** 37-48
[7] Krivoshchekov S N, Vyatkin K A, Ravelev K A and Kochnev A A 2020 Influence of geological and technological parameters on effectiveness of hydrochloric acid treatment of carbonate reservoirs *International Journal of Engineering* **33(10)** 2113-2119
[8] Putilov I, Krivoshchekov S, Vyatkin K, Kochnev A and Ravelev K 2020 Methods of predicting the effectiveness of hydrochloric acid treatment using hydrodynamic simulation *Applied Sciences (Switzerland)* **10(14)** 4828
[9] Yarkeeva N R and Khaziev A M 2018 The use of hydraulic fracturing for the intensification of oil inflow in wells *Petroleum engineering. Oil and gas business* **5** 30-36
[10] Kochnev A A, Kozyrev N D, Kochneva O E and Galkin V I 2020 Development of a comprehensive methodology for predicting the effectiveness of geological and technical measures based on machine learning algorithms *Georesources* **3** 79-86
[11] Ashena R and Moghadasi J 2011 Bottom hole pressure estimation using evolved neural networks by real coded ant colony optimizationand genetic algorithm *Journal of Petroleum Science and Engineering* **77** 375-385

[12] Redikultsev S A, Liplyanin A V and Paliy A O 2010 Using the neural network method for predicting the parameters of well operation after hydraulic fracturing *Drilling, development, production* **1** 33-37
[13] Anifowose F and Abdulraheem A 2011 Fuzzy logic driven and SVM driven hybrid computational intelligence models applied to oil and gas reservoir characterization *Journal of Natural Gas Science and Engineering* **3** 505–517
[14] Demyanov V, Soltani S, Kanevski M, Canu S, Maignan M, Savelieva E, Timonin V and Pisarenko V 2001 Wavelet analysis residual kriging vs. neural network residual kriging *Stoch Environ Res Risk Assess* **15(1)**18-32

IOP Conf. Series: Earth and Environmental Science 1021 (2022) 012045 doi:10.1088/1755-1315/1021/1/012045

Laboratory experiments on oil displacement using complexes based on enzyme solutions on a reservoir model

E M Veliev[1], V V Mukhametshin[2*], L S Kuleshova[1], M M Veliev[3], V I Shchetnikov[3], Vu Viet Thanh[3] and L Z Samigullina[1]

[1] Ufa State Petroleum Technological University, Branch of the University in the City of Oktyabrsky, 54a, Devonskaya St., Oktyabrsky, Republic of Bashkortostan, 452607, Russian Federation
[2] Ufa State Petroleum Technological University, 1, Kosmonavtov St., Ufa, Republic of Bashkortostan, 450064, Russian Federation
[3] JV "Vietsovpetro", 105, Le Loi St., Thang Nhi Ward, Vung Tau, Ba Ria – Vung Tau City, S.R. Vietnam

E-mail: _elshad_@mail.ru, info@of.ugntu.ru

Abstract. It is shown that the oil recovery of all tested samples increases after treatment with the enzyme solution; the oil displacement coefficient varies within 7.23–10.59%. The change in values is associated with various characteristics of core samples in terms of porosity, permeability, initial oil saturation, as well as lithological composition. The model clearly demonstrates that the differential pressure decreases after enzyme solution treatment. This indicates that the presence of the enzyme solution changes the wettability of oil with oil-bearing rock. Oil is easily displaced from the core sample and, therefore, the permeability of the core sample increases. The outcomes of studying water permeability of core samples before and after injection of the enzyme solution reveal that the sample permeability increases when using the enzyme solution (the recovery coefficient of the core sample permeability is 1.34-1.68).

1 Introduction

Improving the efficiency of oil field development is based, primarily, on making the right management decisions [1–6]. The basis of these decisions is the scientific and methodological base, formed on the framework of laboratory, geophysical, hydrodynamic studies, as well as geological and field analysis [7–11].

One of the most promising methods of influencing the bottomhole zone of wells in order to increase its productivity is the application of enzymes [12].

Enzymes are intended to increase the permeability of the bottomhole zone of production wells by reducing the surface tension between fluids and solid rock. The use of biotechnology based on enzymes allows enhancing the production rate of oil wells, reducing the rate of growth in water-cut, and obtaining a positive technical and economic effect.

Well stimulation using enzyme solutions has a number of advantages over other technologies. First, enzyme solutions do not decompose oil when it penetrates into the rock. Second, the enzyme layer is able to exist for a long time on the rock surface. This increases the duration of using enzyme solutions when stimulating the bottomhole zone. Third, the technology of applying enzymes is considered a

IOP Conf. Series: Earth and Environmental Science 1021 (2022) 012045 doi:10.1088/1755-1315/1021/1/012045

clean technology, since it does not affect the corrosion of pipes, instruments, operating installations, human health, and environment.

In practice, the permeability of the production formation deteriorates due to the combination of organic and inorganic matter (precipitation of salt, paraffin, asphaltenes, etc.). It is required to combine them with preparations for cleaning the rock from inorganic sediments in order to increase the enzymes efficiency. The enzyme characteristics do not change when the corresponding reagents are added to the solution.

The technology efficiency in practice depends on specific conditions: oil properties, reservoir characteristics (temperature, water permeability, permeability), and reservoir exploitation methods.

2 Methods and Materials

Experimental studies are carried out according to the following methodology in order to determine the ability to displace oil from rocks under the influence of the enzyme complex solutions.

Core samples are taken from the Miocene deposits of the "White Tiger" field, and a crude oil sample taken from Miocene deposits of the "White Tiger" field is used as a fluid. The main parameters of the core are presented in Table 1.

Table 1. Main parameters of the core sample taken from Miocene deposits of the "White Tiger" field

Model number	Length, 10^{-3} m	Diameter, 10^{-3} m	Porosity, %	Pore volume, 10^{-6} m^3	Gas permeability, 10^{-3} μm^2
1	7.8	50	37.19	28.8	270
2	218.9	50	20.8	83.26	253
3	207.3	50	19.53	69.18	293
4	175	49.9	19.1	63.53	429

A crude oil sample taken from Miocene deposits of the "White Tiger" field is used as a fluid.

The oil viscosity at a temperature of 120° C is 1.051 mPa·s, density is 0.805 kg/m^3, and the water viscosity is 0.59 mPa·s.

Experimental studies were carried out under the following laboratory conditions:

- temperature – 120° C;
- compression pressure – 13 MPa;
- effective pressure – 10 MPa;
- oil displacement rate – $0.5\cdot10^{-6}$ m^3/min.

Testing process:

- determination of the sample initial permeability;
- sample saturation with oil;
- displacement of oil by sea water at a rate of $0.5\cdot10^{-6}$ m^3/min, until oil appears;
- injection of the enzyme solution back into the sample in a volume equal to 0.2 V_{por};
- injection continuation of injected water in a volume of 0.2 V_{por} to move the enzyme solution into the sample;
- after 24 hours - continuation of oil displacement by seawater until oil appears;
- determination of sample permeability, calculation of the recovered oil amount before and after treatment with an enzyme solution.

Oil displacement outcomes are shown in Figures 1-4.

Simulation outcomes are summarized in Table 2.

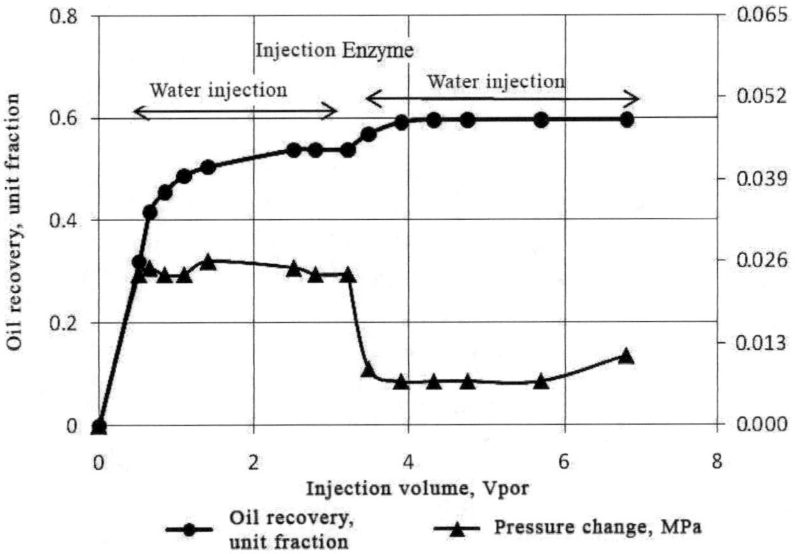

Figure 1. Dynamics of oil recovery and pressure changes before and after treatment with enzyme solution (model 1)

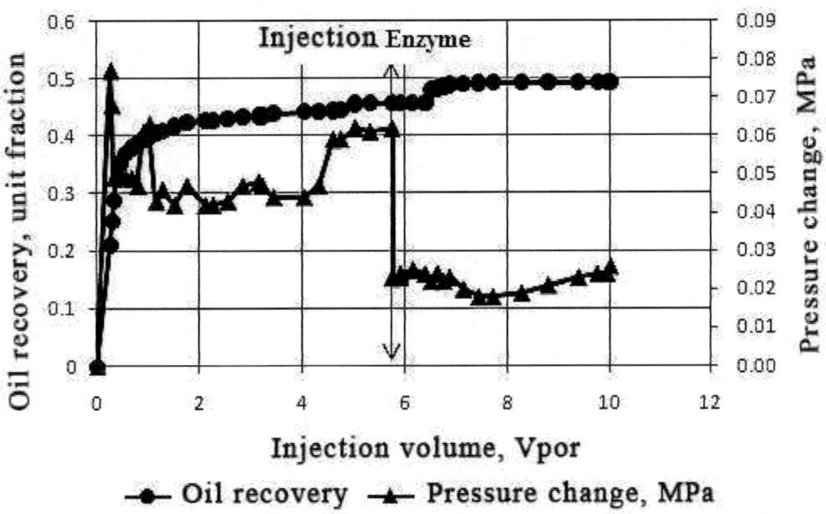

Figure 2. Dynamics of oil recovery and pressure changes before and after treatment with enzyme solution (model 2)

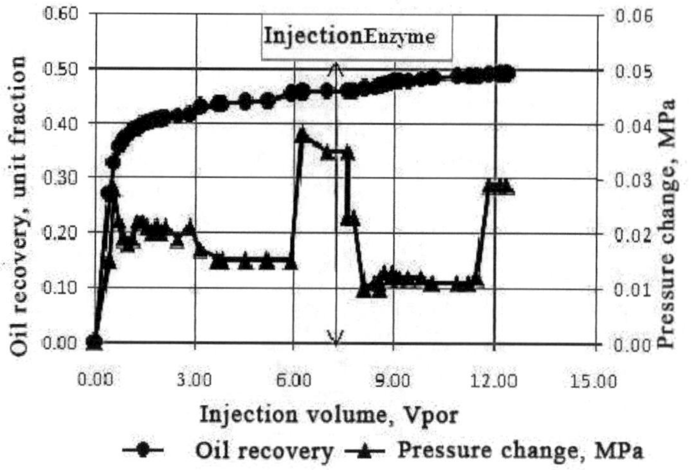

Figure 3. Dynamics of oil recovery and pressure changes before and after treatment with enzyme solution (model 3)

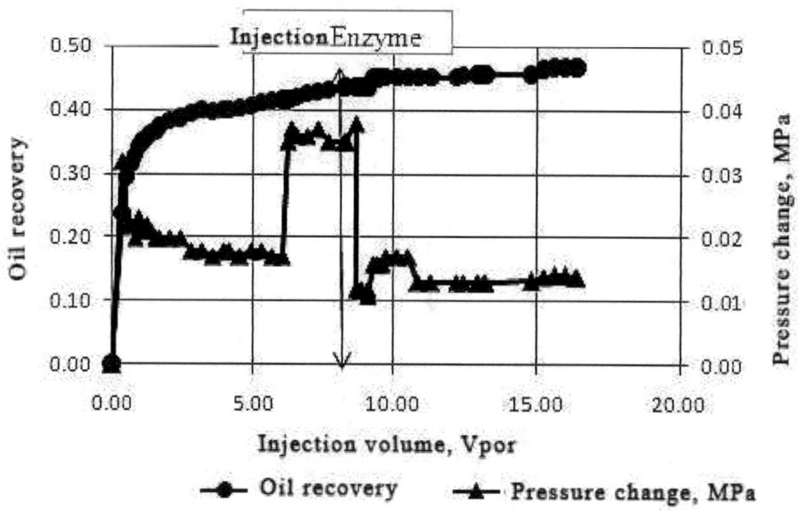

Figure 4. Dynamics of oil recovery and pressure changes before and after treatment with enzyme solution (model 4)

Table 2. Simulation outcomes

Parameters	M 1	M 2	M 3	M 4
Water absorption prior to enzyme injection, 10^{-3} μm ^{2}	8.5	9.63	12.75	15.86
Water absorption after enzyme injection, 10^{-3} μm ^{2}	14.3	12.89	21.31	18.23
Permeability recovery coefficient after enzyme injection	1.68	1.34	1.67	1.48
Oil recovery coefficient	10.59	7.23	8.73	7.28

The outcomes of laboratory tests and evaluation of research results have showed that the enzyme solution meets the following criteria:
- thermal stability;
- decline in the surface tension of water and oil;
- change in reservoir rock wettability;
- increase in the permeability of reservoir rocks in reservoir conditions.

3. Conclusion

Experimental studies of the ability to displace oil using complexes based on enzyme solutions on a reservoir model have revealed that enzyme solutions contribute to an increase in core permeability when simulating Miocene deposites on core samples. The oil recovery coefficient after the enzyme solution injection has increased by 7.23-10.59% during simulation. The permeability coefficient has grown by 1.34-1.68 times.

References

[1] Economides J M, Nolte K I 2000 *Reservoir stimulation* (West Sussex, England, John Wiley and Sons).
[2] Soloviev N N, Mukhametshin V Sh, Safiullina A R 2020 Developing the efficiency of low-productivity oil deposits via internal flooding, *IOP Conf. Ser.: Mater. Sci. Eng.* **952,** 012064 DOI: 10.1088/1757-899X/952/1/012064
[3] Zeigman Yu V, Mukhametshin V V, Kuleshova L S 2020 Management of flooding of low-production deposits according to geological and field data, I*OP: Earth and Environmental Science* **579** 012019 DOI: 10.1088/1755-1315/579/1/012019
[4] Nur Wijaya, Sheng J J 2020 Comparative study of well soaking timing (pre vs. post flowback) for water blockage removal from matrix-fracture interface, *Petroleum* **6(3)** DOI: 10.1016/j.petlm.2019.11.001
[5] Mukhametshin V Sh 2020 Rationale for the production of hard-to-recover deposits in carbonate reservoirs, *IOP Conf. Ser.: Earth Env.* **579,** 012012 DOI: 10.1088/1755-1315/579/1/012012
[6] Mukhametshin V Sh, Andreev V E, Yaskin S A 2020 Designing measures to increase oil recovery based on the identification and grouping of deposits, *IOP: Earth and Environmental Science* **579** 012013 DOI: 10.1088/1755-1315/579/1/012013
[7] Zhang Sh, Fang Zh 2020 Permeability damage micro-mechanisms and stimulation of low-permeability sandstone reservoirs: A case study from Jiyang Depression, Bohai Bay Basin, China, *Petroleum Exploration and Development* **47(2)** DOI: 10.1016/S1876-3804(20)60054-4
[8] Mukhametshin V Sh 2020 Justification for increasing the performance of hydrochloric acid treatment in wells of fields with carbonate reservoir, *IOP Conf. Ser.: Mater. Sci. Eng.* **905** DOI: 10.1088/1757-899X/905/1/012083
[9] Back M J, Kirk G 2012 An Integrated Portfolio Management Approach for More Effective Business Planning, SPE Hydrocarbon Economics and Evaluation Symposium (Calgary, Alberta, Canada, 24-25 September 2012). DOI: 10.2118/162748-MS
[10] Hu W, Wei Y, Bao J 2018 Development of the theory and technology for low permeability reservoirs, P*etroleum Exploration and Development* **45(4)** DOI: 10.1016/S1876-3804(18)30072-7
[11] Akhmetov R T, Mukhametshin V V, Kuleshova L S, Grezina O A, Malyshev P M 2020 The generalized correlating function of capillary curves and the relationship of the filtration-capacitive parameters of reservoirs in Western Siberia with the size distribution of pore channels, *Journal of Physics: Conference Series* **1661** 012016 DOI: 10.1088/1742-6596/1661/1/012016
[12] Shchetnikov V I, Veliyev E M 2019 Criteria for the selection of an operational object for the application of oil production intensification technology based on enzyme complexes, High-tech

technologies in solving the problems of the oil and gas complex (Proceedings of the IX International Youth Scientific Conference). (2019).

Determination of cutting process parameters for tooth of excavator active bucket

A K Muravskiy

Perm National Research Polytechnic University, 29, Komsomolsky prospect, Perm, 614990, Russia

E-mail: Mak2000mak@yandex.ru

Abstract. Excavators with the active buckets are able to produce digging frozen, durable soils and hard rocks. The widespread introduction of active buckets is constrained by a number of reasons, including an incomplete theoretical description of the active bucket operation and the process of interaction of the active bucket with soil and rock. In this work, the absolute speed of movement of the cutting edge of the tooth, the cutting angle and the clearance angle of cutting are mathematically determined.

1. Introduction

An excavator active bucket increases the technological capabilities of the excavator and the efficiency of its operation. This machine can dig frozen soils and hard rocks using non-explosive technology, increase labor productivity, reduce the cost of excavation and loading operations, decrease the losses of minerals, reduce the emissions of harmful substances into the atmosphere, increase the safety of mining operations, expand the boundaries of enterprises and mining operations [1, 2, 3].

By now, working equipment with an active bucket for hydraulic excavators and mechanical shovels has been created and successfully tested. In Russia company Uralmash designed and manufactured an EKG-5V excavator with an active bucket, capable of developing hard rocks with a strength of up to 100 MPa without the use of drilling and blasting preparation. This machine mined rock with low specific power consumption. Conventional excavator dug blasted rock with a higher specific power consumption per 1 m^3 of mined rock [4]. The use of impact teeth in the construction of an excavator bucket makes it possible to obtain forces on the tooth tip of up to 3500-4500 kN and more, which is several times more than the digging force of modern hydraulic excavators with a conventional bucket [5]. Daedong Industrial Co. Limited (South Korea) currently manufactures vibrating buckets capable of working with frozen soils and hard rocks, which are used for excavators weighing from 18 to 42 tons.

The widespread introduction of active buckets is constrained by a number of reasons, including an incomplete theoretical description of the active bucket operation and the process of interaction of the active bucket with soil and rock. The tooth speed, the cutting angle and the clearance angle of the active bucket affect the process of interaction of the active bucket with the soil or rock, in particular, these parameters affect the cutting force.

2. Methods

The active bucket of an excavator includes movable teeth *1* and a bucket body *2* (Figure 1). When using an active bucket, force F periodically acts on the tooth. Force F is created in various ways, for example, with a hammer *3*. As a result, the tooth moves relative to the body of the active bucket. When digging the soil or rock with the active bucket, the teeth *1* make a force effect on the massif, which causes the destruction of the soil or rock, and a "zone with broken bonds" appears [1]. This makes it possible to fill the active bucket with soil or rock. After penetration into the soil or rock, the teeth should make a reverse movement. Usually, the return stroke is carried out due to the influence of the soil or rock on the tooth during the movement of the active bucket body. Then the cycle is repeated.

The operation of the active bucket can be described by the vector of the speed of the bucket body $\vec{v_b}$ in the absolute XY coordinate system and the vector of the speed of the moving tooth relative to the body of the active bucket $\vec{v_t}$. The process of interaction of a movable tooth and soil or rock is characterized by the cutting angle, that is the angle between the front surface of the tooth and the tangent to the cutting direction [6]. When the bucket moves and the tooth is fixed relative to the bucket body (v_t =0), the cutting angle of the tooth is α_b (Figure 2).

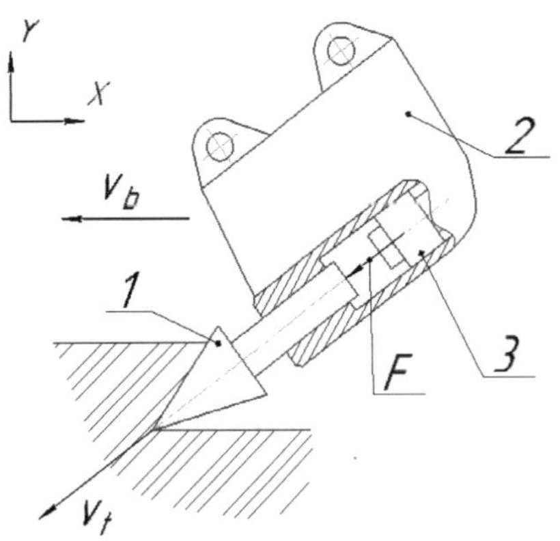

Figure 1. Design of the bucket

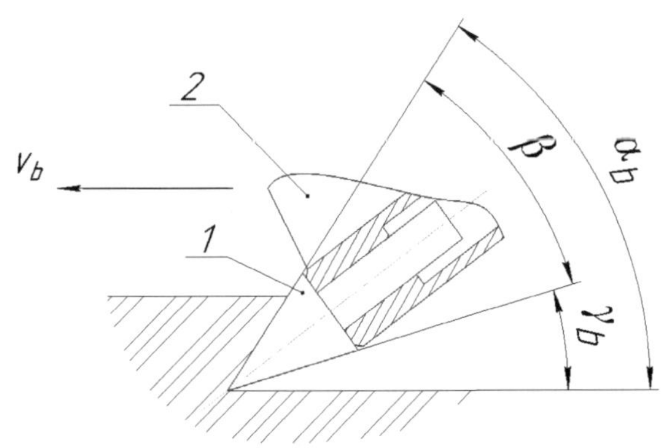

Figure 2. Parameters of the bucket

The cutting angle α_b affects the value of the clearance angle γ_b, that is the angle between the back surface of the tooth and the tangent to the cutting direction. The clearance angle γ_b is related to the wedge angle β of the tooth and the cutting angle α_b. The clearance angle is:

$$\gamma_b = \alpha_b - \beta. \quad (1)$$

Using the methods of vector algebra, geometry, trigonometry, we will determine of the speed, the cutting angle and the clearance angle of the tooth of the active bucket with the joint movement of the bucket body and the tooth.

3. Results

Determination of tooth speed.

With the joint movement of the bucket body and the tooth relative to the bucket body, the vector of the absolute speed of the cutting edge of the active bucket tooth is determined (Figure 3):

$$\overrightarrow{v_{b+t}} = \vec{v_b} + \vec{v_t}. \quad (2)$$

Using the cosine theorem, we determine the value of the absolute speed of the cutting edge of the active bucket tooth:

$$v_{b+t} = \sqrt{v_b^2 + v_t^2 - 2v_b v_t \cos(\pi - \theta)} \quad (3)$$

θ is the angle between vectors $\vec{v_b}$ and $\vec{v_t}$.

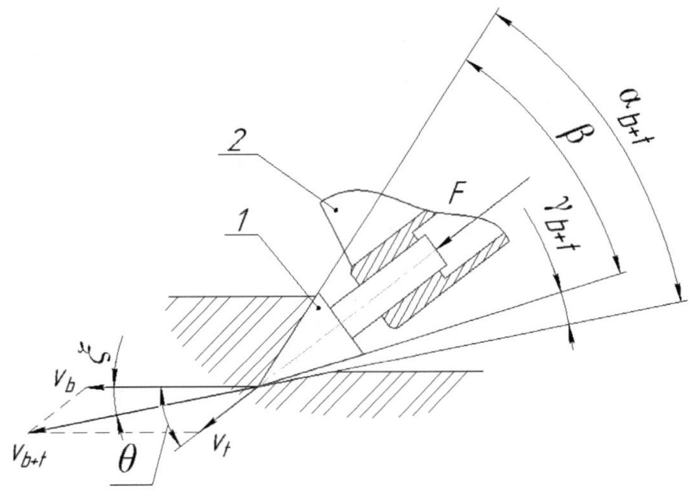

Figure 3. The joint movement of the bucket body and the tooth

Determination of cutting angle.
With the joint movement of the bucket body and the tooth, the cutting angle is:

$$\alpha_{b+t} = \alpha_b - \xi \quad (4)$$

ξ is the angle between the vector of the speed of the bucket body $\vec{v_b}$ and the vector of the absolute speed of the tooth $\overrightarrow{v_{b+t}}$.

Using the theorem of sines, we determine the angle ξ:

$$\frac{v_t}{\sin\xi} = \frac{v_{b+t}}{\sin(\pi - \theta)}. \quad (5)$$

Respectively:

$$\xi = \arcsin\left[\frac{v_t}{v_{b+t}}\sin(\pi - \theta)\right]. \quad (6)$$

Taking into account the equation (3):

$$\xi = \arcsin\left[\frac{v_t \sin(\pi - \theta)}{\sqrt{v_b^2 + v_t^2 - 2v_b v_t \cos(\pi - \theta)}}\right]. \quad (7)$$

Taking into account the equations (6) and (7) with the joint movement of the bucket body and the tooth, the cutting angle is:

$$\alpha_{b+t} = \alpha_b - \arcsin\left[\frac{v_t}{v_{b+t}}\sin(\pi - \theta)\right] \quad (8)$$

$$\alpha_{b+t} = \alpha_b - \arcsin\left[\frac{v_t \sin(\pi-\theta)}{\sqrt{v_b^2 + v_t^2 - 2v_b v_t \cos(\pi-\theta)}}\right]. \quad (9)$$

Consequently, the value of the cutting angle of the active bucket with the joint movement of the bucket body and the tooth is less than the cutting angle of the bucket with a fixed tooth ($\alpha_{b+t} < \alpha_b$). And it depends, in particular, on the ratio of the speeds of the bucket and the tooth (v_t / v_{b+t}) and angle θ.

Determination of clearance angle.

The clearance angle of the tooth with the joint movement of the bucket body and the tooth:

$$\gamma_{b+t} = \alpha_{b+t} - \beta. \quad (10)$$

Taking into accountthe equations (8) and (9), the clearance angle is:

$$\gamma_{b+t} = \alpha_b - \beta - \arcsin\left[\frac{v_t}{v_{b+t}} \sin(\pi-\theta)\right] \quad (11)$$

$$\gamma_{b+t} = \alpha_b - \beta - \arcsin\left[\frac{v_t \sin(\pi-\theta)}{\sqrt{v_b^2 + v_t^2 - 2v_b v_t \cos(\pi-\theta)}}\right]. \quad (12)$$

For common excavator buckets, the presence of a minimum positive clearance angle from 5 to 7^0 allows one to almost completely eliminate friction of the back surface of the tooth against the soil or rock, which reduces tooth wear and does not affect the tractive force [7].

4.Conclusion

We mathematically determined cutting process parameters for tooth of the excavator active bucket with a joint movement of the bucket body and teeth. There is the absolute speed of movement of the cutting edge of the tooth; the cutting angle of a tooth; the clearance angle of cutting of the tooth. The results of the work can be useful in the design of active buckets for excavators.

References

[1] Shishaev S V, Fedulov A I, Mattis A R 1989 *Calculation and creation of an active bucket* (Novosibirsk: IGD)

[2] Mattis A R, Chskidov V I, Labutin V N Zajcev G D 2006 Prospects for the use of non-explosive technologies in quarries *Mining journal* **6** 87–91

[3] Liberman Y L, Gorbunova L N, Letnev K Y 2021 Progressive designs of the most active action for earthmoving machines *Bulletin of the Kuzbass State Technical University* **1** 87–96

[4] Bojko G X 2003 *Mining equipment of Uralmashzavod* (Ekaterinburg: Uralskij rabochij)

[5] Mattis A R, Zajcev G D 2003 Problems of creating excavators for explosion-free technology and ways to solve them *Mining information and analytical bulletin* **5** 132–135

[6] Vetrov Y A 1971 *Cutting of soils by earthmoving machines* (Moscow: Mashinostroenie)

[7] Zelenin A N et al. 1975 *Machines for earthworks* (Moscow: Mashinostroenie)

Paleogeographic reconstructions of Emsian – Lower Frasnian formation for oil and gas prospecting

L D Pleshkov, E E Kozhevnikova

Perm State University, 15 Bukireva St., Perm, 614990, Russia

E-mail: levpleshkov@gmail.com, eekozhevnikova@bk.ru

Abstract. The reconstruction of sedimentation conditions is very important for understanding the geological structure of the region, for assessing the generative potential of the formation, and for the design of prospecting works for oil and gas. In this work we reconstruct the paleogeographic conditions of Emsian – Lower Frasnian formation identifying zones with conditions for the formation of reservoirs, seals and hydrocarbon bearing rocks. The accumulated data on the oil and gas content of the territory were summarized what made it possible to divide the oilfields into several groups with different sources of oil generation. In the conclusion, we allocated zones promising for oil deposits in Devonian Terrigenous formation on the territory of Perm Krai and Udmurt Republic.

1. Introduction

Paleogeographic studies are often undeservedly forgotten by petroleum geologists and not used when planning prospecting works. In many practical situations, paleogeographic data can significantly improve the efficiency of exploratory drilling, which is very important for the deepest oil and gas formations with a low reserves approval rate.

One of the most deeply buried and promising formations for new discoveries is the Emsian – Lower Frasnian oil and gas formation (also called Devonian Terrigenous). The latest regional works on paleogeographic reconstruction in the north of Volga-Ural region for Devonian Terrigenous were performed in the 1970s [1, 2, 3, 4]. Some researchers later paid attention to restoration of lithofacial conditions within the framework of thematic works, but on local sites. The works of N. E. Sosnin [5] are devoted to Devonian Terrigenous formation on the territory of the Udmurt Republic. Much more studies are executed for the territory of Perm region which are partially based on paleogeographic reconstructions [6, 7, 8, 9, 10].

The considerable accumulated amount of well logging data and core sample studies made it possible to reconstruct in detail the paleogeographic conditions of the Emsian – Lower Frasnian sedimentation basin. In this work we consider the data of wireline logs from more than 850 wells penetrated the whole Devonian Terrigenous formation. In addition to reconstructing the paleogeographic conditions, we correlate them with the hydrocarbon content of the formation. This approach allowed us not only to divide the territory into different geological zones, but also to formulate recommendations for the potentially successful areas of prospecting works for oil and gas.

2. Research Methods

The basis of a facies study and paleogeographic conditions reconstruction is the detailed correlation of the well logs. Comparison and adjustment of stratigraphic picks made by different authors over a long period of time (when opinions of stratigraphic partitioning were changing) resulted in a catalogue of updated stratigraphic picks for the Emsian – Lower Frasnian formation for the territory of Perm Krai, the Udmurt Republic and the eastern part of the Kirov region.

The correlation was carried out with the wireline logs and biostratigraphy data. Radioactive logging (natural gamma ray and neutron-gamma ray logs) is the most informative in most cases, but Pashiysky horizon is better determined by laterlog data. In other wells where these methods were not recorded; correlation was made by spontaneous potential (SP) data.

3. Origins of Emsian – Lower Frasnian Rocks Sediments

The Emsian – Lower Frasnian sedimentation was preceded by a long period of continental development of the area. During the investigated time, the transgression in the northern Volga-Ural was unstable and the basin periodically shallowed until the generation of sediment ceased. The changes from transgressive to regressive phases led to formation of cycles in the geological record.

3.1. Emsian Age

During the Emsian age, rocks of the Takatinsky, Vyazovsky and Koyvenksy horizons were formed. At the beginning of the Takatinsky time, intensified tectonic activity led to the advance of the Ural geosyncline from the east of the area into the continental interior in the direction of Izhevsk, but the maximum transgression only reached Votkinsk (Fig. 1). By the end of Emsian age, there was a separation of the Kama Arch (KamA) from the Perm Arch (PA). At the same time in the east, the North-Tatar Arch (NTA) was united with the Perm and Bashkir Arches, and in the north — with the Kama Arch. Sediments with terrigenous varieties (mainly sandstones) formed under the epicontinental plain conditions. Only in the area of the Krasnokamsk-Chusovoy paleodepression the carbonate layers of low thickness deposited under the influence of a shallow sea conditions. In general, the thickness of Emsian stage ranges from 0 m on the arches and up to 100 m in the Krasnokamsk-Chusovoy paleodepression.

3.2. Eifelian Age

The Eifelian age was preceded by a hiatus in sediment generation, further transgression from the east became more active and the area of sedimentation significantly increased in comparison to the Emsian basin. The Eifelian sediments are represented by the Biysky and Afoninsky horizons. During the Eifelian age, landforms rearranged — the Kama Arch separated from the North-Tatar Arch, both remaining the main source of terrigenous material. Sediment in the Krasnokamsk-Chusovoy paleodepression generated in open sea conditions resulting in rocks composed of carbonate varieties. In the direction of the arches, carbonate origin of sediments changed to terrigenous. The thickness of sediments varies from 0 to 50 m within the Krasnokamsk-Chusovoy paleodepression.

3.3. Givetian Age

The Givetian age is subdivided into the Starooskolsky and Pashiysky time. Three cycles of sedimentation are distinguished in the Starooskolsky time, each corresponding to the formation of the Vorobyovsky, Ardatovsky and Mullinsky horizons. The alternation of transgression and regression is confirmed by gravel packs at the base of these horizons. The cycles are not fundamentally different in terms of sediment generation and areal extent. At the same time, the Starooskolsky basin occupied a larger area than the Emsian – Eifelian one reducing the land area of the Bashkir and Kama Arches. During the tectonic rearrangement of the region the North-Tartar and Kama Arches were united by an island line into a single paleosurface. A distinctive feature of this time is the change in the direction of sea transgression.

IOP Conf. Series: Earth and Environmental Science 1021 (2022) 012047 doi:10.1088/1755-1315/1021/1/012047

During the Emsian – Eifelian ages, the sea was actively advancing onshore from the east and since the Starooskolsky time the main direction of the transgression changed to the north-east (from the present position of the Republic of Tatarstan). In the most submerged part of the basin sedimentation occurred in shallow sea conditions, in some places with deposition of clayey carbonates. The shallow-water zone extended from the south-west of the area along the western slope of the Perm and Bashkir Arches, northwards to the position of Perm, turning westwards and along the position of the towns of Kez, Votkinsk and Izhevsk. On the eastern slope of the North-Tatar Arch and on the western slopes of the Perm and Bashkir Arches sediments formed in coastal accumulative plain conditions with formation of channels and detrital cones. Here, mainly well sorted sandstone beds were accumulated with subordinate clayey and silty varieties. The thickness of the deposits varies from 0 m on the arches to 80 m in the shallow sea area.

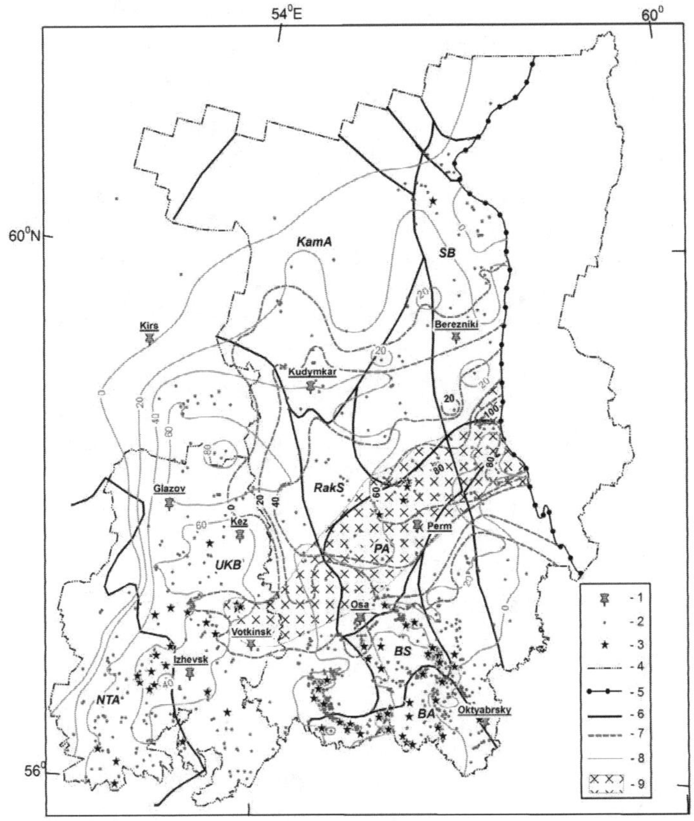

Figure 1. Oil and gas content and capacities of Devonian Terrigenous formation
Legend: 1 — cities; 2 — wells; 3 — oil deposits in Devonian Terrigenous; 4 — administrative boundaries of Perm region and Udmurtia Republic; 5 — boundary of West-Ural folding zone; 6 — boundaries of large tectonic elements: KamA — Kama Arch, SB — Solikamsk Basin, RakS — Rakshinskaya Saddle, PA — Perm Arch, BA — Bashkir Arch, BS — Babka Saddle, UKB — Upper-Kama Basin, NTA — North-Tatar Arch; 7 — Emsian stage thickness; 8 — Starooskolsky superhorizon thickness; 9 — Devonian Terrigenous formation with predominance of shallow sea conditions.

Pashiysky time was again preceded by a regressive period and sediments formed under the differentiated tectonic movements. Active tectonics resulted in the final separation of the North-Tartar and Kama Arches. This separation was accompanied by volcanic activity as evidenced by beds tuff at the base of Pashiysky horizon described in the core from several boreholes of the Bashkir Arch [11]. On the sides of the arches sediments generated in coastal plain conditions, while in the most submerged central part the conditions corresponded to a shallow sea. Depositional thickness varies from 0 to 40 m.

3.4. Frasnian Age

Timansky time is characterised by increasing sea transgression, with all areas of paleosurface sinking below the sea level. By the end of the Timansky time the marine sediment generation conditions have stabilised with the formation of carbonate sediments. These conditions continued without interruption until the Visean age. In the entire area, the lower layers are composed of terrigenous formations, while the carbonate interlayers are found upwards ranging from centimeters to 1 m in thickness. In the east, open sea conditions existed in the area of modern Berezniki under the influence of the geosyncline. From east to southwest, the area dominated by shallow marine conditions, and where the arches dipped the sediments generated in coastal-marine conditions. The total Timansky sediment thickness varies from the several meters in the area of the Bashkir Arch to 40 m in the central part of the Upper-Kama Basin, near the town of Kez.

As a result of updating the stratigraphic picks and reconstructing the sedimentation conditions the following has been established. The most complete section with maximum thickness of Devonian Terrigenous formation (up to 180 m) is in a zone of Upper-Kama Basin and Rakshinskaya Saddle junction, as well as the Perm Arch. A zone has been identified by the predominantly shallow sea sedimentation conditions with the accumulation of mainly carbonate rocks during the Emsian – Early Frasnian ages (Fig. 1).

4. Hydrocarbon Bearing Capacity of Emsian – Lower Frasnian Formation

Within the extents of the Devonian Terrigenous formation that has been studied, there are 64 discovered oil deposits. Beds with production potential are confined to formations from Starooskolsky to Timansky horizons — 'D0' bed in Timansky sediments, 'D1' bed in Pashiysky and 'D2' bed in Starooskolsky. Most of the deposits — 36 — are in the Perm and Bashkir Arches, 15 are discovered in the Upper Kama Basin, 12 are within the North-Tatar Arch, and one in the Solikamsk Basin (Fig. 1). Reservoir beds are sandstones except for a group of oilfields in the central part of Upper-Kama Basin (Vukoshurskoe, Kykvinskoe, Staroyaginskoe) where the reservoirs are confined to carbonates of Starooskolsky superhorizon. Due to high lithological heterogeneity of the reservoir rocks, lithologically screened deposits prevail. There is no consistent fluid seal within the Emsian – Lower Frasnian formation; the regional seal is Timansky mudstones with thicknesses varying over the area from several to 10–15 m. Sometimes, the seals are the overlying clayey carbonates of Sargayevsky superhorizon, thus in Oblastnovskoye and Dorokhovskoye oilfields, the deposits are found in Timansky sandstones directly under Sargayevsky rocks [12]. In some fields the whole thickness of the Timansky sandstones acts as a fluid source and the deposits are found only in Pashiysky reservoirs (Andreevskoye, Rassvetnoye and others). Most often oil deposits are confined to Timansky reservoirs — more than 50 ones. There are multilayer fields with the deposits present in several reservoirs (Andreevskoye, Arkhangelskoye and others). The least number of deposits is confined to the Starooskolsky reservoirs — three in the Udmurt Republic and seven are within Perm Krai. Generally, the Starooskolsky sandstones are not isolated from the overlying reservoirs and have a single oil-water contact with the Pashiysky deposits (Andreevskoye and others).

The basic parameters of physical and chemical properties of oils are presented in Table 1. Terrigenous Devonian oils vary from light to heavy in density, from low-sulfur to high-sulfur in sulfur content, from low to medium in paraffin content. Table 1 shows properties of oils with both typical characteristics for Devonian Terrigenous formation and the oils that differ significantly from the main bulk of terrigenous Devonian oils. The fields of Perm Arch and oil of Chuboysky field are distinguished by low sulfur content, and the highest sulfur content is characteristic to the oil occurring in carbonates of Starooskolsky superhorizon. The oils with the highest wax content are found in the Perm Arch. The special position is occupied by oils of Chuboyskoye field — they are greater in density, thus the minimum content of sulphur. Oils with such combination of the parameters are met only in Riphean – Vendian rocks [13]. It can be assumed that the groups of deposits distinguished by physical and chemical properties of oil have different generation sources.

Table 1. Physical and chemical oil properties of Devonian Terrigenous formation

No.	Field	Bed	Density (kg/m^3)	Content of (%)		
				sulphur	paraffins	resins and asphaltenes
	Andreevskoe	D1	898	2.14	3.98	21.89
1	Andreevskoe	D0	898	2.37	3.10	23.28
	Andreevskoe	D2	903	2.54	3.05	29.32
2	Arkhangelskoe	D0, D1	848	2.40	3.30	16.07
3	Vekoshurskoe	D2, carb. *	896	4.16	4.37	19.32
4	Garankinskoe	D0	888	2.42	1.69	15.46
5	Dorokhovskoe	D0	837	1.50	1.53	7.17
6	Yeseneyskoe	D0	869	2.37	3.68	18.18
7	Zorinskoe	D0	840	0.69	5.36	11.26
8	Kamyshlovskoe	D0	871	1.43	2.02	20.63
9	Krasnokamskoe	D0	846	1.15	5.16	8.36
10	Kuznetsovskoe	D0	857	0.72	5.67	15.32
11	Kuliginskoe	D0, D1	895	2.01	3.28	15.18
12	Kykvinskoe	D2, carb.	883	3.36	1.78	28.68
13	Oblastnovskoe	D0	881	2.20	3.20	3.42
14	Rassvetnoe	D1	860	1.65	4.52	13.92
15	Severokamskoe	D0	835	0.44	5.13	8.38
16	Sosnovskoe	D0	863	1.37	4.55	24.93
17	Staroyaginskoe	D2, carb.	878	3.68	1.70	26.80
18	Cherdynskoe	D0	855	1.12	2.92	10.72
19	Chuboyskoe	D1	972	0.44	1.01	26.49

* carb. — carbonate reservoir bed

5. Conclusions

Around the world, oil and gas prospecting is based on oil system analysis, an approach described in detail in the literature [14, 15]. This approach increases the efficiency of prospecting. The process of identifying an oil system in a region consists in determining oil and gas suites [16], the stage of organic matter catagenetic transformation and tracing reservoirs and seals. The oil system is usually named by the age of the oil and gas source formations. A similar approach in a simplified form is applied in this work. Based on the paleogeographic reconstructions and generalization of the generative potential of the formation [17, 18, 19] we came to the following conclusions.

During the Emsian – Eifelian ages, the potential source rocks were deposited. In the Starooskolsky time, in the greater part of the territory the deposition occurred in conditions of the coastal accumulative plain leading to the formation of terrigenous reservoirs and seals. In a small area in shallow sea conditions Starooskolsky carbonate strata were deposited — also potential oil and gas reservoirs. During the Pashiysky time, sedimentation conditions facilitated to the formation of terrigenous reservoirs with widely distributed sandstone lenses. Timansky rocks were formed in similar conditions.

In terms of oil and gas bearing potential, it can be assumed that the formation of Devonian Terrigenous oil reservoirs was due to three sources of generation. Terrigenous Devonian deposits in the Chuboyskoe field were formed due to migration from Riphean – Vendian rocks. Deposits confined to Perm Arch and Starooskolsky carbonates of Udmurt Republic were formed by source rocks of Devonian Terrigenous themself. Probably, a part of deposits in Devonian Terrigenous relates to Upper Devonian – Tournaisian formation. Proceeding from the received data the most perspective area for searching of oil deposits in Devonian Terrigenous rocks is the Perm Arch where the deposits could be formed due to source rocks of both Devonian Terrigenous and Upper Devonian – Tournaisian formations. In Udmurt Republic, the eastern slope of the North-Tatar Arch is the most promising as the reservoirs are widely distributed in the Devonian Terrigenous and the source of generation is probably the Upper Devonian – Tournaisian formation.

References

[1] Mikhaylova N A 1968 *Paleogeography of Middle and Upper Devonian of Kirov and Perm Regions and Udmurtian ASSR* (Moscow: Nauka)
[2] Kutukov A V 1982 Conditions of manifestation of oil-formation processes in Vendian, Devonian and Viseian terrigenous formations of Perm Prikamie *Geology of oil and gas* **2** 35–9
[3] Chepikov K R 1977 *Lithologo-facial and lithologo-paleotectonic conditions of formation of oil-and-gas-bearing formations on the east of the Russian platform* (Moscow: IG&RGI)
[4] Filippova M F 1958 *Stratigraphy, lithology and facies of the Devonian terrigenous part of the northern regions of the Ural-Volga region* (Moscow: Gostoptekhizdat)
[5] Sosnin N E 2009 Structure and prospects of oil-bearing capacity of terrigenous Devon in territory of Udmurt Republic *Geology, geophysics and development of oil and gas fields* **11** 21–4
[6] Denisov A I 1973 Distribution and lithological and facial characteristics of Vorobiev layers of Perm Prikamie *Geology and oil and gas bearing capacity of Volga-Ural region north. Trudy VNIGNI.* **123** 164–71
[7] Vilesov A P, Putilov IS 2009 Complex use of lithologo-facial analysis and seismo-facial modelling for study of Devonian terrigenous strata structure *Petroleum Economy* **4** 23–7
[8] Putilov I S 2007 Application of probabilistic statistical analysis for study of facies zonation of Turnaisian – Famensky carbonate complex of Siberian oil field *Oil Farming* **9** 112–4
[9] Provorov V M 2003 History of geological development of Perm Oblast territory *General and regional geology, geology of seas and oceans, geol. mapping: Review.* (Moscow: OOO Geoinformcentr)
[10] Stashkova EK, Frik M G, Araslanova RM, Bagaev A N 2015 *Terrigenous Devonian of the Perm Region* (Perm: AO KamNIIKIIGS)
[11] Yekhlakov Y A 2001 Making a catalogue of stratigraphic breakdowns of Pre-Paleozoic and Paleozoic deposits from prospecting and exploration wells in Perm region *Research report,* (Perm, FSUDP KamNIIKIGS)
[12] Karaseva T V, Kozhevnikova E E 2019 *Devonian terrigenous oil-and-gas-bearing complex in the southern districts of Perm region. New ideas about geological structure and formation of oil-and-gas bearing capacity* Monograph, (Perm State National Research University)
[13] Giniyatova L F, Bashkova S E, Karaseva TV 2017 Analysis of the development of Riphean-Vendian deposits of the north-eastern part of the Volga-Ural GGP in connection with their oil and gas content *Bulletin of Perm University. Geology* **16(3)** 275–82
[14] Magoon L B, Dow W G 2012 The petroleum system — from source to trap *AAPG Memoir* **60**
[15] Hantschel Th, Kauerauf A I 2009 Fundamentals of basin and petroleum systems modelling (Springer – Verlag)
[16] 1988 *Dictionary of Oil and Gas Geology* (Leningrad: Nedra)
[17] Rodionova K F 1967 *Organic matter and oil-bearing rocks of Devonian Volga-Ural oil and gas-bearing region* (Moscow)
[18] Larskaya E S 1983 *Diagnostics and methods of study of oil-and-gas-bearing strata* (Moscow: Nedra)
[19] Kozhevnikova E E 2020 Geologic Structure and Hydrocarbon Generation Potential of Devonian Terrigenous Strata in the Perm Territory *Russian Geology and Geophysics* **61(8)** 906–16

IOP Conf. Series: Earth and Environmental Science 1021 (2022) 012048 doi:10.1088/1755-1315/1021/1/012048

Indirect determination of reservoir pressure by changing operating modes of exploration and recovery wells

V A Grishchenko[1], M N Kharisov[1], R F Yakupov[2], V V Mukhametshin[3], L S Kuleshova[2]

[1] LLC «RN-Bashneft», 86/1, Lenina St., Ufa, Republic of Bashkortostan, 450006, Russian Federation
[2] Ufa State Petroleum Technological University, Branch of the University in the City of Oktyabrsky, 54a, Devonskaya St., Oktyabrsky, Republic of Bashkortostan, 452607, Russian Federation
[3] Ufa State Petroleum Technological University, 1, Kosmonavtov St., Ufa, Republic of Bashkortostan, 450064, Russian Federation

E-mail: rustem_geolog@mail.ru, info@of.ugntu.ru

Abstract. The results obtained in the article enable to draw a conclusion about the sufficient accuracy of assessing the energy state and the possibility of using an algorithm for the indirect determination of reservoir pressure for solving practical problems of geological exploration and geological recovery research in selecting and planning geological and engineering operations and monitoring the work of the base well stock. However, algorithms for indirect determination of reservoir pressure should not fully replace well testing but they can be considered as an additional tool for assessing the reservoir pressure and their results should be analyzed considering the experience of the reservoir development in specific geological conditions.

1. Introduction

Key oil fields in Russia were discovered in the middle of the last century. The main part of the largest and most productive deposits has been developed for several decades and is at the final stage of development. The most effective use of the existing fund is necessary to maintain oil production levels amid the gradual depletion of fixed assets and transition to the development of areas with deteriorated filtration and reservoir properties [1–10]. A decrease in reliable information about the current energy state caused by the need to increase the duration of shutdowns during hydrodynamic studies of wells is an attendant problem [11]. In such conditions, mathematical models for the indirect determination of reservoir pressure aimed to reduce oil production losses during research gain ground.

Considerable measures aimed to maintain the efficient development of oil reserves at the final development stage are being implemented. They include transfers of wells between sites, waterflooding organization, injection redistribution, unprofitable wells shutdown, hydraulic fracturing, etc. [12–15] This leads to local changes in the energy state. Thus, one of the main types of geological and engineering operations at production facilities is the optimization of pumping equipment or well stimulation. The annual number of well stimulation measures is massively more than the number of production tests. The use of well performance indicators when optimizing pumping equipment to assess the energy state of the reservoir by indirect methods can significantly increase the coverage of data on the current reservoir pressure. This will help to improve the efficiency of geological and engineering operations on planning and monitoring basic wells stock operation.

Thus, the objective of the work is to analyze the results of indirect determination of reservoir pressure by changing the wells operating modes, which is currently rather relevant.

2. Materials and methods

To achieve this objective, the methods of system analysis, regression analysis, and multi-parameter optimization were used. The scientific literature was mainly taken from the abstract databases of peer-reviewed publications like Web of Science and Scopus, as well as from the scientific search engine Google Scholar.

Partial solution of the generalized Darcy's law for the case of radial steady flow to the well has been used.

Well operation indicators are associated with reservoir pressure by a particular solution of the generalized Darcy's law, which is described by the Dupuis formula for a radial steady-state single-phase inflow [16]:

$$q = \frac{2\pi kh(\overline{P}_r - P_{wf})}{B\mu(ln\frac{r_e}{r_w} - \frac{1}{2} + S_{tot})}, \tag{1}$$

where q is fluid flow rate, m^3/s; k is reservoir absolute permeability, m^2; h is effective collector capacity, m; μ is fluid viscosity, Pa s; B is coefficient of volumetric expansion of fluid, units; $\overline{P}_r$ is average pressure on the supply circuit, Pa; P_{wf} is bottomhole pressure in the well, Pa; r_e is radius of the feed loop, m; r_w is borehole radius, m; S_{tot} is total skin factor, units.

Thus, the flow rate of the fluid at the well is related to the bottomhole pressure by the following dependance:

$$q = \frac{2\pi kh\overline{P}_r}{\mu B\left(\ln\left(\frac{r_e}{r_w}\right) - \frac{1}{2} + S_{tot}\right)} - \frac{2\pi kh}{\mu B\left(\ln\left(\frac{r_e}{r_w}\right) - \frac{1}{2} + S_{tot}\right)} \cdot P_{wf}. \tag{2}$$

In a standardized form, this dependence takes the following form:

$$\frac{q_0}{q_{0max}} = 1 - \left(\frac{P_{wf}}{\overline{P}_r}\right), \tag{3}$$

where $q_{0max} = \frac{2\pi kh\overline{P}_r}{\mu B\left(\ln\left(\frac{r_e}{r_w}\right) - \frac{1}{2} + S_{tot}\right)}$ is fluid flow rate at zero bottomhole pressure, m^3/s.

In 1968 Vogel showed that if bottomhole pressures is below the bubble point pressure, the dependence of the relative fluid flow rate in the well on the relative bottomhole pressure deviates from the linear form and is described by the following empirical relationship [17]:

$$\frac{q_0}{q_{0max}} = 1 - 0{,}2 \cdot \left(\frac{P_{wf}}{\overline{P}_r}\right) - 0{,}8 \cdot \left(\frac{P_{wf}}{\overline{P}_r}\right)^2, \tag{4}$$

The indicator curve proposed by Vogel was obtained when processing the results of computer modeling of wells behavior in several fictitious fields with a wide range of PVT properties and relative phase permeabilities (provided that the oil phase flows into them) from which the dissolved gas is released.

Alternative empirical dependences as well as corrections to Vogel's dependence have been proposed by other authors. Thus, in 1970, Standing proposed an adjustment to the Vogel dependence to consider the total skin factor [18]. Later, in 1973, Fetkovich proposed an alternative empirical relationship based on the analysis of 40 different wells in 8 fields under similar condition of the oil phase inflow, from which the dissolved gas was released [19]. Wiggins in 1991 derived an analytical relationship for a radial steady-state three-phase inflow of oil, water, and gas to a well [20]. Despite the large number of works in this area, the empirical dependences proposed in them have drawbacks leading to significant deviations of the calculated fluid flow rates from the actual flow rates in wells of fields with excellent physical and chemical properties [21, 22]. As a solution to this problem, an algorithm for determining the empirical dependence of the indicator curve at bottomhole pressure below the bubble point pressure was proposed in [23].

Formula (1) for determining the flow rate of the fluid in the well for a radial steady-state single-phase inflow can also be written as follows:

$$q = PI(\overline{P}_r - P_{wf}), \tag{5}$$

where $PI = \frac{2\pi kh}{B\mu(\ln\frac{r_e}{r_w}-\frac{1}{2}+S_{tot})}$ is well productivity index, m^3/(Pa·s).

Regarding the results of Vogel's work, at the bottomhole pressure below the saturation pressure, formula (5) takes the following form:

$$q = q_b + \frac{PI \cdot P_b}{1.8}[1 - 0.2\left(\frac{P_{wf}}{P_b}\right) - 0.8(\frac{P_{wf}}{P_b})^2], \tag{6}$$

where $q_b = PI(\overline{P}_r - P_b)$ is inflow to the well at bottomhole pressure equal to the saturation pressure, m^3/s; P_b is oil saturation pressure with gas, Pa.

Together, formulas (5) and (6) form the following system of equations (Figure 1):

$$\begin{cases} q = PI \cdot (\overline{P}_r - P_{wf}), P_{wf} > P_b \\ q = q_b + \frac{PI \cdot P_b}{1,8}\left[1 - 0,2 \cdot \left(\frac{P_{wf}}{P_b}\right) - 0,8 \cdot \left(\frac{P_{wf}}{P_b}\right)^2\right], P_{wf} \le P_b \end{cases} \tag{7}$$

In practice, in addition to reservoir pressure, the productivity index is also an unknown parameter. When carrying out the well stimulation, the performance indicators are measured in two modes of well operation, specifically, before and after the optimization of pumping equipment, which leads to an analytical solution to the system of equations (7) with two unknowns.

In the case of several sequential well stimulations during a period of time when changes in reservoir pressure and productivity index can be neglected, the problem of determining the values of reservoir pressure and productivity index is solved by the method of regression analysis [24]:

$$Y = f(X_1, \dots, X_i, \dots, X_n, \beta) + e, \tag{6}$$

where X_i is i-th random variable, β is the vector of unknown parameters of the equation, e is the vector of residuals.

One of the methods for solving the regression equation is to determine the values of unknown parameters using a genetic algorithm (Figure 2) [25].

Application of the method of regression analysis with the application of a genetic algorithm enables to refine the value of reservoir pressure based on the results of several sequential well stimulations under conditions of uncertainty in the measurement error of well operation indicators (Figure 3).

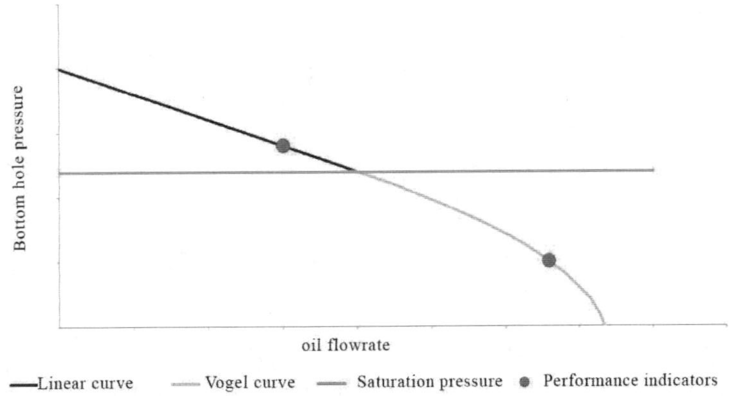

Figure 1. Determination of functional dependence of oil flow rate on bottomhole pressure using two well operation modes

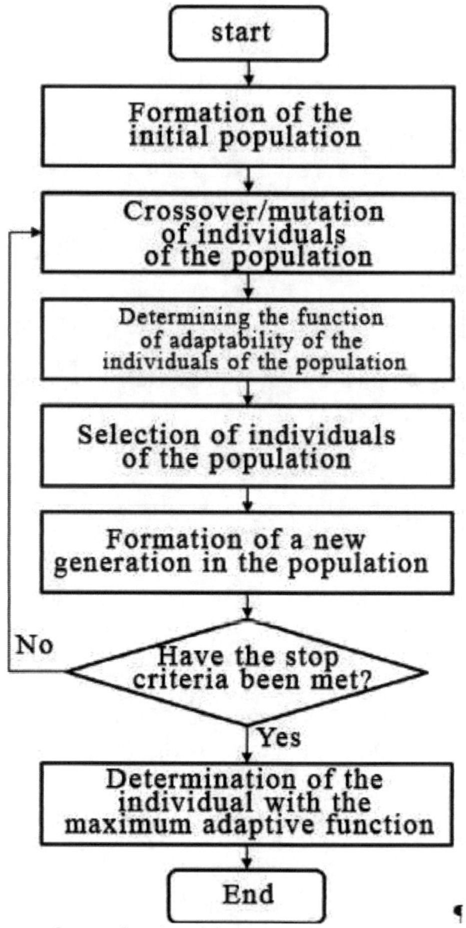

Figure2. Block diagram of genetic algorithm

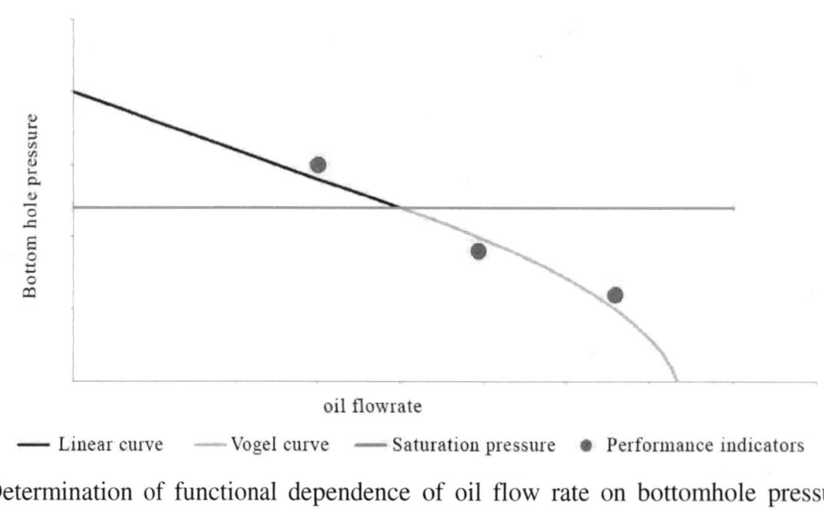

Figure3. Determination of functional dependence of oil flow rate on bottomhole pressure using a variety of well operation modes

To analyze the results of the indirect determination of reservoir pressure by changing the operating modes of wells, the authors used an algorithm described in detail in [26].

3. Results and Discussion

The accuracy of the results obtained and the possibility of using the algorithm for the indirect determination of reservoir pressure in practice was assessed by the method of retrospective analysis of the actual data for wells with the well stimulation. For the analysis, we used wells with reservoir pressure measurements before or after the well stimulation, with a satisfactory technical condition of the production string and cement stone, and with the absence of technical problems during well-workover operation (accidents, significant process liquids intake, etc.). For each well from the final list, reservoir pressure was estimated using the accepted algorithm and the results were compared with actual measurements (Figure 4, 5).

Figures 4 and 5 show that the results of the assessment using the algorithm for indirect determination of reservoir pressure have a fairly high convergence with the actual data. Specifically, the median of the absolute value of the relative deviation is 8%; deviation in 88% of wells does not exceed 20%. The largest contribution to the deviations obtained is probably due to errors in reservoir pressure measurements, which is especially important for low-productivity deposits with a complex geological structure, and for converting bottomhole pressure through a dynamic level.

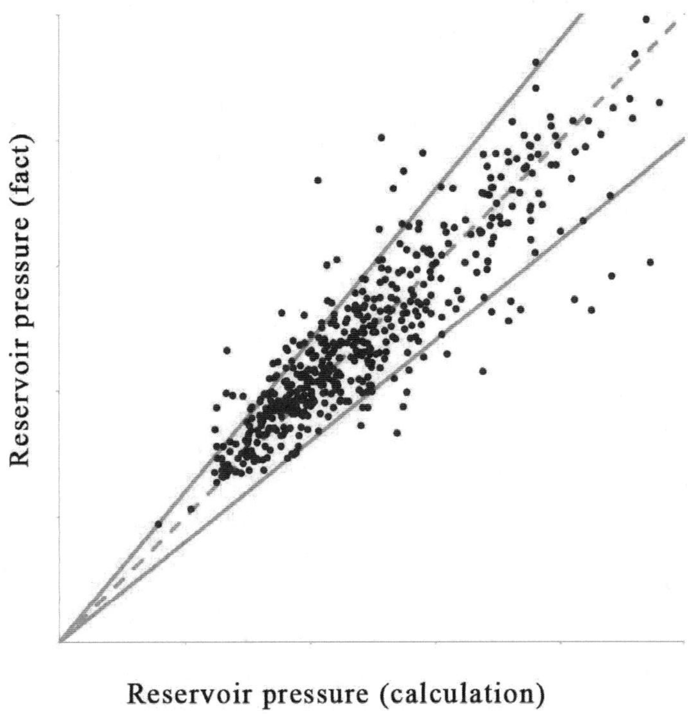

Figure 4. Cross-plot of calculated and actual values of reservoir pressure for analyzed wells

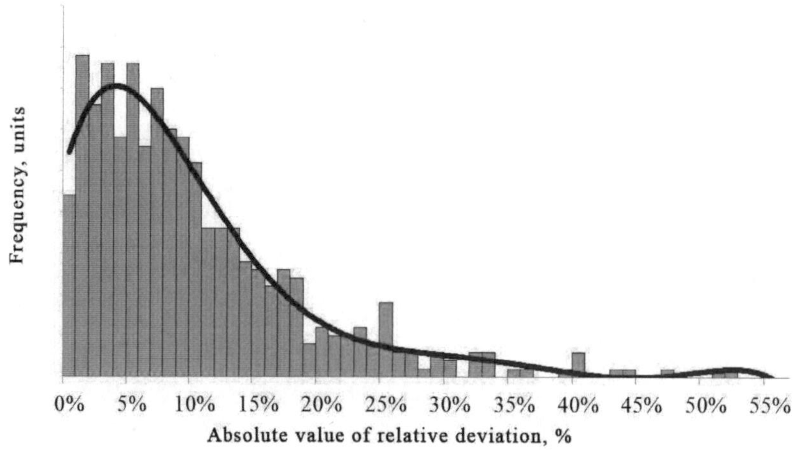

Figure 5. Histogram of deviations of calculated reservoir pressure values from actual values for analyzed wells

4. Conclusion

In general, the results obtained enable to conclude that there is a sufficient accuracy in assessing the energy state and the possibility of using an algorithm for indirect determination of reservoir pressure to solve practical problems of field geological analysis in the selection and planning of geological and technical measures and monitoring the operation of the basic well stock.

It should be noted that algorithms for indirect determination of reservoir pressure should not fully replace hydrodynamic testing of wells, but they can be considered as an additional tool for assessing reservoir pressure, whose results should be analyzed considering the experience of reservoir development in specific geological conditions.

References

[1] Dmitrievsky A N 2017 Resource-innovative strategy for the development of the Russian economy *Oil Industry* **5** 6-7

[2] Gabzalilova A Kh, Batalov D A, Mukhametshin V Sh, and Andreev V E 2021 Geological and technological justification of the parameters of acid-clay treatment of wells *IOP Conference Series: Materials Science and Engineering (MEACS 2020 – International Conference on Mechanical Engineering, Automation and Control Systems)* **1064** 012058 1–9 DOI: 10.1088/1757-899X/1064/1/012058

[3] Shpurov I V, Zakharenko V A, and Fursov A Ya 2015 A differentiated analysis of the degree of involvement and the depletion of stocks of jurassic deposits in the Western Siberian oil-and-gas province *Nedropolzovanie XXI vek* **1 (51)** 12-19

[4] Khuzin R R, Mukhametshin V Sh, Salikhov D A, Andreev V E, Pepelyaev D V, Stefanovich Yu N 2021 Improving the efficiency of horizontal wells at multilayer facilities *IOP Conference Series: Materials Science and Engineering (MEACS 2020 – International Conference on Mechanical Engineering, Automation and Control Systems)* **1064** 012066 1-9 DOI: 10.1088/1757-899X/1064/1/012066

[5] Dmitrievsky A N, and Eremin N A 2015 Modern scientific-technical revolution (str) and the shift of paradigm of hydrocarbon resources development *Problems of Economics and Management of Oil and Gas Complex* **6** 10-16

[6] Malyarenko A M, Bogdan V A, Blinov S A, Kotenev Yu A, Mukhametshin V Sh, and Kotenev M Yu 2021 Improving the reliability of determining physical properties of heterogeneous clay reservoir rocks using a set of techniques *Journal of Physics: Conference Series* **1753** 012074 1-12 DOI: 10.1088/1742-6596/1753/1/012074

[7] Muslimov R Kh 2016 A new strategy for the development of oil fields in modern Russia - optimization of production and maximization of KIN *Oil. Gas. Innovations* **4** 8-17

[8] Batalov D A, Soloviev N N, Mukhametshin V Sh, Andreev V E, Dubinskiy G S, and Safiullina A R 2020 Forecasting the use of non-stationary waterflooding in the conditions of oil deposits in Western Siberia *IOP: Earth and Environmental Science* **579** 1–7 DOI: 10.1088/1755-1315/579/1/012020

[9] Eremin A N, Eremin An N, Eremin N A 2012 *Management of the development of intelligent oil and gas localities: an account for the body in 2 books* (Moscow: Catalog of Gas)

[10] Valeev A S, Kotenev Yu A, Mukhametshin V Sh and Sultanov Sh Kh 2019 Substantiation of the recovery of residual oil from low-productive and heterogeneous formations in Western Siberia by improving the waterflood system using gas and water-gas impacts *IOP Conference Series: Materials Science and Engineering* **560** 012204 1-6 DOI: 10.1088/1757-899X/560/1/012204

[11] Ipatov A I, and Kremenetsky M I 2008 *Hydrodynamic and field-technological studies of wells: a textbook* (Moscow: MAKS Press)

[12] Galkin V I, Rastegaev A V, Kozlova I A, and Krivoshchekov S N 2014 Probability evaluation of a formation hydraulic fracturing based on mining-geological features (deposits of Kogalym region are taken as an example) *Geology, Geophysics and Development of Oil and Gas Fields* **7** 100-104

[13] Yakupov R F, Mukhametshin V Sh, Khakimzyanov I N, and Trofimov V E 2019 Optimization of reserve production from water oil zones of D3ps horizon of Shkapovsky oil field by means of horizontal wells *Georesursy* **21(3)** 55-61 DOI: 10.18599/grs.2019.3.55-61

[14] Alvarado V, Reich E-M, and Yunfeng Yi K 2006 Potsch Integration of a Risk Management Tool and an Analytical Simulator for Assisted Decision-Making in IOR *SPE Europec/EAGE Annual Conference and Exhibition (Vienna, Austria, 12-15 June 2006)* 6 p DOI: 10.2118/100217-MS

[15] Stenkin A V, Kotenev Yu A, Mukhametshin V Sh, and Sultanov Sh Kh 2019 Use of low-mineralized water for displacing oil from clay productive field formations *IOP Conference Series: Materials Science and Engineering* **560** 012202 1-5 DOI: 10.1088/1757-899X/560/1/012202

[16] Deeva T A, Kamartdinov M R, Kulagina T E, and Shevelev P V 2006 *Modern methods of field development at late stages: training manual* (Tomsk)

[17] Vogel J V1968 Inflow Performance Relationships for Solution-Gas Drive Wells *Journal of Petroleum Technology* **20(1)** 83-92

[18] Standing M B 1970 Inflow Performance Relationships for Damaged Wells Producing by Solution-Gas Drive *Journal of Petroleum Technology* **22(11)** 1399-1400

[19] Fetkovich M J 1973 The Isochronal Testing of Oil Wells *Fall Meeting of the Society of Petroleum Engineers of AIME*

[20] Wiggins M L 1992 Analytical Inflow Performance Relationships for Three-Phase Flow in Bounded Reservoirs *SPE Western Regional Meeting*

[21] Gallice F, and Wiggins M L A 1999 Comparison of Two-Phase Inflow Performance Relationships *SPE Mid-Continent Operations Symposium*

[22] Fattah K A 2012 New Inflow Performance Relationship for Solution-Gas Drive Oil Reservoirs *Oil and Gas Business* **5** 343-366

[23] Yunusova E A 2017 Analysis of the effectiveness of the use of the Vogel indicator curve in predicting the oil flow rate *Problems of gathering, treatment and transportation of oil and oil products* **2** 23-31

[24] Shashkov V B 2003 *Applied regression analysis. Multivariate regression: training manual* (Orenburg: GOU VPO OSU)

[25] Panchenko T V 2007 *Genetic algorithms* (Astrakhan: Astrakhan University Publishing House)

[26] Kharisov M N, Yunusova E A, and Maisky R A 2018 Algorithm for indirect determination of reservoir pressure using Data Mining methods *Problems of gathering, treatment and transportation of oil and oil products* **3** 40-48

Study of the reliability of the determination of reservoir characteristics of productive formations in the Stretenskoye field using correlation analysis

A S Chukhlov[1], V I Galkin[2], I N Ponomareva[2], V I Chernykh[3]

[1] LLC LUKOIL PERM, 62, Lenina ave., Perm, 614068, Russia
[2] Perm National Research Polytechnic University, 29, Komsomolskiy avenue, Perm, 614990, Russia
[3] LLC LUKOIL-Engineering PermNIPIneft in Perm, 29, Sovetskoy Armii street, Perm, 614000, Russia

E-mail: vgalkin@pstu.ru

Abstract. Storage properties (porosity) of productive strata at Stretenskoye field were determined using several methods of geophysical studies of wells. The filtration properties (permeability) are determined according to the interpretation of the hydrodynamic study materials under unsteady conditions. At the same time, the methods used in a number of cases showed different results. In order to avoid misinterpretation of the structure of void space and its properties, it is advisable to conduct studies to assess the reliability of all the methods. For this purpose, the paper used the tools of mathematical statistics such as correlation and multivariate regression analysis. As a result of the research, the authors gave the recommendations for the application of methods to determine the filtration and storage properties of reservoirs differentially for terrigenous and carbonate reservoirs of the studied field.

1. Introduction

The determination of reliable values of filtration and storage properties of reservoirs is the most important stage in the study of the geological structure of hydrocarbon deposits [1, 2]. For this purpose we use the materials from geophysical and hydrodynamic studies of wells (GIS, well testing) as well as data from laboratory core studies. Moreover, each of the methods is characterized by its own advantages and disadvantages and has conditions for effective application in the solution of the assigned problems [3-6]. Several methods of geophysical studies were implemented in almost all wells of the tula terrigenous and tournaisian carbonate objects at the Stretenskoye field studied in this paper. Highly informative hydrodynamic studies were carried out at each well at the initial period of operation [7-9]. As a result of the interpretation of these types of studies, sometimes significantly different values of filtration and storage properties were obtained, which made it difficult to use them in the solution of field problems, such as calculation of reserves, creation and adaptation of geological and hydrodynamic models and design in field development in general. In this regard, the assessment of the reliability of the use of each of the well survey methods is an urgent task.

2. Materials and methods

Storage properties of rocks - oil and gas reservoirs at the Stretenskoye field were determined by four methods, including data from core studies (m_1) and methods of geophysical studies of wells (gamma-gamma logging (m_2), taking into account the volumetric model of the rock skeleton (m_3); acoustic logging (m_4)). The filtration properties (permeability) of the reservoir were determined from the data of the interpretation of the pressure recovery curves (PRC) by the tangent method (k_1), in the SAPHIR PC (KAPPA Workstation) (k_2), by the method of accounting for the variable post-inflow by G.I. Barenblatt et al. (K_3). Permeability data characterize the condition of remote areas of productive formations. To assess the state of the bottom-hole areas, the skin factor values determined by the tangent (S_1) methods and SAPHIR (S_2) software, as well as the values of the dimensionless diagnostic feature *d*, determined during PRC processing by the method of deterministic pressure moments.

In order to assess the reliability of the results obtained for each of the listed methods, the well-known methods of mathematical statistics were used: correlation and multivariate regression analysis [10, 11]. According to the well-known theoretical concepts, the reservoir characteristics of productive formations should largely control the flow of fluids during the operation of wells. In this regard, the main characteristic of the inflow - liquid flow rate (Q_l), is used as a dependent feature (predicted parameter) in multivariate regression analysis. Similarly, correlation analysis is reduced to the assessment of the statistical relationship between the reservoir parameters of productive formations and well flow rates. At the same time, the significance of the statistical relationship was assessed in accordance with the level generally accepted in engineering practice, which is $p = 0.05$.

The conclusion of the reliability of the method to determine the reservoir properties of reservoirs was made based on the analysis of the developed multiple regression equations - multidimensional statistical models of well flow rates. During the development of these models, step-by-step regression analysis was used. The flow rate of the liquid acted as a dependent feature, the filtration-storage properties determined by one method or another, as well as a complex of additional geological and field information characterizing the inflow in conjunction with porosity and permeability, were used as independent factors: reservoir (P_{st}) and bottom hole pressure (P_{bp}), reservoir thickness (h), oil viscosity (μ). The models were created individually for two deposits of the studied field, confined to the C_1t and C_2tl reservoirs.

The principle of the valuation assessment is based on the well-known property of multivariate statistical models. It means that the most reliable parameter is used in more models and is included in these models at earlier steps.

3. Results

Table 1. Correlation matrix between production rates and porosity factors (deposit C_2tl)

	Q_l	m_1	m_2	m_3	m_4
Q_l	1.00	0.31	0.29	0.89	0.84
m_1		1.00	0.98*	1.00*	1.00*
m_2			1.00	1.00*	1.00*
m_3				1.00	1.00*
m_4					1.00

Nota bene: * – statistically significant relationship

Table 2. Correlation matrix between production rates and permeability factors (deposit C_2tl)

	Q_l	k_1	k_2	k_3
Q_l	1.00	0.91*	0.89*	0.66
k_1		1.00	0.99*	0.39
k_2			1.00	0.35
k_3				1.00

Nota bene: * – statistically significant relationship

Tables 1 and 2 show the correlation matrices characterizing the presence/absence of a statistical relationship between all values of porosity and flow rates, as well as between all values of permeability and flow rates, created for the C_2tl reservoir.

According to the data in Tables 1 and 2, it follows that the reservoir storage properties do not control the fluid inflow. The highest, although statistically insignificant, relationship was noted between the flow rates and porosity factors, determined from gamma-gamma logging, taking into account the volumetric model of the rock skeleton. The values of the permeability factors, on the contrary, are rather closely correlated with the fluid flow rates, but the maximum value of the correlation coefficient *r* is typical for the tangent method. Statistically significant relationship between the porosity and permeability factors was not established. The maximum, but statistically insignificant values of the coefficient *r* are noted when the porosity factors determined from the core are correlated with the permeability factors according to the Barenblatt method.

The correlation matrix reflecting the results of the assessment of the statistical relationship between flow rates and parameters characterizing the hydrodynamic state of the bottom hole area is shown in Table 3.

Table 3. Correlation matrix between production rates and parameters of bottom-hole areas (deposit C_2tl)

	Q_l	S_1	S_2	d
Q_l	1.00	0.73	0.49	0.78*
S_1		1.00	0.72	0.90*
S_2			1.00	0.58
d				1,00

Nota bene: * – statistically significant relationship

The most closely controlled fluid inflow is the state parameter *d* of bottom hole area, determined by the method of deterministic pressure moments.

The multidimensional model for the determination of the flow rate of the liquid, developed during the mathematical processing of the reservoir and reservoir characteristics of the wells of the C_2tl deposit, has the following form:

$$Q_l^{M\text{-}Tl} = 46.658k_1 + 197.608k_3 + 305.227k_2 + 1.043m_2 + 4.8770d - 28.2;$$

when R=0.998, p<0.105, standard error 1.61 m^3/day

The model was formed in the sequence shown in the regression equation. The values of the coefficients R, describing the strength of statistical relationships, changed in the following way: 0.914; 0.971; 0.990; 0.994; 0.998.

Tables 4 and 5 shows the correlation matrices characterizing the presence/absence of a statistical relationship between all values of porosity and flow rates, as well as between all values of permeability and flow rates, built for C_1t reservoir.

Table 4. Correlation matrix between production rates and porosity factors (deposit C_1t)

	Q_l	m_1	m_2	m_3	m_4
Q_l	1.00	0.075	-0.40	-0.50	-0.37
m_1		1.00	0.21	0.28	0.19
m_2			1.00	0.97*	0.87*
m_3				1.00	0.91*
m_4					1.00

Nota bene: * – statistically significant relationship

Table 5. Correlation matrix between production rates and permeability factors (deposit C_1t)

	Q_l	k_1	k_2	k_3
Q_l	1.00	0.76*	0.86*	0.47
k_1		1.00	0.71*	0.09
k_2			1.00	0.34
k_3				1.00

Nota bene: * – statistically significant relationship

Table 6 shows the correlation matrix characterizing the statistical relationship between flow rates and parameters characterizing the hydrodynamic state of the bottomhole area.

Table 6. Correlation matrix between production rates and parameters of bottom hole area (deposit C_1t)

	Q_l	S_1	S_2	d
Q_l	1.00	-0.05	-0.09	0.45
S_1		1.00	0.79*	0.18
S_2			1.00	0.25
d				1.00

Nota bene: * – statistically significant relationship

The multidimensional model for the determination of the fluid flow rate, built during the mathematical processing of the reservoir and reservoir characteristics of the wells of the C1t reservoir, has the following form:

$$Ql^{M\text{-}T} = 343.034k_1 + 0.419S_2 - 0.826m_2 + 21.32;$$
when $R=0.912$, $p<0.004$, standard error 3.05 m^3/day

4. Conclusion

This correlation and multivariate regression analysis allowed substantiating the reliability of the determination of the reservoir properties of terrigenous and carbonate reservoirs of the Stretensky field based on the materials of geophysical and hydrodynamic studies of wells. It was found that in both terrigenous and carbonate reservoirs, porosity did not control the inflow, and the permeability determined by any of the studied methods, on the contrary, controlled it rather strongly. This conclusion is quite logical, since the well-known analytical formulas of production rates also contain only filtration properties.

It is also known that porosity and permeability also do not control each other. The state of the bottom-hole area in terrigenous reservoirs is most reliably assessed by the method of deterministic moments of pressure, and in carbonate reservoirs - by the value of the skin factor determined in the SAPHIR software. Permeability of terrigenous reservoirs is consistently determined by all three methods, permeability of carbonate is determined by the tangent method. The inclusion of the porosity coefficient values determined by gamma-gamma logging into the resulting multidimensional flow rate models allows giving the highest assessment of the reliability of this method of logging wells.

References

[1] Abrosimov A A, Shelyago E V, Yazinina I V 2018 Substantiation of a representative volume of reservoir properties data to obtain statistically reliable petrophysical relationships *Journal of Mining Institute* **233** 487-491, 10.31897/PMI.2018.5.487

[2] Dragunov A A, Mukhamadiev R S, Chernov S V 2017 Influence of geodynamic processes on reservoir properties of geological environment (on the example of the Romashkino field) *Georesursy* **19(4)** 319-322, 10.18599/grs.19.4.3

[3] Ghareb Hamada, Veronique Joseph 2020 Developed correlations between sound wave velocity

and porosity, permeability and mechanical properties of sandstone core samples *Petroleum Research*, **5(4)** 326-338, 10.1016/j.ptlrs.2020.07.001
[4] Ponomareva I N, Martyushev D A, Akhmetova M I 2016 Evaluation of the optimal duration of the hydrodynamic studies of low-productivity wells on the example of Ozernoye field *Neftyanoe Khozyaystvo - Oil Industry* **3** 36-38
[5] Martyushev D A 2020 Rock stress state influence on permeability of carbonate reservoirs *Bulletin of the Tomsk Polytechnic University. Geo Assets Engineering* **331(8)** 24-33, 10.18799/24131830/2020/8/2765
[6] Ponomareva I N, Martyushev D A 2020 Evaluation of hydraulic fracturing results based on the analysis of geological field data *Georesursy* **22(2)** 8-14, 10.18599/ grs.2020.2.8-14
[7] Cherepanov S S, Martyushev D A, Ponomareva I N 2013 Evalution of filtration-capacitive properties of fracture carbonate reservoir of Predural'skogo edge deflection *Neftyanoe Khozyaystvo - Oil Industry* **3** 62-65.
[8] Davydova A E, Shchurenko A A, Dadakin N M, Shutalev A D, Kvesko B B 2019 Well testing design development in carbonate reservoir *Bulletin of the Tomsk Polytechnic University. Geo Assets Engineering* **330(6)** 68-79, 10.18799/24131830/2019/6/2128
[9] Qiguo Liu, Heng Lu, Longxin Li, Aiting Mu 2018 Study on characteristics of well-test type curves for composite reservoir with sealing faults *Petroleum* **4(3)** 309-317, 10.1016/j.petlm.2018.03.011
[10] Ponomareva I N, Martyushev D A, Cherny K A 2021 Research of interaction between expressive and producing wells based on construction of multilevel models *Bulletin of the Tomsk Polytechnic University. Geo Assets Engineering* **332(2)** 116-126, 10.18799/24131830/2021/02/3048
[11] Ponomareva I N, Galkin V I, Martyushev D A 2021 Operational method for determining bottom hole pressure in mechanized oil producing wells, based on the application of multivariate regression analysis *Petroleum Research*, 10.1016/j.ptlrs.2021.05.010

Technogenic impact on the environment during leaching

Yu I Razorenov[1], R V Klyuev[2], E R Guzueva[3]

[1]Platov South-Russian State Polytechnic University, 132 Prosveshcheniya St., Novocherkassk, 346428, Russian Federation
[2]Moscow Polytechnic University, 38b Semenovskaya St., Moscow, 107023, Russian Federation
[3]Chechen State University named after A. H. Kadyrov, 17a, Dudaev Boulevard, Groznyy, 364015, Russia

E-mail: yiri1963@mail.ru

Abstract. The results of the analysis of the theory and practice of leaching metals from substandard raw materials are presented. A brief historical reference is given on the leaching of metals from experimental leaching in the middle of the last century to industrial development in modern times. The direction of combining traditional and innovative technologies within a single technological process is formulated. The mechanism of underground block leaching of ores is described. A complex method of multi-factorial assessment of the existing natural and man-made factors of development is proposed. The parameters of the proposed technology and some aspects of its application are substantiated. Using the method of engineering forecasting, the prospects for leaching metals from substandard raw materials are assessed. The main results of the development of underground leaching of metals from sub-standard raw materials and the prospects for improving the method with the rational combination of underground leaching technologies with traditional mining technologies are formulated. It is concluded that the leaching of metals from substandard raw materials has prospects for development.

1. Introduction

In the coming years, open-pit mining will no longer play a decisive role, therefore, the importance of combined mining of deposits with the extraction of the richest areas of deposits from preparatory rifling operations with minimal losses and dilution and leaching of the remaining reserves in underground blocks increases.

Priority is given to the introduction of underground leaching development systems with unlimited extraction of reserves that ensure the involvement in the operation of raw materials unsuitable for extraction by traditional methods.

The history of leaching metals from substandard reserves begins with the experimental leaching of uranium in the deposits of Central Asia and Kazakhstan and functions in the status of industrial leaching at the deposits of JSC PPGHO, where about half of the annual production of metals is extracted by leaching.

The greatest development of the technology of underground leaching of metals was obtained during the extraction of uranium. On an industrial scale, the method of underground leaching at the site of occurrence is used for poor ores.

Ore blocks are prepared by mining operations. To do this, a store filled with ore is created in the ore body, where the ore is irrigated with a solution of sulfuric acid. A collection of production solutions is equipped in the lower part of the chamber. The resulting semi-finished uranium is delivered to the hydrometallurgical plant.

The combination of technologies begins with the extraction and delivery of different-grade ore reserves. About 40% of the balance reserves are extracted using traditional technology, and the reserves of the second stage are processed by leaching at the ore deposit site.

The method of leaching separated from the array and crushed ore in the chambers - shops allows you to increase output with a decrease in the volume of laying work and waste on the earth's surface. The main parameter determining the leaching efficiency is the average size of the piece.

The parameters of metal leaching from poor and lost ores are determined taking into account the laws of the mutual influence of the natural and man-made factors involved, including the capabilities of the technologies used, the productivity of enterprises, and environmental and economic efficiency.

The purpose of research in this area of mining is achieved by combining the possibilities of technologies with obtaining an ecological and economic effect.

2. Materials and methods

The complex method of solving the problem includes analysis and generalization of literary sources, research results and production experience, processing of experimental data using statistical and theoretical methods [3–4].

The methodology for substantiating the expediency and technological implementation of it includes a multifactorial assessment of positive and negative factors, technological qualities, the effectiveness of averaging mineral mass; the possibility of selective processing, the size, and the necessity of grinding, etc.

The substantiation of the parameters of the proposed technology is carried out within the framework of an industrial experiment, during which the parameters of the elements of the problem under study are obtained by known direct and indirect ones. According to the results of the experiment using the method of engineering forecasting, technological innovations are recommended.

The performance indicators of underground leaching are determined by the conditions of occurrence and properties of ores, technology parameters, and safety requirements.

3. Results and discussion

The most common mining system is with collapse, storage, and leaching of ore by infiltration flow of reagent.

Blocks of underground leaching are prepared by sinking workings and equipment of irrigation horizons and acceptance of production solutions (Figure 1).

The ore is beaten off by exploding charges in wells into a cutting slot. In order to create a compensatory space, ore is released (up to 30%) during the stripping process. In fractured rocks, the movement of solutions is provided without crushing.

The quality of crushing was determined by photoplanimetry and evaluated by the output of classes: 0–50 mm; +50–100 mm; +100–150 mm; +200 mm.

Approximate results of the crushing quality study are given in Table 1.

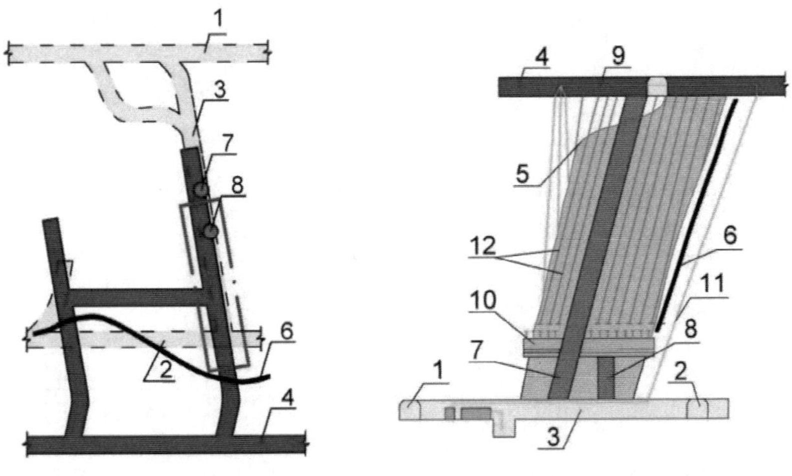

Figure 1. Leaching block: 1, 2, 3, 4 – mining workings; 5 – orebody boundary; 6 – geological fault; 7 – cut-off rising; 8 – ore outlet; 9 – irrigation horizon with solutions; 10 – cutting layer; 11 – well for feeding solutions; 12 – wells for intensification of the leaching process.

Table 1. Indicators of explosive ore breaking

Indicators	Units	Fans of wells					Average value
		1	2	3	4	5	
Volume of beaten ore	m^3	580	630	530	670	250	–
Specific consumption of granulite	kg/m^3	1.0	0.9	1.6	1.4	1.6	1.27
Convergence coefficient	–	1.1	0.7	0.8	0.7	0.9	0.83
Output of beaten ore	m^3/m^3	3.4	4.2	3.0	3.7	3.6	3.38
Fraction yield 0–50mm	%	66.6	52.4	64.1	70.6	66.9	64.1
Fraction yield (+100) –150 mm	%	8.5	4.6	8.8	6.6	3.8	6.5
Fraction yield (+150)–200 mm %	%	10.3	10.4	3.2	2.3	2.6	5.8
Fraction yield (+200) mm	%	2.6	19.6	9.6	3.1	2.3	7.4
Fraction yield (0-50) + (+150–200) mm	%	97.4	80.4	90.4	96.9	97.7	92.6
Average piece size	mm	60.3	115.4	76.9	53.8	52.3	72.3
The criterion of fragmentation	–	3.9	3.3	3.7	4.0	4.5	–

The share of class 0–50 was 64%, and the share of oversized 400mm pieces was only 1.5%. The medium-weighted piece had dimensions of 52–78 mm.

The best crushing was achieved by blasting in a "clamp" with a loosening coefficient of 1.35–1.4. In the range of classes 0-200 mm, the regularity obeys the Gauss law. The average metal content in the 200–400 mm classes turned out to be 5 times less than in the 0–200 mm class (Figure 2).

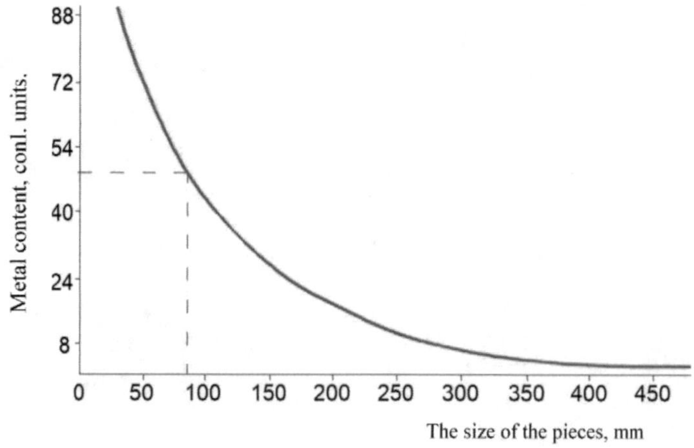

Figure 2. Distribution of metal content in ore by granulometric classes.

Intensification of fading with time of leaching is carried out by the discovery of new active surfaces repulsed and crushed ore through the detonation of explosives in boreholes, as well as the influence of compressed air, vibration, air currents, etc [20, 21].

Underground leaching of uranium mined in the fields of the East and the Star (Northern Kazakhstan), as well as "Belogorskoe" (North Caucasus).

The prospects of underground leaching technologies are associated with the development of continuous unbroken mining with a decrease in the volume of penetration of preparatory-rifled workings due to the combination of irrigation operations of ore collection of production solutions.

The disadvantage of traditional mining and ore processing technologies – the loss of metals and the accumulation of processing tailings can be reduced by combining traditional and innovative technologies within a single technological process. An element of the combination is a complex of technologies with metal leaching (Figure 3).

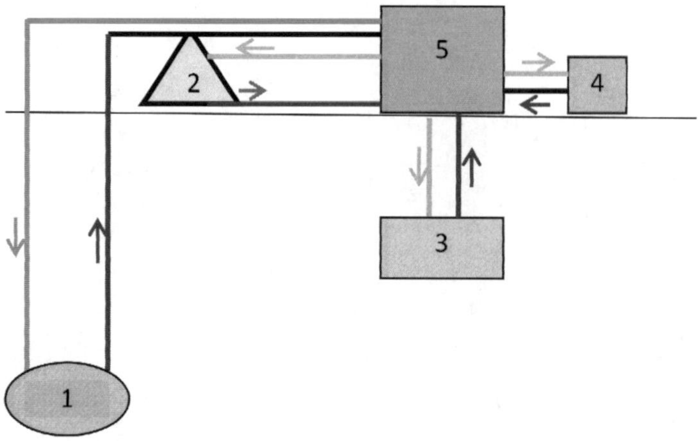

Figure 3. Complex of metal leaching technologies: 1 – borehole; 2 – heap; 3 – underground; 4 – mechanochemical; 5 – solution processing workshop.

The complex method of multifactorial evaluation of interacting factors in the process of metal leaching allows optimizing the parameters of the proposed technology and using the engineering forecasting method to realistically assess the prospects for leaching metals from substandard raw materials to obtain an ecological and economic effect.

A rational combination of underground leaching and traditional technologies is achieved by taking into account the geological and mining conditions of deposits, the content of metals in ores, occupational safety, and other factors [16-19].

One of the aspects of using the possibilities of leaching metals from substandard ores in terms of metal content is the rational use of leaching tailings for the reclamation of disturbed lands. At the exit of the rock collapse zone to the earth's surface, off-balance ore is poured into the formed funnel and leached by irrigation from the surface. At the same time, the completeness of subsurface use increases, and the quality of disturbed lands is restored.

In the case of a chamber system for mining ore bodies in stable host rocks, after the ore is launched, the chambers are filled with substandard ore for leaching in terms of metal content.

4. Conclusion

The main results of the development of underground leaching of metals from substandard raw materials include the following provisions:

- A methodology for assessing the ecological and economic efficiency of underground leaching technology and combined mining technologies has been developed.
- Leaching technologies make it possible to obtain metals of higher quality, which is important when using them in related industries [5–6].

Improvement of the method of underground leaching of substandard ores in terms of metal content includes details of the combination conditions, the use of multifactorial mathematical models, development of forecasting methods.

The data obtained can be used in the design of development systems and for teaching at a university [7–12].

The results of the study are consistent with the concepts of foreign specialists in this field of mining [13–15].

References

[1] Drebenstedt K, Golik V I and Dmitrak Yu V 2018 Prospects for diversification of metal mining technology in the RSO-Alania *Sustainable Development of Mountain Territories* **1 (35)** 125–131

[2] Komashchenko V I, Golik V I, Belin V A and Gaponenko A L 2014 Improving the efficiency of explosive rebounding on the basis of new methods for initiating borehole charges in quarries *Mining Informational and Analytical Bulletin* **9** 293–304

[3] Golik V I and Lyashenko V I 2016 The new technologies and engineering tools for the recovery of the tailings of the ore concentration *Ferrous Metallurgy. Bulletin of Scientific, Technical and Economic Information* **8** 10-19

[4] Shchurov N I, Myatezh S V, Malozyomov B V, Shtang A A, Martyushev N V, Klyuev R V and Dedov S I 2021 Determination of inactive powers in a single-phase ac network *Energies* **14 (16)** 4814

[5] Eremeeva Zh V, Sharipzyanova G H, Nitkin N M, Krikhtin V V, Ter-Vaganyants Yu S and Dakhnova T V 2016 Influence of technology factors of mixing and the nature of nanodimensional particles on mechanical properties of the powder alloyed steel SP60HGS *Nanotechnology: Science and Production* **3** 57–76

[6] Shchurov N I, Dedov S I, Malozyomov B V, Shtang A A, Martyushev N V, Klyuev R V and Andriashin S N 2021 Degradation of lithium-ion batteries in an electric transport complex *Energies* **14 (23)** 8072

[7] Valiev N G, Propp V D and Vandyshev A M 2020 The 100th anniversary of the department of mining engineering of UrSMU *News of Higher Educational Institutions. Mining Magazine* **8** 130–143

[8] Dushin A V, Valiev N G, Lagunova Yu A and Shorin A G 2018 Ural Mining University and

Moscow Mining University: interaction of higher education institutions *Mining Journal* **4** 4–10

[9] Rybak J M, Kongar-Suryun Ch, Tyulyaeva Yu, Khairutdinov A M and Akinshin I 2021 Geomechanical substantiation of parameters of technology for mining salt deposits with a backfill *Mining Science* **28** 19–32

[10] Gavrishev S E, Kornilov S N, Pytalev I A and Gaponova I V 2017 Enhancing mine production efficiency through waste management *Mining Journal* **12** 46–51

[11] Lyashenko V I, Dyatchin V Z and Franchuk V P 2017 Technical means for extraction and processing of ore materials. Message 2 *Mining Information and Analytical Bulletin* **4** 33–41

[12] Kaplunov D R and Radchenko D N 2017 Design principles and selection of subsoil development technologies that ensure the sustainable development of underground mines *Mining Journal* **11** 121–125

[13] Bosikov I I, Klyuev R V, Aimbetova I O and Makhosheva S A 2021 Development of methods and management tools aerogasdynamics processes at mining sites *Sustainable Development of Mountain Territories* **3** 397–405

[14] Doifode S K and Matani A G 2015 Effective industrial waste utilization technologies towards cleaner environment *International Journal of Chemical and Physical Sciences* **4** 536–540

[15] Vrancken C, Longhurst P J and Wagland S T 2017 Critical review of real-time methods for solid waste characterisation: Informing material recovery and fuel production *Waste Management* **61** 40–57

[16] Konyuhov V Yu, Gladkih A M, Chemezov A V 2019 Evaluation of the economic feasibility of the introduction of plasma hardening technologies in the Far North enterprises. *IOP Conference Series: Materials Science and Engineering* **560** 012147

[17] Balanovsky A E, Shtayger M G, Kondrat'Ev V V, Van Huy V, Karlina A I 2018 Plasma-arc surface modification of metals in a liquid medium. *IOP Conference Series: Materials Science and Engineering* **411(1)** 012013

[18] Konyuhov V Y, Gladkih A M, Zott R S 2020 Accelerator as an effective replacement of a business incubator in the Irkutsk region. *Journal of Physics: Conference Series* **1582** 012045

[19] Nemarov A A, Lebedev N V, Kondrat'ev V V, Kornyakov M V, Karlina A I 2016 Theoretical and experimental research of parameters of pneumatic aerators and elementary cycle flotation. *International Journal of Applied Engineering Research* **11(20)** 10222–10226

[20] Gladkih A M, Konyuhov V Yu, Galyautdinov I I, Shchadova E I 2019 Green building as a tool of energy saving. *IOP Conference Series: Earth and Environmental Science* **350** 012032

[21] Konyuhov V Yu, Gladkih A M, Galyautdinov I I, Shchadova E I 2020 Calculations of efficiency in implementing progressive mold forming methods. *IOP Conference Series: Materials Science and Engineering* **760** 012027

IOP Conf. Series: Earth and Environmental Science 1021 (2022) 012051 doi:10.1088/1755-1315/1021/1/012051

The role of deep faults in the structure formation of the TCFD, the distribution of hydrocarbon accumulations, and thermomineral sources

A A Daykaev[1], E A Abubakarova[1,2], A S Elzhaev[2], M A Khasanov[2], T B Ezirbaev[2], D U Germakhanova[2], Kh A Alieva[2], A A Shaipova[2]

[1] Kh. Ibragimov Complex Institute of the Russian Academy of Sciences, 21 a, Staropromyslovskoe highway, Grozny, 364051.
[2] Grozny State Oil Technical University named after Academician M.D. Millionshchikov, 100, Kh.Isaev avenue, 364905, Grozny, Russia.

E-mail: eliza_ggni@mail.ru

Abstract. The work purpose was to assess the prospects of oil and gas potential of the western part of the Tersko-Caspian trough (TCT) concerning the development of deep-sunk areas in the light of modern ideas about the deep-seated formation of hydrocarbons. The main tasks are to analyze the ideas about the formation mechanism of the tectonic structure and formation of hydrocarbon accumulations; to study the relationships between the formation processes, migration, and the areal distribution of hydrocarbon accumulations; to determine the main prerequisites for the deep genesis of oil and gas. The scientific novelty lies in a comprehensive approach to the study of the formation processes of hydrocarbons, migration, the formation of traps, and patterns of placement of hydrocarbon accumulations. The article describes the mechanisms of formation of the Tersk and Sunzha composite anticline as the main zones of oil and gas accumulation. It is suggested that the formation of high-amplitude anticline folds in the upper cretaceous deposits resulted from the vertical migration of high-pressure fluids along deep faults with the breakthrough of their upper horizons of the sedimentary cover into the late-orogenic folding stages. The main prerequisites for the deep genesis of oil and gas are listed as pronounced uneven distribution of hydrocarbon accumulations, AHFP, hydrochemical anomalies, etc.

1. Introduction

Tectonically, the study area covers mainly the western part of the Tersko-Caspian foredeep (TCFD), and administratively, the territories of the Republics of the North Caucasus District of Russia (except for the Republic of Dagestan). The area is characterized by a complex fault-block structure. Deep faults of the sub-lateral, north-east, and northwest orientations are revealed here. The effectiveness of geological research on hydrocarbon raw materials is largely determined by the state of scientific developments in the field of petroleum geology, in particular, studies of the role of deep faults in the formation of tectonic structures, hydrocarbon accumulations, and their spatial distribution.

IOP Conf. Series: Earth and Environmental Science 1021 (2022) 012051 doi:10.1088/1755-1315/1021/1/012051

2. Brief information about deep faults and criteria for their identification

The concept of "deep fault" was introduced into scientific circulation by academician A.V. Peyve in 1945 [1]. According to the definition given in the geological dictionary, deep faults are understood as "zones of mobile insulation joint of unitized packages of the earth's crust and the underlying part of the upper mantle, with a length of up to many hundreds and thousands of km." [2]. By D. G. Azhgireyu: "the deep fault is represented by a relatively narrow, and extending along the mobile belt tectonic zone of concentration of long-term developing deformations of the earth's crust" [3]. Other definitions are given in the literature. At the same time, in almost all definitions, deep faults are marked as zones characterized by a significant length (with small width), the development duration, and the foundation depth.

The doctrine of deep faults was developed by Russian scientists A.V. Peive, A. N. Zavaritsky, N. S. Shatsky, A. I. Suvorov, V. E. Hain, E. E. Milanovsky, N. M. Sinitsyn, G. D. Azhgirei, V. A. Aprodov et al. and foreign researchers R. Zonder, Stile, Kloos, etc.

The accumulation of a large volume of actual geological material contributed to the appearance of many publications and many generalizing works on the role of deep faults in the Earth development, their classification by various characteristics, etc. A.V. Peive (1956) classified them by large tectonic elements, distinguishing deep faults of geosynclines, platform areas, and fore deeps. V. E. Hain subdivided deep faults according to the foundation depth and the order of the structures they share [2]. According to the depth of the deposit, super-deep (with a depth of penetration of 400-7000 km), medium-deep (penetrating to a depth of 300 km), sub-crustal (penetrating the mantle), and crustal, spreading only in the earth's crust, were distinguished [4]. According to the differentiation of the deep structures of the earth's crust, the faults are pericratonic, perioceanic, intra-geosynclinal, inter-block, etc., according to the classification of V.A. Aprodov. Based on the history of the structures delineated by the faults and the relation of the faults to their orientation, two groups of faults are distinguished - continental and oceanic.

There are many criteria by which deep faults are identified: the geophysical and geodesic features of faults are divided into dynamo-kinematic and structural [1].

Dynamo-kinematic features are divided into seismological, slope-measuring, geodesic, and geothermal.

Seismological features are zones of faults of different orders from planetary to intraregional faults that are active at the present stage and are recorded by bands of increased seismicity.

Tilt-measuring features are expressed in changes in the slopes of the earth's surface during earth-quakes in remote areas, which allows us to consider the deep fault in the zone of which this earthquake occurs, according to the degree of impact on the Earth's crust, and to establish the planetary nature of the movements.

Geodesic features. The accumulated factual material of repeated precision levelings and triangulations allowed us to reveal the existence of modern vertical and horizontal movements along the discontinuities. The faults are marked by bands of an increased horizontal gradient of changes in the velocities of vertical movements.

Geothermal features. The increased heat flow in the zone of some currently active deep faults indicates their role as vertical heat conduits. And allows us to judge the possible genesis of tectonic forces, the action of which leads to the formation of the fault.

Structural geophysical features of faults are divided into gravimetric, magnetic, seismic, electrometric, and radiometric.

Seismic features. The identification of the structural features of the Earth's crust and mantle associated with the activity of faults using seismic features is based on the consideration of the - dynamic and kinematic features of the wave pattern.

Magnetic features. Magnetic anomalies that mark the position of deep faults are almost always closely related to the influence of igneous rocks or rocks processed by deep solutions and melts. The increased magnetization of basic and ultrabasic rocks is known. The presence of basic and ultrabasic rocks indicates the depth of the fault.

Electrometric features. The use of direct current methods (DES, VES) for the detection of deep faults have not been widely developed due to the small depth of common modifications of electrical exploration and the high cost of deep methods of research.

Radiometric features. Radiometric methods of fault detection are largely similar to magnetic methods since the radioactivity of rocks is determined by the conditions of magmatism and metamorphism in a given region.

Gravimetric features are used to identify faults at the boundary between two dissimilar blocks with different densities and faults. A universal and frequent sign of a fault is the "gravity stage".

The works of V. V. Belousov, G. D. Azhgirey, M. F. Mirchink, M. V. Muratov, B. G. Sokratov, N. Yu. Uspenskaya, A. I. Letavin, B. K.Lothiev, M. N. Smirnova, Yu. A. Sterlenko, R. S. Bezborodov have reflected the questions related to the nature of the distribution of deep faults, their influence on the tectonic features of the foundation, and sedimentary cover, as well as on the formation of oil and gas accumulations within the Caucasus. The formation of supra-fault anticlinorium, in particular the Sunzhenskaya one over the eastern part of the Przekis-Tyrnyauz deep fault, is explained by the transpression of rocks in the foredeep and the arrival of an increased thermal flux along the deep fault zone. G. D. Azhgiray identified two deep faults extending sub-parallel to the previously identified Przekish-Tyrnyauzsky fault. These are the Southern deep fault located 40 km south of the Przekis-Tyrnyauz fault and the Khasaut fault located 30 km north of the Przekis-Tyrnyauz fault. [2] Thus, G. D. Azhgiray identified the main deep faults of the sub-lateral orientation within the Precaucasia. Based on the analysis of the features of the geological structure of deep fault zones, he noted the different ages of activation of deep faults. Later, V. A. Stanulis, T. V. Yakovleva M. N. Smirnova, I. M. Krisyuk, V. D. Talalaev et al., based on the analysis of geological and geophysical materials and data on the seismicity of the region, revealed systems of differently oriented deep faults: Sub-lateral – Kraevoy, Tersky, Sunzhensky, Przekish – Tyrnyauzsky, Chernogorsky; Northwestern – Gudermessko – Mozdoksky Benoy – Eldarovsky, Datykhsko – Akhlovsky, Nalchik; Northwestern – Gudermessko – Mozdoksky Benoysko – Eldarovsky, Datykhsko - Akhlovsky, Nalchik; Northwestern – Gudermessko-Mozdoksky Benoysko-Eldarovsky, Datykhsko-Akhlovsky, Nalchik; Northwestern-Gudermessko-Mozdoksky, Benoysko-Eldarovsky, Datykhsko-Akhlovsky, Nalchik; Northwestern-Gudermessko-Mozdoksky, Benoysko-Eldarovsky, Datykhsko-Akhlovsky, Nalchik; The eastern ones are the Novogroznensky, Argunsky, Tskhinvali-Kazbek, Ardonsky, Chereksky, Malkinsky, and other faults. Most of them are attributed to the suture zones-systems of converging faults that merge at depth into tectonic seams. The development and nature of the placement of anticline zones and individual structures in the sedimentary cover within the Tersko-Caspian foredeep are mainly due to the presence of suture zones and the fault-block structure of the foundation.

3. On the role of deep faults in the formation of the structure of the Tersko-Sunzha folded zone (TSFZ), the distribution of hydrocarbon accumulations and thermal water sources

About the geological history of the development and the mechanisms of the formation of the TSSZ structure, there are different, often opposite, points of view [5]. As a study result of the deep structure was noted the fault-block structure of the foundation, and the leading role of differently oriented deep faults in the formation of the sedimentary cover structure. A number of researchers (M. I. Zhemerichko, N. V. Koronovsky, B. A. Sokolov, N. Sh. Yandarbiev, V. V. Dotsenko et al.), denying the sedimentary nature of the folding of the Tersko-Sunzha region, consider it a rift structure, linking the formation of anticline supra-fault structures of the rootless folding with fluid-geodynamic processes (the injection mechanism of anticlinorium formation) [6]. That is, it is assumed that the formation of high-amplitude anticline folds in the Upper Cretaceous sediments overlying the layer of the Maikop clay rocks-tires, under the influence of vertical migration of high-pressure fluids and their breakthrough from deep faults into the upper horizons of the sedimentary cover in the late-orogenic phases of tectonic activation. Thus, the most significant hydrocarbon fields are concentrated within the supra-fault anticline zones (Terskaya and Sunzhenskaya). The area of the entire considered part of the TCFD is approximately 30-35 thousand km^2. The total area is occupied by the main large deposits,

IOP Conf. Series: Earth and Environmental Science 1021 (2022) 012051 doi:10.1088/1755-1315/1021/1/012051

with reserves of more than 90% of the total reserves of all deposits in the region – 350-400 km^2, which indicates a concentrated flow of fluid flows from bottom to top along deep faults. The upright migration of oil and gas is also indicated by the multi-layer nature of the main fields. Industrial oil and gas content in them is established in the Miocene sand layers, in the Upper Cretaceous limestone thickness, in the Lower Cretaceous Aptian layers, and some deposits in the carbonate rocks of the Valanginian and Malm. That is, the range of oil and gas content of the section increased as the depth of exploration increased. One of the founders of the theory of the inorganic origin of oil N. A. Kudryavtsev, noting the role of deep faults in the formation of oil and gas fields, stated that they create prerequisites for the accumulation of thick sedimentary rocks, create traps for oil and gas, and at the same time ensure that the traps are filled with these minerals [7;8].

Some researchers call the geological processes associated with the breakthrough of high-pressure thermal reservoir fluids through various kinds of disjunctive dislocations hydrovolcanic [9]. Hydro-volcanic processes are usually timed to the active phases of folding in different regions. With the intensive growth of anticline folds with the formation of various discontinuous disturbances, fluid flows in the fragmented area produced physical and chemical effects on the bedrock, that is, natural hydraulic fracturing, thereby creating hydrocarbon traps. Similar statements are found in many works. For example, in his report at the All-Union Meeting on the genesis of oil and gas, I. M. Sukhov noted: "oil and gas, under favorable conditions, create their own structures and are in them in a complex system of faults, discharges, cracks, cavities and caverns" [10]. In another work [11], it is assumed that the vertical migration of fluids in the presence of layers of rock tires on the path can have a structure-forming effect. Namely, under the thick layer of the Maikop clays, high-amplitude anticline structures were formed in the Upper Cretaceous limestone column, complicated by discontinuous disturbances, which are associated with the main highly productive hydrocarbon deposits. Within the same structures, numerous small oil and gas deposits were formed in the Middle Miocene sand layers, due to the partial breakthrough of fluids through the tire in areas with relaxed shielding properties. In the presence of discontinuous faults that open a certain part of the fluids to the Earth's surface, sources of oil and high-thermal waters were formed by migrating up them. Foci of groundwater discharge in the form of thermal water outlets to the Earth's surface are known within the Tersk and Sunzhensk anticline zones (Bragun, Goryachevod, and Sernovod springs), the juvenile nature of which was noted in several works. In particular, the above-mentioned N. A. Kudryavtsev noted: "... the signs of the deep origin of oil entails the need to review the question of the genesis of underground waters and their salt composition from the point of view of the wide participation of juvenile waters in it" [7].

4. Conclusion

Thus, many factors indicate the formation of large highly productive deposits in the central part of the TCFD due to the upright migration of fluid flows from the deep subsurface. Deep faults served as how the hydrocarbons formed in the deep zones penetrated the sedimentary strata[13; 14]. The determining role in the formation of hydrocarbon deposits was played by vertical migration at the stage of active and short-term manifestations of pre-Pliocene and partially anthropogenic tectonic genesis. The relationship is traced between the migration processes, the formation of traps, and the nature of the distribution of oil and gas accumulations. A comprehensive analysis of neotectonic movements, modern geodynamics, and other factors that determine the conditions for the formation and location of oil and gas accumulations is one of the most important criteria for successful forecasting of oil and gas potential and planning GEW of oil and gas exploration. The presented materials substantiate the deep views of the formation of hydrocarbons and provide a basis for a new assessment of the prospects for the oil and gas content of the deep-submerged zones of the studied territory.

References

[1] Peive A V, Sinitsyn V M 1950 Some basic questions of the theory of geosynclines *News of the AS USSR. Ser. geol.,* **4** 28-53

[2] *Geological Dictionary* 1978 volume 2. (M.: Nedra)

[3] Azhgirei et al. 1974 *General geology. OL range.* (Moscow: "Enlightenment")
[4] Smirnova M N 1984 *Fundamentals of Geology of the USSR* (Moscow: Higher School)
[5] Koronovsky N V, Gushchin A I, Nikitin M Yu et al 1990 Geological development and formation of the modern structure of the Tersko-Caspian foredeep *Tectonics of orogenic structures of the Caucasus and Central Asia.* pp. 4-35. (Moscow: Science)
[6] Kerimov I A, Dotsenko V B, Daukaev A A 2014 The change of geotectonic paradigms and the evolution of ideas about the formation of the Tersko-Sunzha fold area *Bulletin of the Academy of Sciences of the Czech Republic* **1** 85-92
[7] Kudryavtsev N A 1964 Deep faults and oil and gas potential *Deep faults. Collection of articles* pp. 191-201 (Moscow: Nedra)
[8] Daukaev A A, Bachaeva T Kh et al 2018 Deep structure and oil and gas content of submerged zones of western part of Terek-Caspian trough. *Advances in Engineering Research* **453-457,** 2352-5401
[9] Ivanchuk P P 1967 Some features of upright migration of hydrocarbons in the manifestations of hydro volcanism *Genesis of oil and gas.* pp. 484-490 (Moscow: Nedra)
[10] Sukhov I M 1967 On the search and exploration of oil and gas in Bessarabia *The genesis of oil and gas* pp. 350-356. (Moscow: Nedra)
[11] Timofeev A A, Timofeev V A 2004 Oil and gas content of deformed and metamorphosed sedimentary covers *Scientific notes of the Faculty of Geology and Geography* pp. 83-100 (Rostov-on-Don, 2004)
[12] Daukaev A A 2012 On the relationship of abnormally high formation pressures with the nature of the development of anticline structures of the Tersko-Sunzhenskaya oil and gas-bearing area *Oil and gas geology* **2** 51-52
[13] Dotsenko V V 2014 Indicators of modern upright migration of hydrocarbons and the process of formation of oil and gas accumulations in Southern Russia *Bulletin of the Academy of Sciences of the Czech Republic* **3 (24)** 79-85
[14] Abubakarova E A 2020 Morphology of geophysical fields of Terek-Caspian trough and placement patterns of oil and gas structures IOP Conference Series: Materials Science and Engineering **905** 012075

Results of identification of ineffective oil objects in carbonate reservoirs

O A Grezina

Ufa State Petroleum Technological University, Branch of the University in the City of Oktyabrsky Oktyabrsky, Republic of Bashkortostan, Russian Federation

E-mail: ugntu.grezina@mail.ru,

Abstract. The article presents data on the selection of oil analogous objects using multidimensional exploration analysis of geological and physical parameters. The main components and ineffective objects in the carbonate reservoirs of the Tournaisian stage confined to tectonic elements within the inner and outer zones of the Cis-Ural passive margin of the East European paleocontinent were analyzed. The study showed that the use of codes of the tectonic elements to which the facilities belong as a grouping variable makes it possible to identify groups with common characteristics. It is recommended to apply the successful experience of stimulating oil production by acid-implosion in the wells of the Tournaisian stage of the Almetyevsk peak field to the facilities of the same and adjacent (Belebeevsko-Shkapovskaya peak) groups with close axe coordinates for the main components.

1. Introduction

A successful geological and technical measure provides a normative technological effect and profitability of works performed to intensify oil production [1–4].

The result of an event depends on the choice of an object (deposit, well) similar to the one at which the event was successfully replicated [5–8].

The task of selecting identical objects is most relevant for ineffective deposits, where a large amount of work is being performed to regulate the oil recovery. These deposits can be characterized by a significant variety of bedding conditions, properties of productive formations and fluids saturating them [9–11].

To solve the problem of identifying objects (groups with shared characteristics), the principal component method and discriminant analysis can be applied [12, 13].

2. Methods and materials

The geological and physical characteristics of 130 oil objects in the carbonate deposits of the Tournaisian stage of the Volga-Ural oil and gas province were analyzed.

The objects are confined to the following large tectonic elements:

- within the inner zone of the Cis-Ural passive margin of the East European paleocontinent - the Tatar arch (Nizhnekamsk system of linear dislocations, Sarailinskaya saddle, Almetyevskaya peak, Belebeevsko-Shkapovskaya peak), Verkhnekamskaya depression, Birskaya saddle, Bashkir peak;
- Abdulinsky deflection;
- within the outer zone of the Cis-Ural passive margin of the East European paleocontinent - the slope of the East European paleocontinent.

In the wells of the Kizelovsky horizon of the Bavlinskoye field (Almetyevskaya peak) and Alekseevskoye fields (Belebeevsko-Shkapovskaya peak), acid-implosion was applied to stimulate oil production [14, 15]. The long-term technological effect and a significant additional volume of oil production were achieved at low costs.

The technology involves applying multiple hydrochloric acid baths. The implosive effect is created by a shock-wave device, which is a deep sucker rod pump with a plugged bottom part and holes in the casing; the pump plunger is moved using a lifting unit. Implosion can be performed after an acid bath, for pumping out the reaction products of an acid solution with a rock, or during an acid bath, for a dynamic reaction of an acid solution with a rock.

To identify objects similar to the Kizelovsky horizon of the Bavlinskoye field, the method of principal components was used [16].

Each component (new variable, factor) is represented by a linear combination of the initial parameters (variables) that correlate with it. The number of main components does not exceed the number of initial parameters, while the first two to four components contain the main part of the variance; this reduces the number of variables. Values of the main components calculated for the objects are combined into relatively homogeneous groups, that is, a hidden data structure is revealed. The objects are represented in the coordinate axes of the main components (factor axes).

3. Results

Each of 130 objects had 13 initial geological and physical parameters (main variables): roof depth, total and effective oil-saturated thicknesses, porosity coefficients, initial oil saturation and permeability; volumetric coefficient of oil, density of degassed oil, viscosity of formation oil, content of sulfur, paraffins and resins, initial formation temperature, °C (Table 1).

Table 1. Values of geological and physical parameters of objects

Parameter	Parameter values		
	minimum	average	maximum
Roof depth, m	1011	1429	2260
Total oil-saturated thickness, m	0.9	13.7	90.0
Effective oil-saturated thickness, m	0.9	4.8	21.3
Porosity coefficient, units	0.06	0.12	0.16
Oil saturation coefficient, units	0.50	0.75	0.89
Permeability coefficient, μm^2	0.002	0.078	0.935
Volumetric coefficient of oil, units	1.01	1.06	1.21
Density of degassed oil, kg / m^3	840	900	968
Viscosity of reservoir oil, mPa · s	1.9	29.3	203
Sulfur content,%	1.5	3.1	5.2
Paraffin content,%	1.4	3.5	7.3
Resin content,%	11.8	22.7	60
Initial reservoir temperature, ° C	18	27	48

The initial parameters were transformed into 13 main components (factors), the first four of which account for 63.2% of the total variance, respectively 32.5, 13.8, 8.8 and 8.2%. Factorial coordinates of observations and variables were determined.

The main components were interpreted by the factor coordinates of the variables. The first main component (factor) uniformly represents the occurrence conditions (depth and initial reservoir temperature), oil properties (viscosity, volumetric ratio, sulfur and resin content, density) and reservoir (permeability and porosity coefficients); the contribution of nine variables is 94.4 %. In the second factor, the main contribution is made by the thickness properties of the formation (72.4%). In the third factor, the wax content, reservoir temperature and oil saturation account for 82.7% of the total variance. The fourth factor reflects the reservoir properties, the main contribution is made by the permeability and porosity coefficients (72.5%).

To classify objects, projections of observations (objects) were built on the factorial plane (Figure). The number of the group into which the objects were divided according to the tectonic confinement was chosen as a grouping variable. There were nine groups: 1 - Nizhnekamsk linear dislocation system (2 objects); 2 - Sarailinskaya saddle (1 object); 3 - Almetyevskaya peak (45 objects); 4 - Belebeevsko-Shkapovskaya peak (28 objects); 5 - Verkhnekamsk depression (1 object); 6 - Birsk saddle (18 objects); 7 - Bashkir peak (7 objects); 8 - Abdulinsky trough (1 object); 9 - slope of the East European paleocontinent (27 objects).

The relationship between the objects was determined. The objects belonging to each of the tectonic elements were visually grouped. The right side of the zone of the Almetyevsk peak objects was combined with the zone of the Birsk saddle objects. The zone of the objects of the Belebey-Shkapovskaya peak adjoins the boundaries of the zones.

The object of the Bavlinskoye field, belonging to the Almetyevskaya peak is marked with an arrow; it is located at the boundaries of the zones. Similar objects are located in the factor space close to the target object in the zone of the same tectonic element. Similar objects include closely spaced objects from the zone of an adjacent tectonic element (Table 2).

Table 2. Factor coordinates for correlation-based observations

object	No	Factor 1	Factor 2	Factor 3	Factor 4
Group 3 (Almetyevskaya peak)					
Novoelkhovskoe	4	-0.23905	-0.20334	-0.26321	1.16977
Bavlinskoe	7	-0.09347	0.20666	-0.60210	0.49634
Berezovskoe (RT)	13	-0.37793	0.48198	-0.34072	-3.81254
Group 4 (Belebeevsko-Shkapovskaya peak)					
Alekseevskoe (RT)	75	-0.31637	0,. 9699	-0.29702	-0.14903
Group 6 (Birsk saddle)					
Taimurzinskoe	84	-0.58309	0.01108	0,54845	-0.79834
Group 7 (Bashkir peak)					
Oryebash	96	-0.10766	-0.36060	-0.02073	-0.74565
Naratovskoe	95	-0.63429	0.61504	0.14126	-0.57164
Igrovskoe	99	-0.50643	0.70815	0.52646	0.03385

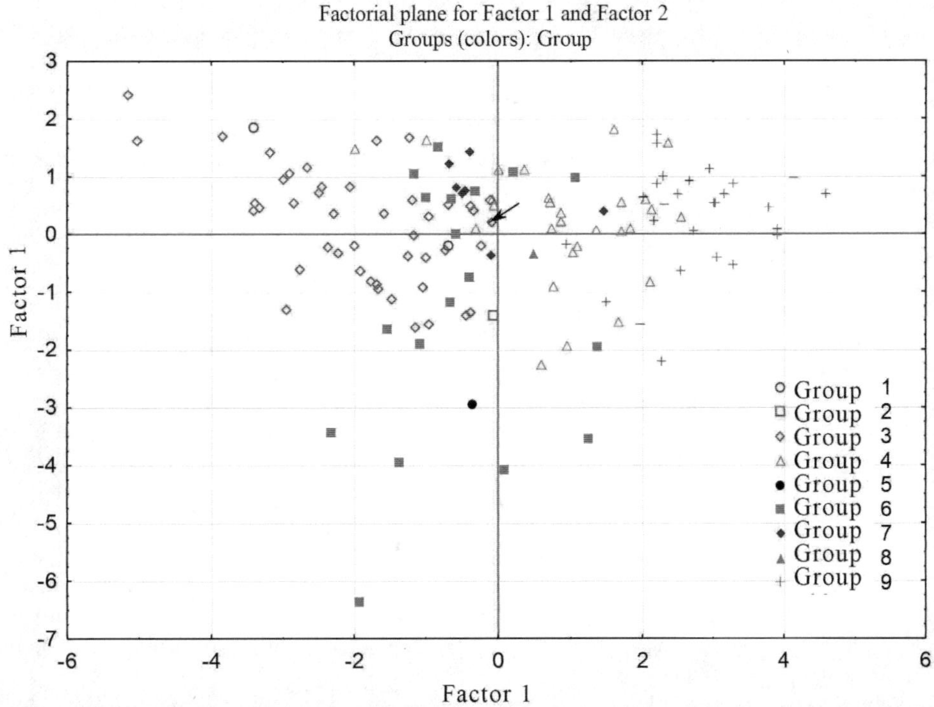

Figure 1. Projection of observations on the factorial plane: Groups: 1 - Nizhnekamsk system of linear dislocations; 2 - Sarailinskaya saddle; 3 - Almetyevskaya peak; 4 - Belebeevsko-Shkapovskaya peak; 5 - Verkhnekamsk depression; 6 - Birsk saddle; 7 - Bashkir peak; 8 - Abdulinsky trough; 9 - slope of the East European paleocontinent. Arrow - for object 7 (Kizelovsky horizon of the Bavlinskoye field, Almetyevskaya peak)

The acid-implosion method can be used at the objects of the Tournaisian stage of the Novoelkhovsky and Berezovsky deposits (Almetyevskaya peak), the Taimurzinsky deposit (Birskaya saddle) and the Oryebashsky, Naratovsky and Igrovsky (Bashkirsky peak) deposits.

4. Conclusion

Geological and physical data for 130 oil objects in the carbonate reservoirs of the Tournaisian stage of the Volga-Ural oil and gas province were reduced by the method of main components; the objects were classified. The grouping variable is the belonging of objects to one of the nine tectonic elements.

As a result, objects similar to the one of the Tournaisian stage of the Almetyevsk peak field were identified; for this peak, a high success rate of stimulation of oil production by the acid-implosion method was achieved. The selected objects have similar values of the main components, belong to the Almetyevskaya peak. With a lesser likelihood of successful impact, it can be objects belonging to the Belebeevsko-Shkapovskaya peak, the Birskaya saddle and the Bashkir peak.

References

[1] Lungwitz B, Fredd C, Brady M, Miller M, Ali S, and Hughes K 2004 Diversion and Cleanup Studies of Viscoelastic Surfactant-Based Self-Diverting Acid *SPE International Symposium and Exhibition on Formation Damage Control (Lafayette, 18-20 February 2004)* 10 p DOI: 10.2118/86504-MS

[2] Andreev A V, Mukhametshin V Sh, Kotenev Yu A 2016 Deposit Productivity Forecast in Carbonate Reservoirs with Hard to Recover Reserves *SOCAR Proceedings* **3** 40-45 DOI: 10.5510/OGP20160300287

[3] Mukhametshin V V 2017 Eliminating uncertainties in solving bottom hole zone stimulation tasks *Bulletin of the Tomsk Polytechnic University. Geo Assets Engineering* **328 (7)** 40–50

[4] Zeigman Yu V, Mukhametshin V Sh, Khafizov A R, and Kharina S B 2016 Prospects of Application of Multi-Functional Well Killing Fluids in Carbonate Reservoirs *SOCAR Proceedings* **3** 33–39 DOI: 10.5510/OGP20160300286

[5] Nasr-El-Din H A, Van Domelen M S, Sierra L, and Welton Th D 2007 Optimization of Surfactant-based Fluids for Acid Diversion *European Formation Damage Conference (Scheveningen, The Netherlands, 30 May-1 June 2007)* 11 p DOI: 10.2118/107687-MS

[6] Mukhametshin V V, Andreev V E, Dubinsky G S, Sultanov Sh Kh, and Akhmetov R T 2016 The Usage of Principles of System Geological-Technological Forecasting in the Justification of the Recovery Methods *SOCAR Proceedings* **3** 46–51 DOI: 10.5510/OGP20160300288

[7] Tyncherov K T, Mukhametshin V Sh, Paderin M G, Selivanova M V, Shokurov I V, Almukhametova E M 2018 Thermoacoustic inductor for heavy oil extraction *IOP Conference Series: Materials Science and Engineering* **327 (4) (042111)** 1–8 DOI:10.1088/1757-899X/327/4/042111

[8] Mukhametshin V V 2018 Rationale for trends in increasing oil reserves depletion in Western Siberia cretaceous deposits based on targets identification *Bulletin of the Tomsk Polytechnic University. Geo Assets Engineering* **329(5)** 117–124

[9] Anderson V J J, Pearson R A, and Boek E S 2006 The rheology of worm-like micellar fluids *Rheology Reviews* 217-253

[10] Yakupov R F, Mukhametshin V Sh, and Tyncherov K T 2018 Filtration model of oil coning in a bottom water-drive reservoir *Periodico Tche Quimica* **15 (30)** 725-733

[11] Akhmetov R T, Kuleshova L S and Mukhametshin V V 2019 Application of the Brooks-Corey model in the conditions of lower cretaceous deposits in terrigenous reservoirs of Western Siberia *IOP Conference Series: Materials Science and Engineering* **560** 012004 DOI: 10.1088/1757-899X/560/1/012004

[12] Fjelde I 2008 Sulfate in Rock Samples from Carbonate Reservoirs *International Symposium of the Society of Core Analysts, SCA2008-19 (Abu Dhabi, UAE, 29 October-2 November 2008)* 12 p

[13] Rogachev M K, Mukhametshin V V and Kuleshova L S 2019 Improving the efficiency of using resource base of liquid hydrocarbons in Jurassic deposits of Western Siberia *Journal of Mining Institute* **240** 711-715 DOI: 10.31897/PMI.2019.6.711

[14] Grezina O A 2018 Assessment of the results of acid implosion stimulation of the near-wellbore area based on statistical data *IOP Conference Series: Earth and Environmental Science* **194** 082016

[15] Grezina O A 2017 Efficiency evaluation for nanomaterials applied in well bottom zone treatment after completion of well construction *Nanotechnologies in Construction* **9(6)** 65-86 DOI: 10.15828/2075-8545-2017-9-6-65-86

[16] Mukhametshin V V, and Kuleshova L S 2020 On uncertainty level reduction in managing waterflooding of the deposits with hard to extract reserves *Bulletin of the Tomsk Polytechnic University. Geo Assets Engineering* **331(5)** 140–146 DOI 10.18799/24131830/2020/5/2644

The influence of reservoir rock fractures on the efficiency of geotechnical jobs on the Yamal oilfields

A S Sharapov

Industrial University of Tyumen, 38, Volodarsky Str., Tyumen, 625000, Russia

E-mail:Sharapov.anatoliy.sergeevich@gmail.ru

Abstract. At present, the majority of Yamal oilfields are characterized by low debits. That is why geotechnical jobs (GTJ) are conducted here to increase oil production. Thus, this article analyzes predicted and actual efficiency of GTJ on nineteen Yamal oilfields in 2010-2017. Therefore, the purpose of the article is to provide information on the quantity and efficiency of GTJ on Yamal oilfields as well as to interpret the results obtained. The question is how reservoir rock fractures influence GTJ. So, much attention is given to the efficiency of hydraulic fracturing operation (HFO) in relation to the rate of field tectonical diversity. In fact, our analysis shows that physical and chemical methods take the leading place among the quantity of actual operations. It was also found that the biggest incremental oil production was obtained from drilling horizontal drain holes, but the smallest quantity – from remedial cementing. Furthermore, the results presented here show the dependence of HFO efficiency on the rate of field diversity. In conclusion, actual data prove that double permeability of reservoir rock increases the efficiency of well perforation, HFO and physical and chemical methods as well as its dependence on the rate of field diversity. The findings are of direct practical relevance.

1. Introduction

Now oil and gas formations on the Yamal are presented by terrigenous fracture, porous, porous and fracture and fracture and porous rock reservoirs influencing geotechnical jobs. In fact, the majority of oil and gas fields are on the third and fourth development stages and are characterized by low debits. Therefore, this article analyzes the efficiency of predicted and actual geotechnical jobs on nineteen Yamal oilfields in 2010-2017.

2. Methods

The following methods were used to obtain the results, i.e. dialectical and retrospective ones, induction and deduction as well as comparative analysis and synthesis of given data.

3. The study of incremental oil production from geotechnical jobs

Actually, the biggest incremental oil production was obtained while drilling horizontal drain holes. It should be mentioned that well perforation and hydraulic fracturing operation gave approximately the same level of incremental oil production. By the way, the Lukoil company in Perm region got similar results for the Kungur oilfields there the greatest incremental oil production levels were typical of hydraulic fracturing operation, but simultaneously the level of incremental oil production during well perforation was equal to hydraulic fracturing operation [1]. It should be noted that reservoir rocks of Kungur oilfields were carbonate ones characterized by exfoliation. Therefore, reservoir rocks fractures were the reason for equal levels of incremental oil production during hydraulic fracturing operation and well perforation. At the same time, such an incremental oil production during well perforation and hydraulic fracturing operation exceeds that obtained during sidetracking . Also, the smallest quantity of incremental oil production was obtained during remedial cementing. Anyway, table 1 shows actual and predicted incremental oil production.

Table 1. Incremental oil production from geotechnical jobs

Geotechnical jobs	Forecast, thousand tons	Fact, thousand tons	Change of meaning, %
Horizontal drain-hole	19.212	13.232	-31
Sidetracking	6.044	4.229	-30
Hydraulic fracturing operation	4.812	4.942	3
Physical and chemical methods (CBZ)	621	1.099	77
Remedial cementing	580	130	-78
Conformance control	500	534	7
Diverter technology	418	511	22
Well perforation	389	4814	1.136
Pumping equipment optimization	281	621	121

4. The study of quantity of geotechical jobs

In fact, physical and chemical methods of cleaning the bottomhole formation zone take the leading place among the quantity of actual geotechnical jobs. So, the biggest increase in the quantity of actual geotechnical jobs was observed during well perforation and pumping equipment optimization (five and two times correspondingly). Also, we observed a 67% decrease in the actual quantity of remedial cementing jobs and 29% - in hydraulic fracturing operation. Figure 1 shows the guantity of operation.

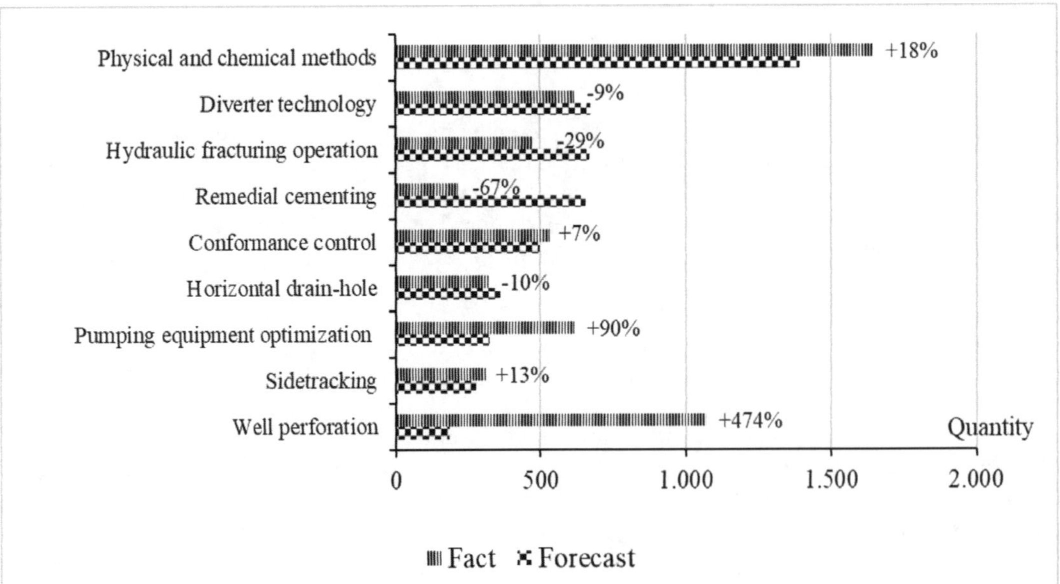

Figure 1. Quantity of operation

5. The study of incremental oil production from one well job

Also, the efficiency of geotechnical jobs is characterized by incremental oil production from one well job (see table 2). The biggest increase was obtained during well perforation (114%), physical and chemical methods of cleaning the bottomhole formation zone (75%). The biggest decrease was obtained during sidetracking (38%), remedial cementing (25%), horizontal drain holes (23%). Overall, low efficiency of remedial cementing can be explained by the fact that they were conducted in water cut marginal wells, while low efficiency of sidetracking and horizontal drain holes is explained - clogging of the bottomhole formation zone.

Table 2. Incremental oil production from one well job

Geotechnical jobs	Forecast, tons	Fact, tons	Change of meaning, %
Horizontal drain-hole	52.800	40.600	-23
Sidetracking	21.500	13.300	-38
Hydraulic fracturing operation	7.200	10.400	44
Remedial cementing	800	600	-25
Pumping equipment optimization	900	1.000	11
Physical and chemical methods (CBZ)	400	700	75
Well perforation	2.100	4.500	114
Diverter technology	800	1.000	25
Conformance control	700	900	29

6. The study of influence of fractures on geotechnical jobs

To study the influence of fractures on geotechnical jobs it is necessary to observe the use of hydraulic fracturing operation in relation to the rate of field tectonical diversity (see Gorbunov P [2]) for nine oilfields. Table 3 shows that the actual efficiency of hydraulic fracturing operation for one well job is 111.500 tons in Vostochno-Tarkosalinsky oilfield with the high tectonical diversity oil field, while in low tectonical diversity oilfields the actual efficiency doesn't exceed 20.000 tons.

Table 3. Incremental oil production for hydraulic fracturing operation for one well job

Field	Forecast, tons	Fact, tons	Change of meaning, %	Rate of field tectonical diversity
Vostochno-Tarkosalinsky	9.500	111.500	1.071	High
Krajnee	4.200	10.600	154	Low
Vyngayakhinskoe	8.000	15.800	99	Low
Kholmogorskoe	7.600	14.800	94	Low
Romanovskoe	11.400	19.700	73	Low
Sredne-Iturskoe	2.300	13.300	478	Low
Muravlenkovskoe	1.600	700	-58	Low
Karamovskoe	2.600	7.600	196	Low
Tarasovskoe	4.300	8.000	87	Low

In fact, high diversity zones were subjected to stronger different vector diastrophism and as a result, have a more developed fracture network [4]. As for low diversity zones, they were subjected to less tectonical diversity resulting in a less developed fracture network. Therefore, new fractures were made in a well during hydraulic fracturing operation resulting in the reservoir to well connectivity

IOP Conf. Series: Earth and Environmental Science 1021 (2022) 012053 doi:10.1088/1755-1315/1021/1/012053

with the fracture volume capacity of reservoir rocks [3]. Thus, it is quite effective in oilfields with high tectonical diversity (see table 3 and figure 2). As for the Vostochno-Tarkosalinsky oilfield, 1000% excess in actual data compared to forecast shows that fractures of reservoir rocks were not taken into account while calculating forecast efficiency data.

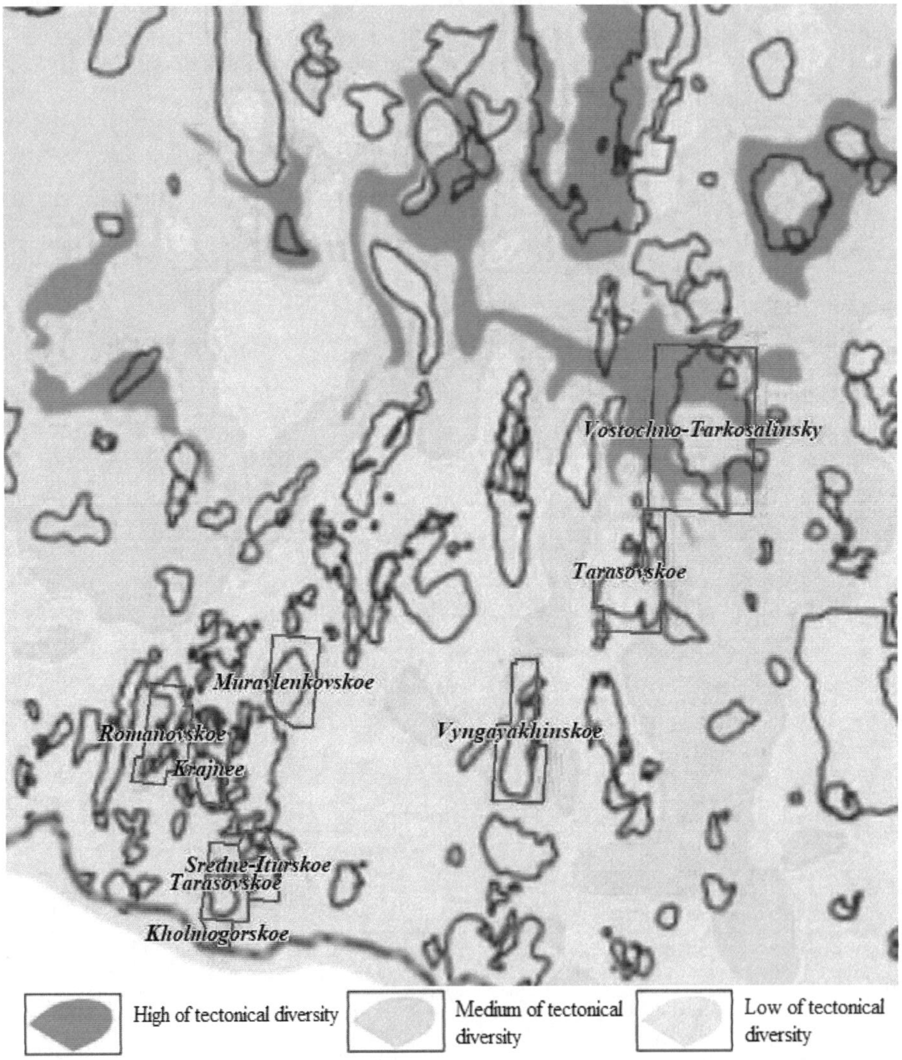

Figure 2. Map of the distribution of rate tectonical diversity (according to P Gorbunov, 2020)

7. Conclusion

The Yamal oilfields are characterized by double permeability i.e. the fluids are contained in fractures and in porous channels between which exchange processes take place [5]. It leads to high actual indexes of incremental oil production during well perforation, hydraulic fracturing operation and physical and chemical methods proved by obtained actual data. The article shows the efficiency of hydraulic fracturing operation depending on the rate of field tectonical diversity.

References

[1] Ilyushin P 2015 Analysis of geotechnical jobs to increase the productivity of producing wells in the oil fields of the Perm region *Vestnik of PNRPU* **15** 81–89

[2] Gorbunov P, Vorobyov S and Maksimenko O 2020 Tectonical diversity of Mesozoic-Cenozoic sediments as one of the main oil and gas control features in the northern part of the West Siberian plate *Geology of oil and gas* **1** 57–68
[3] Khanal A and Weijermars R 2019 Visualization of drained rock volume (DRV) in hydraulically fractured reservoirs with and without natural *Petroleum Science* **16** 550–577
[4] Andrews B, Shipton Z, Lord R and McKay L 2020 The growth of faults and fracture networks in a mechanically evolving, mechanically stratified rock mass: a case study from spireslack surface coal mine *Solid Earth* **11** 2119-2140
[5] Popov I 2013 *New technologies in oil and gas geology and field development* (Tyumen: Industrial university of Tyumen press)

IOP Conf. Series: Earth and Environmental Science 1021 (2022) 012054 doi:10.1088/1755-1315/1021/1/012054

Microbiological and physicochemical treatment of oil to increase field resource extraction

M M Veliev[1], E M Veliev[2], V V Mukhametshin[3], L S Kuleshova[2], Le Viet Zung[1] and D I Akhmetshina[2]

[1]JV "Vietsovpetro", 105, Le Loi St., Thang Nhi Ward, Vung Tau, Ba Ria – Vung Tau City, S.R. Vietnam
[2]Ufa State Petroleum Technological University, Branch of the University in the City of Oktyabrsky, 54a, Devonskaya St., Oktyabrsky, Republic of Bashkortostan, 452607, Russian Federation
[3] Ufa State Petroleum Technological University, 1, Kosmonavtov St., Ufa, Republic of Bashkortostan, 450064, Russian Federation

E-mail: _elshad_@mail.ru

Abstract. It was shown that the use of a physicochemical microbiological composition (FCMC) revealed a slight change in the amount of sulphate-reducing bacteria. The mineralization and density of the fluid taken from production wells before and after injection of this composition almost did not changed. After the FCMC injection, the watering of the pilot site slowed down. The calculations showed that the accumulated additional oil production of the pilot site during the period under study will be 856 tons and is increasing. These initial results suggest further effective application of the technology to other similar facilities.

1. Introduction

The problem of enhanced oil recovery is one of the most important and complex problems of scientific and technological progress. A successful solution to this problem is a basic prerequisite for meeting oil needs, but this is only possible on the basis of a high level economy, as well as the developed production in many industries [1-8].

Due to a number of social, political, techno-economic, market and other circumstances, new methods for enhanced oil recovery are being actively studied today in only a few countries. However, it is already becoming quite obvious that the field of study and implementation of new methods is gradually expanding [9-12].

Since the current state of the raw material base of Vietsovpetro is characterized by the deterioration in the geological and physical parameters of the deposits, a significant increase in the share of hard-to-recover reserves, the development of technologies with enhanced oil recovery coefficient seems especially relevant. They include technologies of complex oil depositsdevelopment, technologies using physical phenomena in filtration of fluids in a porous medium of reservoirs (vibration wave, electric and acoustic impact, hydraulic fracturing), biotechnology [13, 14].

The tasks of technologies using physical phenomena are to intensify production and ensure enhanced oil recovery with complex geological and physical conditions, in particular, at the final stages of field development, when traditional production methods turn out to be ineffective [15].

2. Methods and Materials

Every year microbiological methods of enhanced oil recovery are increasingly recognized in the world as highly effective with their small investment needs and eco-friendliness.

As far back as 1926, J. W. Beckham determined that most of the oil in fields around the world would remain nonrecoverable and that it was necessary to develop methods for enhanced oil recovery. He suggested that microorganisms may be used to solve the problem [16].

Along with physical and chemical methods of enhanced oil recovery, microbiological technologies for enhanced oil recovery (EOR), which have now been introduced in various oil producing regions of Russia and abroad, are becoming increasingly widespread in the world.

The prospects of EOR microbiological methods are specified ин their low investment need, high efficiency and environmental safety. The practical application of biotechnologies in oil production allows increasing the final oil recovery factor (ORF) by 5-7%.

The core of EOR microbiological methods is the effect of metabolites – chemical products of microorganisms – on oil and reservoir properties, as a result of which oil displacement and sweep coefficients increase, and hence, the ORF. According to the mechanisms of this effect, microbiological methods to some extent are similar to physicochemical methods.

The purpose of this study is the use of technology using a physicochemical and microbiological composition (FCMC) and obtaining a scientific and practical justification for the expanded use of technology at other Vietsovpetro facilities.

The microbiological method of enhanced oil recovery is used in two ways:

- technology of microbes cultivation outside the deposit with subsequent injection of their products into the deposit;
- injection of nutrient solutions (molasses) to activate microbial activity directly in the formation (in situ).

The technology of molasses injection to activate the activity of microbes directly in the formation is only effective at low temperature of the formation, and in the conditions of Vietsovpetro it is not effective. Therefore, the technology of microbe cultivation outside the deposit with subsequent injection of their products into the deposit is considered.

An enzymatic liquid was injected into the deposit – biopolymers, biosurfactants, bio cosurfactants, bioantioacidants, biostabilizers, enzymes, as well as chemical surfactants, chemical cosurfactants and others. Especially, impressive results were obtained with the integrated application of the microbiological method with physicochemical factors of treatment.

3. Results and Discussion

One injection well (No. 74) and two production wells (No. 117, 705) were chosen as pilot sites.

For enhanced oil recovery at the Lower Miocene deposits of the White Tiger field the FCMC was prepared meeting the corresponding requirements for quantity and quality.

FCMC was injected into the formation as follows.

The formation-pressure maintenance was stopped for 3 h 20 min and at P=0 MPa 67 t of FCMC was injected into the tubing in two parts: first 11.8 t, and then another 55.2 t. The total injection duration was 4 h 40 min.

IOP Conf. Series: Earth and Environmental Science 1021 (2022) 012054 doi:10.1088/1755-1315/1021/1/012054

For several months, the wells were monitored and a slight change in the number of sulphate-reducing bacteria was revealed. The mineralization and density of the liquid extracted from production wells before and after the FCMC injection has not changed much.

Prior to the FCMC injection, the pressure and acceleration of the well No. 74 were unstable and were at a high level. The stabilization of these parameters after the FCMC injection reflects a noticeable improvement in the technical parameters of the well. In production wells 117 and 705, the watering growth rate slowed down, oil production stabilized.

The general performance of the pilot site is shown in Table 1, and the operation dynamics of the pilot site are shown in Figure 1.

The table and the figure show that in March the injection volume of the injection well No. 74 increased dramatically, which affected the well pressure. Over the next month, the total oil production of wells increased markedly and reached its maximum value in May.

Table 1. General performance of the pilot site

Month	Intake capacity, m^3/day	Well pressure, MPa	Sum of oil flowrates in two wells, t/day
1	349.8	8.09	69.6
2	361.2	7.67	63
3-intensive injection	655.7	9.08	56
4	358.8	5.82	52
5	363.4	5.77	80
6	360.2	5.64	59
7-FCMC injection	357.9	5.43	63
8	364.2	5.34	52
9	356.9	5.14	49.2
10	360.4	5.55	49
11	357.3	5.63	50
12	359.9	5.74	50

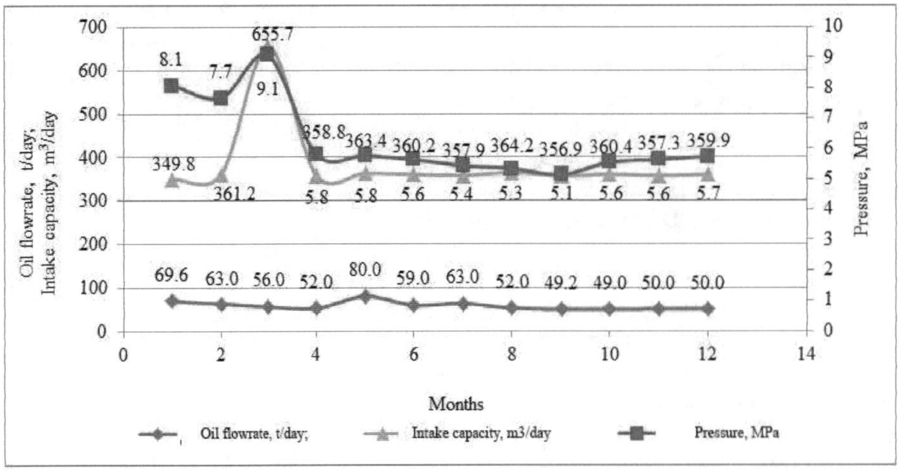

Figure 1. Dynamics of pilot site performance

To objectively assess the FCMC efficiency, let us divide the process of deposit development into three stages (Table 2):

1. Normal injection stage, from January to March. The total current oil production decreased at a rate of 6.6-6.7 t/day.

2. Development stage under the influence of a sharp increase in the volume of injection (May-July and partially August). The total current oil production increased reaching the maximum in May, then sharply decreased at an average speed of 9.3 t/day.

3. Development stage under the influence of the FCMC injection: since September the total current oil production has stabilized.

The dynamics of the total current oil flowrate in the pilot area over three stages of oil deposit development is given in Figure 2.

Table 2. Total oil flowrate of production wells over three periods

Month	Total flowrate by months 1, 2, 3, 4	Total flowrate by months 5, 6, 7, 8	Total flowrate by months 8, 9, 10
1	69.6		
2	63.0 ↓ 7 t/day		
3-intensive injection	56.0		
4	52.0 – transition point		
5		80.0	
6		59.0 ↓ **9** t/day	
7-FCMC injection		63.0	
8		52.0 – transition point	
9			49.2
10			49.0
11			50.0
12			50.0

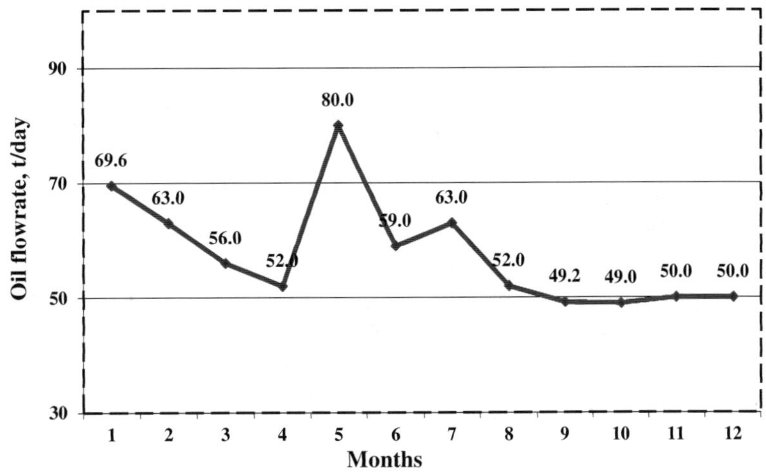

Figure 2. Dynamics of the total current oil flowrate in the pilot area over three stages of oil deposit development

From January to March of the second year: the tendency to decrease the total current oil production was stable – the change in the total oil production was a steep curved inclined line characteristic of the decreasing exponential function (Figure 3).

The actual and calculated curves almost coincide. The compatibility or regression factors (R^2) between actual and calculated data is very high and amounts to 0.9929, which is an ideal compatibility factor between actual and calculated data.

At the stage under the influence of intensive water injection, the oil flowrate increased sharply, then also decreased sharply. The rate of flowrate decline was unstable.

At the third stage under the influence of the FCMC injection, the total current oil flowrate in production wells was stabilized until December.

Mathematical processing of data on the total current oil flowrate is required for an objective assessment of the FCMC efficiency. There are only two suitable types of mathematical functions or regressions to simulate the current oil flowrate from May to August (Figure 4).

- power function tending to zero

$$y = 75.472 \cdot x^{-0.2286} \quad (R^2 = 0.8644),$$

where y – oil flowrate, x – month.

- exponential function tending to zero

$$y = 72.577 \cdot e^{-0.0587x} \quad (R^2 = 0.69).$$

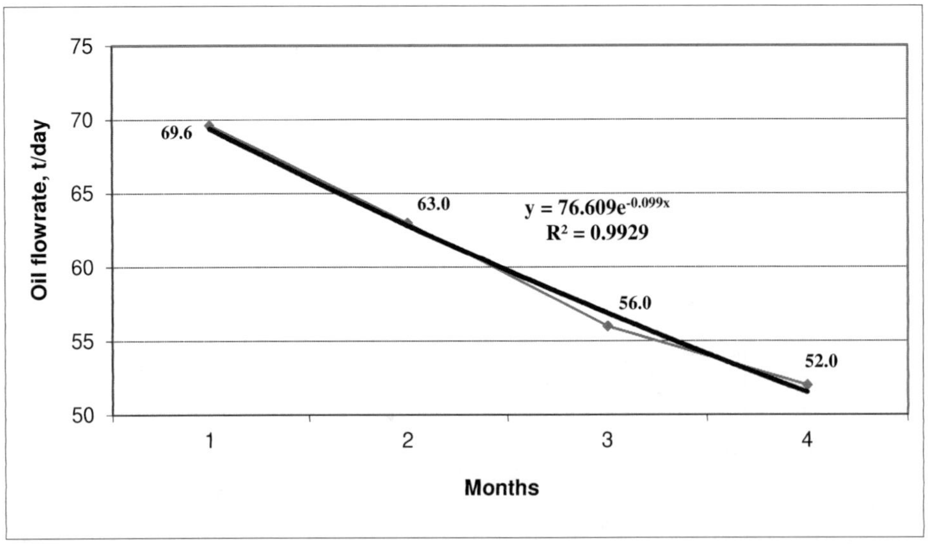

Figure 3. Dynamics of the total current oil flowrate in the pilot site over three stages of oil deposit development

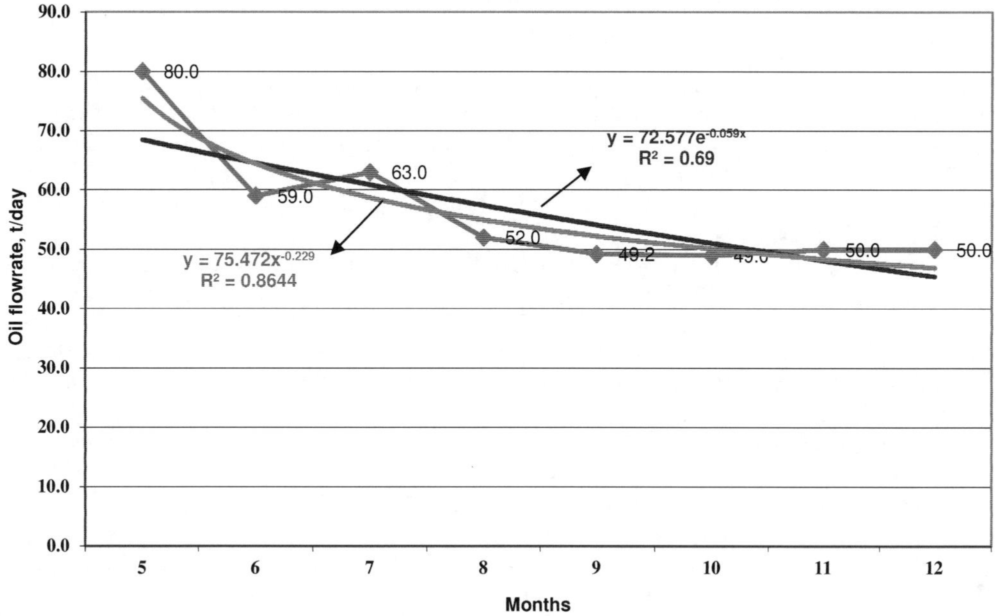

Figure 4. Dynamics of the total current oil flowrate in the pilot site over the third stage of oil deposit development

For the total current oil flowrate from January to March, there is an exponential zero-tending function $y = 76{,}609 \cdot e^{-0{,}9992x}$ with compatibility factor $R^2 = 0.9929$ (Figure 3) that is very similar to the real data. As is known, the most reliable trend line is for which the value of the compatibility factor is equal to or close to 1. It can be seen that the power function has a higher compatibility factor than the exponential function. Therefore, using the power function, let us calculate the additional oil produced from September.

The calculations showed that the accumulated additional oil flowrate over the seven months in the pilot site will amount to 856 tons and is increasing.

4. Conclusions

1. A slight change in the amount of sulphate-reducing bacteria was revealed after the FCMC injection. The mineralization and density of the liquid extracted from production wells before and after the FCMC injection has not changed much.
2. The growth in watering at the pilot site slowed down after the FCMC injection.
3. The calculations show that the accumulated additional oil flowrate in the pilot site during the period under study will amount to 856 tons and is increasing.
4. These initial results allow further effective application of the technology at other facilities.

References

[1] McLachlan G J 2004 *Discriminant Analysis and Statistical Pattern Recognition* (John Wiley & Sons, Hoboken, New Jersey)

[2] Soloviev N N, Mukhametshin V Sh and Safiullina A R 2020 Developing the efficiency of low-productivity oil deposits via internal flooding *IOP Conf. Ser.: Mater. Sci. Eng.* **952** 012064 DOI: 10.1088/1757-899X/952/1/012064.

[3] Kadyrov R R, Mukhametshin V V, Galiullina I F, Kuleshova L S and Safiullina A R 2020 Prospects of applying formation water and heavy brines derived therefrom in oil production and national economy *IOP Conference Series: Materials Science and Engineering* **905** 012081 DOI: 10.1088/1757-899X/905/1/012081.

[4] Muslimov R Kh 2014 Oil recovery: past, present, future (production optimization, maximization of oil recovery (FEN, Kazan)

[5] Mukhametshin V Sh 2020 Rationale for the production of hard-to-recover deposits in carbonate reservoirs *IOP Conf. Ser.: Earth Env.* **579** 012012 DOI: 10.1088/1755-1315/579/1/012012.

[6] Kuleshova L S and Mukhametshin V V 2020 Justification for the selection of distances between wells to increase the efficiency of flooding by deposits in terrigenous reservoirs of Western Siberia, *Journal of Physics: Conference Series* **1582** 012057 DOI: 10.1088/1742-6596/1582/1/012057.

[7] Back M J and Guercio C 2010 Portfolio Management for Strategic Planning and Operational Optimization, SPE Annual Technical Conference and Exhibition (Florence, Italy, 19-22 September), DOI: 10.2118/134339-MS.

[8] Mukhametshin V V and Kuleshova L S 2020 On uncertainty level reduction in managing waterflooding of the deposits with hard to extract reserves *Bulletin of the Tomsk Polytechnic University. Geo Assets Engineering* **331(5)** 140–146 DOI 10.18799/24131830/2020/5/2644.

[9] Mukhametshin V Sh, Andreev V E and Yaskin S A 2020 Designing measures to increase oil recovery based on the identification and grouping of deposits, *IOP: Earth and Environmental Science* **579** 012013 DOI: 10.1088/1755-1315/579/1/012013.

[10] Back M J and Kirk G 2012 An Integrated Portfolio Management Approach for More Effective Business Planning, *SPE Hydrocarbon Economics and Evaluation Symposium* (Calgary, Alberta, Canada, 24-25 September) DOI: 10.2118/162748-MS.

[11] Malyarenko A M, Kotenev Yu A, Mukhametshin V Sh, Bogdan V A, Kotenev A Yu, Blinov S A and Umetbaev V G 2020 The influence of the rocks material composition on the reservoir properties of the reservoir *Journal of Physics: Conference Series* **1582** 012061 DOI: 10.1088/1742-6596/1582/1/012061

[12] Khokhlov V I, Galimov Sh S, Devyatkova S G, Kotenev Yu A, Sultanov Sh Kh and Mukhametshin V Sh 2019 Justification of impact and planning of technology efficiency on the basis of limy-emulsion formulation in low-permeability highly-rugged reservoirs of Tyumen deposits, IOP Conference Series: Earth and Environmental Science **378** 012114 DOI: 10.1088/1755-1315/378/1/012114

[13] Feng Zhang and Daoyong Yang 2014 Determination of minimum permeability plateau and characteristic length in porous media with non-Darcy flow behavior, *Journal of Petroleum Science and Engineering* **119** 8-16 DOI: 10.1016/j.petrol.2014.04.018.

[14] Xinlei Shi, Yunjiang Cui, Wankun Xu, Jiansheng Zhang and Yeqin Guan, Formation permeability evaluation and productivity prediction based on mobility from pressure measurement while drilling, Petroleum Exploration and Development **47(1)** 146-153 DOI: 10.1016/S1876-3804(20)60013-1.

[15] Mukhametshin V Sh 2020 Justification for increasing the performance of hydrochloric acid treatment in wells of fields with carbonate reservoir *IOP Conf. Ser.: Mater. Sci. Eng.* **905** 012083 DOI: 10.1088/1757-899X/905/1/012083.

[16] Beckham J W 1926 The action of bacteria on mineral oil, *Ind. Eng. Chem. News* **4** 3

IOP Conf. Series: Earth and Environmental Science 1021 (2022) 012055 doi:10.1088/1755-1315/1021/1/012055

Use of acid fracturing to increase the production potential of subsoil users

G M Siraeva[1], I G Fattakhov[2], L S Kuleshova[2], R R Stepanova[2]

[1] PJSC TATNEFT named after V.D. Shashin, 75, Lenin St., Almetyevsk, Republic of Tatarstan, 423450, Russian Federation
[2] Ufa State Petroleum Technological University, Branch of the University in the City of Oktyabrsky, 54a, Devonskaya St., Oktyabrsky, Republic of Bashkortostan, 452607, Russian Federation

E-mail: i-fattakhov@rambler.ru, info@of.ugntu.ru

Abstract. The paper reflects the main problems of the search and development of Domanic deposits, including the following: complexity of the area search for promising sites; difficulty in extracting reservoir beds from well geophysical survey (GIS) data; low permeability of formations; low profitability of oil production; difficulty in reserve estimate and and registration. Oil and gas companies face problems such as the lack of reliable information on Domanic deposits from specific GIS studies, core samples, a significant risk of obtaining unprofitable flow rates, and a low degree of adaptation of existing oil production technologies to the products of Domanic deposits. The formation of reserves with mobile hydrocarbons is influenced by such parameters as: degree of maturity of source rocks; composition of substances; amount of minerals that determine the maturation of the organic substance and formation of the reservoir; presence of intervals with increased fragility, presence of impermeable layers that ensure the safety of deposits, etc. It was shown that the Domanic deposits should be developed using horizontal wells and multi-stage acid hydraulic fracturing treatment (HFT).

1. Introduction

Wide variation intervals of oil deposit parameters reflecting geological, physical and physicochemical properties of formations and their saturating fluids require a differentiated approach to analysis, control, regulation and design of development systems [1-6].

With the discovery and introduction of new objects with specific features in terms of geological structure into development, it is required not only to adapt the established technologies, but also to create innovative approaches to oil production [7-10].

One of these promising objects is the deposits of the Dankov-Lebedyan horizon of the Volga-Ural oil-and-gas province.

For the purpose of expanding the production potential of PJSC Tatneft on Bavlinsky oil field the company drilled a pilot inclination well No. xxx7G with oriented core sampling to study the lithological section and the geological structure of beds and a horizontal well on the DankovoLebedyan horizon using multizone HFT through ports with TAM packers [11].

These developments will increase the share of recoverable oil reserves in the oldest oil producing area of the Volga region due to the development of new productive horizons and will ensure the enhanced oil recovery of these reservoirs. The methods used in the work will reduce the specific costs of enhanced oil recovery and will make it possible to achieve a profitable performance at such wells.

Domanic deposits are characterized by a small grain size – fine grains. Rocks do not represent a typical reservoir either in lithology or in the structure of the void space – there is no connection between porosity and permeability, as well as quite low reservoir properties in the presence of industrial oil reserves, which indicates the unconventional nature of the reservoir. But it is worth taking into account that these deposits can be represented by mixed-type reservoirs.

Filtration properties are often caused by subhorizontal fracturing with a branched configuration. The deposits of the DankovoLebedyan and Yelets horizons within the South Tatar Suite (STS) are mainly represented by carbonate rocks. According to the mineral composition, the deposits of the Famennian tier include 86% limestones and 14% dolomites. The DankovoLebedyan horizon is mainly composed of fine-grained limestones, sometimes with recrystalized areas. In transitional areas, the horizon changes from limestone to dolomite limestone, from dolomite limestone to dolomite with different degrees limestone, which has undergone secondary changes. The promising sections of the DankovoLebedyan horizon are confined to the pore-cavern-fractured sections. The variability of reservoir properties is determined by both the primary sedimentation conditions and the secondary processes.

One of the promising directions for hard-to-recover reserves is the industrial development of the above type reservoirs.

Matrix treatment of the multi-zone HFT was carried out in the first port (15% HCL injection V=12 m^3, aqueous polysaccharide gel V=6 m^3, technical fluid with y=1.0 g/cm^3, V=10 m^3, P_{in}=13.0 MPa, P_{av}=11.0 MPa, P_{fin}=11.0 MPa) and the fourth port (15% HCL V=35.6 m^3, aqueous polysaccharide gel V=12 m^3, technical fluid with y=1.0 g/cm^3, V=7.3 m^3, P_{in}=11.0 MPa, P_{av}=10.5 MPa, P_{fin}=8.2 MPa), while in the second (15% HCL V=10.3 m^3, aqueous polysaccharide gel V=10 m^3, technical fluid with y=1.0 g/cm^3, V=10 m^3, P_{in}=11.0 MPa, P_{av}=10.5 MPa, P_{fin}=11.0 MPa) and third (15% HCL V=50.8 m^3, aqueous polysaccharide gel V=24 m^3, technical fluid with y=1.0 g/cm^3, V=10 m^3, P_{in}=11.5 MPa, P_{av}=10.5 MPa, P_{fin}=8.6 MPa) – acid HFT.

2. Materials and methods

The work (Fig. 1) was accompanied by some features, such as rather long fracture closure time in Domanic deposits during the pressure loss registration was one of the difficulties in interpreting the Mini-Frac HFT (Meyer software). During the mini-frac jobs the analysis of the Nolte G Time Test graph showed that the crack did not close, therefore, it is impossible to determine the closing pressure (fracture pressure gradient) with reliable accuracy. Therefore, the HFT model was calibrated by the actual (effective) fracturing pressure after the process was completion (Fig. 2) [12].

HFT control at well No. xxx7G was carried out via the method of microseismic monitoring with continuous recording of seismic noise by high-sensitivity stationary seismic sensors.

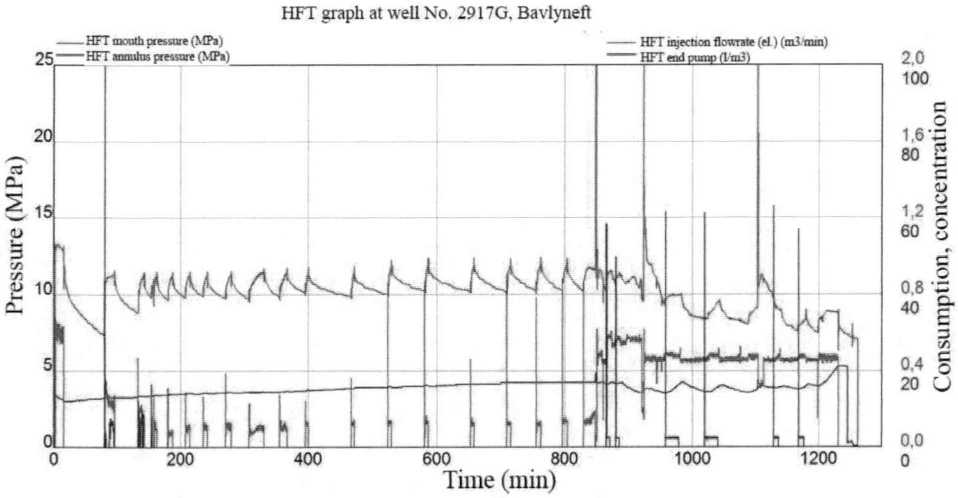

Fig 1. Multi-stage HFT graph

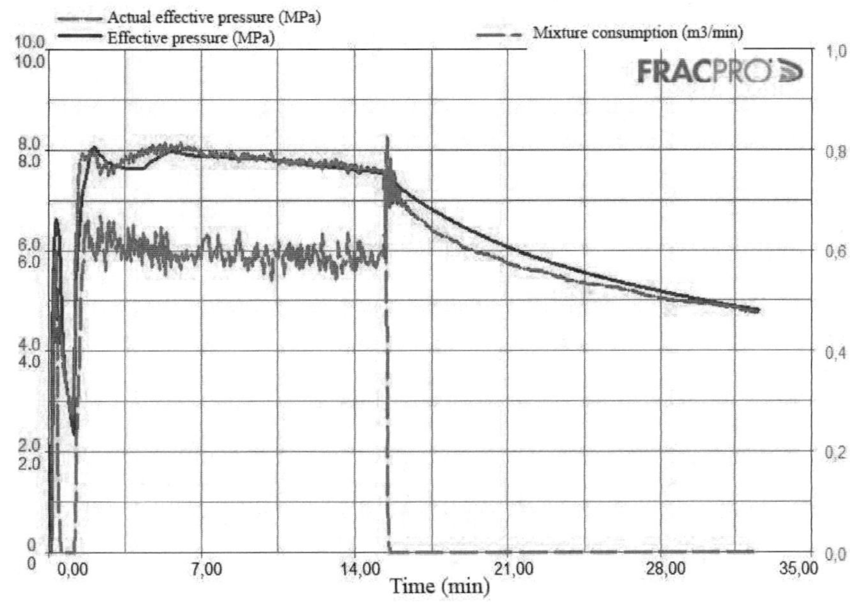

Fig 2. Calibration of the HFT model according to actual pressure

At the same time, before the treatment of port No. 1 (Fig. 3), a mini-HFT and a test with increased fluid consumption were carried out. The graphs of the mini-HFT interpretation are given in Fig. 2, 4, 7. The analysis of the mini-HFT was carried out in the MinFrac program (Meyer software). The following dependencies were used: dP on dT and on G function.

The duration of fracture closure in unconventional objects behaves unpredictably and can be very long. So, in case of a mini-frac at well No. xxx7g, the crack did not close at the expected time.

IOP Conf. Series: Earth and Environmental Science 1021 (2022) 012055 doi:10.1088/1755-1315/1021/1/012055

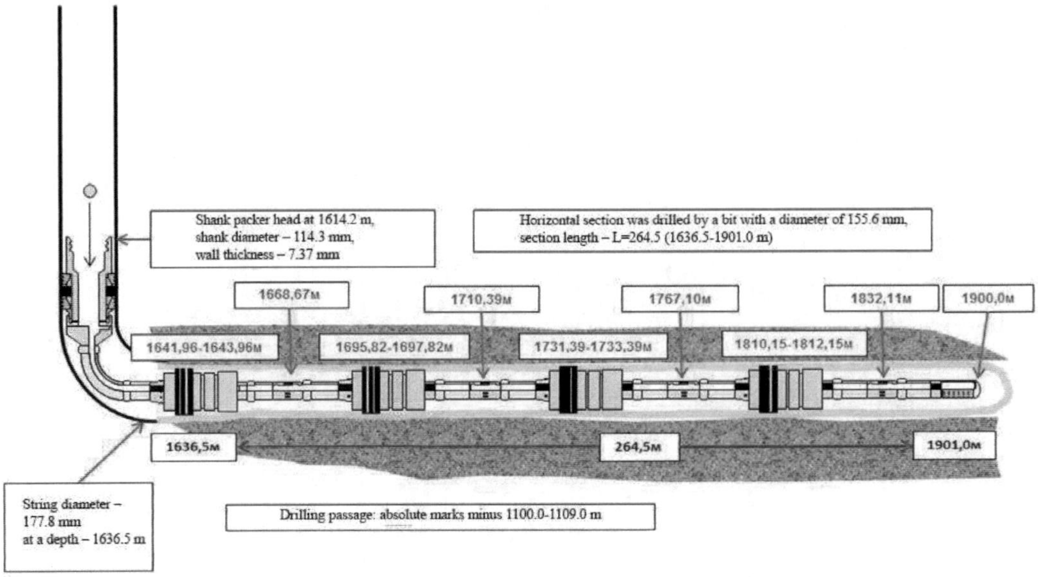

Fig 3. Layout of downhole equipment in well xxx7G

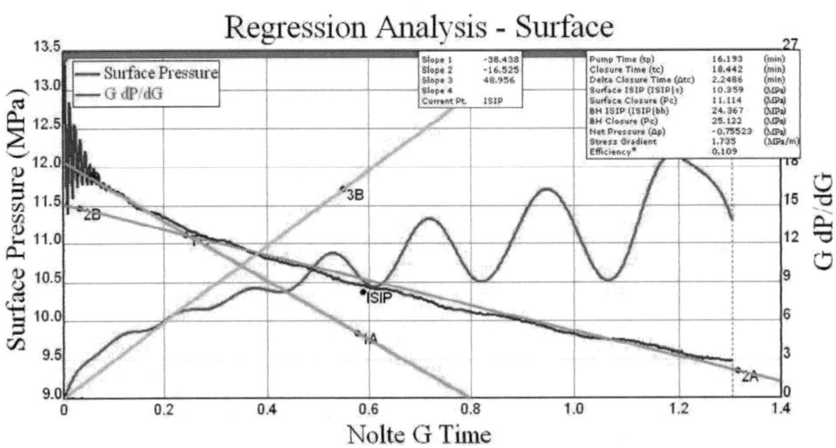

Fig 4. Wellhead pressure depending on Nolte G Time

Multi-stage HFT is recommended only in case of consolidated rocks.

Based on the previous numerical simulation of fracture propagation during fracturing it was determined that fracturing occurs intermittently over time, which leads to pressure surge, crack opening, and displacement, and that fracture propagation is sensitive to factors such as the initial state of gas saturation, the type of injected fluid, inhomogeneity, ultimate strength, elastic modulus, and model permeability. The increase of fracture pressure is also affected by permeability. High intrinsic relative permeability results in rapid movement of the injected fluid. High elastic modulus and high Poisson coefficient are not favorable for fracture propagation, although they are not very sensitive. The effective permeability for such reservoirs is low – the matrix permeability is from several $1\cdot10^{-3}$ to $600\cdot10^{-3}$ μm^2 (in the presence of natural cracks).

As part of the work on reserves estimates, a full lithological-petrographic justification of the belonging of the Middle Franian-Famennian deposits to "unconventional" reservoirs was carried out in accordance with the temporary methodological recommendations of the State Committee on Natural Reserves.

Carbonate Devonian deposits are distributed throughout the territory of Tataria of the Franian-Famennian complex. Based on facial and tectonic confluence in terms of lithological features, deposits containing organic matter have different stages of oil formation. According to the features of oil content, three zones are distinguished by area: southeast, central and north-western; two groups of deposits are distinguished along the section: group I (D3ev – D3zg) – carbonate dense weakly permeable rocks with a content of Corg = 0.5-5% and group II (D3sr – D3mn) – silicon-carbonate weakly permeable rocks with a content of Corg = 5-20%.

Domanic deposits have traditionally been considered as oil source rocks, but they are at the same time unconventional reservoirs that include industrial oil deposits. The deposits have no bottom waters, water oil contacts, are not structurally controlled and limited only to replacement of reservoir rocks with rocks having lower reservoir properties. The development of such deposits is impossible without the use of special modern technologies, including horizontal drilling and hydraulic fracturing.

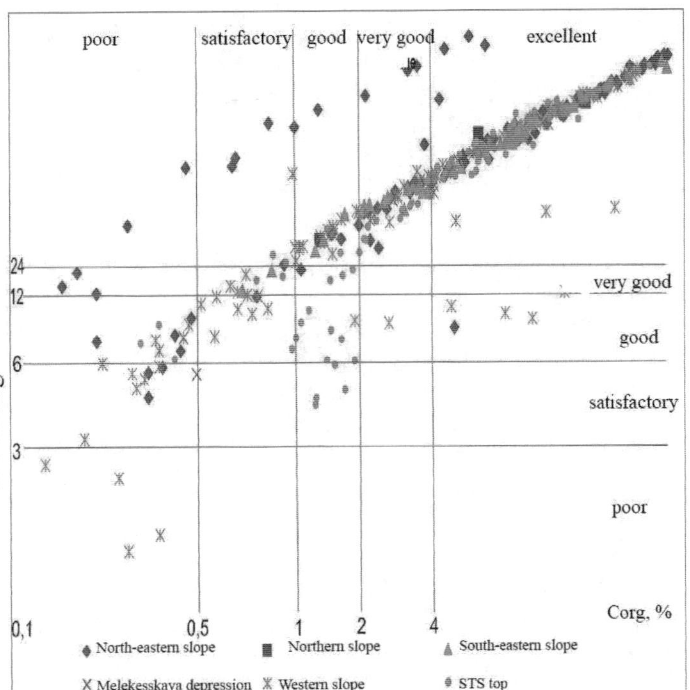

Fig 5. Generation potential of domanikites. World development technologies: Shell ICP, Chevron CRUSH, ExxonMobil ElectroFrac, AMSO, EGL Technology, TGT (OJSC RITEK)

PJSC TATNEFT distinguishes both domanicites and domanicoids belonging to Domanic deposits:

- domanicites, which in the Volga-Ural and Timan-Pechersk oil and gas-bearing provinces are represented by limestones and siliceous-clay dolomites containing organic matter, which act as the main oil source rocks. The reservoirs are characterized by very low porosity and very low Corg > 5% permeability. Deposits of the Voronezh, Rechitsa, Domanic, Sargaev horizons are represented by limestones with the development of secondary processes, with carbonate nodules, as well as brecciated carbonate rocks with a carbonaceous aggregate. The development of dark gray, almost black high-carbonaceous carbonate-siliceous rocks, aged in thickness, is characteristic of the Domanic rocks.

Such a complex of deposits is interpreted as an oil source thickness with a high generation potential (Fig. 5, 6) [13].

- domanicoids – in the Ural-Volga region cover the entire section from the Upper Franian to the Upper Turney rise – deposits of Rechitsk- (Mendym) – of the Volga age, developed in the axial zones of the Kama-Kinel system of uncompensated deflections.

Reservoirs are characterized by low porosity and very low permeability, Corg < 5%. Domanicoids reached the conditions of the main oil formation zone – the protocatagenesis zone.

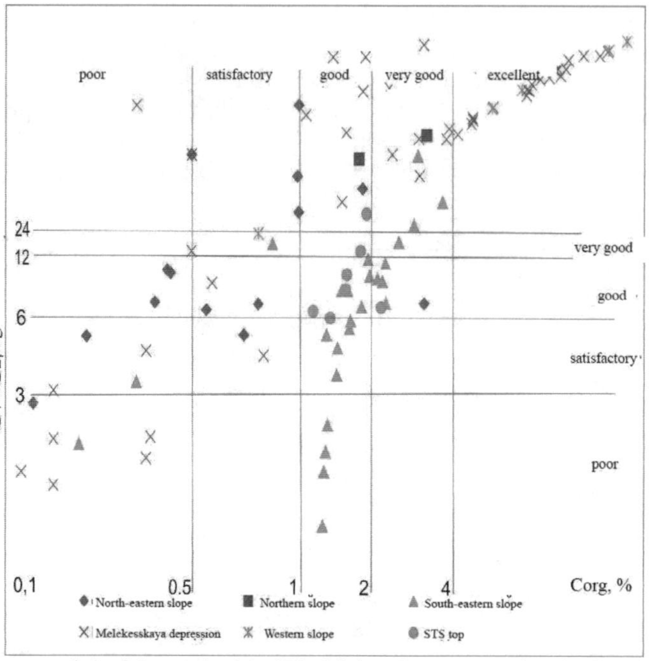

Fig 6. Generation potential of domanicoids. World development technologies: horizontal drilling, HFT, multi-stage HFT, bottomhole treatment, multi-scale bottomhole treatment

3. Results and Discussion

To start the facility for pilot operation, the 25-175RHAM-12-4-2-2 pump was lowered to a depth of 1355 m with an upper drive SK8-3.5-4000. The stroke length is 3.5 m, and the pumping speed is 3.5. The theoretical capacity of the pump is 26 m^3/day. At these process parameters of pumping equipment the initial parameters of well operation were as follows: fluid flowrate – 24.4 m^3/day (at constant decreasing dynamic level), oil flowrate – 17 t/day, water production – 22%.

Ten days after the well was put into operation, hydrodynamic studies were carried out, which were carried out using an echo sounder. The conversion of the fluid level values in the annular space during the level restoration curve study to the values of the bottom ole pressure was carried out by calculation on the basis of the PJSC TATNEFT methodology. When interpreting the hydrodynamic study results, the best match between the experimental and design curves was achieved using a horizontal well model at a homogeneous reservoir.

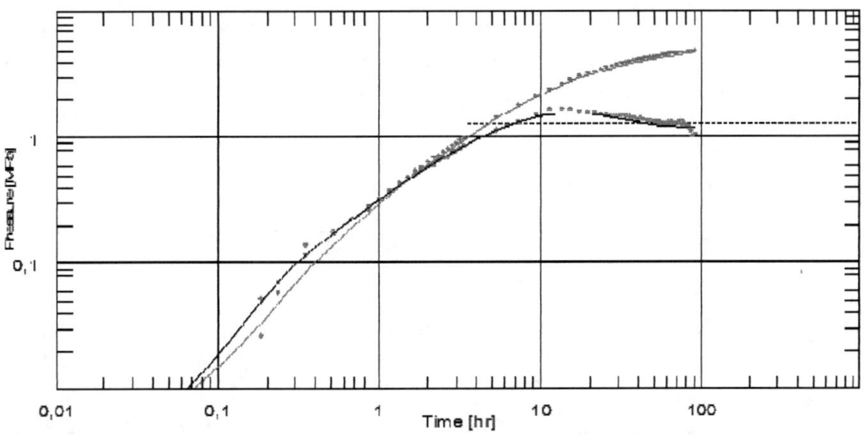

Fig 7. Pressure buildup curve and its derivative in bilogarithmic coordinates

Table. Results of level restoration curve interpretation of well No. xxxG

Parameter	Value
Bottomhole pressure, MPa	6.3
Formation pressure, MPa	12.2
Length of horizontal section, m	251.1
Productivity, m^3/day MPa	3.14
Hydraulic conductivity, $\mu m^2 \cdot m/Pa \cdot s$	15.8
Inflow factor, m^3/MPa	2.24
Summary skin effect	-5.04
Skin effect of bottomhole zone	1.38
Geometric skin effect (for a horizontal well this parameter is the equivalent of a vertical well with a crack passing through it)	-6.42
Radius of productive horizon survey, m	207.0

The graph (Fig. 7) shows the maximum derivative, which for a horizontal well with cracks the HFT may indicate clearly deteriorated properties of the bottomhole zone. At the same time, the deterioration of bottomhole zone may be connected with weak opening of created cracks [14].

Well performance as of 01.02.2020: $q_{f\ in}$/March 2020 – 24.4/5.73 m^3/day, $q_{o\ in}$/March 2020 – 17/3.6 t/day, 22/29.4% water initial/March 2020. The accumulated oil production through the well amounted to 9.080 thousand tons of oil.

4. Conclusion

1. The main problems of search and development of Domanic deposits are as follows:
   - complexity of the area search for promising sites;
   - difficulty in extracting reservoir beds from well geophysical survey (GIS) data;
   - low permeability of formations;
   - low profitability of oil production;
   - difficulty in reserve estimate and registration.

2. Oil and gas companies face problems such as the lack of reliable information on Domanic deposits from specific GIS studies, core samples, a significant risk of obtaining unprofitable flow rates, and a low degree of adaptation of existing oil production technologies to the products of Domanic deposits.

3. The formation of reserves with mobile hydrocarbons is influenced by such parameters as: degree of maturity of source rocks; composition of substances; amount of minerals that determine the

maturation of the organic substance and formation of the reservoir; presence of intervals with increased fragility, presence of impermeable layers that ensure the safety of deposits, etc.

4. It was shown that the Domanic deposits should be developed using horizontal wells and multi-stage acid hydraulic fracturing treatment.

References

[1] Nurgaliev R Z, Kozikhin R A, Fattakhov I G, Kuleshova L S, and Gabbasov A Kh 2019 Prospects for the use of new technologies in assessing the impact of geological and technological risks *IOP Conference Series: Earth and Environmental Science* **378** 012117 1-9 DOI: 10.1088/1755-1315/378/1/012117

[2] Mukhametshin V V 2020 Oil Production Facilities Management Improving Using the Analogy Method *SOCAR Proceedings* **4** 42-50 DOI: 10.5510/OGP20200400464

[3] Shoupeng Zhang and Zhengwei Fang 2020 Permeability damage micro-mechanisms and stimulation of low-permeability sandstone reservoirs: A case study from Jiyang Depression, Bohai Bay Basin, China *Petroleum Exploration and Development* **47(2)** 374-382 DOI: 10.1016/S1876-3804(20)60054-4

[4] Gabzalilova A Kh, Batalov D A, Mukhametshin V Sh, and Andreev V E 2021 Geological and technological justification of the parameters of acid-clay treatment of wells *IOP Conference Series: Materials Science and Engineering* **1064** 012058 1–9 DOI: 10.1088/1757-899X/1064/1/012058

[5] Fattakhov I G, Kuleshova L S, Kochetkov A V, Galiullina I F, Andaryanov I G, Safiullina A R, and Gizatullin I R 2020 The use of a liquid destructor with flow into the stream during the injection of the fracturing mixture *IOP Conference Series: Materials Science and Engineering* **952** 012044 DOI: 10.1088/1757-899X/952/1/012044

[6] Rogachev M K and Mukhametshin V V 2018 Control and regulation of the hydrochloric acid treatment of the bottomhole zone based on field-geological data *Journal of Mining Institute* **231** 275-280 DOI: 10.25515/PMI.2018.3.275

[7] Nitika Kalia and Vemuri Balakotaiah 2009 Effect of medium heterogeneities on reactive dissolution of carbonates *Chemical Engineering Science* **64(2)** 376-390

[8] Khuzin R R, Mukhametshin V Sh, Salikhov D A, Andreev V E, Pepelyaev D V, and Stefanovich Yu N 2021 Improving the efficiency of horizontal wells at multilayer facilities *IOP Conference Series: Materials Science and Engineering* **1064** 012066 DOI: 10.1088/1757-899X/1064/1/012066

[9] Xiangdong Qiua, Gallyam Aidagulova, Mehdi Ghommem, Eric Edelman, Dominic Brady and Mustapha Abbad 2018 Towards a better understanding of wormhole propagation in carbonate rocks: Linear vs. radial acid injection *Journal of Petroleum Science and Engineering* **171** 570-583

[10] Mukhametshin V V, and Andreev V E 2018 Increasing the efficiency of assessing the performance of techniques aimed at expanding the use of resource potential of oilfields with hard-to-recover reserves *Bulletin of the Tomsk Polytechnic University. Geo Assets Engineering* **329(8)** 30–36

[11] Ambrish Singh, Ansari K R, Quraishi M A, Hassane Lgaz and Yuanhua Lin 2018 Synthesis and investigation of pyran derivatives as acidizing corrosion inhibitors for N80 steel in hydrochloric acid: Theoretical and experimental approaches *Journal of Alloys and Compounds* **762** 347-362

[12] Nurgaliev R Z, Kozikhin R A, Fattakhov I G, and Kuleshova L S 2019 Application prospects for new technologies in geological and technological risk assessment *Mining Journal* **4(2261)** 36–40 DOI: 10.17580/gzh.2019.04.08

[13] Malyarenko A M, Bogdan V A, Blinov S A, Kotenev Yu A, Mukhametshin V Sh, and Kotenev M Yu 2021 Improving the reliability of determining physical properties of heterogeneous clay reservoir rocks using a set of techniques *Journal of Physics: Conference Series* **1753** 012074 DOI: 10.1088/1742-6596/1753/1/012074

[14] Lei Wang, Jianye Mou, Siqi Mo, Bing Zhao, Zhiyuan Liu and Xuxin Tian 2020 Modeling matrix acidizing in naturally fractured carbonate reservoirs *Journal of Petroleum Science and Engineering* **186** 106685.

Gas content and gas-dynamic characteristics of rocks of the IV-p horizon of the Starobin deposit of potash salts

D A Bobrov[1], **N A Litvinovskaya**[1], **E A Nesterov**[2]

[1] Mining Institute of the Ural Branch of the Russian Academy of Sciences, 78a, Sibirskaya St., Perm, 614007, Russia
[2] Perm National Research Polytechnic University, 29, Komsomolsky prospect, Perm, 614990, Russia

E-mail: camdobr@yandex.ru

Abstract. The article presents the results of studies of the gas content and gas-dynamic characteristics of the free gas contained in the salt rocks of the IV – p horizon of the Starobin deposit of potash salts. The research was carried out in the workings of the Petrikovsky ore enrichment plant. The IV-p horizon was discovered for the first time and data on the content of free gas in the rocks have not existed until now. Given that mining operations in other potash horizons and areas of the field are regularly accompanied by various gas-dynamic phenomena, all commissioned horizons and new areas should be examined for the content of free gas in the rocks and at the rock contacts. For the research, mine instrumental observations were used according to the methodology developed at the Mining Institute of the Ural Branch of the Russian Academy of Sciences. The obtained data serve as the basis for the development of recommendations for the safe conduct of mining operations in the rocks of the IV – p horizon.

1. Introduction

In spite of the fact that sudden emissions of rock and gas are more typical of coal mines [1-5] this dangerous natural phenomenon can also be found in the extraction of potash ore. Mining operations at all the mines of the Starobin deposit of potash salts are accompanied by gas-dynamic phenomena of various nature. Even on the II potash horizon, which was previously considered safe for rock and gas emissions, a strong gas-dynamic phenomenon occurred in 2020 in the form of roof rock collapse, accompanied by gas release. Some of the foci of gas-dynamic phenomena, such as those on the I potash horizon, are of a man-made nature due to the redistribution of free gas as a result of subsurface mining [6]. And some, as those on the III potash horizon, have a natural origin and need a special approach to predicting and preventing gas-dynamic phenomena [7, 8].

At the Petrikovsky ore enrichment plant under construction, the main horizon planned for the development is the IV-p horizon. The rocks of this horizon are similar in structure and conditions of occurrence to the rocks of the III potash horizon. In view of this similarity of the two horizons, the fact that the III potash horizon at the entire Starobin deposit was recognized as dangerous for gas-dynamic phenomena, and the fact that no studies of the gas content and gas-dynamic characteristics of the rocks of the IV-п horizon were carried out in the course of geological studies, these rocks could also, predictably, be classified as dangerous for gas-dynamic phenomena. However, any measures to prevent gas-dynamic phenomena are associated with additional economic expenses and are justified only if there is really a danger of sudden emissions of rock and gas. Since no studies of the gas content and gas-dynamic characteristics of the rocks were carried out in the course of geological studies, it became necessary to study further the content of free gas in the rocks and its characteristics. The results of the research are the basis for the development of recommendations for the safe conduct of mining operations in the pit field of the Petrikovsky ore enrichment plant.

2. Materials and methods

Studies of free gas in salt rocks were carried out according to the method developed at the Mining Institute of the Ural Branch of the Russian Academy of Sciences [9]. For the research, the workings were selected that were carried out no longer than a day in order to avoid natural degassing of rocks.

The research methodology involves two stages. At the first stage, mine instrumental measurements of the initial gas pressure in the rock mass and the pressure increment with a fully sealed hole (well) are carried out. For this purpose, boreholes and wells are drilled, in which measurements are carried out. After this stage, the gas content of the rock mass (1) and the initial rate of free gas release from the walls (2) of the research hole (well) are calculated [10].

The function of the dependence of the gas content of rocks on free gases on the value of the initial gas pressure has the following form:

$$X = (4.4 + 12.9Pr) /71.94Pr0.16. \quad (1)$$

The initial rate of gas release of free gas from the walls of the well is calculated based on the increment of pressure for 30 seconds when the well is completely sealed. The formula for the calculation has the following form:

$$Ur=0.027\Delta P0.638 \quad (2)$$

Boreholes and wells are drilled into the walls of mine workings, as well as into the soil and roof. Thus, a comprehensive study of the rocks is carried out. In addition, point-to-point drilling and sealing of boreholes and wells allow us to find out the gas content and gas-dynamic characteristics of each layer separately. In the future, this will allow one to identify the most gas-bearing layers of rocks and give recommendations for their degassing.

3. Research results

Currently, more than 200 studies have been conducted in various layers of the IV-p potash horizon in different workings of the mine's near-barrel yard, workings of the main directions, as well as preparatory workings of lavs 1-1, 2-1 and 3-1. The results obtained allow us to draw the first conclusions about the gas content and gas-dynamic characteristics. The generalized, average indicators for each layer of rocks of the IV-p potash horizon are presented in Figure 1.

As can be seen from the Figure 1, the most gas-bearing is the middle halite layer with the carnallite layer contained in it. Here, in the carnallite layer, the highest values of the initial gas release rate are also found. In addition, the halite layer 1-2 is distinguished in the section. Here, too, the gas content and the initial rate of gas release are greater than in other rocks. In all silvinite layers, the gas-carrying capacity and gas-dynamic characteristics are not significant, within the margin of error.

Table 1 shows the quantitative indicators of gas content and gas-dynamic characteristics for each layer, obtained in the course of all conducted studies.

The analysis of the data in the table shows that, in general, both the gas content and the gas-dynamic characteristics (the initial rate of gas release and the pressure of free gas in the rock mass) for the planned formations fluctuate within insignificant limits. This points out to background indicators, and does not imply the presence of foci of gas-dynamic phenomena in the rocks.

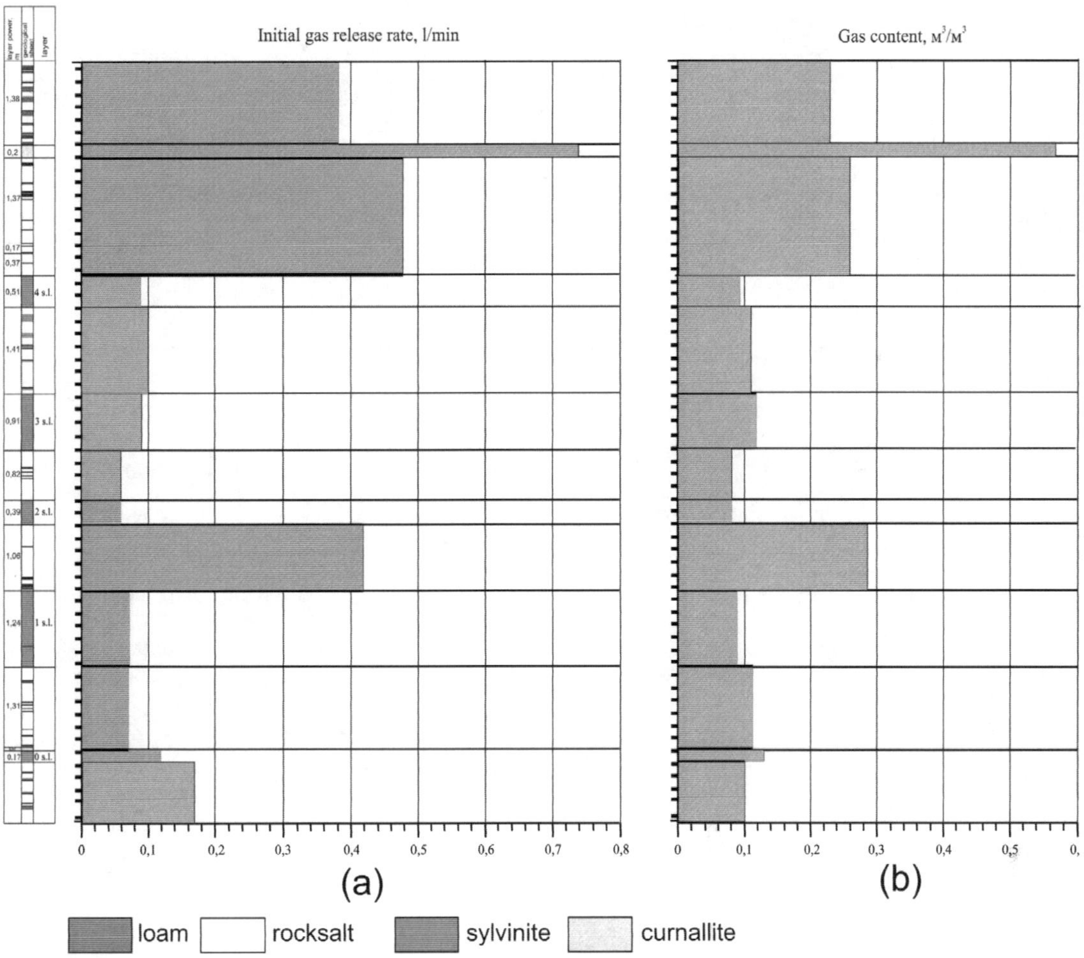

Figure 1. The initial rate of gas release (a) and gas content (b) of rocks of the IV-p potash horizon.

All the maximum values in the middle halite layer and halite layer 1-2 were recorded in the zone of substitution of silvinite layer 3 by halite rocks. This is a local replacement zone, intersected by the mounting and adjacent, auxiliary mounting drift of lava 2-1 (figure 2). The studies conducted within the replacement zone showed that the rocks do not contain a large amount of free gas, the indicators of gas content and gas-dynamic characteristics here are the same as in the rest of the sites. Thus, it is the substitution zone that is characterized by an increased free gas content. Apparently, in this zone, the rocks were recrystallized with the dissolution of sylvinite and the deposition of halite, which was accompanied by the release of intracrystalline gas and the accumulation of existing free gas in microcracks and various cavities. However, the values of both the gas content and the gas-dynamic characteristics at this point, although large, are still not sufficient for the development of gas-dynamic phenomena. This only indicates the potential danger of such replacement zones and requires further study.

Table 1. Gas content and gas-dynamic characteristics of rocks

Layer	Gas content, m^3/m^3			Initial gas release rate, l/min			Gas pressure in the array, MPa		
	Minimum value	Maximum value	Average value	Minimum value	Maximum value	Average value	Minimum value	Maximum value	Average value
Middle halite formation	0.02	1.16	0.25	0.01	2.99	0.43	0.190	0.215	0.193
Carnallite layer	0.11	1.86	0.57	0.05	4.74	0.74	0.190	0.22	0.196
4 sylvinite layer	0.02	0.23	0.1	0.01	0.13	0.09	0.189	0.191	0.190
3-4 halite layer	0.02	0.44	0.11	0.03	0.35	0.1	0.189	0.194	0.190
3 sylvinite layer	0.02	0.35	0.12	0.01	0.13	0.09	0.190	0.193	0.191
2-3 halite layer	0.02	0.12	0.08	0.01	0.13	0.06	0.190	0.190	0.190
2 sylvinite layer	0.02	0.12	0.08	0.01	0.13	0.06	0.190	0.190	0.190
1-2 halite layer	0.02	1.46	0.28	0.03	2.35	0.42	0.190	0.211	0.193
1 sylvinite layer	0.02	0.15	0.09	0.03	0.7	0.07	0.189	0.190	0.190
0-1 halite layer	0.02	0.14	0.11	0.06	0.07	0.07	0.190	0.190	0.190
0 sylvinite layer	0.02	0.19	0.13	0.03	0.21	0.11	0.190	0.190	0.190
Lower halite formation	0.02	0.17	0.1	0.11	0.22	0.17	0.190	0.190	0.190

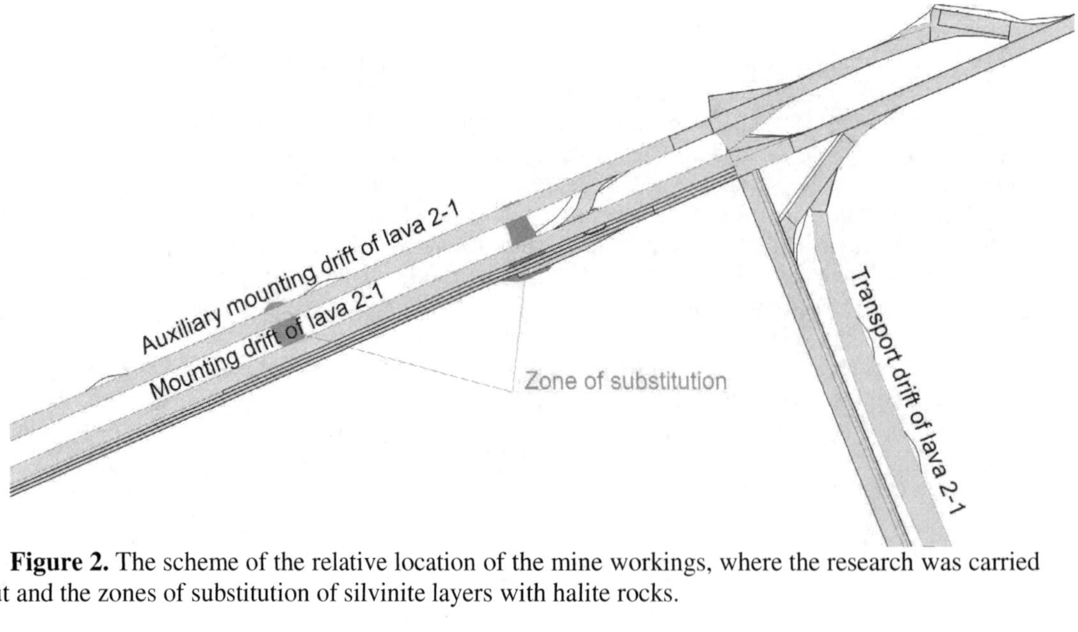

Figure 2. The scheme of the relative location of the mine workings, where the research was carried out and the zones of substitution of silvinite layers with halite rocks.

4. Conclusion

The results of experimental studies have shown that the rocks of the exposed IV-p potash horizon have low gas content and gas-dynamic characteristics and do not contain a large amount of free gas. Smooth and sustained occurrence without folding and any geological disturbances does not imply the occurrence of foci of gas-dynamic phenomena. However, the substitution zones found in the rocks can carry a potential danger, since it is in them that an increased value of free gas is observed.

In general, it can be concluded that the workings on the IV-п horizon can be carried out in the usual mode and do not require special measures for rock degassing. At the same time, when conducting mining operations through the zones of substitution of silvinite rocks with halite, it is necessary to monitor the gas content and gas-dynamic characteristics of the rocks in order to take measures for degassing the formation in advance in case of danger.

5. Acknowledgments

The research was supported by RFBR grant No. 20-45-596017.

References

[1] Zhou Peiling, Zhang Yinghua, Huang Zhi'an, Gao Yukun, Wang Hui, Luo Qiang 2017 Coal and gas outburst prevention using new high water content cement slurry for injection into the coal seam *International Journal of Mining Science and Technology* **27(4)** 669-673

[2] Zakharov V N, Malinnikova O N 2018 The emission hazard of the bottom-hole zone of gas-saturated coal seams during mining operations *Industrial safety and labor protection* **12** 81-85

[3] Odintsev V N, Makarov V V 2020 Gas-dynamic destruction of a coal seam in the process of its zonal disintegration *Physical and technical problems of mineral resources development* **6** 61-77

[4] Liang Chen, Enyuan Wang, Jianchun Ou, Jiangwei Fu 2017 Coal and gas outburst hazards and factors of the No. B-1 Coalbed, Henan, China *Geosciences Journal* **22** 171-182.

[5] Shadrin A V 2016 Geophysical criteria for the pre-release development of cracks in a coal seam *Physical and technical problems of mineral resources development* **4** 48-62.

[6] Andreiko S S, Ivanov O V, Litvinovskaya N A 2015 *Forecasting and prevention of gas-dynamic phenomena from the soil during the passage of preparatory workings in the sub-worked array of salt rocks* (Perm: Publishing House of Perm. Nat. Research. Polytech. University)

[7] Andreiko S S 2007*Gas-dynamic phenomena in potash mines: methods of forecasting and methods of prevention* (Perm: Publishing House of Perm State Technical University)

[8] Tarakanov V A, Golovaty I I, Beresnev S P, Andreiko S S, Ivanov O V 2010 Studies of rocks gas content of the formation of the Third potash horizon of the Starobin deposit of potash salts *Mining Journal* **8** 25-27.

[9] Zemskov A N, Kondrashev P I, Travnikov L G 2008 *Natural gases of potash deposits and measures to combat them* (Perm: Merchant Tarasov's printing house)

[10] Leibenzone L S 1947 *Movement of natural liquids and gases in a porous medium* (Moscow: State Publishing House of Technical and Theoretical Literature)

Forecasting the hydraulic fracturing efficiency when selecting wells for an event

A I Valeev, V Sh Mukhametshin and V V Anisimov

Ufa State Petroleum Technological University, Branch of the University in the City of Oktyabrsky, 54a, Devonskaya St., Oktyabrsky, Republic of Bashkortostan, 452607, Russian Federation

E-mail: vsh@of.ugntu.ru

Abstract. To this date, most of the oil and gas reserves are hard-to-recover, that is, they are located in deposits confined to low-permeability reservoirs. Hydraulic fracturing treatment (HFT) is one of the methods of production stimulation that allows making the oil recovery profitable from such reservoirs. The given article discusses regression analysis to predict HFT efficiency and select candidate wells.

1. Introduction

Improving the efficiency of managing the development of oil deposits is based on the study and formalization of various processes aimed at extracting liquid hydrocarbons [1–10]. One of the given processes is the intensification of oil production with the application of impact on the bottomhole zone using hydraulic fracturing treatment.

Predicting HFT efficiency, assessing the factors to one degree or another influencing the event, as well as a detailed and substantiated selection of design wells for impacts are fundamental processes that need to be carried out and coordinated at the early stages of field development. An early assessment of possible risks before the event and the identification of various factors influence on the potential for the projected event efficiency will minimize the probability of an unfavorable outcome.

According to the published data, HFT efficiency varies within a fairly wide range in many fields. Increasing HFT efficiency is associated with both deeper laboratory studies and the experience generalization of applying various technologies for such treatments in diverse field conditions. Identification of parameters that have a prevailing effect on efficiency, and the subsequent formalization of the process based on statistical modeling, allow implementing a scientifically grounded choice of wells and selecting the optimal stimulation technology in order to increase HFT efficiency [11–16].

2. Materials and methods

Geological factors that need to be taken into account (total, effective and perforated thickness of the formation, permeability, compartmentalization, thicknesses of interlayers entering the highly connected zones of the formation, etc.) affect the efficiency of the hydraulic fracturing method. The given factors are considered for each specific well, and in relation to the adjacent area.

Geological factors are regarded both for an individual well and for the space around the well.

The procedure forecast is compiled in accordance with the types of geological section available in the well. Many of the treated wells are located in the discontinuous zone of AB formation. Two lithotypes can occur in such a part, considering the parameters of the geological model: a thin-layered reservoir and semi-monoliths.

According to the obtained results, the hydraulic fracturing efficiency directly correlates with the identified geological and physical factors. It has been determined that in most cases the efficiency of combining several wells when opening a single interlayer is directly designated by the thickness of the latter.

The analysis of performed HFT operations was carried out on the operating well stock of the investigated field in order to study the influence of geological and technological parameters on the success of HFT in the conditions of the investigated field.

The regression model was built in order to identify the mathematical dependence of technological and geological parameters. Twenty five wells, belonging to the AB_{1-3} object, were identified on the basis of the production data on hydraulic fracturing treatment of the reservoir in the considered field. HFT operations were performed using a linear gel (Zeta-Gel).

The component structure is a low-viscosity, two-component, and SAS-based fracturing gel system. The gel includes polydiallyldimethylammonium chloride, less sodium lauryl sulfonate, and lauryl alcohol. A distinctive feature of the gel is friction reduction. Zeta-Gel application is required to regulate the goals of selective stimulation of production in the oil-saturated area of the reservoir with the proximity of water-saturated and washed horizons.

Using the company data that uses this Zeta-Gel polymer-free gel system, it can be said to aid rapid gelation and is successful up to 80°C. Compared to guar-type fluids, which tend to form an insoluble residue after final degradation, the Zeta-Gel system does not deposit any sediment. The disadvantage of the Zeta-Gel component is a very large foaming of the liquid during mixing at high speeds, low sanding capacity.

The Zeta-Gel system is prepared from fresh water, to which cationic SAS WGA305, clay and shale stabilizer NLC 100, biocide BioClear 1000, and demulsifier NE 201 are added. The anionic component WGA300 is added at low stirring speed (to avoid foaming). The solution gradually becomes cloudy and a gel with significant elasticity and viscosity is formed after a few minutes.

The influence of various factors on the event success was defined by means of regression analysis. Well data are shown in Table 1.

The main geological and production indicators were considered by wells.

If a close relationship is found between the given parameters, one of them is excluded. The final correlation matrix is presented in Table 2, where h is an effective formation thickness, k is the permeability, M is the proppant weight, f is the water-cut up to HFT, q is the liquid flow rate up to HFT, and N is the multiplicity of productivity increase.

The correlation matrix demonstrates that the effective formation thickness and proppant mass, as well as the formation permeability and water-cut of wells' production, strongly affect the productivity growth.

Further, the regression model is built by the method of multiple linear regression. The general view of the regression equation and the equation obtained after calculations in Microsoft Excel are presented below.

$$Y = \beta_0 + \beta_1 \cdot X_1 + \beta_2 \cdot X_2 + ... + \beta_n \cdot X_n, \quad (1)$$

$$N^` = 2{,}021 + 0{,}58 \cdot h + 0{,}0048 \cdot k - 0{,}222 \cdot M + 0{,}03 \cdot f - 0{,}043 \cdot q, \quad (2)$$

In the resulting equation, N' is the estimated multiplicity of the increase in well productivity after hydraulic fracturing treatment. The N' values for 25 wells are designated by substituting parameters into the obtained equation. Comparative graph of calculated and actual values is built for the given group of wells (Figure 1).

Table 1. Input data for regression analysis

Well number	Effective thickness, m	Permeability, 10^{-3} μm²	Proppant weight, t	Water-cut to HFT, %	Liquid flow rate to HFT, t/day	Actual rate of increase in productivity, un.
	Factors					
1888	17.8	71.8	12	90.6	14.1	9.5
1895	7.6	120.4	6	45.8	10.8	3.9
1954	14.8	59.7	16.4	8.7	1.7	10.0
1956	6.5	175.5	18.7	7.0	3.2	3.0
2031	4.2	21.6	12.3	39.6	7.2	4.6
2114	6.4	118.5	15	10.7	3.7	7.2
2155	16.9	322.3	12	54.9	15.2	8.1
2156	16.2	4.1	11	10.5	3.5	7.8
2161	19.6	54.4	8.1	24.2	2.9	7.4
2265	31.6	52.1	3	73.6	1.0	24.9
2459	12.1	48.7	7.8	21.2	5.3	5.2
2518	28.4	296.3	5	11.5	6.8	16.2
2656	16.2	69.2	32.8	70.8	3.5	5.4
2721	21.6	175.3	8	9.8	4.3	17.8
2787	13.8	223.1	13	9.7	5.7	10.5
3033	13.6	77.3	10	54.9	1.8	5.0
3167	12.8	68.7	5	60.0	1.7	14.7
3169	16.0	237.8	5	72.6	5.4	16.5
3208	15.0	57.6	5	80.3	9.1	13.2
3553	13.7	2.1	5	6.7	11.5	6.8
3577	6.6	6.9	8.7	89.8	10.9	5.2
6076	11.7	19.2	6.3	17.6	12.4	8.4
8709	23.1	24.8	10.1	9.0	14.8	14.0
2938U	20.0	17.5	12.2	97.1	47.6	14.0
3551U	9.2	7.5	6.4	11.8	8.5	4.1

Table 2. Correlation matrix

Parameter	h	k	M	f	q	N
h	1	-	-	-	-	-
k	0.210576	1	-	-	-	-
M	-0.21066	0.025855	1	-	-	-
f	0.10344	-0.09805	0.008537	1	-	-
q	0.096658	-0.1427	-0.03093	0.386109	1	-
N	0.799729	0.219994	-0.39952	0.213239	0.061171	1

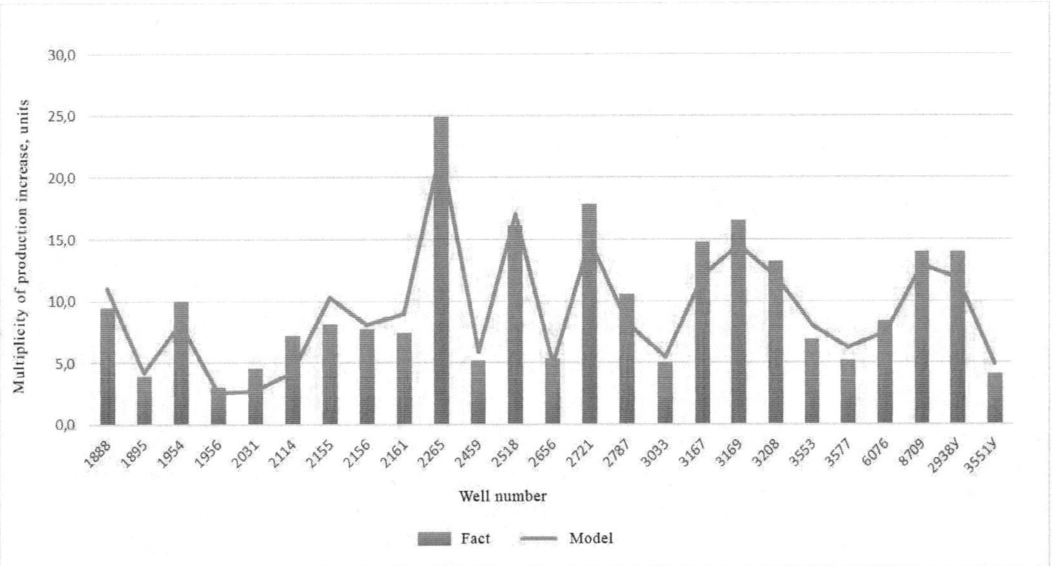

Figure 1. Comparison of the calculated and actual rate of production increase

The obtained regression model can be applied to predict the success of the hydraulic fracturing operation.

The determination coefficient according to calculations in Microsoft Excel is 0.823. This indicates the model applicability for forecasting. It takes into consideration the difference in wells' geological parameters (effective reservoir thickness, permeability coefficient), production data for wells (liquid flow rate, water-cut), and the amount of proppant used to fix HFT fractures is included in the model.

The selection of candidate wells for hydraulic fracturing treatment was carried out among the operating production wells of the third section of geological analysis of AB_{1-3} object of the considered field.

Well numbers of 2387, 2329, 3206, and 2837 have been chosen. According to the geological and production data, the design technological effect of hydraulic fracturing treatment was designated for the given four wells for each of them based on the constructed regression model. The calculation outcomes are presented in Table 3.

Table 3. Technological success forecast for HFT

Well number	2329	2387	2837	3206
Effective thickness, m	11.3	10.3	9.0	7.8
Permeability, 10^{-3} μm ^{2}	5.7	27.6	19.8	43.9
Proppant weight, t	6	4	10	5
Water-cut to HFT, %	58.3	67.2	88.1	79.5
Liquid flow rate to HFT, t/day	5.0	3.4	19.0	6.9
Estimated rate of increase in productivity, un.	5.6	6.5	5.1	6.2

According to the regression model, the predicted increase in the productivity factors of the wells under consideration is essential. It can be concluded that hydraulic fracturing treatment in wells No. 2329, 2387, 2837, 3206 is expedient and technologically successful.

A typical computation of HFT parameters (determination of the length and opening of fractures, liquid flow rate after the operation), and a regression model that allows considering the geological and production characteristics of each well, can be applied in the further selection of wells for hydraulic fracturing treatment in the field.

3. Results and Discussion

Forecasting HFT efficiency, and a detailed and substantiated selection of design wells for the event is a fundamental process that must be carried out and coordinated at the early stages of design. An early assessment of possible risks before the event and the identification of the various factors influence on the success potential of the projected event will minimize the probability of an unfavorable outcome.

A fairly high technological success of the events is meant by comparing the outcomes of calculating the efficiency of the hydraulic fracturing operation and the predicted data obtained by the regression model for the four proposed wells. Hydraulic fracturing in wells No. 2329, 2387, 2837, and 3206 will further contribute to an increase in oil production and the most complete working-out of reserves.

4. Conclusion

The regression model has been built on the basis of statistical data on operations in the wells of AB object to predict the stimulation method success. Linear gel technology has been considered. The post-HFT productivity gain has been estimated for four candidate wells applying the generated model.

The regression model allows taking into consideration the cumulative effect of both geological and technological parameters on the final HFT result.

It contributes to the rational selection of candidate wells from the total stock of operating wells. A high effect is feasible to be achieved while reducing the cost of financial resources, evaluating well current performance and adjusting technological characteristics.

References

[1] Gabzalilova A Kh, Batalov D A, Mukhametshin V Sh, and Andreev V E 2021 Geological and technological justification of the parameters of acid-clay treatment of wells. *IOP Conference Series: Materials Science and Engineering* **1064** 012058 DOI: 10.1088/1757-899X/1064/1/012058

[2] Rodrigues V F, and Medeiros A C R 2011 Limits of Fracture Conductivity Correlation Improvement Through Acid Fracture Surface Characterization. *OTC Brasil, Rio de Janeiro, Brazil, October 2011* DOI: 10.4043/22337-MS

[3] Mukhametshin V V, and Kuleshova L S 2021 Using the method of canonical discriminant functions for a qualitative assessment of the response degree of producing wells to water injection during the development of carbonate deposits. *IOP Conference Series: Materials Science and Engineering* **1064** 012069 DOI: 10.1088/1757-899X/1064/1/012069

[4] Economides M J, and Martin A N 2010 How to decide between horizontal transverse, horizontal longitudinal and vertical fractured completion *SPE Annual Technical Conference and Exhibition, held in Florence, Italy* DOI: 10.2118/134424-MS.

[5] Khuzin R R, Mukhametshin V Sh, Salikhov D A, Andreev V E, Pepelyaev D V, Stefanovich Yu N 2021 Improving the efficiency of horizontal wells at multilayer facilities. *IOP Conference Series: Materials Science and Engineering* **1064** 012066 DOI: 10.1088/1757-899X/1064/1/012066

[6] Litvin V T, Strizhnev K V, Shevchuk T N, and Roshchin P V 2018 The acid treatment of the well on Bazhenov oil formation after hydraulic fracturing operation *Oil Industry* **4** 70-73 DOI: 10.24887/0028-2448-2018-4-70-73.

[7] Batalov D A, Andreev V E, Mukhametshin V V, and Kuleshova L S 2021 Development regulation of oil and gas reservoirs based on effective geological and geophysical information *IOP Conference Series: Materials Science and Engineering* **1064** 012055 DOI: 10.1088/1757-899X/1064/1/012055

[8] Dikshit A, Woiceshyn G, Agnihotri V, Chochua G, and Md. Noor M N B 2021 Quantifying Erosion of Downhole Solids Control Equipment during Openhole, Multistage Fracturing *SPE Drill & Compl.* **36 (02)** 413–426. DOI: 10.2118/203096-PA

[9] Malyarenko A M, Bogdan V A, Blinov S A, Kotenev Yu A, Mukhametshin V Sh, and Kotenev M Yu 2021 Improving the reliability of determining physical properties of heterogeneous clay reservoir rocks using a set of techniques. *Journal of Physics: Conference Series* **1753** 012074 DOI: 10.1088/1742-6596/1753/1/012074

[10] Ghaleb Al Habsi, Soheil Ghanbarzadeh, Siyavash Motealleh, and Badar Al Busafi. 2019. Hydraulic Fracturing: Best Remedy for Condensate Banking Effects in Tight Gas-Condensate Reservoirs. *Abu Dhabi International Petroleum Exhibition & Conference, Abu Dhabi, UAE, November 2019.* DOI: 10.2118/197422-MS

[11] Mukhametshin V G, Dubinskiy G S, Andreev V E, Mukhametshin V V, and Kuleshova L S 2021 Geological, technological and technical justification for choosing a design solution for drilling wells under different geological conditions. *IOP Conference Series: Materials Science and Engineering* **1064** 012061 DOI: 10.1088/1757-899X/1064/1/012061

[12] Kaiyi Zhang, Bahareh Nojabaei, Kaveh Ahmadi, and Russell T. Johns. 2020. Effect of Gas/Oil Capillary Pressure on Minimum Miscibility Pressure for Tight Reservoirs *SPE J.* **25 (02)** 820–831 DOI: 10.2118/199354-PA

[13] Zeigman Yu V, Mukhametshin V Sh, Khafizov A R, Kharina S B, Abutalipova E M, Avrenyuk A N 2017 Peculiarities of selecting well-killng fluids composition for difficult conditions. *Oil Industry.* **1** 66–69.

[14] Piyush Pankaj, Alhad Phatak, and Sandeep Verma 2018 Evaluating Natural Gas-Based Foamed Fracturing Fluid Application in Unconventional Reservoirs *SPE Asia Pacific Oil and Gas Conference and Exhibition, Brisbane, Australia, October 2018* DOI: 10.2118/192042-MS

[15] Yaskin S A, Mukhametshin V V, Andreev V E, and Dubinskiy G S 2018 Forecasting the parameters of non-stationary water flooding of oil deposits. *IOP Conference Series: Earth and Environmental Science* **194** 062037 DOI: 10.1088/1755-1315/194/6/062037

[16] Jin F, Shunyuan Z, Bingshan L, Chen C, and Kedi M 2017 Green Fracturing Technology of Shale Gas: LPG Waterless Fracturing Technology and its Feasibility in China. *SPE Latin America and Caribbean Petroleum Engineering Conference, Buenos Aires, Argentina, May 2017* DOI: 10.2118/185500-MS

Justification for proppant concentration choice during hydraulic fracturing

A V Kochetkov[1], I G Fattakhov[2], L S Kuleshova[2], R R Stepanova[2] and A R Safiullina[2]

[1] PJSC Tatneft named after V.D. Shashin, 75, Lenin St., Almetyevsk, Republic of Tatarstan, 423450, Russian Federation
[2] Ufa State Petroleum Technological University, Branch of the University in the City of Oktyabrsky, 54a, Devonskaya St., Oktyabrsky, Republic of Bashkortostan, 452607, Russian Federation

E-mail: i-fattakhov@rambler.ru, info@of.ugntu.ru

Abstract. The article presents a mathematical model of linear and nonlinear increase in proppant concentration during hydraulic fracturing. For the first time, an analytical solution using direct calculation for determining the parameters of the main hydraulic fracturing stages including the transition concentrations for given masses of proppant of various types is presented. It is shown that the use of a mathematical model in the development of a processing plan provides the correct distribution of the proppant mass by fractions, which facilitates the introduction of information and analytical systems and transferring data directly from the operating program to the database. The spreadsheet forms used in the production have been elaborated, which enables to apply the mathematical model of the operating program at each hydraulic fracturing without additional labor costs. The obtained mathematical model can be applied to improve the software used in the design, modeling and engineering support of hydraulic fracturing.

1. Introduction

The main purpose of using methods for intensifying production and increasing the degree of hydrocarbon reserves development is to increase the efficiency of oil deposits development [1–8]. However, it is important to correctly select the technological parameters of impacts on the bottomhole zone and formation [9, 10]. Against the background of a center-of-gravity shift in oil production towards the formations with hard-to-recover reserves, solving the issues of substantiating the adoption of control decisions becomes especially relevant due to a significant increase in the cost of errors [11, 12]. Hydraulic fracturing plays a special role in the development of deposits with hard-to-recover reserves, which quite often ensures cost-effective deposits development.

A significant part of stimulation technologies by applying hydraulic fracturing involves the use of proppant. To place the granules in the fracture, a uniform supply of proppant of a given concentration to the fracture fluid is ensured.

The target proppant concentration in a hydraulic fracturing project is always higher than the concentration used in the initial stages. This is technologically necessary because an instant increase in concentration from 0 to the maximum value will lead to complications during hydraulic fracturing. At the initial stages, the proppant plays the role of an abrasive material, and the perforation interval and the bottom-hole part of the fracture are prepared for the injection of proppant with the target concentration. To enhance the abrasive effect, reduce the risks of technological complications and level the filtration profile of the fracturing fluid during the injection process, at the initial stages, proppant is used with a smaller average particle diameter, and at the same time with a larger variation in diameter. In practice, this is achieved by using proppant of two fractions: an insignificant amount of the first one and then the main mass of a larger fraction [13].

2. Materials and methods

Supervising during the hydraulic fracturing includes a reconciliation of the actually injected volumes of fluids and the corresponding values of the proppant concentration. The transparency of the process control necessitates the separation of the total mass of the proppant at a stage, as a rule, with a uniform increment of the proppant concentration. The increment step remains constant at the stages with the main proppant fraction. However, during the transition from the abrasive proppant to the main proppant subject to the planned linear increase in concentration, a problem arises with three interdependent variables: the proppant mass of the abrasive fraction, the rate of concentration increase during injection and the transitional proppant concentration being the maximum for abrasive and minimal for the main fraction. Of the three indicated variables, the mass is set by the design value directly, and the rate of concentration increase is determined by the total proppant mass and the final concentration, which is also set by the project.

If a strict uniform step of concentration increment is applied from stage to stage during the injection of two proppant fractions, this will lead to one of two errors depending on the initial data. If the correct mass is preserved in stages, the linear nature of the concentration increase will be violated, Figure 1, a. In the opposite case, while maintaining the linear nature of the concentration increase, an incorrect distribution of the proppant mass by stages will be obtained, Figure 1, b.

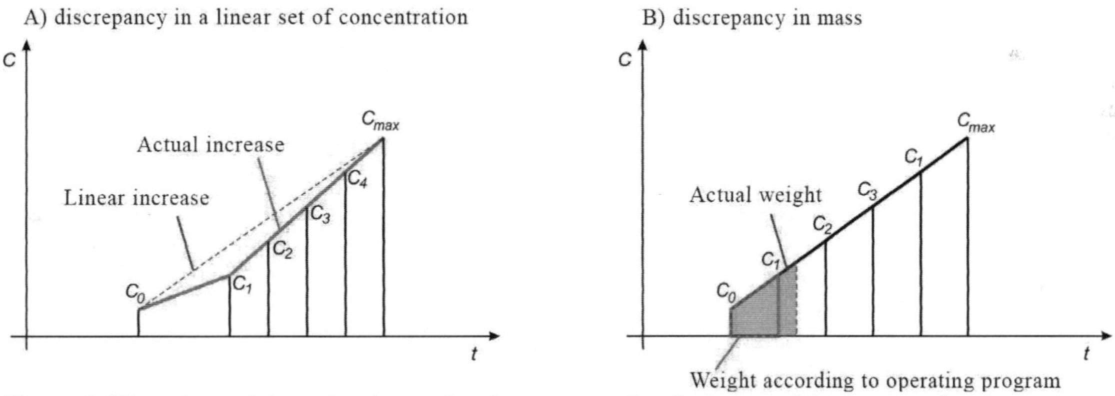

Figure 1. Distortions arising when increasing the concentration during transition between fractions

Thus, provided that the mass by fractions is correctly regarded, a mathematically accurate linear set of concentration is possible only if the transition concentration is calculated for the given conditions. This will accordingly violate the multiplicity of increasing the concentration by stages but will ensure the correct display of the mass of each fraction of the proppant while maintaining the overall linear set of concentration. The deviation shown in Figure 1, b benefits a technically correct hydraulic fracturing process. However, this approach will lead to errors in case of modeling and automated metering systems. The proposed approach enables to correctly display the moment of transition of the proppant type in the operating program and, thus, both the mass and the concentration set will correspond to the given technology.

3. Results and Discussion

To solve the problem, a mathematical model of the change in the proppant concentration during hydraulic fracturing was built. When constructing a mathematical model, it was revealed that to ensure the linear nature of the proppant concentration increase, the following condition must be met for stages with an increase in proppant concentration:

$$\frac{v_s}{C_2 - C_1} = const = K, \quad (1)$$

where C_1, C_2 are the initial and final concentration of the considered section of the processing plan, v_s is volume of mixture of fracturing fluid with proppant, *K* is concentration line slope coefficient.

To maintain a linear set during the transition between proppant fractions in the operating program based on a mathematical model, it is necessary to calculate the C_a transition concentration, at which the stage with fraction No.1 ends and fraction No.2 begins for a given proppant mass by fractions and maximum concentration C_{max}.

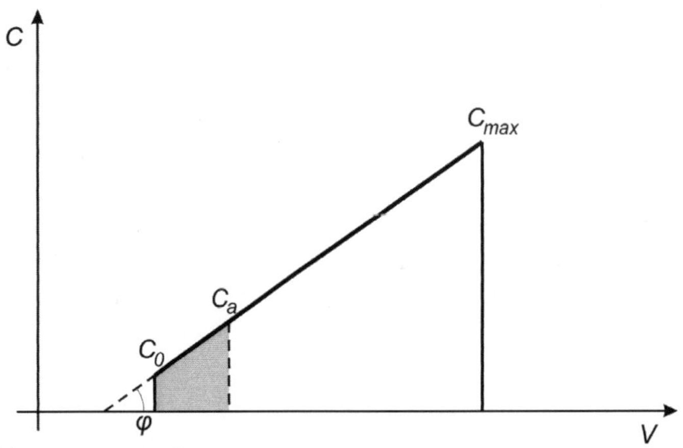

Figure 2. Linear concentration set

Figure 2 shows a linear concentration set. If we represent the proppant concentration *c* as a function of the fracturing fluid volume *V*, then the condition for a linear increase in concentration will remain at the same slope angle φ for straight lines C_0C_a and C_aC_{max} as well as in standard graphs with a time axis *t* at a constant injection rate *r*. In this case, the proppant mass can be represented as the trapezoid area bounded by the concentration line, a projection on the V axis and perpendiculars of the initial and final concentration to the projection.

$$m = S_{trapezium} = \frac{C_1 + C_2}{2} V, \quad (2)$$

where C_1, C_2 is initial and final concentration in the area under consideration, volume *V* of clean liquid.

Thus, knowing the proppant mass by fractions, the planned minimum and maximum concentration, as well as the absolute density of the proppant, it is possible to calculate the required volume of fracturing fluid for the condition of a linear increase in concentration:

$$V = \frac{2M}{C_0 + C_{max}}, \quad (3)$$

where M is total mass of proppant of all fractions, C_0, C_{max} are initial and maximum concentrations.

To calculate the volume of the mixture of the fracturing fluid with the proppant, it is necessary to add the volume occupied by the proppant particles regarding the location of the fracturing fluid in the space between the particles to the volume of the fracturing fluid (3):

$$V_s = \frac{2M}{C_0+C_{max}} + \frac{M}{\rho}, \quad (4)$$

where ρ is absolute concentration of proppant.

Considering the constancy of the coefficient K in the formula (1) and the application of formula (4) for the proppant fraction No.1, a system of equations was obtained:

$$\begin{cases} \frac{V_s}{C_{max}-C_0} = \frac{V_{sa}}{C_a-C_0}; \\ V_{sa} = \frac{2m_a}{C_0+C_a} + \frac{m_a}{\rho}. \end{cases} \quad (5)$$

There are two unknowns in the equation: C_a and V_{sa}. First of all, for practical purposes, the transition concentration C_a is of interest. Solving the system of equations, we come to the analytical solution:

$$C_a = \frac{\sqrt{(C_0m_a-C_{max}m_a)^2-4\rho V_s(C_0^2m_a-C_0^2\rho V_s-C_0C_{max}m_a+2C_0m_a\rho-2C_{max}m_a\rho)}-C_0m_a+C_{max}m_a}{2\rho V_s}. \quad (6)$$

Thus, the minimum number of stages required to describe the injection of two types of proppant is reduced to two stages with a transitional C_a stage. Table 1 shows the principle of forming a work program according to the presented mathematical model [14].

Table 1. Principle of forming operating program

Proppant type	Mass	Discharge	Liquid volume	Mixture volume	Time	Concentration from	Concentration to
No.1	m_a	r	$V = \frac{2m_a}{C_0 + C_a}$	$V_s = V + \frac{m_a}{\rho}$	$t = \frac{V_s}{r}$	C_0	C_a
No.2	m_b	r	$V = \frac{2m_b}{C_a + C_{max}}$	$V_s = V + \frac{m_b}{\rho}$	$t = \frac{V_s}{r}$	C_a	C_{max}

If necessary, further division into stages is possible according to the same principle. In this case, there will be no need to calculate the transient concentrations as they can be set equal for the main proppant fraction.

If the above-mentioned distortions occur, the proppant amounts by fractions are distorted. As well, to control the consumption of materials and store data in the database, engineers need to manually adjust the indicators. In contrast to the standard approach, the work program obtained on the basis of a mathematical model can be directly loaded into information systems and databases since it will correspond to the actual download at the process termination.

Some of the fracturing technologies imply a non-linear increase in the proppant concentration during the injection process. This reduces the required volume of the gel while maintaining the proppant mass and concentration. To ensure the condition of nonlinearity and smooth transition between proppant stages, changes are made to the treatment plan.

The coefficient K in formula (1) for a non-linear increase in concentration is not constant for the entire injection but varies from stage to stage. In order to set the degree of curvature, the coefficient ω taking values in the range from 1 to 1.5 is introduced. When ω=1, a linear increase in concentration is obtained, and a larger coefficient corresponds to a greater degree of nonlinearity in the curve of proppant concentration growth.

If we express the volume of the stage in terms of the coefficient K in formula (1), then with a known K and a given initial and final concentration, the volume of the mixture can be determined as follows:

$$V_s = (C_{max} - C_{min}) \cdot K \quad (7)$$

For the nonlinear case, $K=K_n$ for each stage starting from stage 2, n is determined using an arithmetic progression. Preliminarily, we introduce an intermediate factor K' also determined by the progression.

$$K'_1 = 1;\ K'_n = K'_{n-1} + \frac{\omega-1}{N-1}, \tag{8}$$

where N is number of stages. Then, we introduce the desired K_n

$$K_1 = K;\ K_n = K_{n-1} \cdot K'_n \tag{9}$$

Accordingly, the volumes of the stages can be determined by the progression.

$$V_{sn} = \left(C_n^{max} - C_n^{min}\right) \cdot K_n. \tag{10}$$

Figure 3 shows a nonlinear increase in concentration at ω=1,37.

The formula for calculating the transient concentration (6) can be introduced into the forms of spreadsheets used by hydraulic fracturing engineers, which will enable to apply the presented approach to all hydraulic fracturing processes without the need to perform calculations manually. When developing and implementing new specialized information and analytical systems, the presented mathematical model can be built into the appropriate modules to simplify the system management. The introduction of nonlinearity through the coefficient ω not only simplifies the preparation of the operating program for the appropriate technologies but also indicates the aggressiveness of the injection in a digital equivalent for being accounted in the database and analyzed in the corporate integrated AS.

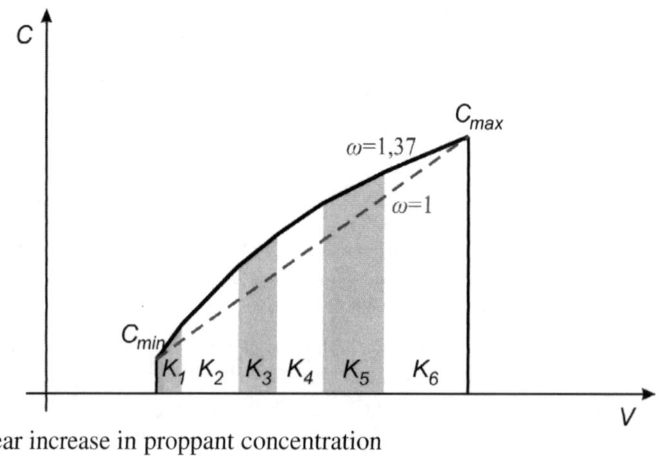

Figure 3. Non-linear increase in proppant concentration

4. Conclusion

1. A mathematical model of linear and nonlinear increase in proppant concentration has been obtained during hydraulic fracturing. For the first time, an analytical solution using direct calculation for determining the parameters of the main hydraulic fracturing stages including the transition concentrations for given masses of proppant of various types has been presented.

2. The use of a mathematical model in the development of a processing plan enables to maintain the correct distribution of the proppant mass by fractions, which facilitates the introduction of information and analytical systems transferring data directly from the operating program to the database.

3. Refinement of the spreadsheet forms used in the production will enable to apply the mathematical model of developing the operating program at each hydraulic fracturing process without additional labor costs.

4. The obtained mathematical model can be used to improve the software used in the design, modeling and engineering support of hydraulic fracturing processes.

References

[1] Seeyub Yang, Prashanth Siddhamshetty, Joseph Sang-Il Kwon 2017 Optimal pumping schedule design to achieve a uniform proppant concentration level in hydraulic fracturing *Computers & Chemical Engineering* **101** 138-147

[2] Mukhametshin V G, Dubinskiy G S, Andreev V E, Mukhametshin V V, and Kuleshova L S 2021 Geological, technological and technical justification for choosing a design solution for drilling wells under different geological conditions *IOP Conference Series: Materials Science and Engineering* **1064** 012061 DOI: 10.1088/1757-899X/1064/1/012061

[3] Gabzalilova A Kh, Batalov D A, Mukhametshin V Sh, and Andreev V E 2021 Geological and technological justification of the parameters of acid-clay treatment of wells *IOP Conference Series: Materials Science and Engineering* **1064** 012058 DOI: 10.1088/1757-899X/1064/1/012058

[4] Prashanth Siddhamshetty, Shaowen Mao, Kan Wu, Joseph Sang-Il 2020 Kwon - Multi-Size Proppant Pumping Schedule of Hydraulic Fracturing: Application to a MP-PIC Model of Unconventional Reservoir for Enhanced Gas Production *Processes* 8 570 DOI: 10.3390/pr8050570

[5] Khuzin R R, Mukhametshin V Sh, Salikhov D A, Andreev V E, Pepelyaev D V, Stefanovich Yu N 2021 Improving the efficiency of horizontal wells at multilayer facilities *IOP Conference Series: Materials Science and Engineering* **1064** 012066 DOI: 10.1088/1757-899X/1064/1/012066

[6] Tong S, Mohanty K K 2016 Proppant transport study in fractures with intersections *Fuel* **181** 463–477

[7] Malyarenko A M, Bogdan V A, Blinov S A, Kotenev Yu A, Mukhametshin V Sh, and Kotenev M Yu 2021 Improving the reliability of determining physical properties of heterogeneous clay reservoir rocks using a set of techniques *Journal of Physics: Conference Series* **1753** 012074 DOI: 10.1088/1742-6596/1753/1/012074

[8] Jiaxiang Xu, Yunhong Ding, Lifeng Yang, Zhe Liu, Rui Gao, Hanxuan Yang, Zhen Wang 2019 Effect of proppant deformation and embedment on fracture conductivity after fracturing fluid loss *Journal of Natural Gas Science and Engineering* **71** 102986

[9] Andreev V E, Chizhov A P, Chibisov A V and Mukhametshin V Sh 2019 Forecasting the use of enhanced oil recovery methods in oilfields of Bashkortostan *IOP Conference Series: Earth and Environmental Science* **350** 012025 DOI: 10.1088/1755-1315/350/1/012025

[10] Batalov D A, Andreev V E, Mukhametshin V V, and Kuleshova L S 2021 Development regulation of oil and gas reservoirs based on effective geological and geophysical information *IOP Conference Series: Materials Science and Engineering* **1064** 012055 DOI: 10.1088/1757-899X/1064/1/012055

[11] Mingulov Sh G and Mukhametshin V Sh 2020 Restoring the intake capacity of injection wells by means of shock-wave exposure *IOP Conf. Ser.: Mater. Sci. Eng.* **952** 012060 DOI: 10.1088/1757-899X/952/1/012060

[12] Mukhametshin V V, and Kuleshova L S 2021 Using the method of canonical discriminant functions for a qualitative assessment of the response degree of producing wells to water injection during the development of carbonate deposits *IOP Conference Series: Materials Science and Engineering* **1064** 012069 DOI: 10.1088/1757-899X/1064/1/012069

[13] Fattakhov I G, Kuleshova L S, Kochetkov A V, Galiullina I F, Andaryanov I G, Safiullina A R, and Gizatullin I R 2020 The use of a liquid destructor with flow into the stream during the injection of the fracturing mixture *IOP Conference Series: Materials Science and Engineering* **952** 012044 DOI: 10.1088/1757-899X/952/1/012044

[14] Nurgaliev R Z, Kozikhin R A, Fattakhov I G, Kuleshova L S, and Gabbasov A Kh 2019 Prospects for the use of new technologies in assessing the impact of geological and technological risks *IOP Conference Series: Earth and Environmental Science* **378** 012117 DOI: 10.1088/1755-1315/378/1/012117

The comparison of methods for modeling seismic impact on buildings and structures

G M Kravchenko[1], E V Trufanova[1], K Kh Mazhiev[2,3], Adam Kh Mazhiev[4], Aslan Kh Mazhiev[4,5], A Kh Mazhieva[2]

[1] Don State Technical University, 1, pl. Gagarina, Rostov-on-Don, 344003, Russia
[2]Grozny State Oil Technical University named after Academician M.D. Millionshchikov, 100, pr. Kh.A. Isaeva, Grozny, 364051, Russia
[3] Complex Research Institute named after Kh. I. Ibragimov, Russian Academy of Sciences, 21A, Staropromyslovskoe shosse, Grozny, 364906, Russia
[4] Academy of Sciences of the Chechen Republic, 13, pr. M. Esambaeva, Grozny, 364024, Russia
[5] Youth government of the Chechen Republic, 4, prospect named after V.V. Putin, Grozny, 364024, Russia

E-mail: galina.907@mail.ru; seismofund@mail.ru

Abstract. The article compares methods for modeling seismic effects on buildings and structures to improve the structural safety of design solutions. A linear-spectral method of accounting for seismic impact and an integral model are considered. Seismic loads are determined along the main axes of the building under study, taking into account translational and rotational effects. The orientation of the cosines of the seismic impact on the building is calculated. The assessment of participation factors for each form of natural fluctuations is carried out. Combinations of seismic loads are compiled, their influence on the stress-strain state of the building frame structures is analyzed. Recommendations for modeling the seismic impact on buildings and structures by the finite element method have been developed. The proposed algorithm for modeling seismic effects by the finite element method allows determining the most dangerous propagation of seismic effects in accordance with the modal analysis of the frame of a multi-storey building.

1. Introduction

Modern methods of calculating buildings and structures should ensure operational safety throughout the entire life cycle, taking into account the random nature of seismic effects. When updating the basic building codes and regulations, the seismic resistance criteria are defined implicitly [1]. The requirement is stated on the need to prevent the loss of operational properties of the structure or collapse, which poses a threat to the safety of people. The presented criteria are not related to limiting states, which are determined by the current state standard for the reliability of building structures.

In modern building codes, the calculation of the building structures of buildings and structures is performed for the design basis earthquake and the maximum design earthquake. However, for all declared levels of calculation, one seismic zoning map and one calculation method is used—the spectral method with the coefficient of inelastic work of the foundation [2].

The foregoing allows asserting that the presented method for calculating seismic effects does not rely on the basic provisions of existing design standards. The accumulated experience in predicting the behavior of buildings under special influences and the probabilistic approach in calculating seismic loads allow simulating the seismic effects in accordance with the required level of reliability of building structures [3-4].

2. Statement of the Problem

To improve the accuracy of engineering calculations of frame buildings in seismically hazardous areas, it is necessary to compare the calculation methods for seismic effects using modern software systems, recommend a seismic action model that takes into account the complexity of the structural scheme of the building.

To solve the problem, it is necessary:

- to develop a computational model of the investigated object by the finite element method;
- to perform a dynamic calculation of the building frame, to evaluate the forms and frequencies of natural vibrations;
- to determine the category of complexity of the building's structural system;
- to determine seismic loads according to two theories of calculation;
- to analyze the stress-strain state of the building frame, taking into account the obtained seismic loads;
- to elaborate recommendations on the method of seismic impact modeling in the design of multi-storey buildings.

3. Materials and Methods

Two theories for calculating the seismic impact on the building frame are considered:

- linear-spectral method, which takes into account the response of a structure to an arbitrary direction of seismic action;
- integral model of seismic action, taking into account the translational and rotational motion of the base.

The linear spectral method is used in the regulatory documents of most countries and provides the required level of seismic resistance of structures with simple design solutions.

When using an integral model of seismic action for calculating buildings and structures with a height of more than 50 m, it is recommended to apply a direct dynamic method of numerical integration of the equations of motion.

To assess the advantages and disadvantages of the theories for determining seismic loads, the object of study was chosen—a residential building with 23 above-ground floors, a basement and 3 underground ones. Load-bearing vertical structures are walls made of monolithic reinforced concrete. Foundation is a 1700-mm thick slab grillage. The floors of the building are bezel-less, rigidly connected with monolithic reinforced concrete walls. The concrete class of load-bearing structures is B25. Seismicity of the site is 9 points, soil category is II.

The modeling of the spatial frame of the building was carried out by the finite element method in the Stark_ES software package [5]. Slab grillage, floor slabs and coverings, walls were modeled with modern high-precision hybrid finite elements. The main step of the finite element mesh of the computational model is 900x900 mm (Figure 1).

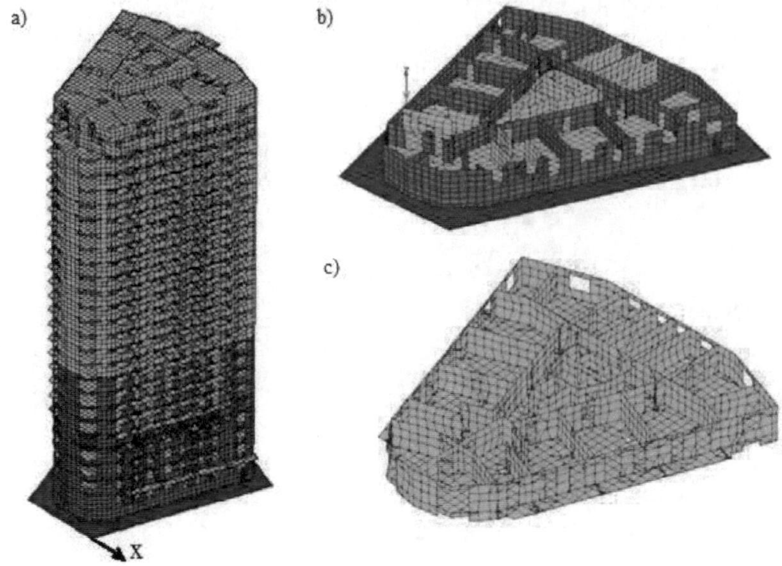

Figure 1. Design model of the building: a) spatial diagram; b) the basement of the building; c) typical floor

When calculating the parameters of free vibrations, the soil foundation of the building was not specified. Parametric foundations approximately took into account the rigidity and did not take into account the significant inertial properties of soils [6]. In this case, the vibration modes obtained taking into account the compliance of the parametric base moved to the beginning of the spectrum, where "false" forms appear. Such a distortion of the spectrum of natural frequencies was aimed at reducing the dynamic effects on the building structure [7-11].

When using an absolutely rigid base, the dynamic effects on the topside were increased, which increased the safety margin of the system.

Constant, long-term and snow loads were converted into masses. The coefficients for loading were used equal to the ratio of the combination coefficients for special seismic loading and load reliability.

To control the correctness of the determined parameters of natural vibrations, epy Sturm sequence check was performed to exclude zero frequencies.

As a result of the dynamic calculation, the frequencies and forms of natural vibrations of the system were determined (Figure 2).

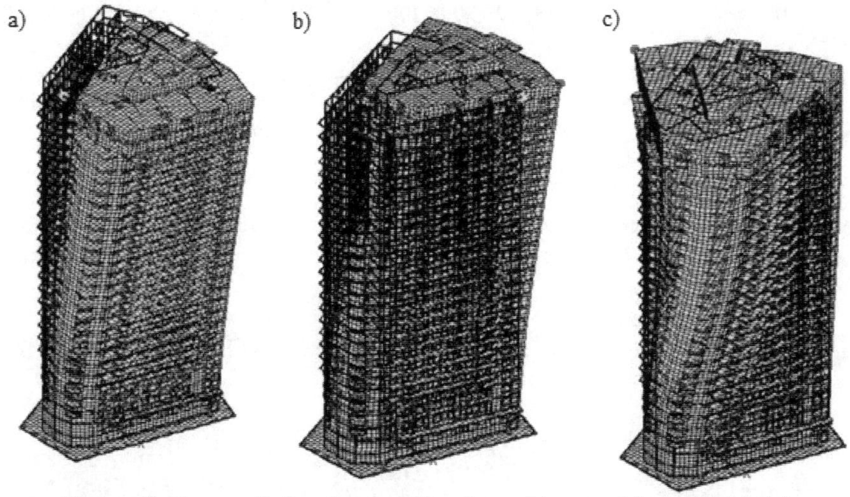

Figure 2. Forms of vibrations: a) first form; b) second form; c) third form

The frequencies of the first ten modes of vibration vary in the range from 1.04 to 8.41 Hz. The periods of fluctuations differ sequentially by 7.7–133%. Consecutive ten-percent excess of the oscillation periods indicates a complex structural and planning solution of the building frame [12]. In accordance with the requirements of regulatory documents, it is necessary to take into account the most dangerous directions of seismic effects from the point of view of the maximum values of the dynamic response of a structure or its parts [13–15].

Initial data for the first theory of calculation are: characteristics of the construction site, allowable damage to the building, taking into account the ability of the building to dissipate energy.

The calculation of seismic loads was carried out according to the first calculation theory. Two directions of seismic impact were accepted: along and across the X axis.

As a result, 10 load cases (corresponding to 10 modes of vibration) were added for seismic with an angle between the direction of the seismic action and the X-axis of 0 degrees. When setting the angle of seismic action at 90 degrees, 10 additional loads were obtained.

To analyze the effect of seismic loads on stresses in load-bearing structures, the load-bearing wall of the first parking level was selected.

The results of the calculation were obtained taking into account the seismic effect according to the first model for the investigated fragment (Table 1). The maximum and minimum values of normal stresses, bending and torques in finite elements with a combination of seismic loads were considered.

Table 1. Forces in a fragment of the building frame (model 1)

	Sr [kN/m^2]	Ss [kN/m^2]	Srs [kN/m^2]	Mr [kN/m]	Ms [kN/m]	Mrs [kN/m]
Min	-3948	8434	-2728	-24	-47.5	-7.9
Max	3101	13539	2048	17.5	45.7	9.4

The calculation of the seismic impact by the integral model was carried out taking into account the translational and rotational effects of the foundation. In accordance with the initial data in the software package, the worst direction of the seismic action for the translational action was determined: the angle with the OX axis was 65.974 degrees, the angle with the XOY plane was 0.0 degrees. The worst direction of seismic action for rotational action: angle with the OX axis was 89.665 degrees, the angle with the XOY plane was 10.777 degrees.

Based on natural fluctuations, the participation factors were assessed for each form. The number of vibration modes taken into account made it possible to collect at least 90% for translational directions and 75% for rotational action. The calculation took into account the forms with a participation factor of more than 1%.

The direction (orientation) of the action cosines was calculated for each form of natural vibrations. The initial data for determining the seismic impact were: the vibration amplitude, the number of taken into account eigenvalues, the forms with a participation factor of more than 1%, the direction cosines of the seismic impact, the seismicity of the construction site, the coefficients normalizing the possible damage to the building.

The calculation for the first and second hazardous directions of seismic action added 8 loads to the design model of the building. Taking into account the rotational component of the seismic action corrected three additional loads.

The calculation results in the Stark_ES software package demonstrated the stress-strain state of the building frame structures, taking into account the most dangerous directions of seismic effects. The efforts in the considered fragment of the structure are summarized in Table 2.

Table 2. Forces in a fragment of the building frame (model 2)

	Sr [kN/m^2]	Ss [kN/m^2]	Srs [kN/m^2]	Mr [kN/m]	Ms [kN/m]	Mrs [kN/m]
Min	-4448	8658	-2639	-21.6	-64.3	-11.2
Max	6048	10216	3986	34	68.4	15

4. Results and Discussion

The paper analyzed the methods of modeling seismic impact on buildings and structures developed in the Stark_ES software package based on the finite element method [16-17].

The first model was linear spectral. The advantages of the model are simplicity of calculation; ensuring the required level of reliability of the results for buildings with simple design solutions. The disadvantages are the number of seismic impacts taken into account and their direction is set at the discretion of the calculator; the rotational movement of the base is not taken into account.

The second model is integral. The advantages of the model are consideration of the translational and rotational motion of the base; high level of reliability of results for buildings with simple and complex structural solutions. The disadvantages are additional calculations to determine the dangerous direction of exposure with processing the results for a sample of considered forms with a participation factor of more than 1%.

The analysis of the results of calculating a fragment of the building frame with a minimum superposition of combinations of loads showed that when calculating within model 2, the minimum values of stresses in the horizontal direction increased by 13%, the remaining stresses remained practically unchanged. The values of the momenta increased to 42%.

When calculating seismic actions within model 2, with the maximum superposition of combinations of loads, the stresses in the vertical direction increased by 25%, horizontal stresses increased by more than 95%. The values of the momenta increased to 95%.

5. Conclusion

In the study of the stress-strain state of the frame of a multi-storey building, a computational model has been developed that makes it possible to determine the dynamic characteristics by the finite element method. Modal analysis showed that the first and second modes of natural vibrations of the building have a translational direction, the third has a torsional one. This nature of the vibration modes shows the correctness of the adopted structural scheme of the building during construction in an area with seismic activity.

The horizontal displacements of the building from seismic effects, calculated using the integral model, are 2 times greater than similar characteristics determined using the linear spectral method.

To ensure the reliability of design solutions, it is recommended to design buildings according to an integral model of seismic action, taking into account the translational and rotational motion of the base.

When determining the most dangerous direction of seismic impact, it is necessary to take into account the amplitude of vibrations, the number of considered eigenvalues, forms with a participation factor of more than 1%, direction cosines of the impact, seismicity of the construction site, coefficients normalizing possible damage to the building. Taking these characteristics into account enables additional combinations of loads from both translational and rotational influences.

An increase in the level of reliability in the field of earthquake-resistant construction is impossible without a modern regulatory framework based on the achievements of Russian and foreign science. Methods for calculating modern buildings and structures for seismic resistance should take into account the complexity of structural schemes and ensure the reliability of the adopted design solutions.

The proposed algorithm for modeling seismic effects by the finite element method allows determining the most dangerous propagation of seismic effects in accordance with the modal analysis of the frame of a multi-storey building.

References

[1] Aisenberg Ya M 2017 Seismic intensity scale. Analysis and suggestions for improvement *Natural and man-made risks. Safety of structures* **5-6 (31)** 17-21

[2] Panasyuk L, Kravchenko G, Trufanova E 2017 Researching design solutions for frames of buildings in case of increased seismic intensity in specific zones, *in collection: MATEC Web of Conferences* **106** 02027

[3] Abakarov A D, Zainulabidova Kh R 2009 Probabilistic models of seismic impacts *Bulletin of the Dagestan State Technical University. Engineering Science* **15** 92-97

[4] Murtazaliev G M, Dibirgadzhiev A M 2015 Variational principles of structural mechanics *Proceedings of the XXXVI final scientific and technical conference of teachers, staff, graduate students and students of Dagestan State Technical University* pp. 118-119 (Makhachkala: Dagestan State Technical University)

[5] Yakushev V 2011 Analysis of numerical methods for building structures in STARK ES *ABSE-IASS-2011 Symposium, London, 20th - 23rd September* pp. 54-56

[6] Abdurazakov G M, Abakarov AD 2015 *Construction of computational models for assessing the survivability of frame systems under seismic impact. Questions of modern technical sciences: A fresh look and new solutions. Conference Proceedings* (Issue 2) pp. 58-61 (Makhachkala: Innovation center for the development of education and science)

[7] Mazhiev Kh N, Zaalishvili V B, Mazhiev K Kh and others 2018 Seismic and Wind High-Rise Building Design Proceedings of the International Symposium *Advances in Engineering Research* pp 258-266 (Amsterdam: Atlantis Press)

[8] Koss H, Sahlmen J 2002 Methods in pedestrian wind comfort assessment; theoretical and practical comparison, *in Impact of Wind and Storm on City life and Built Environment Proceeding of the Workshop, June 3rd – 4th NANTES (France)*

[9] Tieleman H W, Akins R E 1998 Mean and fluctuating pressure distributions on rectangular prisms immersed in a variety of turbulent shear flows *AIAA/ASME/SIAM/APS , Ist National Fluid Dynamics Congress, july 25-28, ,vol. 2* pp. 1749-1756

[10] Stathopoulos T, Dumitrescu-Brulotte M 1988 Design recommendations for loading on buildings of intermediate height *ANNUAL Conference, 25-27 May, 1988, vol.1* pp. 275-293

[11] Mazhiev Kh N, Bataev D K.S, Mazhiev K Kh et al 2018 The Construction of the Dolphinarium Building in the Seismic Area *Advances in Engineering Research,* (Amsterdam: Atlantis Press) pp 240-245

[12] Kravchenko G M, Korobkin A P, Trufanova E V, Lukyanov V I 2014 The criteria for evaluating dynamic models of the reinforced concrete frame of a building *Science Time* **12** 255-260

[13] Wacker J, Friedrich R, Plate E J, Bergdolt U 1991 Fluctuating wind load on cladding elements and roof pavers. *J. of Wind Engineering and Industrial Aerodynamics* **38** 405-418
[14] Wisse J A, Krus H W, Willemsen F 2002 Wind comfort assessment by CFD *in Impact of Wind and Storm on City life and Built Environment, Proceeding of the Workshop, June 3rd – 4th NANTES (France)*
[15] Cook N J 1990 *The designer's guide to wind loading of building structures. Part 2. Static structures* (London: Butterwords)
[16] Batht K-J 1996 *Finite Element Procedures* pp. 95-97 (New Jersey: Prentice Hall)
[17] Simbirkin V 2004 Analysis of Reinforced Concrete Loadbearing Systems of Multistorey Buildings *Modern Building Materials, Structures and Techniques: CD-ROM Proceedings of the 8th International Conference, Vilnius, May 19-21,* pp. 98-99

IOP Conf. Series: Earth and Environmental Science 1021 (2022) 012060 doi:10.1088/1755-1315/1021/1/012060

Initial strength analysis of reinforced concrete monolithic supports for spherical and cylindrical mine workings

A V Zaitsev[1], A V Kutergin[1], A A Fukalov[1], A S Votinov[2], and I Yu Kolytchev[1]

[1] Perm National Research Polytechnic University, 29 Komsomolsky prospekt, Perm, 614990, Russian Federation
[2] PermNIPIneft branch of LUKOIL-Engineering LLC in Perm, 3a Permskaya st., Perm, 614015, Russian Federation

E-mail: a-zaitsev@mail.ru

Abstract. By using decomposition of hoop and radial components of the displacement vector to the trigonometrical and generalized power series, we obtained the new exact analytical solutions to problems of the equilibrium state of thick-walled heavy anisotropic central- and axial-symmetric bodies fixed on their external surfaces and subject to uniform internal lateral pressure. The estimation of an initial strength of cylindrical and spherical reinforced concrete monolithic mine working supports for a long-term storage of gas and oil is carried out on the basis of a multicriteria approach taking into account real damage mechanisms. They include damage from tension or compression in radial, hoop and axial directions, and from transversal and antiplane shear) of anisotropic axial- and central-symmetric bodies.

1. Introduction

Cylindrical and spherical structures and facilities produced of anisotropic materials (the weight of these materials is an important factor), i.e. reinforced concrete mine working supports, coal slurry pipelines, and tunnel coatings, are widely applied in different industrial spheres, construction, geology and at oil and gas chemical enterprises. Thus, there is a real necessity in obtaining exact analytical solutions of the equilibrium state of thick-walled heavy anisotropic elastic central- and axial-symmetric bodies which are subject to uniform internal lateral pressures. These solutions will be used for engineering methods of the refine strength analysis. Reinforced concrete, cement stone, and rocks also contain reinforcement elements, combined into a frame, and mineral particles in their structures. Therefore, these materials demonstrate natural anisotropy of strength and deformation properties under loading. The refine strength analysis of cylindrical and spherical structures and facilities located at rocks or sedimentary rocks should be carried out on the basis of the multicriteria approach [1], considering various mechanisms of partial or complete loss-of-strength, which is characteristic for anisotropic damaged media. This approach can also be used to estimate the stress state for horizontal oil wells, to determine the conditions for their safe operations in order to maximize the carrying capacity reserves. The obtained analytical dependences will also be used to identify material constants and material functions of anisotropic materials (reinforced concrete supports are created together with the mining facility itself) and to test numerical algorithms worked out to solve more complicated tasks, when certain elements of structures and facilities have similar geometry and boundary conditions.

Also, the obtained data will make it possible to test experimental methods with heavy bodies of elementary geometries.

2. Exact analytical solutions

Let us consider the equilibrium of the thick-walled elastic heavy anisotropic central- and axial-symmetric bodies (i.e. fixed on the external surfaces and subject to uniform internal lateral pressure p) in the spherical (ρ, θ, and φ) and cylindrical (ρ, θ and z) co-ordinate systems. Let us assume that the uniform material, used to produce the bodies, is spherically transversely isotropic on any radius vector drawn from the center or has cylindrical orthotropy. The displacements and strains in the points that belong to the sphere (due to the body asymmetry and external load) are not dependant on component φ, and if the points belong to the cylinder (due to infinite length) from z meet the following Cauchy relations:

$$\varepsilon_{\rho\rho} = \partial_\rho u_\rho, \quad \varepsilon_{\theta\theta} = (\partial_\theta u_\theta + u_\rho)/\rho, \quad \varepsilon_{\varphi\varphi} = (u_\theta \operatorname{ctg}\theta + u_\rho)/\rho, \quad \gamma_{\rho\theta} = \partial_\rho u_\theta + (\partial_\theta u_\rho - u_\theta)/\rho, \tag{1}$$

and equilibrium equations in spherical coordinates [2]:

$$\partial_\rho \sigma_{\rho\rho} + (\partial_\theta \tau_{\rho\theta} + 2\sigma_{\rho\rho} - \sigma_{\varphi\varphi} - \sigma_{\theta\theta} + \tau_{\rho\theta} \operatorname{ctg}\theta)/\rho + F_\rho = 0, \tag{2}$$

$$\partial_\rho \tau_{\rho\theta} + \left[\partial_\theta \sigma_{\theta\theta} + (\sigma_{\theta\theta} - \sigma_{\varphi\varphi}) \operatorname{ctg}\theta + 3\tau_{\rho\theta}\right]/\rho + F_\theta = 0,$$

and cylindrical ones [3]:

$$\partial_\rho \sigma_{\rho\rho} + (\partial_\theta \sigma_{\rho\theta} + \sigma_{\rho\rho} - \sigma_{\theta\theta})/\rho + F_\rho = 0, \quad \partial_\rho \sigma_{\rho\theta} + (\partial_\theta \sigma_{\theta\theta} + 2\sigma_{\rho\theta})/\rho + F_\theta = 0. \tag{3}$$

Here $F_\rho = -\gamma\cos\theta$ and $F_\theta = \gamma\sin\theta$ are the vector components of weight forces, γ is the gravity of the material, and $\partial_x = \partial/\partial x$.

Constitutive equations for the spherically transversely-isotropic body [4]:

$$\sigma_{\rho\rho} = A_{11}\varepsilon_{\rho\rho} + A_{12}\varepsilon_{\theta\theta}, \quad \sigma_{\varphi\varphi} = \sigma_{\theta\theta} = A_{12}\varepsilon_{\rho\rho} + H\varepsilon_{\theta\theta}, \quad \sigma_{\rho\theta} = A_{44}\gamma_{\rho\theta} \tag{4}$$

and cylindrically orthotropic body [5]:

$$\sigma_{\rho\rho} = K_{11}\varepsilon_{\rho\rho} + K_{12}\varepsilon_{\theta\theta}, \quad \sigma_{\theta\theta} = K_{12}\varepsilon_{\rho\rho} + K_{22}\varepsilon_{\theta\theta}, \quad \sigma_{\rho\theta} = G_{\rho\theta}\gamma_{\rho\theta},$$

$$\sigma_{zz} = E_z\left(\mu_{\rho z}\sigma_{\rho\rho}/E_\rho + \mu_{z\theta}\sigma_{\theta\theta}/E_\theta\right) \tag{5}$$

contain the following coefficients [2, 3]:

$$A_{11} = E_\rho(1-\mu_{\theta\varphi})/\nu, \quad A_{12} = E_\theta\mu_{\rho\theta}/\nu, \quad A_{44} = G_{\rho\theta}, \quad A_{22} = E_\theta\left(1-\mu_{\rho\theta}^2 E_\theta/E_\rho\right)/(\nu+\mu_{\theta\varphi}\nu),$$

$$A_{23} = E_\theta\left(\mu_{\theta\varphi}+\mu_{\rho\theta}^2 E_\theta/E_\rho\right)/(\nu+\mu_{\theta\varphi}\nu), \quad \nu = 1-\mu_{\theta\varphi}-2\mu_{\rho\theta}^2 E_\theta/E_\rho, \quad H = A_{22}+A_{23}.$$

$$K_{11} = E_\rho(1-\mu_{\theta z}\mu_{z\theta})/D, \quad K_{22} = E_\theta(1-\mu_{\rho z}\mu_{z\rho})/D, \quad K_{12} = E_\rho(\mu_{\rho\theta}+\mu_{\rho z}\mu_{z\theta})/D,$$

$$D = 1-2\mu_{\rho\theta}\mu_{\theta z}\mu_{z\rho} - \mu_{\rho\theta}\mu_{\theta\rho} - \mu_{\theta z}\mu_{z\theta} - \mu_{z\rho}\mu_{\rho z}.$$

These coefficients are defined by Young's moduli E_ρ, E_θ, and E_z in the directions ρ, θ, and z, Poisson ratios $\mu_{\rho\theta}$, $\mu_{\theta z}$, $\mu_{z\rho}$, and $\mu_{\theta\varphi}$, and antiplane shear modulus $G_{\rho\theta}$ as well.

The sequential substitution of geometrical ratio (1) into the constitutive equations (4) and (5) and by substituting the obtained result into the equilibrium equation (2) and (3) helps to write down non-uniform systems of Lamé partial differential equations.

Let us write down the boundary conditions on the internal and external surfaces of a heavy cylinder and a sphere bounded with radiuses R and r consequently, as follows:

$$u_\rho\big|_{\rho=R}=0,\quad u_\theta\big|_{\rho=R}=0,\quad \sigma_{\rho\rho}\big|_{\rho=r}=-p,\quad \sigma_{\rho\theta}\big|_{\rho=r}=0. \tag{6}$$

For any boundary conditions that do not disrupt the symmetry of tasks by laying out components of displacement vectors on a hoop and radial into trigonometric and generalized power series [6, 7] and taking into account the uniformity of boundary conditions for $i>0$, we obtain exact analytical solutions of the equilibrium state of the thick-walled heavy transversely isotropic sphere [8–11]:

$$u_\rho=D_1\rho^{-1/2-s}+D_2\rho^{-1/2+s}+\left(d_2C_2/\rho+d_3C_3\rho^{-1/2+t}+d_4C_4\rho^{-1/2-t}+H_\rho\rho^2\gamma/H_1-C_1\right)\cos\theta, \tag{7}$$

$$u_\theta=\left(C_1+C_2/\rho+C_3\rho^{-1/2+t}+C_4\rho^{-1/2-t}+H_\theta\rho^2\gamma/H_1\right)\sin\theta,$$

where $s=\sqrt{1/4+2H_4/A_{11}}$, $H_\rho=H_4-H_3-A_{44}$, $H_\theta=2A_{11}-H_2$, $H_2=H+2A_{44}$,

$$t=\sqrt{9/4+[H(A_{11}+2A_{44})-2A_{12}(H_3+2A_{44})]/(A_{11}A_{44})},\quad d_2=-H_2/(H_4+A_{44}),$$

$$d_3=[Y+2A_{44}H_3t]/L,\quad d_4=[Y-2A_{44}H_3t]/L,\quad Y=-A_{44}(H_2+H_4+A_{44}),\quad H_3=A_{12}+A_{44},$$

$$H_1=2\left[A_{11}(H-4A_{44})+2A_{44}(H_4-2A_{12})-2A_{12}^2\right],\quad L=2H_3^2-A_{11}H_2,\quad H_4=H-A_{12}$$

and thick-walled heavy orthotropic cylinder:

$$u_\rho=Q\left(R^{2n}\rho^{-n}-\rho^n\right)+\left[a_1\rho^\beta+a_2\rho^{-\beta}+a_3\ln\rho+a_4+\left(\gamma\Theta\rho^2\right)/G_{\rho\theta}\right]\cos\theta, \tag{8}$$

$$u_\theta=\left\{a_1\alpha_1\rho^\beta+a_2\alpha_2\rho^{-\beta}-a_3\ln\rho-a_4+\left[\gamma(\omega_4\Theta-1)\rho^2\right]/\left(\omega_3G_{\rho\theta}\right)\right\}\sin\theta,$$

where $Q=p/\left(a^{n-1}\Lambda_{11}-a^{-n-1}R^{2n}\Lambda_{21}\right)$, $\beta=\sqrt{K_{22}\left(1/K_{11}+1/G_{\rho\theta}\right)-K_{12}/K_{11}\left(2+K_{12}/G_{\rho\theta}\right)+1}$,

$$\Lambda_{11}=K_{11}n+K_{12},\quad \Lambda_{21}=K_{12}-K_{11}n,\quad \alpha_1=\left[\beta\left(k^2+1\right)+\omega_5\right]/\left(\beta^2-\omega_5\right),$$

$$\alpha_2=\left[-\beta\left(k^2+1\right)+\omega_5\right]/\left(\beta^2-\omega_5\right),\quad \Theta=\left(\omega_3G_{\rho\theta}+\omega_2K_{11}\right)/[K_{11}(\omega_1\omega_3+\omega_2\omega_4)],$$

$$\omega_1=4-m^2-n^2,\quad \omega_2=2q^2+m^2-n^2,\quad \omega_3=3-g^2,\quad \omega_4=2k^2+g^2+3,\quad \omega_5=4-\omega_3,$$

$$m=\sqrt{G_{r\theta}/K_{11}},\quad n=\sqrt{K_{22}/K_{11}},\quad q=\sqrt{K_{12}/K_{11}},\quad g=\sqrt{K_{22}/G_{r\theta}},\quad k=\sqrt{K_{21}/G_{r\theta}}.$$

Integration constants D_1, D_2, C_1, C_2, C_3, C_4, a_1, a_2, a_3 and a_4, included into expressions (7) and (8), are determined from boundary conditions (6) and are not specified because of their complexity.

A consequent substitution of (7) and (8) into Cauchy relations (1) and constitutive equations (4) or (5) makes it possible to write down the expressions for strains and stress, respectively.

3. Estimation of the initial strength

The obtained analytical solutions (7) and (8) can be used to analyze weight forces to the stress states of solid-cast reinforced working supports of underground spherical and level cylindrical mine workings. We can also assess the initial strength of these facilities by the set of criteria [1].

The work [12] was supplemented with independent quantities:

$$J_\sigma^I=\sigma_{\varphi\varphi}=\sigma_{\theta\theta},\quad J_\sigma^{II}=\sigma_{\rho\rho},\quad J_\sigma^{III}=\sigma_{\rho\theta},$$

and

$$J_\sigma^I=\sigma_{\theta\theta},\quad J_\sigma^{II}=\sigma_{\rho\rho},\quad J_\sigma^{III}=\sigma_{zz},\quad J_\sigma^{IV}=\sigma_{\rho\theta}.$$

These quantities refer to the orthogonal transformations that are admissible upon spherically transversal-isotropic and orthotropic bodies, consequently, and describe different damage mechanisms caused by the tension or contraction in a hoop or radial direction, shifts on the covering or isotropy and diametric subspace too.

Figures 1 and 2 show the distribution of non-zero independent invariants of stress tensor in solid-cast reinforced supports ($p=0$ MPa): the spherical support is along meridional coordinate θ and nondimensionalized radial coordinate $\tilde{\rho}=(\rho-r)/(R-r)$, the plane cylindrical support is along hoop and radial component $\tilde{\rho}$.

We have chosen the following parameters of geometry and elastic moduli of the reinforced concrete ($\gamma=40$ kN/m^3): $r=3.0$ m, and $R=6{,}0$ m; $E_\theta=50.0$ GPa, $E_\rho=35.0$ GPa, $G_{\rho\theta}=56.5$ GPa, $\mu_{\theta\varphi}=0.075$, and $\mu_{\rho\theta}=0.15$ (the spherical working); $r=3.0$ m, and $R=5.5$ m; $E_\rho=40$ GPa, $E_\theta=80$ GPa, $E_z=53.3$ GPa, $G_{\rho\theta}=56.5$ GPa, $\mu_{\rho\theta}=0.15$, $\mu_{\rho z}=0.075$, and $\mu_{z\theta}=0.375$ (the plane cylindrical working). If we consider the classical Lamé solution for an anisotropic sphere and cylinder [2, 3], we have nonzero stresses even by internal loads as a new obtained analytical solution (7) and (8) consider the input of weight forces.

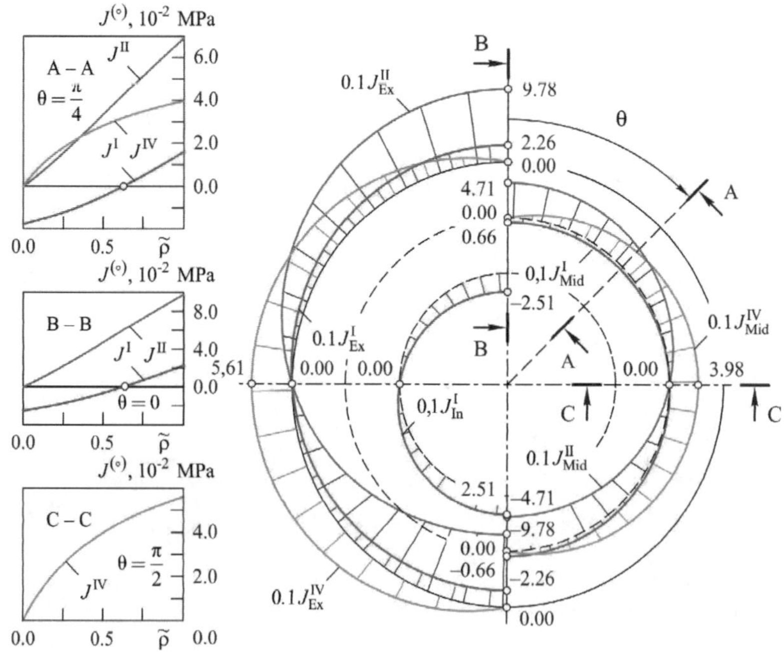

Figure 1. The distribution of non-zero invariants of stress tensor (KPa) on the fixed external ($J^{(\bullet)}_{Ex}$), load-free internal ($J^{(\bullet)}_{In}$) and medial ($J^{(\bullet)}_{Mid}$) surfaces in the concrete monolithic spherical mine working support

4. Conclusion

On the internal free covering (Fig. 1 and 2), only the first invariant J^{I}_{σ} is nonzero; this invariant steadily increases in the upper part of the facilities along $\tilde{\rho}$, and at the bottom it steadily decreases taking zero indexes at $\tilde{\rho}=0.63$ and $\tilde{\rho}=0.87$ by the supports of the spherical and cylindrical mine working. It should be emphasized that the location of these points that dispart the sphere of positive and negative values J_σ^I, does not depend on θ (with the exception of a diametric plane that corresponds to $\theta = \pi/2$, where $J_\sigma^I = 0.00$ kPa). The third invariant J_σ^{III} (Fig. 2) has the same nature of modification along the radial component in the plane working supports. The points where the change of symbol J_σ^{III} takes place are located at the distance $\tilde{\rho} = 0.75$ from the internal covering. Upon the modification of the radial component from the non-fixed surface to the fixed one, J_σ^{II} increases everywhere in magnitude (Fig. 1 and Fig. 2). As the points reach the highest magnitude J_σ^I and J_σ^{II} (the spherical working), J_σ^I, J_σ^{II}, and J_σ^{III} (the cylindrical mine working) in the places of the vertical diametric plane these points are considered as most hazardous as it can damage monolithic supports. The reason can be tension or contraction in hoop and radial directions (the spherical mine working), and in the hoop, radial and axial directions (the plane cylindrical mine working).

The third J_σ^{III} and the forth J_σ^{IV} invariants describing the damage by the shear of the points of the solid-cast supports by the spherical and plane cylindrical mine working, respectively, are equal to zero in the points that are located in the plane area (Fig. 1 and Fig. 2). J_σ^{III} (Fig. 1) and J_σ^{IV} (Fig. 2) grow by the angle and get at their highest amplitude by $\theta = \pi/2$ (the remaining invariants in this diametric plane take zero values). J_σ^{III} steadily changes its factors from zero on the internal covering up to the maximum on the external fixed covering (Fig. 1). J_σ^{IV} (Fig. 2) unsteadily grow, have a local minimum by $\tilde{\rho} = 0.16$ and change the symbol by $\tilde{\rho} = 0.34$. The state of this minimum does not depend on θ; an exception is a sectional area that coincides with the vertical diametrical plane, when all the indexes of the fourth invariant are zero.

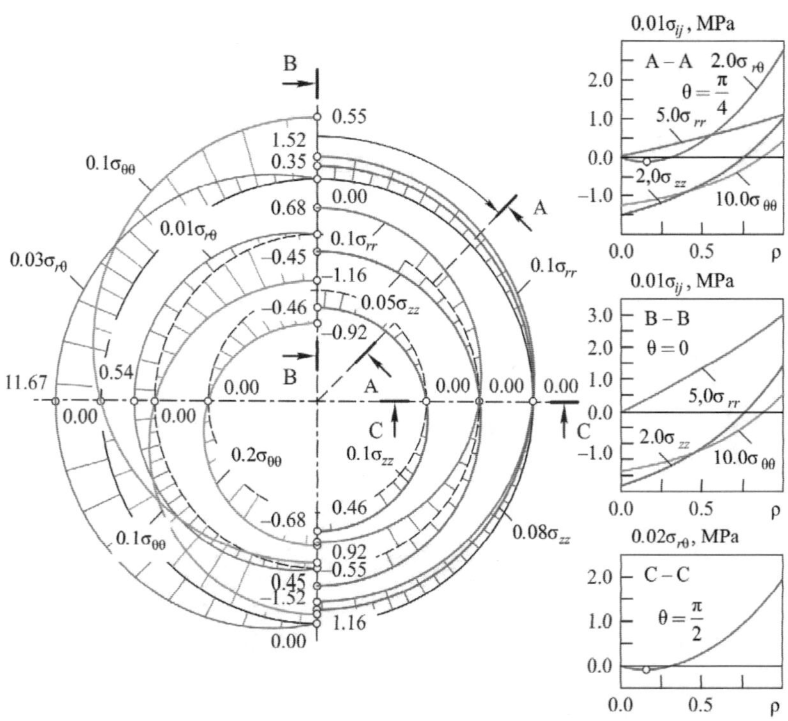

Figure 2. The distribution of non-zero invariants of stress tensor (MPa) on the fixed external ($J_{\mathrm{Ex}}^{(\bullet)}$), load-free internal ($J_{\mathrm{In}}^{(\bullet)}$) and medial ($J_{\mathrm{Mid}}^{(\bullet)}$) surfaces in the concrete monolithic horizontal cylindrical mine working support

Acknowledgements
The work is supported by the Russian Ministry of Science and Education, State Assignment FSNM–2020–0027 for Fundamental Researches in 2020–2022.

References

[1] Wildemann V E, Sokolkin Yu V, and Tashkinov A A 1997 *Mechanics of inelastic deformation and failure of composite materials* (Moscow: Nauka Publishers).
[2] Lekhnitskii S G 1981 *Theory of Elasticity of an Anisotropic Body* (Moscow: Mir Publishers).
[3] Birger I A, Shor B F, Dem'yanushko I V et al 1975 *Thermostability of machine elements* (Moscow: Mashinostroenie Publishers)
[4] Zaitsev A V, Sokolkin Yu V and Fukalov A A 2021 *J. Samara State Tech. Univ., Ser. Phys. Math. Sci.* **25(1)** 83–96 (In Russian)
[5] Zaitsev A V and Kislitsyn A V 2007 *J. Samara State Tech. Univ., Ser. Phys. Math. Sci.* **1(14)** 164–167. (In Russian)
[6] Kozhevnikova L L, Kuznetsov G B, Matveenko V P and Shardakov I N 1974 *Strength of Materials.* **6(9)** 1057–1061
[7] Kuznetsov G B 1979 *Elasticity, viscoelasticity and long-term strength of cylindrical and spherical bodies* (Moscow: Nauka Publishers)
[8] Zaitsev A V, and Fukalov A A 2010 *J. Samara State Tech. Univ., Ser. Phys. Math. Sci.* **5(21)** 85–95. (In Russian)
[9] Zaitsev A V, Sokolkin Yu V, and Fukalov A A 2013 *PNRPU Mechanics Bulletin.* **4** 59–75 (In Russian)
[10] Zaitsev A V, Sokolkin Yu V, and Fukalov A A 2019 Initial Strength Analysis of Anisotropic Concrete Supports for Spherical Mine Workings in a Sedimentary Rock Mass *Physical and Mathematical Modeling of Earth and Environment Processes (2018)* (Springer Proceedings in Earth and Environmental Sciences. Springer, Cham) pp. 463–471. https://doi.org/10.1007/978-3-030-11533-3_46.
[11] Fukalov A A, Zaitsev A V, Sokolkin Yu V, and Bayandin Yu V 2021 *J. Samara State Tech. Univ., Ser. Phys. Math. Sci.* **25(2)** 303–319.
[12] Pobedrya B E 1984 *Mechanics of composite materials* (Moscow: Moscow State Univ. Publishers)

Geological and statistical simulation for assessment of zonal oil and gas potential formation processes in the Visimskaya monocline

V I Galkin, K A Koshkin, O A Melkishev, I A Kozlova

Perm National Research Polytechnic University, Komsomolsky Prospekt 29, Perm, 614990, Russia

Email: vgalkin@pstu.ru

Abstract. The paper considers probabilistic approach application to assess zonal oil and gas potential of the Visimskaya monocline area. The earlier studies have shown that the processes of generation, migration and accumulation of hydrocarbons (HC) shall be most effectually assessed from a probabilistic standpoint. Generation processes are best described by a complex probabilistic criterion – $Pcomp^{G}$, migration and accumulation processes – by $Pcomp^{MAC}$ [1]. These criteria were used to develop a multivariate statistical model – $Pcomp^{zon}$, acting as an estimate of the system emergence regarding the zonal oil and gas potential. The values were ranked from maximum to minimum to assess the effect of $Pcomp^{G}$ and $Pcomp^{MAC}$ indicators on $Pcomp^{zon}$ at different ranges of its values. We successively built and analysed 379 models. The analysis of these models has shown that at $Pcomp^{zon}>0.5$ there is a congruent effect of $Pcomp^{G}$ and $Pcomp^{MAC}$ on this indicator, the effect being different at $Pcomp^{zon}<0.5$. A pattern of its variation over the area was built based on $Pcomp^{zon}$ values, which allowed differentiating the area of the Visimskaya monocline into a high-potential western part and a lower-potential eastern part.

1. Introduction

The Visimkaya monocline is located in the north-east of the Volga-Ural oil and gas province. In the plan, this area has a shape similar to triangle with dimensions of 180x120 km. It should be noted that this monocline plunges to the east in the direction of the Pre-Urals foreland basin. As it follows from the oil deposits location analysis, these are mainly concentrated in the west of the area. It therefore appears relevant to determine the criteria forming the location of HC deposits in this area [1,2]. It should be pointed out that the authors do not consider the impact of the Kama-Kinel trough system on the processes under study. This paper addresses the joint effect of generation, migration and accumulation processes on the oil deposits location character for the first time for this area. The authors believe that the development of probabilistic criteria describing these processes will allow a zonal prediction of oil and gas potential in the Visimkaya monocline.

2. Development of a zonal oil and gas potential criterion

For a zonal oil and gas potential rationale, two probability criteria were used: $Pcomp^{G}$ and $Pcomp^{MAC}$. The probabilistic criteria application technology and the statistical validation methods to address similar problems are provided in [3,4,5,6].

Generation processes are best described by a complex probabilistic criterion – $Pcomp^G$, and migration and accumulation processes by $Pcomp^{MAC}$ [1].

Figure 1 and Figure 2 show a quantity of individual probabilities (m) providing the maximum difference between the mean values of $Pcomp^G$ and $Pcomp^{MAC}$ for two classes of territories.

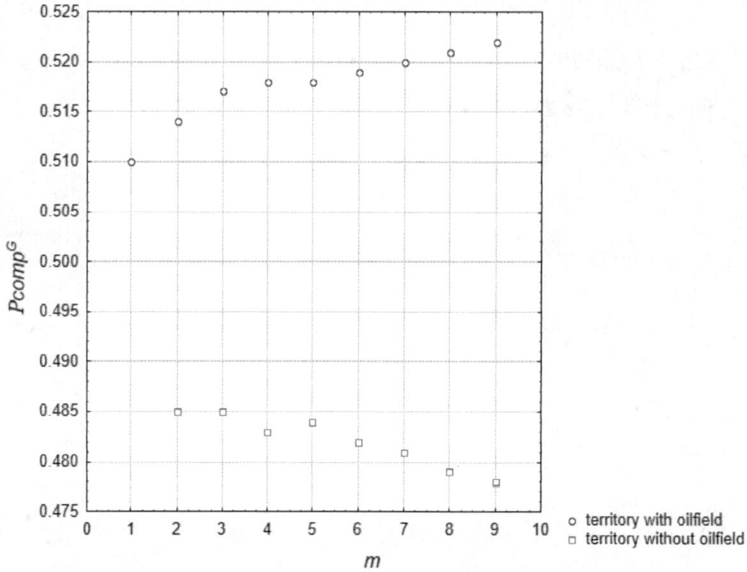

Figure 1. Value $Pcomp^G$ for m

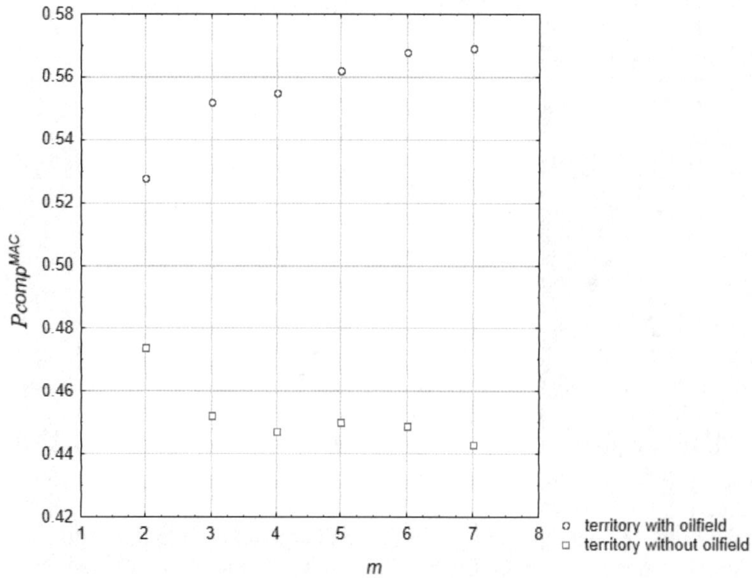

Figure 2. Value $Pcomp^{MAC}$ for m

A complex criterion $Pcomp^{zon}$ was developed based on $Pcomp^G$ and $Pcomp^{MAC}$ criteria, which acts as an estimate of the system emergence in relation to the zonal oil and gas potential of the Visimskaya monocline. The proposed criterion is calculated from the following formula:

$$Pcomp^{zon}=0.4756-1.9517Pcomp^{G}+0.3426Pcomp^{MAC}+1.9886(Pcomp^{G})^2+0.8385Pcomp^{G}*Pcomp^{MAC}+0.1596(Pcomp^{MAC})^2.$$

Using this formula, $Pcomp^{zon}$ values were calculated for 382 deep wells located across the Visimskaya monocline. To estimate the effect of generation, migration and accumulation conditions that quantitatively describe the process of emergence of $Pcomp^{G}$ and $Pcomp^{MAC}$ values on $Pcomp^{zon}$ at different ranges of its values, the values were ranked from maximum ($Pcomp^{zon}$ =0.862) to minimum ($Pcomp^{zon}$ =0.015). This value range was used to estimate the effect of $Pcomp^{G}$ and $Pcomp^{MAC}$ values on $Pcomp^{zon}$in the following order: the first model is constructed with 4 maximum values $Pcomp^{zon}$ (n=4), the next one with n=5, and so on until n=382. To account for the variety of various, and in some cases, divergent effects of the indicators under study on $Pcomp^{zon}$, 379 models were built.

To quantitatively estimate the change in the performance of the system under study and validate the presence of emergence of effect of $Pcomp^{G}$ and $Pcomp^{MAC}$ on $Pcomp^{zon}$, a graph (Figure3) of *t* criterion values variation was plotted that evaluates the 'contribution' of these probabilities to the multivariate models built.

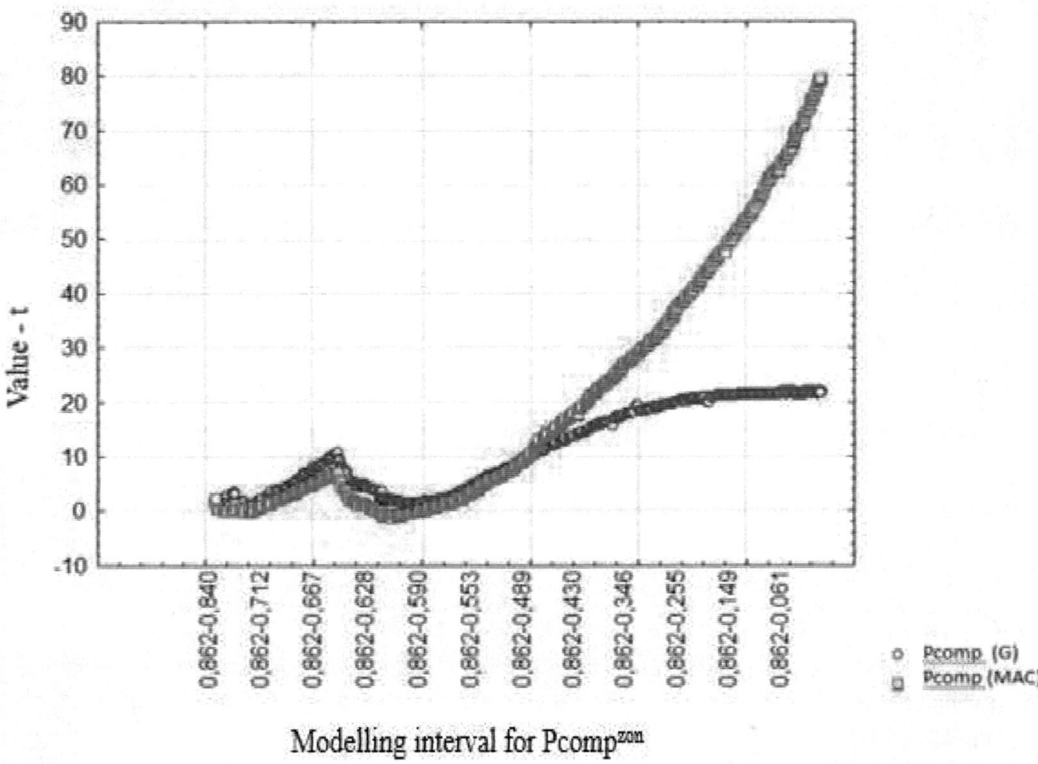

Figure 3.T criterion graph for $Pcomp^{G}$ and $Pcomp^{MAC}$ as a function of $Pcomp^{zon}$

Hence, it can be seen that changes in *t* values are characterised by the emergence of the zonal oil and gas potential system over a certain range of values. Congruent effect of $Pcomp^{G}$ and $Pcomp^{MAC}$ values is observed at $Pcomp^{zon}>0.5$. At this range of values, firstly, there is an insignificant excess of *t* criterion values for $Pcomp^{G}$ over *t* values for $Pcomp^{MAC}$, and secondly, there is a congruent variation in *t* values for these criteria.

At $Pcomp^{zon}<0.5$, when both $Pcomp^{zon}>0.5$ and less than 0.5 values are included in the sample set, the ratios of *t* values for $Pcomp^{G}$ and $Pcomp^{MAC}$ discord; there is a significant excess of *t* criterion for

$Pcomp^{MAC}$ over *t* for $Pcomp^{G}$. This might indicate the prevalence of the factors responsible for the hydrocarbons migration and accumulation processes over their generation. Under these conditions, the hydrocarbon generation processes start to deteriorate in certain areas, resulting in a situation where some of the structures are unfilled with hydrocarbons, i.e. 'empty'. Using this process of transition of quantitative relationships of $Pcomp^{G}$ and $Pcomp^{MAC}$ indicators to $Pcomp^{zon}$ allows one to set a boundary value of $Pcomp^{zon}$ criterion equalling to 0.5, and to determine the emergence system relating to the zonal oil and gas potential. In cases where changes in $Pcomp^{zon}$ are consistent across the area, this criterion can be applied to assess the oil and gas potential of the area under study.

3. Conclusions

To determine the variation of the developed $Pcomp^{zon}$ criterion values, a pattern of its variation for the area was built, as shown in Figure4.

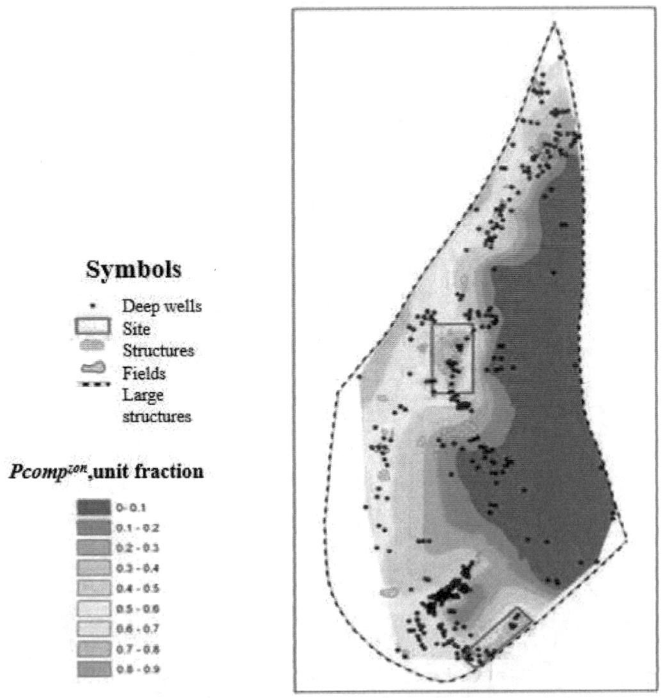

Figure 4.$Pcomp^{zon}$variation pattern across the Visimskaya monocline

From here it follows that the variation in the $Pcomp^{zon}$ values across the Visimskaya monocline has a consistent pattern. The resulting pattern lies in a consistent increase in $Pcomp^{zon}$ values from east to west, and in this case the iso-probabilities of $Pcomp^{zon}$ values are quite complex. If we divide the area under study into two parts based on $Pcomp^{zon}$=0.5, it may be conventionally assumed that in the west of the area with fairly high $Pcomp^{zon}$ values observed there is good correspondence between generation and migration-accumulation conditions of the zonal oil potential formation. This suggests that the presence of the system emergence relating to the zonal oil and gas potential is manifested in this area. For the east of the area, these conditions change significantly with the manifested absence of the system emergence relating to the zonal oil and gas potential. The performed geological and statistical simulation of the formation processes of the Visimskaya monocline's zonal oil and gas potential using the emergence approach allowed one to zone the Visimskaya monocline into a high-potential western part and a lower-potential eastern part. Two sites for detailed geological prospecting works were allocated within the high-potential part of the area, as designated in Figure4.

References

[1] Galkin V I, Koshkin K A, Melkishev O A 2018 The justification of zonal oil and gas potential of the territory of Visimskaya monocline by geochemical criteria *Perm Journal of Petroleum and Mining Engineering* **18(1)** 4-15

[2] Galkin V I, Kozlova I A, Krivoshchekov S N, Melkishev O A 2015 On the justification of the construction of models for oil and gas potential area forecast Visean deposits of Perm region *Oil industry* **8** 32-35

[3] Mikhalevich I M 2006 *Application of mathematical methods in the analysis of geological information (using computer technologies: Statistika)* (Irkutsk: ISU)

[4] Dubrule O 2003 *Geostatistics for Seismic Data Integration in Earth Models* (Tulsa: Society of Exploration Geophysicists)

[5] Isaaks E H, Srivastava R M 1989 *An Introduction to Applied Geostatistics* (New York: Oxford University Press)

[6] Tran D T, Gabbouj M, Iosifidis A 2017 Multilinear class-specific discriminant analysis *Pattern Recognition Letters* **100** 131–136

IOP Conf. Series: Earth and Environmental Science 1021 (2022) 012062 doi:10.1088/1755-1315/1021/1/012062

Development of a mathematical model for the formation of organic deposits based on research on their formation

A V Kozlov, K A Vyatkin, P U Ilushin

Perm National Research Polytechnic University, 29, Komsomolsky ave., Perm, 614990, Russia

E-mail: anton.kozlov@girngm.ru

Abstract. The processes of production and transportation of reservoir fluids are often complicated by various phenomena, among which one can distinguish the formation of organic deposits on the walls of the production tubing or linear oil pipeline. The fundamental problem in the field of oil production is the modeling of the processes of formation of these deposits, for which many models and equations have been developed. Their common disadvantage is high complexity and the need to use a complex mathematical apparatus. This work presents an equation based on laboratory studies and allowing one to model the processes of paraffin formation depending on the temperature gradient and water cut. This model shows high reliability in comparison with the classical Matzein model.

1. Introduction

In the present time, most of the oil fields in Russia, in particular in the Perm Krai, are moving to late stages of development. This process is accompanied by a decrease in production levels, an increase in water cut, and complications in the production of fluids. One of the typical complications arising in the production and transportation of oil is the fallout and subsequent accumulation of asphalt-resin-paraffin deposits (ARPD) [1].

High-molecular components, which are these organic deposits, are in a dissolved state in reservoir conditions [2]. However, during the extraction and transportation of the fluid, it cools down and, when the temperature drops below the crystallization temperature of ARPD, a sharp process of precipitation of solid sediment begins [3]. The formation of these deposits leads to disruptions in the operation of oilfield equipment, an increase in the load on it, and often leads to accidents [4]

An important task of modern oil production is the correct modeling of all processes occurring in the oil field. Modeling paraffin formation processes is a fundamental problem, the solution of which is complicated by different compositions of fluids, mechanisms of formation of deposits and conditions of paraffin deposition [5]. Currently, several mechanisms of the formation of ARPD have been described, but the most recognized of them is molecular diffusion [6]. This mechanism can be described by the following stages:

- formation of the primary layer of deposits;
- creating a concentration gradient of dissolved paraffins and their diffusion to the cold surface;
- growth of the thickness of organic deposits and their compaction due to internal diffusion.

The literature describes many methods for predicting the formation of organic deposits, including the "RRR Model", "Matzain Model" and "Heat Analogy" [7]. The disadvantages of these methods are their extreme complexity and the presence of empirical coefficients, the determination of which is either difficult or completely impossible. For example, in the “Matzain Model” method there is a diffusion coefficient, a change in which can multiply the thickness of the predicted layer of organic deposits. Within the framework of this project, it is proposed to use the results of laboratory studies to predict the formation of organic deposits.

2. Materials and methods

These studies were carried out on a laboratory unit "WaxFlowLoop", which allows one to investigate the process of paraffin formation under specific thermobaric conditions. The principle of operation of this installation is based on separate or joint regulation of the temperatures of the liquid flow and the wall of the test section to create conditions for the formation of wax deposits. The test section is a 1 meter long, 1/4” diameter spiral tube, made of stainless steel. This installation, shown in figure 1, consists of a pump (a), a differential pressure gauge (b), a hydraulic circuit (c) equipped with a cooling jacket (d), a feed tank (h), a cooling jacket for a feed tank (m) and a liquid level sensor (i). Temperature control in jackets d and m is carried out using circulation thermostats e and g, respectively. The flow rate and density of the liquid are monitored with a flow meter f. If it is necessary to check the liquid level in the raw tank, we open valves 1 and 5 alternately. If it is necessary to limit the flow of liquid into the working part of the installation, we turn off the valve 2. Valves 3 and 4 are used for flushing or when it is necessary to organize movement along impulse lines leading to the differential pressure gauge.

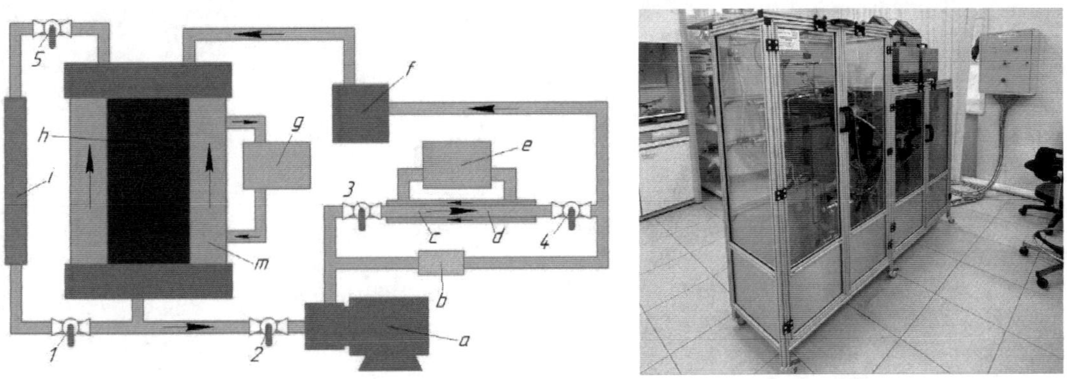

a) b)

Figure 1. Hydraulic diagram of the «WaxFlowLoop» (a); the photo of this installation (b)

The principle of operation of this installation is to use the Poiseuille equation to estimate the diameter of the test section at known values of the flow rate, oil viscosity and the length of the test section. Laboratory tests on this installation last from 8 to 36 hours due to the low rate of formation of organic deposits. At the end of the study, a database is formed, reflecting the change in all the values described above over time. Then, based on the data obtained, the thickness of the organic deposits is determined at each time point. Figure 2 shows an example of a performed laboratory study.

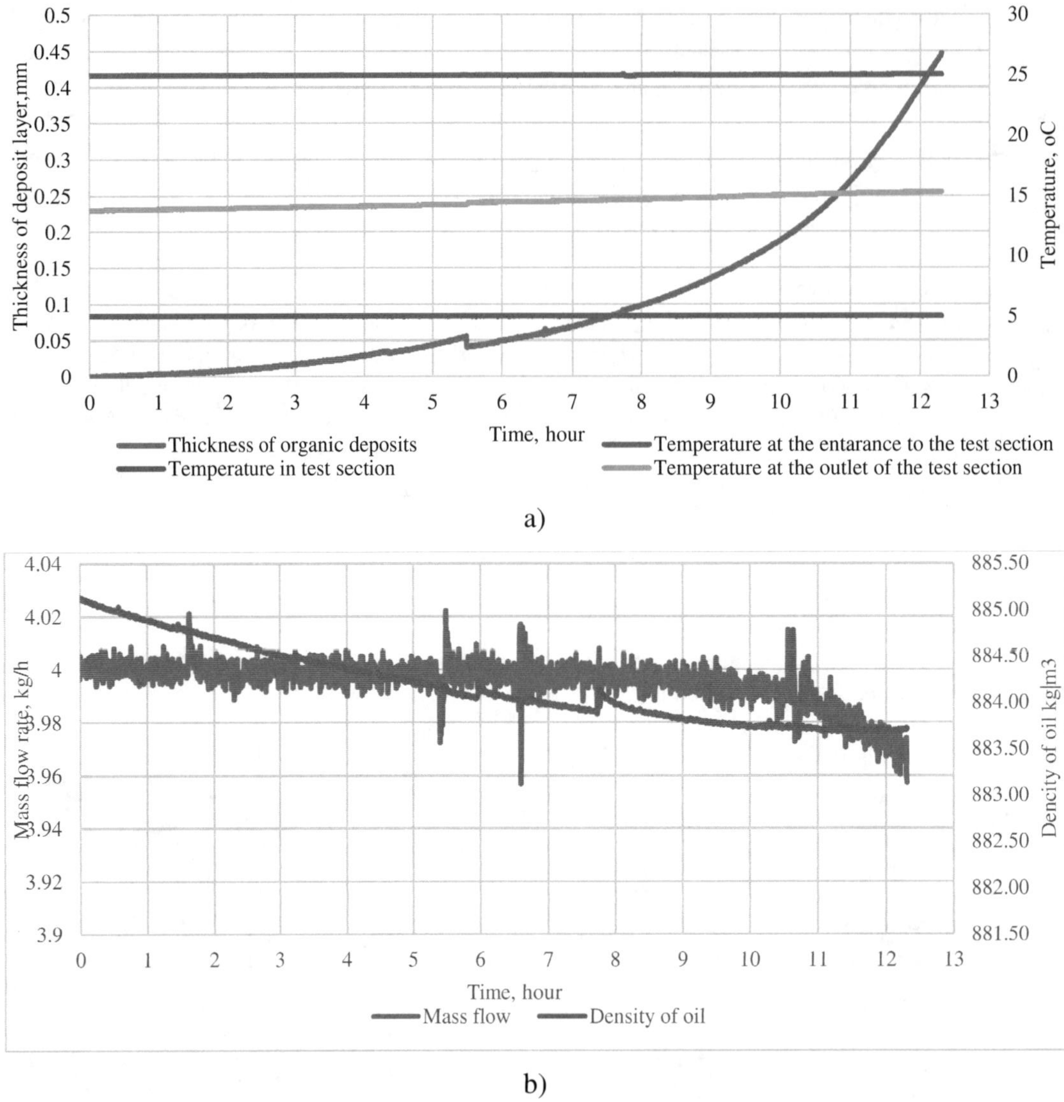

Figure 2. Change in the operating temperature of the installation and the thickness of organic deposits (a), density and flow rate of the liquid (b)

3. Results and Discussion

For the obtained values of the thickness of organic deposits, it is necessary to determine the most suitable law of change. Based on the carried out laboratory studies, the exponential and power laws are most often encountered, implying a sharp increase in the thickness of the ARPD. The resulting equation will be the basis for the expression for further predicting the thickness of the formed organic deposits. For the laboratory study shown in Figure 2, the following law of change in the thickness of organic deposits from the time of the study is determined. The formula for predicting the thickness of organic deposits in this case can be written as expression 1:

$$h = \frac{T_i - T_w}{\nabla T} \cdot C_1 \cdot e^{C_2 \cdot t} \cdot (1 - \omega_t) \tag{1}$$

where h - layer thickness, mm; t – time, sec; ∇T – the difference between the temperature of the wall and the oil during the experiment, °C; T_i – current fluid temperature, °C; T_w – current wall temperature, °C; ω_t – water cut, %; C_1, C_2 – empirical coefficients.

Using this equation, it becomes possible to estimate the thickness of the formed layer of organic deposits at different temperature regimes of their formation and at different water cut of the formation fluid. This model assumes a linear dependence of the rate of formation of organic deposits on the magnitude of the temperature gradient. Also, the assumption of this technique can be considered ignoring the fact of the inclusion of the aqueous phase of the formation fluid in the volume of organic deposits, which can increase their thickness with an increase in the water cut of the fluid. For this fluid, the process of wax formation was modeled using the "Matzain Model" method. Figure 3 shows a comparison of the result of this simulation and the application of the method described above.

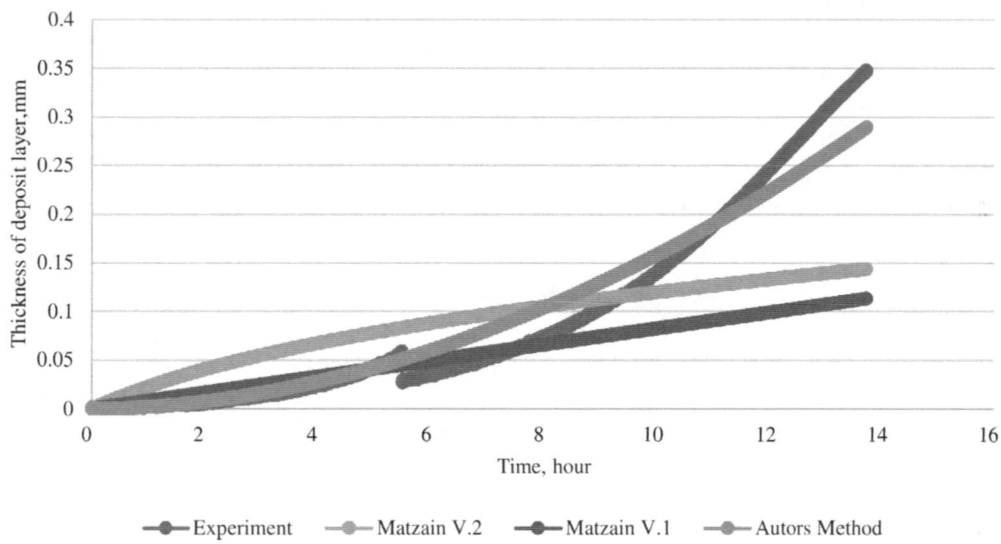

Figure 3. Comparison of the results of modeling the paraffin formation process

4.Conclusion

In the present time, there are many methods for predicting the formation of organic deposits during the production and transportation of fluid. A common disadvantage of these methods can be considered the presence of a set of parameters, the determination of which is either not possible or requires the use of a complex mathematical apparatus. This work presents a method for the interpretation of laboratory studies to simulate the process of paraffin formation under real temperature and pressure conditions. This method makes it possible to estimate the thickness of organic deposits in different temperature conditions and water cut in the fluid. The disadvantage of this method can be conducting research on a degassed sample, which may affect the accuracy of the application of this technique. Figure 3 shows a comparison of the use of this method with the "Matzain Model" method, which shows that the proposed method more accurately describes the process of wax formation of the target fluid.

Acknowledgments:

The work was carried out in the organization of the Lead Contractor as part of the R&D, carried out with the financial support of the Ministry of Science and Higher Education of the Russian Federation (agreement number 075-11-2021-052 of 24 June 2021) in accordance with the decree of the Government of the Russian Federation: 09.04.2010, number 218 (PROJECT 218). The main R&D contractor is Perm National Research Polytechnic University.

IOP Conf. Series: Earth and Environmental Science 1021 (2022) 012062 doi:10.1088/1755-1315/1021/1/012062

References

[1] Krivoshchekov S N, Vyatkin K A, Kochnev A A, Kozlov A V 2021 An approach to estimating the rate of organic deposit formation in a hollow rod string and selection of methods for deposit prevention *Periodico Tche Quimica.* **18(37)** 164-178.

[2] Pedersen K S, Christensen P L, Shaikh J A 2006 *Phase behavior of petroleum reservoir fluids.* (CRC press)

[3] Venkatesan R, Nagarajan N R, Paso K, Yi Y B, Sastry A M, Fogler H S 2005. The strength of paraffin gels formed under static and flow conditions *Chemical Engineering Science.* **60(13)** 3587-98.

[4] Sousa A L, Matos H A, Guerreiro L P 2019 Preventing and removing wax deposition inside vertical wells: are view *Journal of Petroleum Exploration and Production Technology.* **9(3)** 2091-2107

[5] Veiga H M B, Fleming F P, Azevedo L F A 2017 Wax deposit thermal conductivity measurements under flowing conditions *Energy&Fuels.* **31(11)** 11532-47

[6] Aiyejina A, Chakrabarti D P, Pilgrim A, Sastry M K S 2011 Wax formation in oil pipelines: A critical review *International journal of multiphase flow.* **37(7)** 671-94.

[7] Noville I, Naveira L 2012 *Comparison between real field data and the results of wax deposition simulation* In SPE Latin America and Caribbean Petroleum Engineering Conference

Geological and Technological Parameters and Development of Deposits in Carbonate Reservoirs to Solve Exploration Tasks

V Sh Mukhametshin

Ufa State Petroleum Technological University, Branch of the University in the City of Oktyabrsky, 54a, Devonskaya St., Oktyabrsky, Republic of Bashkortostan, 452607, Russian Federation

E-mail: vsh@of.ugntu.ru

Abstract. More than 30 geological and technological parameters to affect the predicted production of reserves in carbonate reservoirs were studied to solve exploration tasks. A physical interpretation of the findings is given. The paper concludes that there is a need for complex geological and statistical modeling. The analysis does not give the parameters that most largely determine oil recovery due to their cross-correlation dependence. The final answer can be provided by a comprehensive oil recovery model involving all parameters and an analysis of cause-and-effect relationships.

1. Introduction

It is important to know what effect various geological and technological parameters have on the depletion of deposits to analyze, manage and design oilfield development [1–8]. This is of particular importance to produce deposits in carbonate reservoirs of the Volga-Ural oil and gas province that contains significant residual oil reserves.

2. Materials and methods

In [9], the deposits are arranged into 14 relatively homogeneous groups. Over 300 productive formations were investigated.

The final oil recovery per formation was determined by the formula

$$\eta = \frac{\frac{n'}{n}\left(\sum_{i=1}^{k} Q^{I}_{rec.i} + \sum_{k}^{l} Q^{II}_{rec.i} + \sum_{l}^{m} Q^{III}_{rec.i} + \sum_{m}^{n} Q^{IV}_{rec.i} + \right)}{Q_{in-place}} \cdot 100, \quad (1)$$

where $Q^{I}_{rec.i}$ is the actual recoverable reserves produced by an *i-th* oil well up to the profitability limit (= 72 t/year); $Q^{II}_{rec.i}$ is recoverable reserves produced by an *i-th* oil well, determined by extrapolating the actual data in coordinates $Q_o - \ln t$ and $\ln Q_o - t$ (depending on a group), provided that there is some data on oil production for the first 5–7 years or more;

$Q^{III}_{rec.i}$ is the recoverable reserves produced by an *i-th* well, determined by equations (4.22) – (4.33) [10], provided that there is no data on oil production for the first 5-7 years and there are productivity values;

$Q^{IV}_{rec.i}$ is the recoverable reserves produced by an *i-th* well, determined by empirical relationships (2) – (13), provided that there is no data on oil production for the first 5-7 years and no early productivity values;

$Q_{in-place}$ is in-place formations in the reservoir;

n' is the total number of wells in operation at the time of calculations;

n is the number of wells for which the recoverable reserves were determined;

k is the number of wells for which $Q^{I}_{rec.i}$ was determined;

$l-k$ is the number of wells for which $Q^{II}_{rec.i}$ was determined;

$m-l$ is the number of wells for which $Q^{III}_{rec.i}$ was determined;

$n-m$ is the number of wells for which $Q^{IV}_{rec.i}$ was determined.

3. Results and Discussion

Given that, no data is available on oil production for the first 5-7 years and no early productivity values for a well, the recoverable reserves were determined based on a close correlation between Q_{rec} and Q^1_o (oil production in the first year of operation) according to the formulas:

$$Q_{rec} = Q^1_o \sum_{t=1}^{t_{tot}} (1 - 0.320 \ln t_i) \quad (2)$$

(for wells of groups 1, 2);

$$Q_{rec} = Q^1_o \sum_{t=1}^{t_{tot}} (1 - 0.303 \ln t_i) \quad (3)$$

(for wells of group 3);

$$Q_{rec} = Q^1_o \sum_{t=1}^{t_{tot}} (1 - 0.315 \ln t_i) \quad (4)$$

(for wells of group 4);

$$Q_{rec} = Q^1_o \sum_{t=1}^{t_{tot}} (1 - 0.308 \ln t_i) \quad (5)$$

(for wells of group 5);

$$Q_{rec} = Q^1_o \sum_{t=1}^{t_{tot}} (1 - 0.316 \ln t_i) \quad (6)$$

(for wells of group 6);

$$Q_{rec} = 1.269 Q^1_o \sum_{t=1}^{t_{tot}} e^{-0.238 t_i} \quad (7)$$

(for wells of group 7);

$$Q_{rec} = Q^1_o \sum_{t=1}^{t_{tot}} (1 - 0.345 \ln t_i) \quad (8)$$

(for wells of group 8);

$$Q_{rec} = Q_o^1 \sum_{t=1}^{t_{tot}} \left(1 - 0.299 \ln t_i\right) \tag{9}$$

(for wells of group 9);

$$Q_{rec} = 1.169 Q_o^1 \sum_{t=1}^{t_{tot}} e^{-0.156 t_i} \tag{10}$$

(for wells of group 10);

$$Q_{rec} = Q_o^1 \sum_{t=1}^{t_{tot}} \left(1 - 0.330 \ln t_i\right) \tag{11}$$

(for wells of group 11);

$$Q_{rec} = 1.149 Q_o^1 \sum_{t=1}^{t_{tot}} e^{-0.139 t_i} \tag{12}$$

(for wells of group 12);

$$Q_{rec} = 1.123 Q_o^1 \sum_{t=1}^{t_{tot}} e^{-0.116 t_i} \tag{13}$$

(for wells of group 13, 14).

The final oil recovery coefficients calculated by the formula (1) for groups of formations are shown in the Table.

Table 1. Final oil recovery by groups of reserves

No. of group	Final oil recovery, %		
	max	medium	min
1	15.6	2.4	0.2
2	31.5	11.6	2.5
3	33.8	9.2	0.2
4	25.1	9.8	1.0
5	31.1	18.7	3.9
6	24.0	5.0	0.4
7	38.9	13.9	24.7
8	31.0	18.7	4.6
9	41.0	33.7	20.0
10	18.5	6.3	0.2
11	31.6	9.2	1.8
12	23.7	6.9	0.3
13	8.7	3.8	0.3
14	4.7	1.0	0.2
1-14	41.0	10.1	0.2

The final oil recovery varies within a fairly wide range. This is due to both natural impacts and different reservoir engineering, namely, different well spacing. With the current reservoir engineering (in primary production of reservoirs), the final oil recovery on average from all productive formations will be about 10%, which indicates the need to find ways to enhance it.

Based on reservoir information available gathered from geophysical, laboratory and hydrodynamic studies, the relationship between oil recovery and the following parameters was studied, namely: oil pay in a formation O_E and in a development zone O_E^{DZ}, standard deviation σ_{O_E}, coefficient of variation W_{O_E}, entropy E_{O_E}, relative entropy $\overline{E}_{O_E}$, oil pay heterogeneity factor P_m; mean value O_F, standard

IOP Conf. Series: Earth and Environmental Science 1021 (2022) 012063 doi:10.1088/1755-1315/1021/1/012063

deviation σ_{O_F}, coefficient of variation W_{O_F}, entropy E_{O_F}, relative entropy $\overline{E}_{O_F}$, heterogeneity factor of oil-bearing interlayers F_{O_L}; average permeability coefficients C_{perm}, oil saturation C_s, laboratory m_L and geophysical m_G porosity, standard deviation σ_m, relative entropy $\overline{E}_m$, geophysical porosity heterogeneity F_m; sand-to-shale ratio R_s, the proportion of reservoir rocks in the total reservoir thickness P_R, a complex heterogeneity index I_{heter}; productivity factor F_{prod} and specific productivity factor F_{prod}^{sp}; viscosity μ_o, relative viscosity μ_{rel}, density of reservoir oil ρ_o, reservoir gas-oil ratio G, bubble-point pressure P_{bbl}; input reservoir pressure P_{res}, temperature t_{res}, reservoir depth O_{depth}; well spacing S.

Paired relationships show that the effect of the above parameters on oil recovery from the selected formations is mostly unified, although in some cases it may be different. A set of parameters that have significant relationships with oil recovery is also different (hereinafter, the significance of relationships was determined using a table of confidence limits for a sample correlation coefficient at 95% significance level), so is their effect, which is attributed to geological and technological features of production.

Some productive formations with higher net pay, thickness of effective oil-bearing interlayers, the proportion of reservoir rocks in the total productive thickness, permeability coefficients, productivity, specific productivity factors and containing less viscous and dense oils are characterized by higher oil recovery rates, which conforms to basic concepts of the effect these parameters have on oil production [11–15].

At first glance, it may seem strange to push up oil recovery by increasing parameters that denote oil pay heterogeneity ($\sigma_{O_E}, W_{O_E}, E_{O_E}, \overline{E}_{O_E}, F_{O_E}$) and thickness of oil-bearing interlayers ($\sigma_{O_L}, W_{O_L}, E_{O_L}, \overline{E}_{O_L}, F_{O_L}$). However, these parameters are closely related to the net pay thickness and reservoir productivity, which strongly affect oil recovery. Besides, the higher the productivity, the greater the effective oil-bearing thickness and indicators of O_E and O_L heterogeneity.

As porosity increases, there is an overall decrease in oil recovery. This is due to the fact that the lower the porosity, the higher the fracturing, permeability and productivity, which contribute to higher oil production. In the most fractured formations of groups 7 and 5, this effect is most noticeable. However, given that porosity exceeds 9%, fracturing of reservoir rocks has no effect, and with a further rise in porosity, oil recovery increases, but from the formations of groups 10 and 11 alone.

In the said groups of formations, the porosity (besides for fractured Famennian limestone of group 7 and, to a lesser extent, fractured limestone of the Tournaisian stage of group 5) is accompanied by increased heterogeneity in this parameter ($\sigma_m, W_m, E_m, \overline{E}_m, P_m$). As porosity increases (starting from 9%), the formations become more productive and the depletion of reserves increases. This explains the way the porous heterogeneity parameters affect oil production. In the formations of group 5, lower porosity is accompanied by a reduced parameter spread against an average, with the fracturing of rocks to go up, productivity to grow and, accordingly, oil recovery to increase. In this regard, lower porosity heterogeneity designated by relative entropy increases the final oil recovery.

The oil-saturation factor has an ambiguous effect. For example, in the conditions of groups 3 and 11, a rise in C_s is accompanied by a corresponding increase in porosity, permeability and productivity, which undoubtedly increases oil recovery. For the formations of group 5 (with porosity being less than 9%), growing oil saturation increases porosity, but decreases fracturing and productivity and, as a consequence, deteriorates the conditions for the development of oil reserves.

The higher the sand-to-shale ratio, the higher the oil recovery. This is due to the fact that the formations that have a higher sand-to-shale ratio are confined to the increased net pay zones. These zones also have better reservoir characteristics, which improves oil recovery with no water injection into the reservoir.

A complex heterogeneity factor, first proposed by M.A. Tokarev, unambiguously affects nearly all groups of formations, Naturally, as it grows, the conditions for the development of reserves deteriorate, which entails a decrease in the final oil recovery.

The input reservoir pressure and temperature grow given that the reservoir depth increases. The viscosity and density of reservoir oil decreases, and due to the compaction of reservoir rocks, fracturing and, accordingly, productivity increases. This explains an increase in oil recovery with an increase in reservoir pressure, temperature and depth.

In the target groups of formations, the deeper the formation, the lower the viscosity and the higher the bubble-point pressure. For sure, this is reflected in the relationships that show that the higher the G and P_{bbl}, the higher the oil recovery.

The production well spacing has a significant effect on oil recovery. As you can see, there is a steady decline in oil recovery for each group of formations, as well as for all formations as a whole, within a varying well spacing from 10 to 250 hectares/well.

4. Conclusion

1. The paper establishes the way various geological and technological parameters influence the final oil recovery from various groups of formations in carbonate reservoirs during primary production.
2. The above analysis, though, does not give the parameters that most largely determine oil recovery due to their cross-correlation dependence. The final answer can be provided by a comprehensive oil recovery model involving all parameters and an analysis of cause-and-effect relationships.

References

[1] Economides J M and Nolte K I 2000 *Reservoir stimulation* West Sussex, (England: John Wiley and Sons)

[2] Mukhametshin V V and Andreev V E 2017 Search and argumentation of decisions aimed at increasing the efficiency of bottom-hole zone stimulation in oil accumulations with challenged reserves *SPE Russian Petroleum Technology Conference (16-18 October 2017, Moscow)* DOI: 10.2118/187785-MS

[3] Yakupov R F, and Mukhametshin V Sh 2013 Problem of efficiency of low-productivity carbonate reservoir development on example of Turnaisian stage of Tuymazinskoye field *Oil Industry* **12** 106–110

[4] Bos C F M 2005 A Framework for Uncertainty Quantification and Technical-to-Business Integration for Improved Decision-Making *SPE Europec/EAGE Annual Conference (Madrid, Spain, 13-16 June 2005)* DOI: 10.2118/94109-MS

[5] Yaskin S A, Mukhametshin V V, Andreev V E and Dubinskiy G S 2018 Forecasting the parameters of non-stationary water flooding of oil deposits *IOP Conference Series: Earth and Environmental Science* **194** 062037 DOI: 10.1088/1755-1315/194/6/062037

[6] Akhmetov R T, Mukhametshin V V, and Andreev A V 2017 A quantitative assessment method of the productive formation wettability indicator according to the data of geophysical surveys *SPE Russian Petroleum Technology Conference (16-18 October 2017, Moscow)* DOI: 10.2118/187907-MS

[7] Collins P M, Dusseault M B, Dorscher D, and Kueber E 2008 Implementing CHOPS in the Karazhanbas heavy oil field (Paper 2008-500) *World Heavy Oil Congress (Edmonton, Canada, 10-12 March 2008)*

[8] Mukhametshin V V, and Kuleshova L S 2020 On uncertainty level reduction in managing waterflooding of the deposits with hard to extract reserves *Bulletin of the Tomsk Polytechnic University. Geo Assets Engineering* **331(5)** 140–146 DOI 10.18799/24131830/2020/5/2644

[9] Mukhametshin V Sh 2003 *Differentiation and grouping of complex deposits in solving*

development problems (Ufa: USPTU)
[10] Mukhametshin V Sh 2003 *Development of low-productive oil deposits under natural conditions* (Ufa: USPTU)
[11] Webb K J, Black C J J, Tjetland G 2005 A Laboratory Study Investigating Methods for Improving Oil Recovery in Carbonates *International Petroleum Technology Conference (Doha, Qatar, 21-23 November 2005)* DOI: 10.2523/IPTC-10506-MS
[12] Batalov D A, Mukhametshin V V, Dubinskiy G S, and Andreev V E 2018 Laboratory grounding of waterproofing sealant based on acrylic polymers *IOP Conference Series: Earth and Environmental Science)* **194** 042003 DOI: 10.1088/1755-1315/194/4/042003
[13] Muslimov R Kh 2005 *Modern methods of oil recovery increasing: design, optimization and performance evaluation* (Kazan': FEN)
[14] Tokarev M A 1990 *Current oil recovery under water-oil displaced comprehensive geological monitoring* (Moscow. Nedra)
[15] Akhmetov R T, and Mukhametshin V V 2018 Estimation of displacement coefficient with due account for hydrophobization of reservoir using geophysical data of wells *IOP Conference Series: Earth and Environmental Science* **194** 062001 DOI: 10.1088/1755-1315/194/6/062001

Assessment of the fracture stress of clay-carbonate rocks of the Vereiskian reservoir for oil fields in the Perm region

A S Votinov[1], A V Zaitsev[2], Yu A Rozhkova [2], B M Osovetsky[3], V V Seredin[3]

[1] PermNIPIneft branch of LUKOIL-Engineering LLC in Perm, 3a, Permskaya st., Perm, 614015, Russia
[2] Perm National Research Polytechnic University, 29, Komsomolsky prospect., Perm, 614990, Russia
[3] Perm State National Research University, 15, Bukireva st., Perm, 614068, Russia

E-mail: votinov14@yandex.ru

Abstract. Currently, one of the most effective stimulating oil production methods from the Vereiskian production zones of oil fields in Perm Krai is proppant hydraulic fracturing (PHF). The effectiveness of the method largely depends on the geometric dimensions of the fracture and the optimal distribution of the proppant in it. The mudstones of the Vereiskian stage limit the growth of the fracture in height, but in certain cases its can be broken in the course of hydraulic fracturing. In this regard, the determination of the conditions for their fracturing is an urgent and important task. In the absence of rock samples of proper quality, the chemical composition of mudstones was determined during the study. The obtained results are compared with the existing data on the minerals of the montmorillonite group of clays. To determine the parameters of the ultimate stress state of the formation, a Mohr's circle of stress is constructed using the known values of the compressive stress and the angle of internal friction of the clays. As a result, the fracture stress of mudstones is obtained, which is in the range of bottom-hole pressures during hydraulic fracturing, that is, there is a possibility of fracture growth outside the target interval.

1. Introduction

In recent years, proppant hydraulic fracturing (PHF) has been widely used in the world practice as a method of stimulating oil production in carbonate reservoirs [1-3]. One of the most promising objects of production is the Vereiskian oil formations, which are an interlayer of carbonate formations of various porosity and clay rocks. The most reliable idea of the structure of the pore space of rocks is given by the method of X-ray tomography of the core [4, 5]. The article [6] presents the results of studying the capacitance characteristics of Vereiskian oil formations by X-ray tomography.

The success and efficiency of the technology largely depends on the geological and physical characteristics of the deposit, as well as on the quality of modeling the hydraulic fracturing process. The maximum values of the productivity gains of wells are achieved not only by creating a fracture of the largest geometric dimensions, but also by the most effective distributing of the propping agent and achieving the optimal value of the dimensionless conductivity of the fracture and the dimensionless number of proppant [7].

The lower part of the Vereiskian formation of oil fields of Perm Krai is characterized by the presence of dense clay barriers (mudstones) above and below the productive reservoirs that limit the growth of the fracture in height. However, if a bottom-hole pressure is created that exceeds the breakdown pressure of the mudstone, these barriers break and the crack grows in height, which leads to inefficient placement of proppant and suboptimal sizes of the fracture. Assessment of the fracture stress of tight barriers is an important task for improving the quality of hydraulic fracturing process planning.

2. Materials and methods

In the research work, the properties of clay rock (mudstone) of the Vereiskian formation of the Moskudinskoe oil deposit of Perm Krai were studied, its chemical composition was determined by the fluorescence X-ray spectral method and losses during calcination by the gravimetric method were also determined. The equipment used is a wave-dispersive X-ray fluorescence spectrometer of the sequential type of action S8 Tiger. The elasto-mechanical and physical properties of the rock were also determined by comparative analysis and a systematic approach. To improve the quality of hydraulic fracturing process planning, the fracture stress and tangential stresses of mudstone were determined using the Coulomb-Mohr theory of failure and using the graphical method for constructing Mohr's circle of stress. The correlation between stress and fracture for many rocks can be determined graphically using circular Mohr stress diagrams. Due to the low quality of the rock samples and the limited data, a single Mohr's circle of stress is constructed in the graph of normal and tangential stresses.

3. Estimation of rock stresses by constructing a stress diagram

With a decrease in the tension (reservoir pressure), mudstones change their strength properties, which leads to their delamination when raised to the surface, so it is impossible to do studies with a tension on samples of this rock. The main petrogenic elements for this type of rocks can be determined by the fluorescent X-ray spectral method. In the research the experiments were carried out for core samples of wells of the Moskudinskoe oil deposit, as a result, it was found that the composition of the studied mudstones includes minerals of the montmorillonite group of clay, the structural packages of which are interconnected mainly by potassium and sodium ions. The obtained results are compared with the data from the research described in article [8].

The toughness of the mudstones structure, their crack resistance, is connected not only with their mineral composition of the rock, but also with the tenacity of the structural bonds between colloids and aggregates. In the scientific articles [9, 10], the possibility of clays to change their structure and composition at a stress of even 5-10 MPa is noted.

Taking this fact into account, experiments were carried out in this work to determine a number of physical and elastic-mechanical characteristics of mudstones of the Moskudinskoe oil deposit. As a result, the following experimental results were obtained: moisture content – 1.19 %; density – 2.51 g/ cm^3; uniaxial tensile strength along the bedding – 2.2 MPa; uniaxial tensile strength across the bedding – 3.7 MPa; uniaxial compressive strength along the bedding – 50 MPa; uniaxial compressive strength across the bedding (τ_c) - 69 MPa; Young's modulus along the bedding – $0.065 \cdot 10^5$ MPa; Young's modulus across the bedding – $0.075 \cdot 10^5$ MPa; Poisson's ratio along the bedding – 0.29; Poisson's ratio across the bedding is 0.33; the adhesion along the bedding – 12.5 MPa; the adhesion across the bedding – 17.3 MPa; the angle of internal friction (φ) is 36°.

To describe the ultimate stress state of the rock at the moment of its destruction, the Coulomb-Mohr strength theory is most widely used. According to it, the destruction of a rock, that is, the transition from an elastic state to a plastic one, is caused by the combined action of normal and tangential stresses. At the same time, it is necessary to calculate and determine the relationship of stresses acting in the rock.

Figure 1 shows a graphical justification for calculating the normal stress (σ) on the sites of maximum tangential stresses (τ). When the maximum ultimate stresses are reached ($\sigma_1 \rightarrow$ max at σ_3 = const), the effective stress σ_n can be considered as a criterion for the formation of a main fracture of destruction in the area of the site of maximum tangential stresses (τ_1). In addition, it is possible to construct a tangent line (envelope) to the Mohr's circle at an angle to the horizontal axis equal to the angle of internal friction (φ). The point of intersection of the Mohr's enveloping curve with the vertical axis (τ_0) is the cohesive strength (adhesion strength) of the rock. Producing this Mohr's enveloping curve to the intersection with the horizontal axis, it is possible to determine the fracture stress of the rock (σ_f).

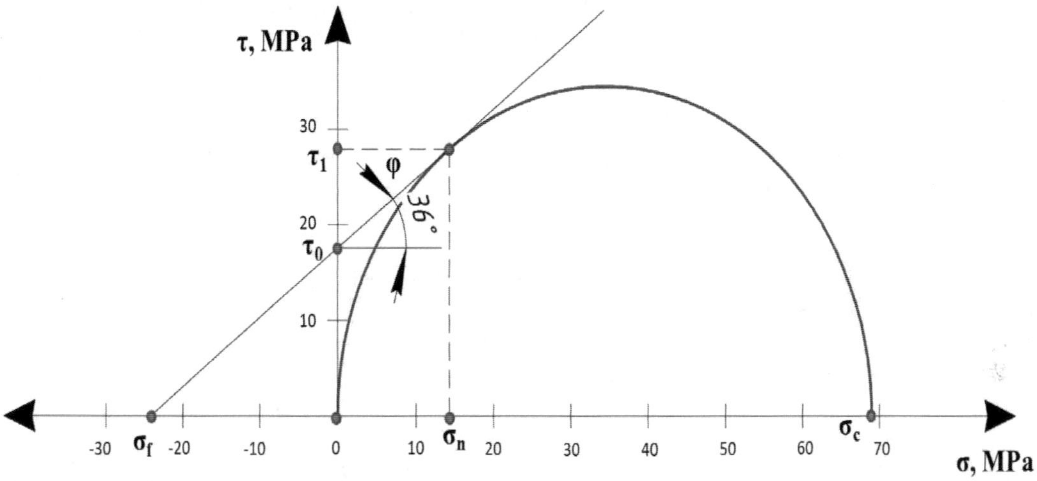

Figure 1. Constructed stress diagram for the mudstone of the Vereiskian formation of the Moskudinskoe oil field.

The shear strength of the body at each specific site is equal to the sum of the cohesion value τ and the normal stress σ, taking into account the angle of internal friction φ. To estimate σ_n, the above-justified Vereiskian mudstones, the strength index for uniaxial compression across their bedding σ_c=69 MPa and the angle of internal friction φ = 36° were used.

The stress assessment was carried out by a graphical method by constructing a failure envelope. It was assumed that the lateral strain σ_3=0; the vertical intensity σ_1=σ_c. Assuming uniaxial compressive strength across the bedding σ_c that is equal to 69 MPa, a circle (Mohr's circle) with a radius R that is equal to $\sigma_c/2$ was deposited. Then a tangent line was drawn to the Mohr's circle at an angle φ that is equal to 36°, and then a straight line to the abscissa axis (the axis of normal stresses) was drawn from the point of contact of the tangent line and the circle. Besides the prolongation of the Mohr's envelope to the abscissa axis was carried out. As a result, for the mudstone of the Vereiskian formation of the Moskudinskoe oil field, the adhesion τ_0, the maximum tangential stress (τ_1), the normal stress σ_n and the fracture stress σ_f, taking into account the internal friction angle 36°, equal 17.6, 27.9, 14.3 and 24.2 MPa, respectively (Figure 1).

The same result for σ_n can be obtained by calculating using the following equation (1):

$$\sigma_n = \frac{\sigma_c}{2} - \cos(90° - \varphi) \cdot \frac{\sigma_c}{2} = \frac{69}{2} - \cos(90° - 36°) \cdot \frac{69}{2} = 14.3\ MPa. \quad (1)$$

Thus, the carried out theoretical calculations show that the studied mudstones can crack and be potentially fractured at the pressure of more than 24.2 MPa.

The average bottom-hole injection pressure during hydraulic fracturing at the Vereiskian reservoir of the Moskudinskoe oil field varies in the range from 19.3 to 27.5 MPa. The upper limit of the range exceeds the fracture stress of mudstones. Therefore, there is a possibility that a hydraulic fracturing crack may develop above and below the perforation interval, involving not only the opened layer V3, but also the layer V4, in the case when it is not opened, as well as the suboptimal geometry of the fracture. In cases where there is a possibility of suboptimal fracture propagation in shoulder beds, the creation of fractures is possible by adjusting the technological parameters of the breakdown fluids and the injection program, as well as by more gentle methods. For the territory of the Perm region, the radial drilling technology has proven itself well for these purposes [11, 12].

The disadvantage of the Mohr's circles method for determining the ultimate fracture stresses of rock are that this method does not take into account the influence of the average main stress, σ_2, and, therefore, it gives results that may not converge with the actual characteristics of rocks [13]. In this regard, there is a need to search for samples of the mudstone core of a satisfactory state and to determine the elastic-mechanical properties (EMP) and stresses of rocks by other methods.

4. Conclusions

As a result of the completed work, the following conclusions were obtained:

- Assessment of the fracture stress of tight barriers that limit the fracture propagation of HF in height is an important task to improve the quality of planning the fracturing process.
- The fracture stress of mudstones of the Vereiskian reservoir of the Moskudinskoe oil deposit is no more than 24.2 MPa.
- There is a possibility that a fracture of HF may propagate above and below the target fracturing intervals in tight clay barriers (mudstones).
- It is necessary to search for samples of the mudstone core of the satisfactory condition and to use other methods for determining the EMP and stresses of rocks.

Acknowledgments

The study was carried out according to the state task of the Ministry of Education and Science of the Russian Federation FSNM-2020-0027 for performing academic studies for 2020 and the planning period of 2021 and 2022.

References

[1] Zhang G, Zhang Y, Xu A, Li Y 2019 Microflow effects on the hydraulic aperture of single rough fractures *Advances in GeoEnergy Research* **3 (1)** 104–114

[2] Kanaun S 2017 Hydraulic fracture crack propagation in an elastic medium with varying fracture toughness *International Journal of Engineering Sciences* **120** 15–30.

[3] Stephansson O, Semikova H, Zimmermann G, Zang A 2019 Laboratory pulse test of hydraulic fracturing on granitic sample cores from Äspö HRL *Rock Mechanics and Rock Engineering* **52 (2)** 629–633

[4] Efimov A A, Savitskii Ia V, Galkin S V, Soboleva E V, Gurbanov V Sh 2016 Study of wettability of reservoirs of oil fields by the method of X-ray tomography core *Neftegaz GNKAR* **4 (4)** 55-63

[5] Galkin S V, Kolychev I J, Savitskiy Ya V 2019 Potentialities of investigation of reservoir hydrophobization by compilation of X-ray core tomography and lateral logging *Russian Geology and Geophysics* **10** 1496-1507

[6] Efimov A A, Galkin S V, Savitckii Ia V, Galkin V I 2015 Estimation of heterogeneity of oil & gas field carbonate reservoirs by means of computer simulation of core x-ray tomography data *Ecology. Environment and Conservation* **21** 79-85.

[7] Economides M, Oligney R, Valkó P 2007 *Unified Fracture Design. Bridging the Gap Between Theory and Practice* (Izhevsk)

[8] Seredin V V, Siteva O S, Alvanyan K A, Andrianov A V 2020 Change in the Physico-Chemical Properties of Clays Subjected to Pressure *Perm Journal of Petroleum and Mining Engineering* **20 (4)** 304-316

[9] Seredin V V, Parshina T Y, Rastegaev A V, Galkin V I, Isaeva G A 2018 Changes of energy potential on clay particle surfaces at high pressures *Applied Clay Science* **155** 8–14

[10] Seredin V V, Khrulev A S, Rastegaev A V, Galkin V I 2020 Procedure of stress state assessment in rocks *Gornyi Journal* **2** 30–34

[11] Kochnev A A, Zotikov V I, Galkin S V 2018 Analysis of the influence of geological technological parameters on the effectiveness of radial drilling technology by the example of operational objects in perm region *Bulletin of Tomsk Polytechnic University, Geo Assets Engineering* **329 (12)** 20–29
[12] Kochnev A, Galkin S, Krivoshchekov S, Kozyrev N, Chalova P 2021 Application of machine learning algorithms to predict the effectiveness of radial jet drilling technology in various geological conditions *Applied Sciences (Switzerland)* **11,** 4487.
[13] Tiab D, Donaldson Erle C 2009 *Petrophysics: Theory and Practice of Measuring Reservoir Rock and Fluid Transport Properties* (Premium Engineering LLC)

IOP Conf. Series: Earth and Environmental Science 1021 (2022) 012065 doi:10.1088/1755-1315/1021/1/012065

Improving the technology of opening productive layers to boost the productivity of exploration and production wells

A P Chizhov, V E Andreev, L S Kuleshova, V V Mukhametshin, A V Andreev and A V Chibisov

Ufa State Petroleum Technological University, 1, Kosmonavtov st., Ufa, Republic of Bashkortostan, 450064, Russian Federation

E-mail: 4ap@list.ru, vv@of.ugntu.ru

Abstract. The article describes the results of improving the well construction completion technology. The factors and processes decreasing technological indicators of new exploration and production wells are analyzed. Research is based on systemic approaches to solving complex scientific and technical problems. The parameters of implementation of the improved technology are presented. The results of field trials intended to implement the technology in the construction of wells in Russian regions are analyzed.

1. Introduction

The efficiency and pace of prospecting, exploration and development of hydrocarbon deposits are determined by the end results at the well completion stage. Therefore, attention is paid to the problems of improving the quality of well completion in Russia and abroad [1-3].

The well completion stage is a complex geological and technical system that includes several sequentially conducted technological stages. All stages were considered as an integral system, which has a mutual influence on each other and the final quality of well construction.

2. Materials and methods

The stage of primary drilling-in of productive deposits is a key one in the well completion. The efficiency of technologies used at this stage determines the final indicators of well completion - hydrodynamic perfection and potential capabilities of productive formations and the degree of oil reserves recovery [4–10]. One of the main technological tasks of primary drilling is to maintain the wellbore in a technically reliable state, determined by the indicators of tightness and strength of the borehole walls in the intervals of permeable formations and unstable rocks. The indicators determine the protection of reservoir properties of the bottomhole zone of productive formations from pollution and the stability of technological processes.

While maintaining the hydraulic connection between the reservoirs and the well during the crossing of the hydrocarbon reservoir by the borehole, the bottomhole zone of oil and gas intervals will be polluted with varying intensity at the stage of primary opening and will continue at the stages of combating complications, testing the reservoirs, casing the well and secondary opening due to the filtration, clogging and physical chemical processes.

The hydraulic conditions of the primary opening of the productive strata is characterized by the regular change in the processes of dynamic and static filtration of fluid through the bottomhole zone of the productive strata under the influence of hydrodynamic pressures of various amplitudes and frequencies of oscillations. This intensifies the processes of reducing the permeability of oil and gas reservoirs at the depth of penetration of the liquid phase and solid particles of the drilling fluid.

If rocks with low permeability (of the order of 0.05 μm^2 and less) are exposed, the predominant effect on the decrease in the reservoir properties of the bottomhole zone will be exerted by the filtrate of the drilling fluid. With an average permeability of reservoirs (approximately from 0.2 to 0.7 μm^2), a change in their properties will occur under the influence of filtration of both phases of the drilling fluid - solid and liquid, but with a limited depth of penetration of the solid phase. With a high permeability of the reservoir (more than 1.2 μm^2), the improvement of filtration properties will be determined by the clogging process, i.e., penetration of the solid phase of the solution into the channels to a depth that depends on the repressions in the well.

Reservoirs of the carbonate type are exposed to the greatest decrease in natural permeability during the initial opening [2]. Fractured rocks plugging occurs when the fracture opening is less than 0.25 mm. Fracture clogging with a solid phase of drilling fluids is typical for channels with openings from 0.25 to 1.2 mm. With large openings of cracks and caverns, continuous filtration of drilling fluids into the depths of permeable formations occurs.

The influence of the above factors on the behavior of the bottomhole zone of productive formations in the initial opening by drilling determines the degree of impact on the hydrodynamic perfection of wells at all subsequent stages.

If at the stage of primary drilling, there is no hydraulic connection between the reservoirs and the well, all the previously listed negative consequences will not arise at the subsequent stages of completion or be minimized.

The stage of the secondary opening of productive formations includes the replacement of drilling fluid with special one, the perforation of the production string against the productive formations and the preparation of the well for development.

The stage of secondary opening of productive formations determines the conditions for maintaining the reservoir properties of the bottomhole formation zone (BFZ) of oil and gas horizons and isolating them from alien formations during operations aimed to establish a hydraulic connection with the well and the bottomhole structure. At this stage, the quality of work depends on the degree of protection of the bottomhole zone of productive formations at the stages of primary opening, borehole fastening and secondary opening technology. The waterproofing of oil and gas formations from other elements will depend on the results of the well casing stage and the methods used to create communication channels between the well and productive formations.

3. Results and analysis

For reliable isolation of productive intervals, it is proposed to treat the walls of the well with jets of drilling mud. The method is used in the process of drilling a well and primary opening of productive intervals. For this purpose, an above-bit sub with 1–3 jet nozzles, through which the borehole walls are treated with jets of drilling mud, is included in the bottom of the drill string.

Kinetic energy of the jet is transformed into a field of high-amplitude and low-frequency pressure pulses in the wellbore zone of the exposed rocks. These processes have the following technological parameters: time of contact with the rock is $(12–17)\cdot 10^{-3}$ s; the force of the dynamic impact of the jet is 1.5–4.0 kN, the temperature of the bridging medium is about 500–600^{0}C. As a result, a waterproofing screen is formed with the depth of the colmatation zone in the void medium of rocks $(10–30)\cdot 10^{-3}$ m. The insulating properties of the screen are characterized as follows: at depressions - $(1.0–1.2)\cdot 10^4$ Pa/m; during repression - $1.8–2.5\cdot 10^4$ Pa/m.

In addition, the technology makes it possible to improve the strength characteristics of the borehole walls, including in unstable rocks, represented by terrigenous and carbonate rocks.

At the same stage of well completion, the formation of the bottomhole structure is completed, which will determine the degree of hydraulic perfection of the production well.

The generalizing indicator characterizing the influence of all stages of well completion on the filtration properties of the bottomhole zone is the coefficient of hydraulic perfection of the well [2, 3]:

$$\delta = \frac{Q_A}{Q_{per}}, \tag{1}$$

where Q_A – actual well production rate, m³/s; Q_{per} – flow rate of a hydrodynamically perfect well with an open bottom and natural reservoir properties of the bottomhole zone, m³/s.

For a hydraulically perfect well, filtration resistance of the bottomhole zone is associated with the curvature and thickening of the flow lines of formation fluids when moving from the formation to the wellbore. The flow rate of a perfect well is determined by the Dupuis formula for a steady radial inflow of incompressible fluid.

For a hydraulically imperfect well, the Dupuis formula can be written as

$$Q_{\Phi} = \frac{2\pi k h \Delta P}{\mu(\ln R_{\kappa}/r_c + C + S)}, \text{ м}^3/\text{с} \tag{2}$$

where $C = C_1 + C_2$ is the coefficient that takes into account additional filtration resistance of the bottomhole formation zone due to imperfections in the degree (C1) and the nature of the opening (C2), determined from the graphs by V.I. Shchurov [4–6]; $S = S_1 + S_2$ is the coefficient taking into account additional filtration resistance in the bottomhole formation zone due to the hydraulic imperfection of the primary opening of the productive formation (S_1) and the secondary opening - perforation (S_2).

If coefficient C characterizes the imperfect geometry of the bottomhole structure, S is the deterioration of the natural reservoir properties of the bottomhole zone during well completion [7, 8].

The hydrodynamic perfection of wells by the degree of preservation of the natural reservoir properties of the bottomhole zone of productive formations is determined by the processes of hydromechanical, physicochemical and chemical interaction with drilling fluid and grouting slurries at all well completion stages.

4. Discussion

Based on the results of analytical studies of a large amount of field information (more than 260 wells), it was found that the hydraulic perfection coefficient can decrease to 0.2.

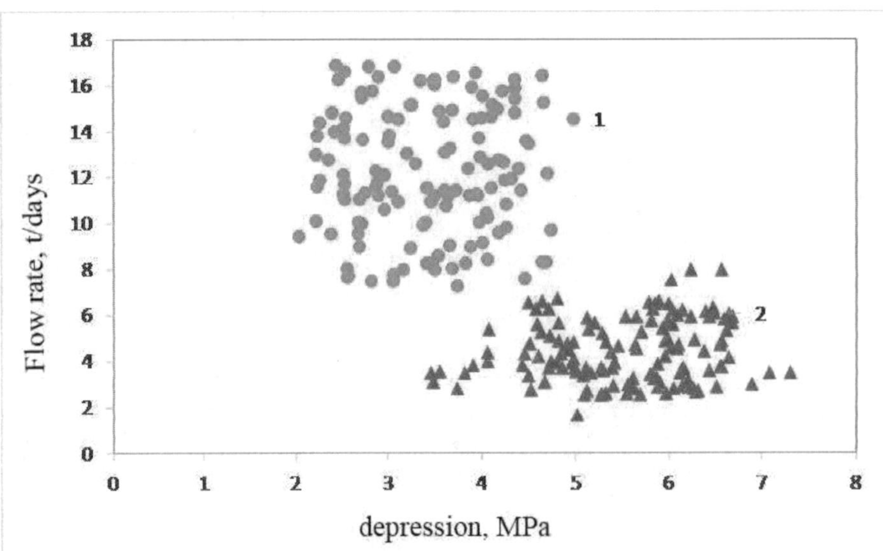

Figure 1. Actual values of production rates introduced from well drilling with the use (1) and without the use of GISS technology (2) during well construction

The generalization of integral indicators of drilling operations revealed an increase in the current flow rate by 1.6-3.0 times (Figure) a decrease in the water cut of the products by 2.6 times (Table).

The GISS technology, causing a non-linear increase in the baseline well construction indicators, creates optimal thermodynamic conditions for improving the design of deep wells and filters, completing wells with an open bottom, hydraulically perfect in nature and degree of penetration.

Table 1. Average technological indicators of completed wells using the GISS technology (more than 150 wells were introduced)

Indicators	Oil flow rate, t / day	Water cut, %	Depression, MPa	Productivity coefficient, m^3/day MPa m
Indicators of pilot wells	11.1	34.4	3.6	1.20
Indicators of base wells	4.3	88.0	5.7	0.33
The ratio of indicators of pilot and base wells	+2.6	-2.6	-1.6	+3.6

In addition, the technical and economic indicators of drilling operations increased: ROP - by 13.6%; drilling technical speed - by 8.6%; commercial drilling rate - by 6.5%; the coefficient of cavernousness of the Verey horizon decreased; the time for additional operations (isolation, borehole reaming) decreased by 97.5%.

5. Conclusion

The results obtained showed that:

- with a non-linear increase in the quality, efficiency, environmental and technical safety of the construction of oil and gas wells, the amount of funds and time for their construction decreased;
- in comparison with traditional ones, the GISS technology meets the key requirements of construction, well operation and development of oil and gas fields - achieving the maximum technical and economic effect with minimal material and financial costs.

References

[1] Dupriest F E and Koederitz W L 2005 Maximizing Drill Rates with Real-Time Surveillance of Mechanical Specific Energy *SPE/IADC Drilling Conference (Amsterdam, The Netherlands, 23-25 February 2005) SPE/IADC 92194.*

[2] Romero S, Monroy R R, Johnson C, Cardenas F and Abraham G A T 2006 Preventing Lost Circulation by Use of Lightweight Slurries With Reticular Systems: Depleted Reservoirs in Southern Mexico *SPE Drilling & Completions Journal* **21(3)** 185-192

[3] Dupriest F E 2006 Comprehensive Drill Rate Management Process to Maximize Rate of Penetration *SPE Annual Technical Conference and Exhibition (San Antonio, USA, 24-27 September 2006) SPE 102210*

[4] Chizhov A P, Andreev V E, Chibisov A V, Mukhametshin V V, and Kuleshova L S 2020 Hydraulically perfect modes of injection of grouting mixtures when isolating absorbing formations *IOP Conference Series: Materials Science and Engineering* **952** 012040 DOI: 10.1088/1757-899X/952/1/012040

[5] Salimov F S, Kotenev Yu A, Mukhametshin V Sh, Sultanov Sh Kh and Safiullina A R 2020 Development of oil deposits in tectonic conditions *IOP Conference Series: Materials Science and Engineering* **905** 012086 DOI: 10.1088/1757-899X/905/1/012086

[6] Mukhametshin V V and Kuleshova L S 2020 On the use of geophysical research data aiming to increase the efficiency impact on the bottomhole zone of wells *IOP Conference Series: Materials Science and Engineering* **905** DOI: 10.1088/1757-899X/905/1/012084

[7] Batalov S A, Andreev V E, Lobankov V M and Mukhametshin V Sh 2019 Numerical

simulation of oil formation with regulated disturbances. Oil recovery quality simulation *Journal of Physics: Conference Series* **1333** 032006 DOI: 10.1088/1742-6596/1333/3/032006

[8] Mingulov Sh G and Mukhametshin V Sh 2020 Restoring the intake capacity of injection wells by means of shock-wave exposure *IOP Conf. Ser.: Mater. Sci. Eng.* **952** 012060 DOI: 10.1088/1757-899X/952/1/012060

[9] Soloviev N N, Mukhametshin V Sh and Safiullina A R 2020 Developing the efficiency of low-productivity oil deposits via internal flooding *IOP Conf. Ser.: Mater. Sci. Eng.* **952** 012064 DOI: 10.1088/1757-899X/952/1/012064

[10] Mukhametshin V Sh 2020 Rationale for the production of hard-to-recover deposits in carbonate reservoirs *IOP Conf. Ser.: Earth Env.* **579** 012012 DOI: 10.1088/1755-1315/579/1/012012

[11] Roth J, Reeves C, Johnson C R, De Bruijn G, Bellabarba M, Le Roy-Delage S and Bulte-Loyer H 2008 *Innovative Hydraulic Isolation Material Preserves Well Integrity (SPE 112715) IADC/SPE Drilling Conference (Orlando, Florida, March 4-6 2008)*

[12] Mukhametshin V Sh 2020 Justification for increasing the performance of hydrochloric acid treatment in wells of fields with carbonate reservoir *IOP Conf. Ser.: Mater. Sci. Eng.* **905** 012083 DOI: 10.1088/1757-899X/905/1/012083

[13] Mukhametshin V Sh, Andreev V E, and Yaskin S A 2020 Designing measures to increase oil recovery based on the identification and grouping of deposits *IOP: Earth and Environmental Science (EES) (International Symposium «Earth sciences: history, contemporary issues and prospects»)* **579** 012013 DOI: 10.1088/1755-1315/579/1/012013

[14] Mukhametshin V Sh, Tyncherov K T and Rakhimov N R 2020 Geological and field control and diagnostics of treatment efficiency of a wellbore zone within carbonate reservoirs *IOP Conf. Ser.: Earth Env.* **579** 012021 DOI: 10.1088/1755-1315/579/1/012021

Estimation of the optimal direction of the horizontal borehole relative to the minimum and maximum formation stress

I I Bosikov[1], E V Egorova[2], A V Karpikov[3]

[1]North Caucasian Institute of mining and metallurgy (State Technological University), 44 Nikolaeva St., Vladikavkaz, 362021, Russian Federation
[2]Astrakhan State Technical University, Institute of Oil and Gas, 20a Tatishcheva St., Astrakhan 414056, Russian Federation
[3] Irkutsk National Research Technical University, 83, Lermontova str., Irkutsk, 664074, Russia

E-mail: igor.boss.777@mail.ru

Abstract. The article discusses the analysis and assessment of the optimal direction of the horizontal wellbore and its spatial location relative to the minimum and maximum reservoir stresses play a significant role in the design of an operation to create multiple hydraulic fractures. Theoretically, two cases are possible: the horizontal wellbore is located strictly in the direction of maximum horizontal stress; the horizontal wellbore is located strictly in the direction of the minimum horizontal stress. As a conclusion based on the hydraulic fracturing design, the well productivity factor was calculated and compared for longitudinal and transverse fractures; longitudinal fractures have a clear advantage. According to the presented calculation, the hydraulic fracturing method will have advantages in a horizontal well in a highly permeable reservoir. The hydraulic fracturing method will increase the vertical permeability. According to the presented calculation, the hydraulic fracturing method will have advantages in a horizontal well in a highly permeable reservoir. The hydraulic fracturing method will increase the vertical permeability and unite the disparate parts of the reservoir. The hydraulic fracturing method will increase the vertical permeability and unite the disparate parts of the reservoir, which will make it possible to more efficiently develop the reserves of the Neocomian deposits in the field.

1. Introduction

Determination of the optimal direction of the horizontal wellbore and its spatial location relative to the minimum and maximum reservoir stresses play a significant role in the design of an operation to create multiple hydraulic fractures.

Theoretically, two cases are possible: the horizontal wellbore is located strictly in the direction of maximum horizontal stress; the horizontal wellbore is located strictly in the direction of the minimum horizontal stress.

2. Materials and methods

When choosing the optimal technology, several criteria were followed – achieving the maximum efficiency of hydraulic fractures due to the maximum interference time; the minimum degree of damage to the created fractures and the productive formation, as well as the achievement of maximum conductivity of the fixed hydraulic fractures [11, 14].

The optimal geometry of multiple hydraulic fractures created in a horizontal wellbore depends entirely on the choice of the applied working fluid, the fixing material, as well as the timeliness and quality of the extraction of the decomposition products of the working fluid.

The most reasonable description of the behavior of a horizontal wellbore with hydraulic fractures can be obtained only from the results of hydrodynamic modeling [1–3].

Determination of the direction of the minimum and maximum horizontal stress is necessary to select the orientation of the horizontal wellbore with the subsequent determination of the type of hydraulic fractures created if there is a need to control their propagation into the water-saturated zone.

When solving such a problem, the primary methods are the available seismic data, built on their basis maps of natural fracturing, the direction of development of the network of natural fractures - as an indicator of the direction of the minimum horizontal stress and possible zones of anisotropy of the reservoir, certain tectonic disturbances - faults, faults, shears and thrusts, data of core studies and tests of MiniFrakov for wells located in the study area and which are typical for it [4–6].

Then it is necessary to conduct a thorough comparison of development options using transverse fractures (the well is drilled in the direction of minimum horizontal stress, and hydraulic fractures will be perpendicular to the well axis) or a longitudinal fracture (the well is drilled in the direction of maximum horizontal stress, and the hydraulic fracture will be parallel to the well axis) [13, 15].

Of course, this issue should be resolved in the context of field development. If waterflooding is planned, then by default it is recommended to place the wellbores along with the maximum horizontal stress, so it is easier to control the waterflooding front [7–8].

Often, with permeability > 10 mD, horizontal fractures gradually begin to dominate over perpendicular ones.

When considering the position of a candidate well (horizontal, cased, and stimulated by the creation of multiple hydraulic fractures of the well) within the reservoir, the permeability of the reservoir is taken into account; required well productivity; economy; placement of proppant and duration of operation, especially in case of depletion of the object to an average pressure below the bubble point pressure. The first and last of the above considerations should be decisive in the choice of well completion with longitudinal fractures [9, 12].

Lack of adhesion between the casing and the formation rock is considered one of the most important factors that can lead to stimulation failure.

3. Results and discussion

3.1. The main parameters of hydraulic fracturing of a highly permeable formation

Since neocom is a highly permeable formation, the TSO (tipscreenout) technology is proposed for hydraulic fracturing - this is end screening, which stops lateral fracture growth and then allows it to increase its opening and fill it with proppant. The result is a short but wide crack.

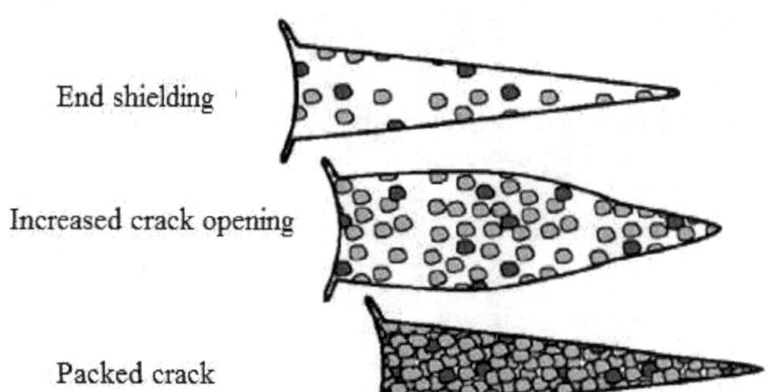

Figure 1. Increased crack opening (inflation) of a crack with end screening technology.

End screening occurs when a proppant concentration is created at the front of the fracture that is sufficient to prevent further fracture elongation. As soon as the growth of the crack is stopped (in this case, we assume that the rate of injection is greater than the rate of fluid leakage into the formation), the continued injection will inflate the crack, i.e., increase its width (Figure 1).

End screening and fracture inflation should be accompanied by an increase in the effective fracture pressure.

Cracking and stalling (i.e., end shielding) is followed by a relatively small "cushion" of clean fluid (no sand) followed by a "slurry" containing 1–4 pounds of sand per gallon of fluid (120–480 kg / m^3).

When fracture growth is stopped, further injection increases fracture opening and allows high concentration slurry to be injected (i.e., 1016 lb/gal = 1200–1920 kg/m^3). Final areal proppant concentrations of 20 pounds per square meter are possible (98 kg/m^2).

Thanks to end shielding, it became possible to increase the crack width without increasing its longitudinal dimension [17-19].

Throughout the fracturing process in highly permeable formations, considerations of effective pressure and fluid leakage are dominant. Firstly, high-permeability formations are usually loose and have low values of the elastic modulus, and secondly, the volumes of fluid are relatively small, while the rates of leakage (filtration) into the formation are high (high permeability, compressible fluids, and fracturing fluids that do not create crusts on the walls of the crack).

For hydraulic fracturing of highly permeable formations, the main issue of paramount importance, in addition to maintaining high permeability at all stresses, is the choice of proppant sizes. While high permeability fractures of course use proprietary proppants such as medium strength and resin-coated proppants, most fracturing operations are performed using standard-sized sand [20-23].

Typical high permeability fracturing uses a 20/40 mesh proppant (sand). Achieving the highest possible fracture conductivity can by itself help prevent sand production by reducing the drawdown.

The choice of fluids for high-permeability fracturing has always been guided by considerations of minimal damage to a high-permeability formation, both due to the formation of a filter cake, and (especially) due to the penetration of the polymer into the formation.

If the penetration of fracturing fluid is minimized, then the degree of damage (i.e., impairment of permeability caused by filter cake or penetration of fracturing fluid) is no longer as important. If the penetration of fluid into the formation during lost circulation is small, then even severe permeability degradations may be permissible without creating a positive skin effect.

3.2. Hydraulic fracturing design for the Western reservoir

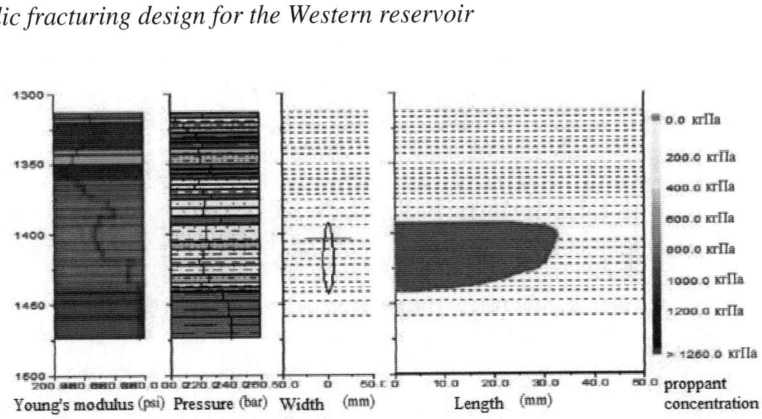

Figure 2. Crack simulation.

Design data was taken from a geomechanical model.

The fluid for hydraulic fracturing was taken by analogy with well R-8: guar-based liquid (carrier liquid) YF130STLUKOIL; displacement fluid WF115. The proppant permeability was calculated based on the following parameters:

- Static temperature: 63 ^{0}C.
- Proppant pressure: 103 bar.
- Proppant concentration: 4.88 kg/m^2.
- Average Young's Modulus: 3.270E + 04 bar.
-

As a conclusion, according to the hydraulic fracturing design, the productivity factor of the wells was calculated and compared for longitudinal and transverse fractures; longitudinal fractures have a clear advantage (Figure 3).

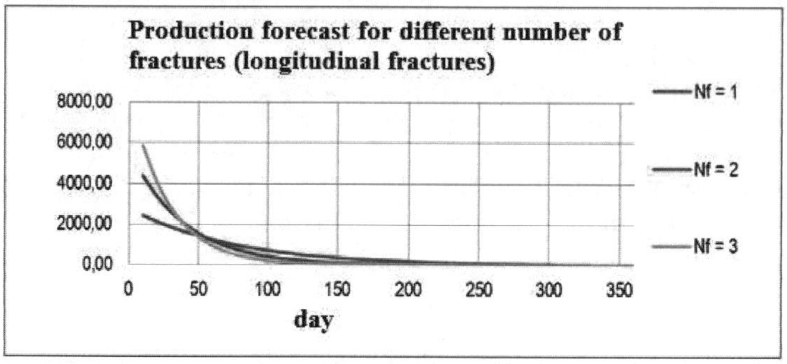

Figure 3. The production forecast for the different number of fractures (longitudinal fractures).

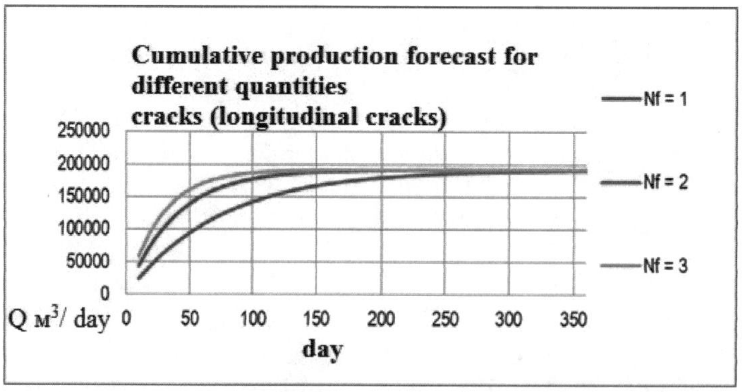

Figure 4. The cumulative production forecast for different quantities cracks (longitudinal cracks).

According to the presented calculation, the hydraulic fracturing method will have advantages in a horizontal well in a highly permeable reservoir. The hydraulic fracturing method will increase the vertical permeability.

According to the presented calculation, the hydraulic fracturing method will have advantages in a horizontal well in a highly permeable reservoir. The hydraulic fracturing method will increase the vertical permeability and unite the disparate parts of the reservoir.

4. Conclusion

In the research work, the analysis of oil recovery methods used, hydraulic fracturing modeling techniques was carried out, and hydraulic fracturing of a highly permeable formation was simulated for two wells 1 and 3 (double-wellbore) in this field.

Hydraulic fracturing design optimization is aimed at creating a fracture of maximum conductivity and minimum length. Since neocom is a highly permeable formation, the TSO (tipscreenout) technology is proposed for hydraulic fracturing – this is end screening. This TSO technology is successfully applied at the Prudhoe Bay field (USA), in the Gulf of Mexico, Indonesia, the North Sea, Canada, Brazil, Venezuela, Vietnam, Saudi Arabia. In Russia, an example of successful design and execution is the hydraulic fracturing operation with TSO in well H-34.

The hydraulic fracturing method will increase the vertical permeability and unite the disparate parts of the reservoir, which will make it possible to more efficiently develop the reserves of the Neocomian deposits in the field.

References

[1] Melnikov N N, Kalashnik A I, Kalashnik N A and Zaporozhets D V 2018 Integrated multi-level geomonitoring of natural-and-technical objects in the mining industry *Journal of Mining Science* **54 (4)** 535–540

[2] Litvinenko V S 2020 Digital economy as a factor in the technological development of the mineral sector *Natural Resources Research* **29 (3)** 1521–1541

[3] Makridin E V, Tyulenev M A, Markov S O, Lesin Y V and Murko E V 2020 Overburden management towards higher safety in coal mining regions *Mining Informational and Analytical Bulletin* **12** 89–102

[4] Petrov Yu S and Sokolov A A 2016 Increase of effective management of technological processes of the mountain enterprise on the basis of the analysis of information on technogenic cycles *2nd International Conference on Industrial Engineering, Applications and Manufacturing* ICIEAM 2016 19–20 May (Chelyabinsk) pp 1716–1721

[5] Bosikov I I, Klyuev R V, Khetagurov V N and Azhmukhamedov I M 2021 Development of methods and man-agement tools aerogasdynamics processes at mining sites *Sustainable Development of Mountain Territories* **1** 77–83

[6] Shchurov N I, Dedov S I, Malozyomov B V, Shtang A A, Martyushev N V, Klyuev R V and Andriashin S N 2021 Degradation of lithium-ion batteries in an electric transport complex *Energies* **14 (23)** 8072

[7] Rylnikova M V, Vladimirov D Y, Pytalev I A and Popova T M 2017 Robotic geotechnologies as way of improving efficiency and ecologization of mineral resource management *Physical and Technical Problems of Mining* **1** 92–101

[8] Katsubin A V, Khoreshok A A, Tyulenev M A and Markov S O 2020 Technology of advance cutting of sloping and steeply pitching coal seams using hydraulic backhoe excavators *Mining Informational and Analytical Bulletin* **11** 27–36

[9] Zhukovskiy Yu L, Korolev N A, Babanova I S and Boikov A V 2017 The prediction of the residual life of electromechanical equipment based on the artificial neural network *IOP Conference Series-Earth and Environmental Science* **87** 032055

[10] Konyuhov V Yu, Gladkih A M, Galyautdinov I I, Shchadova E I 2020 Calculations of efficiency in implementing progressive mold forming methods. *IOP Conference Series: Materials Science and Engineering* **760** 012027

[11] Bolobov V I, Chupin S A, Akhmerov E V, Plaschinskiy V A 2021 Comparative wear resistance of existing and prospective materials of fast-wearing elements of mining equipment. *Materials Science Forum* **1040** 117–123

[12] Martyushev N V, Pashkov E N 2013 Bronze sealing rings defects and ways of its elimination. Applied Mechanics and Materials, **379,** 82–86

[13] Karlina Yu I, Ershov V A, Olentsevich V A, Karlina A I 2021 Justification of the choice of a thermal deburring machine for deburring small parts of microwave electronics. *Journal of Physics: Conference Series* **2032(1)** 012067

[14] Konyuhov V Yu, Gladkih A M, Chemezov A V 2019 Evaluation of the economic feasibility of the introduction of plasma hardening technologies in the Far North enterprises. *IOP Conference Series: Materials Science and Engineering* **560** 012147

[15] Bataev D K S, Khadisov V Kh, Abukhanov A Z, Salamanova M Sh 2019 Application of man-induced raw material resources as a way to solve ecological problems. *IOP Conference Series:*

Materials Science and Engineering **663(1)** 012047
[16] Balanovskiy A E, Astafyeva N A, Kondratyev V V, Karlina A I 2021 Study of mechanical properties of C-Mn-Si composition metal after wire-arc additive manufacturing (WAAM). *CIS Iron and Steel Review* **22** 66–71
[17] Bazhenov Yu M, Murtazaev S-A Yu, Bataev D K-S, Alaskhanov A H, Murtazaeva T S A, Saydumov M S 2021 High-strength concretes based on anthropogenic raw materials for earthquake resistant high-rise construction. *Engineering Solid Mechanics* **9(3)** 335–346
[18] Konyuhov V Y, Gladkih A M, Zott R S 2020 Accelerator as an effective replacement of a business incubator in the Irkutsk region. *Journal of Physics: Conference Series* **1582** 012045
[19] Voropai N, Ukolova E, Gerasimov D, Suslov K, Lombardi P, Komarnicki P 2019 *Simulation approach to integrated energy systems study based on energy hub concept.* 2019 IEEE Milan PowerTech, PowerTech 2019 8810666
[20] Lombardi P, Styczynski Z, Sokolnikova T, Suslov K 2014 *Use of energy storage in Isolated Micro Grids.* Proceedings - 2014 Power Systems Computation Conference, PSCC 2014, 7038361
[21] Stotzer M, Gronstedt P, Styczynski Z, Buchholz B M, Glaunsinger W, Suslov K V 2012 *Demand side integration-A potential analysis for the German power system.* IEEE Power and Energy Society General Meeting, 6345090
[22] Konyuhov V Yu, Gladkih A M, Galyautdinov I I, Kiseleva T Yu 2019 Ecological architecture: The green roofs. *IOP Conference Series: Earth and Environmental Science* **350** 012035
[23] Lombardi P, Sokolnikova T, Suslov K V, Styczynski Z 2012 *Optimal storage capacity within an autonomous micro grid with a high penetration of Renewable Energy Sources.* IEEE PES Innovative Smart Grid Technologies Conference Europe, 6465706

Algorithm for optimization of methanol consumption in the «gas inhibitor pipeline-well-gathering system»

M S Proshutinskiy[1], **I R Raupov**[1], **N M Brovin**[2]

[1] Saint-Petersburg Mining University, 2, 21st Line ave., St. Petersburg, 199106, Russia
[2] Peter the Great St. Petersburg Polytechnic University, 29, Polytechnicheskaya, St. Petersburg, 195251, Russia

E-mail: matvey.proshutinskiy@mail.ru, inzirrr@yandex.ru, Brovin.NM@mail.ru

Abstract. The article proposes an algorithm of actions to optimize the consumption of methanol which is used as an inhibitor of hydrate formation in the process chain of oil and gas production. The proposed approach was developed in the process of analyzing the methanol supply system and planning measures to optimize it for the Eastern section of the Orenburg oil and gas condensate field. The algorithm consists of five sections which to a greater extent affect the efficiency of the inhibition system and are closely related to each other. These are data preparation, determination of equilibrium conditions of hydrate formation, calculation of methanol consumption rates, analysis of the inhibition system flexibility, operational control. The proposed solution in the form of an algorithm makes it possible to carry out a complex analysis of the hydrate formation prevention system, to determine the most optimal solutions to reduce the consumption of methanol.

1. Introduction

Natural hydrates are crystalline solids in which gas molecules (guests) are trapped in the cavity of the crystal lattice formed by water molecules (host) [1, 2].

Hydrate formation is one of the most common problems in the production and transportation of hydrocarbons in both purely gas and oil, gas and condensate fields. The subject of research (the Eastern sector of the Orenburg oil and gas condensate field) in the analyzed period from November 2020 to March 2021 were presented by more than 400 cases of hydrate formation which led to the loss of about 2000 tons of oil and 4 Mcm of gas.

Hydrates are deposited on the inner walls of the tubes sharply reducing their throughput up to a full blockage; the deposition of hydrates inside the valves leads to the loss of its performance. Hydrate formation not only reduces the efficiency of field operation, but also leads to emergency situations, the removal of which makes a significant contribution to the hydrocarbons production cost [3].

The most technologically advanced method of preventing hydrate formation is chemical, namely, the hydrate formation inhibitor injection to sections complicated by hydrates. The greatest practical application in the gas-hydration control was obtained by the method of injecting an inhibitor into a fluid flow. At oil and gas companies of Russia methyl alcohol CH_3OH (methanol) is used as an inhibitor due to the ratio of cost and high technological efficiency [4, 5]. It can be injected into the fluid flow using methanol pipeline from the chemical storage, or through chemical injection skid closely installed to the inhibitor feed point.

E B Buhgalter, Yu F Makogon, A G Burmistrov, G V Ponamorev, F A Kuznetsov, Y A Dyadin, V R Belosludov, V A Istomin, E D Sloan and many others were involved in the development of gas-

IOP Conf. Series: Earth and Environmental Science 1021 (2022) 012067 doi:10.1088/1755-1315/1021/1/012067

hydrates control methods at different times [1, 11, 12-16, 21-27]. In Russia, the document STO Gazprom 3. 1-3-010-2008 «Methodology for calculating the rates of consumption of chemicals for gas production enterprises of Gazprom JSC», which is currently used for rationing methanol consumption, was developed by the Scientific-Research Institute of Natural Gases and Gas Technologies – VNIIGAZ.

Despite the considerable experience in the use of methanol to prevent hydrate formation, the problems associated with the increased consumption of methanol and the production of outputs that do not satisfy the rules on the content of dissolved methanol [8]. The use of methanol as a hydrate inhibitor significantly increases the cost of production of hydrocarbons, so the planning of measures to optimize its consumption is an actual problem.

2. Algorithm for optimization of methanol consumption

The consumption of methanol injected into the system is set based on the consumption rates (units of mass of methanol/unit of volume of gas) which are individual for each technological sections. These rates are calculated using methods and adjusted based on the actual data on objects complicated by hydrate formation.

In the process of analyzing the methanol inhibition system in the Eastern section of the Orenburg oil and gas condensate field, the main directions of optimization of methanol consumption were determined and an algorithm of actions was drawn up (figure 1). In this article, the sections of the proposed algorithm are sequentially discussed, their interrelation and the expected result of each stage are shown [18-20].

The first basic stage is the preparation of information. The main information required for calculations and planning of actions to prevent hydrate formation is: consumption of produced / injected (in the case of gas-lift operation) gas; its component composition; specific humidity of gas; pressure and temperature conditions; salinity of water; the amount and composition of oil, condensate and water produced together with gas. Information collection occurs by sensors installed on the pipes, wellheads and bottom-holes of the wells, well test, laboratory tests for determination of the component composition of fluids, specific humidity of gas, ion composition of the water (figure 1, direction 1). After data collection, it is necessary to analyze their completeness. The result is a list of technological sections for which the necessary data are missing (not equipped with sensors, the sensor is defective, surveys were not carried out or were carried out incorrectly). The quality control of the provided data is carried out in order to determine their reliability and relevance. At this sub-stage, it is necessary to clarify the time of the research and the reliability of the data obtained. The transition from data to information, determination of invalid data can be carried out with filters configured in the program code or installed through the built-in functions of data processing programs. For example, if temperature sensors show approximately the same value on the wells of one cluster, but the temperature of one of the wells at the same flow rate significantly out of this array, then it is advisable to verify the performance of the installed temperature sensor. The missing parameters determined in the data completeness analysis, as well as the data, the reliability of which is in doubt, can be determined by a calculation method based on the available information on the given object. The final result of the data preparation stage is the information ready for use in calculations. To be able to analyze parameters in dynamics (temperature and pressure of the fluid flow, flow rate of produced water, locations of hydrate formation), it is necessary to maintain a database containing information structured by date and type of data.

The next stage is to determine the curve of equilibrium conditions for hydrate formation. There are many engineering techniques for determining the equilibrium conditions of hydrate formation [9, 10, 11]. The choice of a particular technique should be carried out on the basis of information of the component composition of the gas for which it is applicable, as well as on the number of factors taken into account by it, such as the salinity of formation water, the presence of nitrogen, hydrogen sulfide, and carbon dioxide in the gas mixture. In addition to determining the conditions of hydrate formation using engineering techniques, the curve can be determined experimentally by the laboratory tests [6], can be plotted from actual field data if information of hydrate formation depth and the distribution of pressure and temperature is available. The depth hydrate formation can be determined by drift run and the distribution of temperature and pressure using fiber-optic temperature log [17]. The next sub-stage is to determine the technological section at risk of hydrate formation using the curve and information obtained in the first stage. It should be noted that various simulators for mathematical modeling of oil and gas production processes are used to determine equilibrium conditions for hydrate formation and predict sections of hydrate formation: OLGA, CSMHyK, CSMHyFAST and CSMGem [5]. If these areas match with the actual information about the places of formation of gas hydrates, then the curve of hydrate equilibrium conditions can be used for further calculations. If for most of the objects the actual data do not match with the calculated ones, then it is advisable to recheck the initial data used for plotting the curve and, provided they are reliable, choose another method for calculating (figure 1, direction 2). For one field, there can be several curves if the input data for its plotting are different (the component composition of the produced / injected gas, salinity of the produced water, etc.).

The key factor in choosing a method for calculating the consumption rates of methanol is the assessment of its applicability for the conditions of a particular field. If there are recommendations for determining consumption rates for purely gas and gas condensate fields (VRD 39-1.13-010-2000. Instructions on the calculation of methanol consumption standards for calculation of the maximum permissible or temporarily agreed methanol discharges for the facilities of Gazprom JSC), there are no such recommendations for oil and gas condensate fields where oil production is accompanied by a high gas-oil ratio and formation of hydrates. After selecting the method, the methanol consumption rates are calculated which are compared with the current actual values. If the calculated values are higher than the actual ones and at the same time there are no problems with the formation of hydrates in the system, or the calculated values are much less than the actual ones and at the same time the accumulated practical experience indicates the unreliability of the obtained values, it is advisable to choose a different calculation method and check whether all influencing factors were taken into account (figure 1, direction 3). Based on the results of this stage, actions are planned to optimize the actual consumption rates to the calculated ones (figure 1, direction 4). To check the calculated consumption rates, it is necessary to select objects that do not have a high impact on production at the field, and methanol consumption reduction programs should be planned step by step with operational control of the situation.

An important stage is the analysis of the flexibility of the methanol supply system that means:

1. number and location of methanol feed points (for more targeted distribution of methanol throughout the system);
2. speed of response to changes in thermobaric conditions.

Let's analyze the dependence of the methanol consumption on the number and location of the feed points on the example of a section of the gas-lift gas distribution system expose to hydrate formation. Let the gas-lift gas flow rate required to maintain the operating parameters of Well 1 be Q_1, Well 2 – Q_2, and the rates of methanol consumption be G_1 and G_2, respectively, where $G_1 = 2G_2$.

Thus, having one feed point (figure 2, a), to ensure a hydrate-free operation of gas pipelines, it is necessary to inject methanol in the amount:

$$q = G_1 \cdot (Q_1 + Q_2) = G_2 \cdot (2Q_1 + 2Q_2). \quad (1)$$

When adding a feed point to the FL_1 - Inj 1 gas pipeline (figure 2, b), the methanol consumption required to ensure the hydrate-free operation of gas pipelines will be:

$$q = G_1 \cdot Q_1 + G_2 \cdot Q_2 = G_2 \cdot (2Q_1 + Q_2). \quad (2)$$

That is, the savings of methanol will be $G_2 \cdot Q_2$.

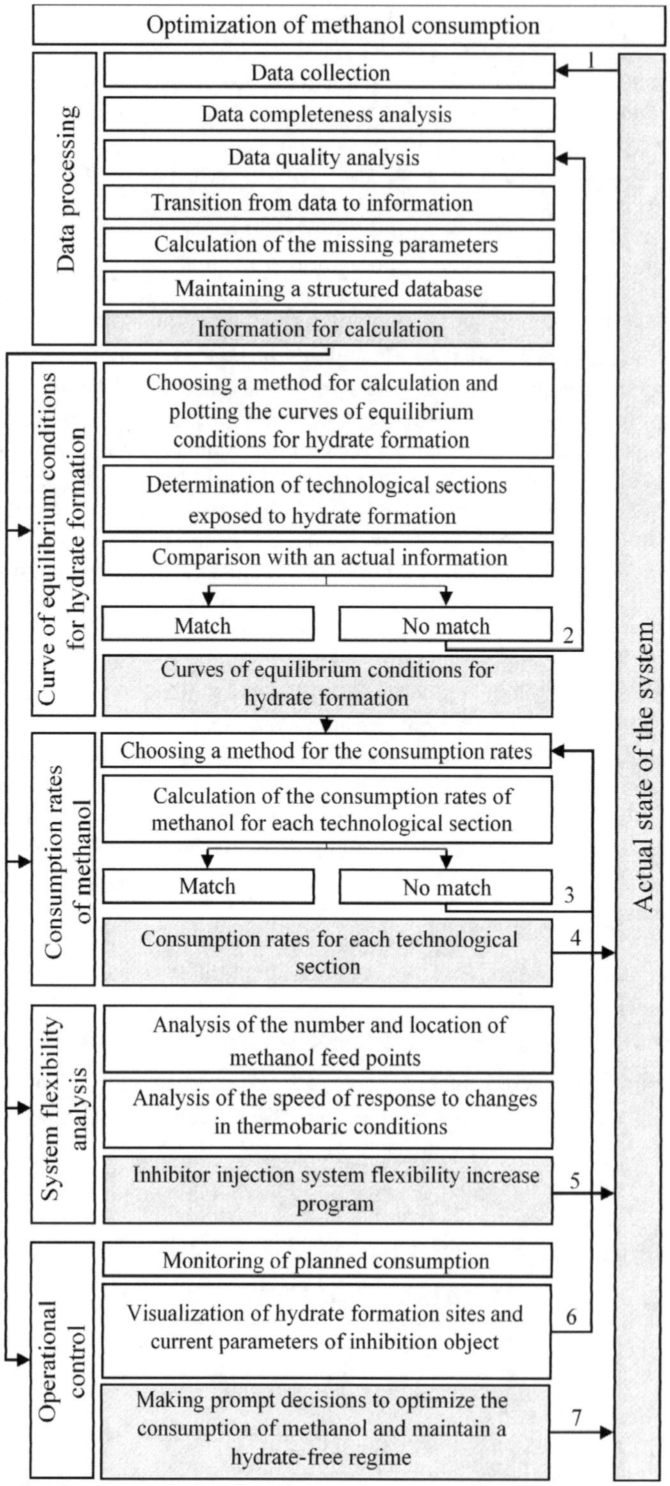

Figure 1. Algorithm of actions to optimize the methanol consumption.

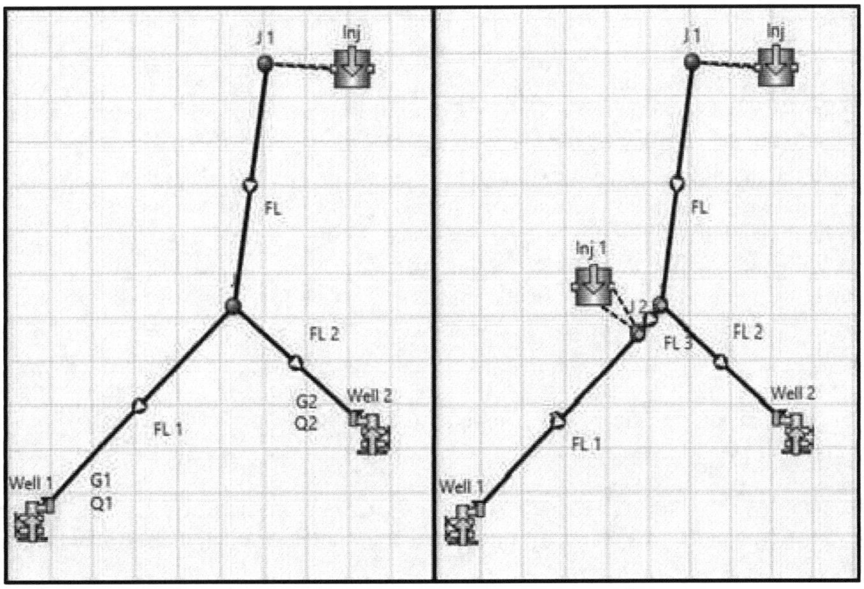

a b

Figure 2. A scheme of a gas-lift gas supply system: a) one inhibitor feed point; b) two inhibitor feed points.

A significant factor in optimizing the consumption of methanol is the response time to changes in thermobaric conditions. The change in temperature and pressure along the wellbore mainly depends on the well operating parameters, but the change in the temperature of the pipeline is significantly influenced by daily and seasonal changes in the ambient temperature, while the degree of influence depends on the pipeline laying method (ground-surface, aboveground, underground). With decreasing temperature and increasing pressure, the required methanol consumption increases. The speed of response to changes in thermobaric conditions means the following chain of actions: a change in thermobaric conditions – data acquisition and processing – recalculation of the methanol consumption rate – a change in methanol consumption. It is worth remembering that the main component of methanol consumption, in addition to the consumption rates, is the gas flow rate, so when it changes it is necessary to adjust the inhibitor injection. Prompt adjustment of the methanol consumption rate allows to adapt to the current conditions that prevents the overrun of the inhibitor and hydrate formation when the air temperature drops sharply. The planning of actions to increase the flexibility of the inhibitor injection system (increasing the number of feed points, installing remote control equipment of methanol consumption) should be carried out with confirmation of their economic and technological feasibility (figure 1, direction 5).

The final stage which should be carried out continuously throughout the entire field development period in the case of the presence of hydrate formation conditions in the system is operational control. First of all, it is the control of the planned consumption of methanol, prompt servicing and maintenance of the feed-control devices / methanol pipelines in workable condition. For a better understanding of the current situation, analysis and timely decision-making it is necessary to visualize the actual parameters of the objects of inhibition, as well as the problems that occur due to hydrate formation which can be carried out using digital twins. For problem areas it is necessary to adapt the methodology for calculating methanol consumption rates and, if necessary, refine the initial data to ensure a hydrate-free well operating parameters (figure 1, direction 6). Operational control of methanol consumption includes a change in the inhibitor injection rate depending on the thermobaric conditions and the current status of development of the production zone (figure 1, direction 7).

3. Conclusion

The algorithm presented in this article allows one to plan the methanol consumption justification program. The main areas of optimization are:

- determination and correction of the curve of equilibrium conditions of hydrate formation;
- determination and correction of methanol consumption rates;
- increasing the inhibition system flexibility;
- operational control over the current state of the system for making quick decisions.

It should be noticed that the basic component of any calculations and decisions is the availability of high-quality information about the state of the system.

Any actions to optimize the supply of the inhibitor must be carefully worked out, technologically and economically justified. In order to confirm the calculated consumption rates of the inhibitor, it is recommended to test the proposed amount of methanol in the pilot areas of the fields by gradually reducing the inhibitor supply, since a sharp change in the consumption of methanol can lead to the formation of hydrates and significant losses in hydrocarbon production.

References

[1] Sloan E D, Koh C A 2008 *Clathrate Hydrates of Natural Gases* (Boca Raton: CRC Press)

[2] Gong J, Shi B and Zhao J 2010 Natural Gas Hydrate Shell Model in Gas-Slurry Pipeline Flow *J. of Natural Gas Chemistry* **19** 261-6

[3] Prakhova M Yu, Krasnov A N, Khoroshavina E A and Shalovnikov E A 2016 Methods and Means to Prevent Hydrate Formation in Gas Production Facilities *Oil and Gas Business* **14(1)** 101–118

[4] Istomin V A 1990 *Prevention and Liquidation of Gas Hydrates in Gas and Oil Systems* (Moscow: VNIIGAZ)

[5] Makwashi N, Zhao D, Ahmed T and Paiko I 2019 *Pipeline Gas Hydrate Formation and Treatment: A Review* 3RD National Engineering Conf. on Bridging the Gap between Academia and Industry (ACICon2018)

[6] Kamari E and Oyarhossein M 2012 *J. of Natural Gas Science and Engineering* **9** 11-15

[7] Prakhova M Yu, Krasnov A N, Khoroshavina E A, Shalovnikov E A and Kolovertnov G Yu 2016 Optimization of Methanol Supply Control in Gas Collection Systems *Oil and Gas Territory* **6** 22-29

[8] Kolchin A V 2019 *Improving the Technology of Using Methanol in Gas Production and Transmission Systems,* PhD Dissertation Ufa

[9] Katz D L 1959 *Handbook of Natural Gas Engineering* (New York: McGraw-Hill)

[10] Makogon Yu F 1974 *Hydrates of Natural Gases* (Moscow: Nedra)

[11] Ponomarev G V 1960 Conditions for the Formation of Hydrates of Natural and Associated Gases (Kuibyshev: NIINP) pp 49-55

[12] Buznikov N A, Istomin V A, Mitnitskiy R A 2016 Influence of the Accumulated Liquid in the Field Pipeline on the Movement of the Hydrate Inhibitor *Vesti gazovoy nauki* **2(26)** 112-16

[13] Buhgalter E B 1984 *Hydrates of natural and petroleum gases (Moscow:* VINITI) pp 63-126

[14] Istomin V A, Kvon V G, Lakeev V P 1986 *Determination of Methanol Consumption at Integrated Gas Treatment Units* (Moscow: VNIIEGazprom) 1-6

[15] Makogon Yu F 1985 *Gas Hydrates, Prevention of Their Formation and Use* (Moscow: Nedra)

[16] Yamamuro O, Suga H J 1989 *J. Therm. Anal.* **35** 2025-64

[17] Rybka V F, Vasyutinskaya S I 2018 Experience in Monitoring the Formation of a Gas Hydrate Plug by Means of Fiber-Optic Well Thermometry *Sci. J. of the Russian Gas Society* **1** 43-46

[18] Prischepa O M, Nefedov Y V and Ibatullin A K 2020 Raw material source of hydrocarbons of the arctic zone of Russia *Periodico Tche Quimica* **17(36)** 506-26

[19] Sabukevich V S, Podoprigora D G, Shagiakhmetov A M 2020 Rationale for selection of an oil field optimal development system in the eastern part of the pechora sea and its calculation *Periodico Tche Quimica* **17(34)** 634-55

[20] Zhenzhen Wei, Shanyu Zhu, Xiaodong Dai, Xuewu Wang, Lis M. Yapanto, Raupov I R 2021 Multi-criteria decision making approaches to select appropriate enhanced oil recovery techniques in

petroleum industries *ENERGY REPORTS* **7** 2751-58
[21] Tcvetkov P, Cherepovitsyn A, Makhovikov A 2020 Economic assessment of heat and power generation from small-scale liquefied natural gas in Russia *Energy Reports* **6** 391-402 DOI: 10.1016/j.egyr.2019.11.093
[22] Ilinova A, Chanysheva A 2020 Algorithm for assessing the prospects of offshore oil and gas projects in the Arctic *Energy Reports* **6** 504 DOI: 10.1016/j.egyr.2019.11.11
[23] Tcvetkov Pavel, Cherepovitsyn Alexey, Makhovikov Alexey 2020 Economic assessment of heat and power generation from small-scale liquefied natural gas in Russia *Energy Reports* **6** 391-402 DOI: 10.1016/j.egyr.2019.11.093
[24] Cherepovitsyn Alexey, Evseeva Olga 2021 Parameters of sustainable development: Case of arctic liquefied natural gas projects *Resources* **1(10)** 1-27. DOI: 10.3390/resources10010001
[25] Rogachev M K, Mukhametshin V V, Kuleshova L S. 2019 Improving the efficiency of using resource base of liquid hydrocarbons in Jurassic deposits of Western Siberia. *Journal of Mining Institute* **240** 711 https://doi.org/10.31897/pmi.2019.6.711
[26] Menshikov S N, Dzhaljabov A A, Vasiliev G G, Leonovich I A, Ermilov O M 2019 Spatial Models Developed Using Laser Scanning at Gas Condensate Fields in the Northern Construction-Climatic Zone *Journal of Mining Institute* **238** 430 https://doi.org/10.31897/pmi.2019.4.430
[27] Shostak N A, Zaporozhets E P 2019 Efficiency Estimation of the Single- and Multicomponent Anti-hydrate Reagents *Journal of Mining Institute* **238** 423 https://doi.org/10.31897/pmi.2019.4.423

Investigation of the influence of the drive shaft speed on the characteristics of a submersible pump when pumping out a gas-liquid mixture

A A Odintsov[1,2], **A N Musinskii**[1,2], **V G Ostrovskii**[1,2]

[1] Perm National Research Polytechnic University, 29, Komsomolskii ave., Perm, 614990, Russia
[2] Novomet-Perm, 395, Kosmonavtov highway, Perm, 630065, Russia

E-mail: Anton.Odintsov@novometgroup.com

Abstract. The article presents the results of small size ESP (electrical submersible pump) tests at a speed of 6000-8500 rpm in the presence of free gas at the inlet to the device. The submersible centrifugal pump is quite sensitive to the presence of free gas in the pumped gas-liquid mixture. Depending on the amount of free gas, the actual characteristics of the pump are deformed, and at a certain gas content, the pump stops supplying the gas-liquid mixture. Stable operation of the ESP is carried out with a free gas content at the pump inlet ranging from 5% to 25%, depending on the pump size. There are several ways to deal with gas in wells operated by ESP: from running the pump into the zone where the intake pressure ensures optimal pump flow and stable operation to the use of separating and dispersing devices at the intake. Results represent pump flow characteristics along with different inlet free gas contents.

1. Introduction

While the artificial lift of oil and gas is moving towards increasing the rotational speed of ESP units [1], and the main attention is paid to the reliability of such units, it is necessary to pay attention to the problem of operating wells with a high content of free gas [2].

In traditional ESP installations, various upstream devices are used to cope with gas: gas separators, gas handlers, dispergators, multiphase pumps. Their use makes it possible to increase the limiting gas content at the pump inlet from 20-30% to 60-80% [3, 4].

At the same time, it is not known for certain about the effect of increasing the rotational speed on the operation of the ESP in the presence of gas at the intake [5].

Thus, the main goal of this work is to study the effect of gas on the operation of an ultra-small ESP [6] with an increase in the rotational speed from 6000 rpm to 8500 rpm. The article presents the results of tests of an ESP and a gas stabilizer of small dimensions with gas at speeds from 6000 to 8500 rpm.

To achieve this goal, bench tests were carried out on a gas-liquid mixture, during which the characteristics of various devices of small size (55 mm) were investigated in the presence of gas.

2. Materials and methods

Investigations were carried out at the test bench for testing devices on the gas-liquid mixture [7, 8].

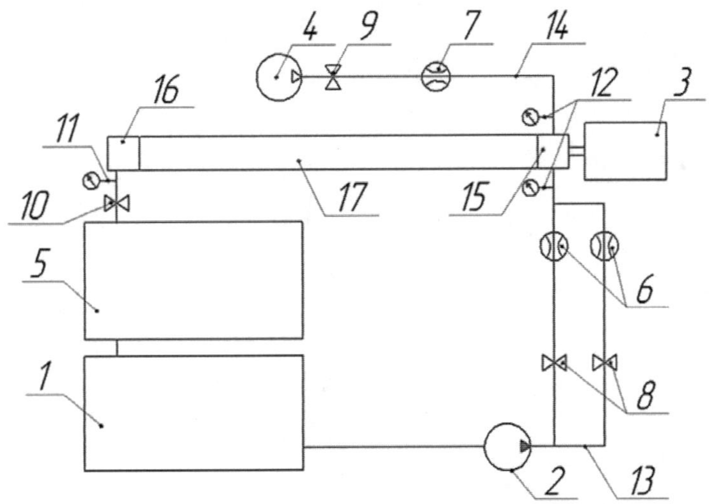

Figure 1. Stand layout: 1 - water tank; 2 - booster pump; 3 - electric drive; 4 - compressor; 5 - shelf separator; 6.7 - flow meters; 8,9,10 - gate valves; 11.12 - pressure sensors; 13 - liquid metering line; 14 - air measurement line; 15 - input module; 16 - output module; 17 - tested pump section

During the experiment, using two gas flow meters, an electromagnetic liquid flow meter, three pressure sensors and a temperature sensor, the following quantities are measured:

- Q_l is volumetric liquid supply, m^3/day;
- T_l is temperature of the mixture in the test object, °C;
- q_l is the volumetric gas supply at the inlet to the test object, measured at a pressure P_0 = 1 atm and a temperature T_0 = 273 K, m^3/day;
- P_1 ispressure at the entrance to the test object, MPa;
- P_2 is pressure at the outlet of the test object, MPa.

The values of these quantities are determined at a confidence level of 0.95 with the errors listed in Table 1.

Table 1. Errors quantities

Value	Relative error	Value	Absolute error
Q_l, m^3/day	0,35 %	t, s	0,00144
T, °C	0,25 %	V, l	0,5
q_l, m^3/day	1%	P^1, MPa	0,00375

3. Results

An ESP2 pump was used as a test device.

The results of the ESP2 operation on gas-liquid mixture were obtained at three different frequencies of 6000, 7200, 8000 rpm.

During the tests, water was supplied to the pump. Then, using a compressor, gas (air) was supplied so that its concentration in the gas mixture was 10%. Using pressure sensors installed at the pump inlet and outlet, and flow meters installed on the gas and liquid lines, the pressures and flow rates of water and gas were recorded. Then the gas content in the mixture was increased to 15%, 20%, etc. as long as the pump created a head. At the next stages, the water supply increased and the test was repeated.

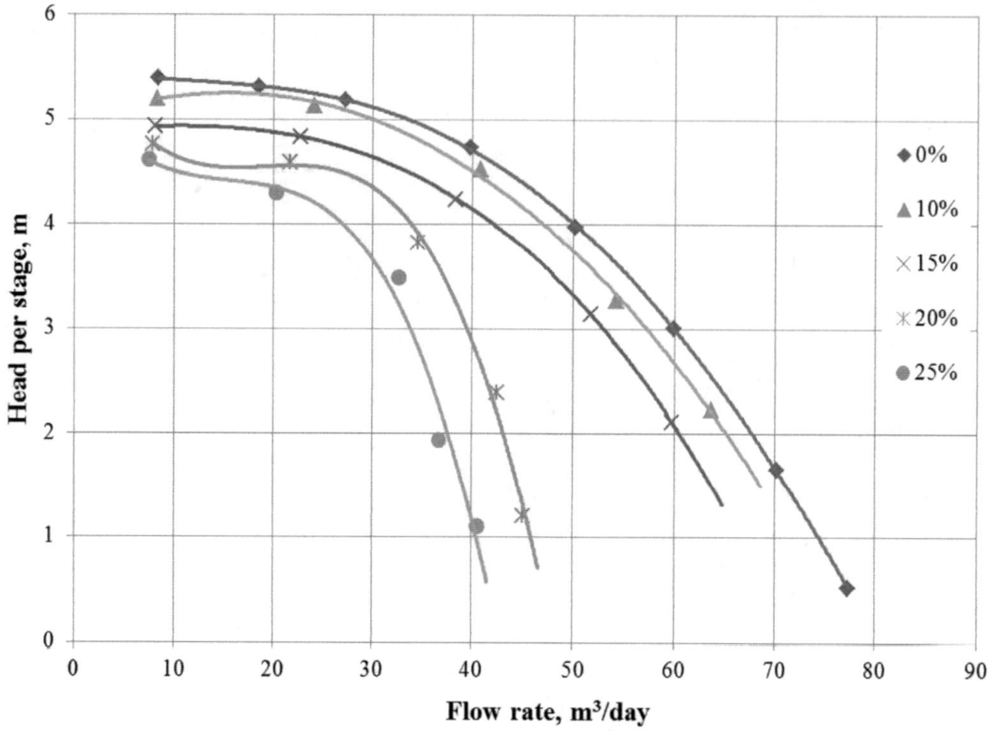

Figure 2. Head-flow characteristic of ESP2 at different values of β_{in} , 6000 rpm

Figure 2 shows the pressure-flow characteristic of ESP2 on a gas-liquid mixture at a speed of 6000 rpm. It can be seen that the maximum gas concentration with which the pump can operate stably is 25%, while the operating range for the supply decreases from 80 to 40 m^3/day.

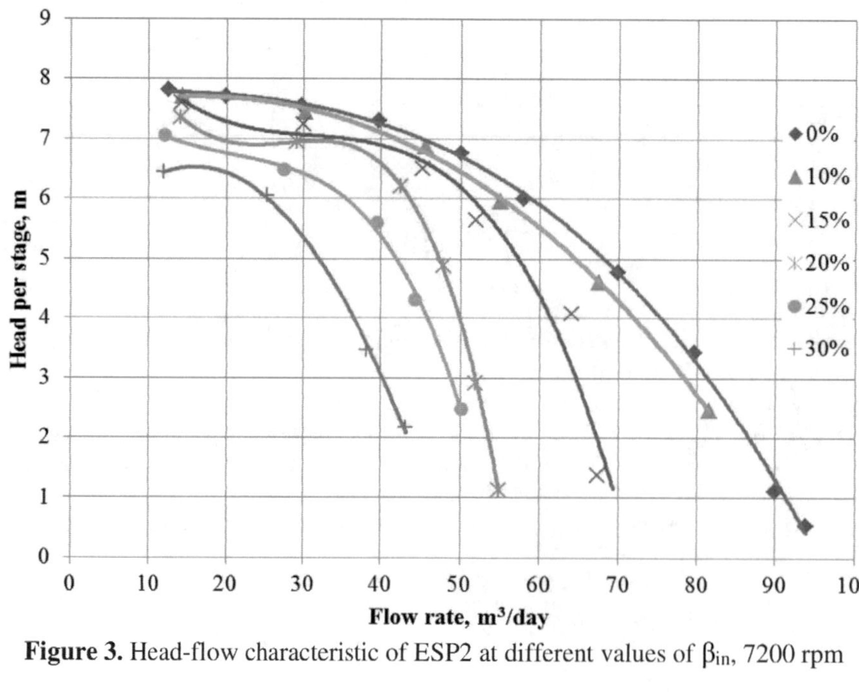

Figure 3. Head-flow characteristic of ESP2 at different values of β_{in}, 7200 rpm

With an increase in the speed of rotation to 7200 rpm, the maximum gas concentration at which the pump operates stably increased to 30% (Figure 3).

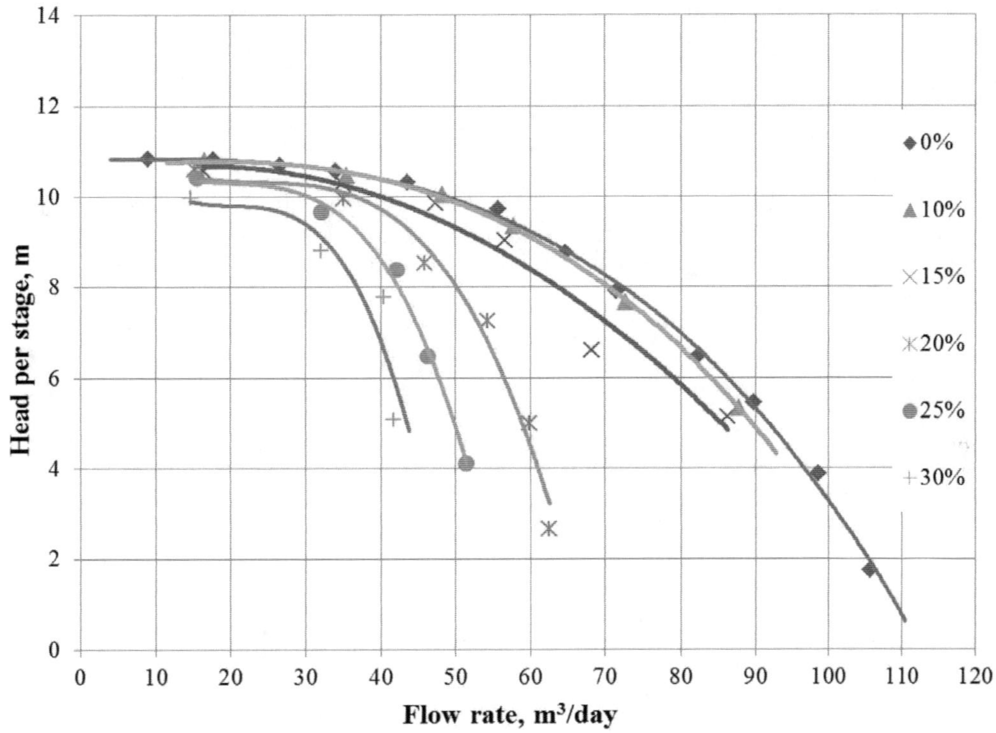

Figure 4. Head-flow characteristic of ESP2 at different values of β_{in}, 8500 rpm

At a speed of 8500 rpm, the maximum gas concentration pumped by the pump did not change and amounted to 30% (Figure 3).

It can be noted that at a gas concentration of more than 20% at all test frequencies, there is a sharp decrease in the pressure curves in the right working zone. This may be due to the fact that a gas lock begins to form in the stage, thereby reducing the area of the flow channel.

Based on the test results, data were obtained, on the basis of which the pressure-flow characteristics of the pump were constructed at different gas contents in the gas mixture (0 - 30%). As can be seen from the test results, with an increase in the rotational speed, the maximum gas concentration at the inlet increases from 25 to 30%.

4. Conclusion

1. An ultra-small pump is capable of pumping a gas-liquid mixture with a gas content of 25%, which corresponds to the test results known in the literature [5].

2. An increase in the rotational speed from 6000 rpm to 8500 rpm slightly increases the maximum permissible gas content at the inlet to the ESP from 25 to 30%.

References

[1] Koropetskii V, Shakirov A 2020 High-speed technologies are changing the rules of the game *Neftegaz.ru* **9** 62-68

[2] Volkov M G, Smolyanets E F 2018 Specifics of oil well operation in the conditions of high free gas content in the production stream *Oil Industry* **11** 120-126

[3] Pescherenko M, Perelman M, Rabinovich A 2014 Increase of ESP efficiency. Multiphase

pumps application *Drilling and oil* **4** 55-57
[4] Pescherenko M, Perelman M, Pescherenko S 2014 Multiphase dispersing pump *Oil Industry* **11** 136-139
[5] Ageev S R, Grigoryan E E, Makienko G P 2007 *Russian vane pump installations for oil production and their application* (Press-master, Perm) p. 144-152
[6] Sidorov M, Savelyev K 2020 Expanding horizons: Novomet's new developments in the field of centrifugal pumps *Oil Production Arsenal* **22** 20-25
[7] Odintsov A A, Musinskii A N, Pescherenko S N 2020 Multiphase axial pump for surface pumping gas-liquid mixture *Oil Industry* **2** 62-64
[8] Ostrovskii V G, Perelman M O, Pescherenko S N, RU Patent No. 2 531 090 *Method of testing gas separators on gas-liquid mixtures and a stand for its implementation* (20.10.2014)

Laboratory studies on the use of surfactant compositions to solve resource extraction problems

Ty Than Ngia[1], M M Veliev [1], V Sh Mukhametshin [2], S S Karimov [1], E M Veliev [2], A V Andreev [2]

[1]JV "Vietsovpetro" Ba Ria – Vung Tau City, S.R. Vietnam
[2]Ufa State Petroleum Technological University, Branch of the University in the City of OktyabrskyOktyabrsky, Russia

E-mail: vsh@of.ugntu.ru

Abstract. Based on studies on the interaction and heat resistance of some heat-resistant polymer additives at 128 °C, a sulfonated polyacrylamide polymer AN-125 is proposed as an additive for improving the viscosity of a surfactant composition. The results of experimental studies of a new surfactant composite prepared based on a mixture of ethoxylated and propoxylated sulfate, α-olefin sulfonate and alkylphenol ethoxylate with a concentration of 500 ppm and additives to improve the viscosity in the reservoir model from the deposit basement samples showed that the surface tension at the oil-solution boundary before the exposure to temperature is 0.19 mN/m, and after exposure to temperature 91 °C – 0.21 mN/m. The experimental studies of water and surfactant compositions injection on 5 prefabricated models composed of core samples of basement rocks covering a wide range of reservoir properties showed that the injection of the oil rim with compositions based on the mixture of ethoxylated and propoxylated sulfate, α-olefin sulfonate and alkylphenol ethoxylate with a concentration of 800 ppm and an additive of polymerasulfonated polyacylate increases the oil displacement ratio.

1. Introduction

The experience of long-term development of fields shows that as oil reserves are depleted, it is necessary to make sound technological decisions based on the study of oil recovery processes in laboratory and field conditions [1-10].

The current stage of operation of the Dragon field is exacerbated by the problems associated with the fact that the field is in the falling stage of development characterized by a further increase in negative factors: uneven production of reserves, increase of water content, increase of hard-to-recover reserves, increase of oil viscosity associated with its oxidation upon contact with oxygen dissolved in water injected into a reservoir, increase of concentration of asphalt-resin substances, aging and wear of well stock, etc. [11, 12].

The limited use of modern enhanced oil recovery technologies leads to the fact that the oil recovery factor (ORF) decreases over the decade by 3-4%. Therefore, today it is necessary to intensively introduce new advanced technologies aimed at involving all types of residual oil in the fields that have entered the final stage of operation, and effectively develop heavy high-viscosity oil fields.

This paper considers the results of the development of surfactant compositions to increase the oil displacement coefficient during the flooding of the Dragon field basement. The paper also proposes the selection, mixing of marketable surfactants for optimal compositions, the evaluation of basic indices, injection of surfactants on the reservoir model of the Dragon field basement rocks.

2. Methods and materials

The subject of the study is the efficiency of displacement ability with surfactant solutions. The purpose of laboratory research is to analyze the possibility of increasing the development efficiency of high-viscosity oil fields using surfactants. The increase of the recovery coefficient of highly viscous oil in conditions of non-uniform reservoirs by permeability should be ensured due to the introduction of the surfactant injection technology.

The study covers the influence of a surfactant composition consisting of a mixture of ethoxylated and propoxylated sulfate, α-olefin sulfonate and alkylphenol ethoxylatane, the process of oil displacement on the reservoir model from the samples of the Dragon field basement with the length of >200 mm and diameter of 50 mm to estimate the oil displacement coefficient with the surfactant composition.

Working fluids consist of a model of izoviscous oil of the Dragon field basement, sea water and surfactant compositions.

The oil displacement ratio (η1) during water injection is determined as follows:

- reservoir model is prepared;
- izoviscous oil is pumped into the reservoir model in the amount of 5 pore volumes and then permeability is determined by oil Ko;
- oil is displaced by injection of seawater into the reservoir model in the amount of 6-10 pore volume at a flowrate of 1-2 m3/day, then 3 pore volume at a flowrate of 3-5 m3/day. Next, the value of displacement coefficient is determined (η1) alongside with the permeability of the reservoir model by water (K1).

The oil displacement ratio during injection of the surfactant solution (η2) is determined as follows:

- oil displacement is continued through the injection of a surfactant solution into the reservoir model in the amount of 0.15 pore volume at a flowrate of 1-2 m3/day, then sea water is pumped into the reservoir model in the amount of 3-5 pore volume at a flowrate of 1-2 m3/day until the displacement coefficient is stabilized and then the value of displacement coefficient (η2) is determined.
- permeability of the reservoir model by water (K2) is determined.

After that, a core holder is cooled to room temperature, the samples are taken out and residual oil saturation of the reservoir model is determined [4].

The results of the experiments are calculated according to the following formulas:

$$\eta_1 = \frac{1-S_{OB}-S_{1OH}}{1-S_{OB}}, \quad (1)$$

$$\eta_2 = \frac{1-S_{OB}-S_{2OH}}{1-S_{OB}}, \quad (2)$$

$$\Delta\eta = \eta_2 - \eta_1, \quad (3)$$

where: SOB – residual water saturation, unit fractions; S1OH – residual oil saturation after water injection, unit fractions; S2OH – residual oil saturation after surfactant injection, unit fractions; η1, η2 – displacement coefficients before and after surfactant injection, unit fractions; K1, K2 – water permeability before and after surfactant injection, 10-3 μm2; KВП – permeability recovery coefficient for water before and after surfactant injection, unit fractions; Δη – increment of the displacement coefficient after surfactant injection, unit fractions.

The following indices were evaluated through the experimental studies of surfactant compositions to increase the oil displacement ratio:

- surface tension between oil and water phases;
- possibility of surfactant connection with sea water;
- heat resistance of surfactant solution systems at 91 °C;
- degree of surfactant adsorption on rock surface;

- effect of surfactant compositions on capillary autopermeability;
- selection of a suitable formulation of surfactant compositions;
- degree of oil displacement coefficient of during surfactant solution injection;
- degree of permeability recovery by the reservoir water after surfactant injection.

To this end, the interaction with sea water and surface tension at the oil-solution boundary at various temperatures are first considered.

3. Results

The results of the study on the interaction of surfactant with sea water at different holding times (1-31 days) and different concentrations (0-1000ppm) show that the colors of the surfactant solution systems remain transparent. The surface tension at the oil-solution boundary фe various temperature is shown in Table 1.

Table 1. Surface tension at oil-surfactant boundary

Surface tension (mN/m) with temperature measurements				
Surfactant concentration, ppm	30oC	50oC	70oC	90oC
0	21.57	21.43	21.3	21.24
100	1.66	1.59	1.57	1.54
250	0.49	0.47	0.46	0.45
500	0.18	0.17	0.17	0.16
1000	0.13	0.13	0.12	0.11

The table shows that the surface tension is inversely proportional to the temperature. This is caused by the fact that as the temperature increases, the solubility of oil components in surfactant solutions increases, which leads to a decrease in oil viscosity and an increase in its mobility, as a result of which the surface tension at the oil-water boundary decreases. However, the change is very small and is within the permissible measurement error. Therefore, the laboratory studies to determine the surface tension are carried out at room temperature (about 30 °C).

All experimental results at 91 °C gave very low surface tension values (from 0.13 to 0.32 mN/m within the total concentration of surfactant systems amounting to 350-1000 ppm), and they slightly changed during the entire thermal exposure for 31 days. All this indicates that the selected surfactant composition for the oil displacement study on the reservoir model of the basement deposit has good thermal resistance, low surface tension and good interaction with sea water.

In order to study the degree of adsorption of surfactants on the surfaces of basement rocks, the samples of basement rocks were crushed, washed several times with distilled water to pH = 7, then dried and crushed several times on a ball crusher to obtain the necessary dimension and then sieved using a 80 μm sieve. Prior to the experiment, the samples were baked in a furnace at 400 °C for 4 hours.

Rock powder and surfactant were placed in heat-resistant tubes at a ratio of 1:20, respectively, and thoroughly shaken for 20 minutes, held at 91 °C for 4 weeks. Thereafter, the surfactant samples were centrifuged, the solution was separated from the rocks, and the concentration of sulfate ions in the solution was determined.

It is necessary to determine the residual surfactant content in the water after injection in the reservoir model based on the ultraviolet (UV) absorption. The process is as follows:

- a reference line is built. UV absorption of AMS-2 surfactant solutions samples prepared at different concentrations in the wave region from 200-400 nm is measured;
- the possibility of absorption of surfactant solutions after injection is determined. The surfactant solutions obtained after injection in the reservoir model are diluted and UV absorption is measured.

The obtained results show that at different times of exposure in sea water, there was almost no adsorption of surfactant solutions on the reservoir rock surface (the concentration of sulfate ions changed very little in temperature during the exposure time). This is caused by the fact that the surface of the reservoir rock is charged with a negative charge, and in sea water the surfactant solutions represent anionic surfactants with negatively charged hydrophilic groups, and all this leads to the interaction to repel the surface of the reservoir rock from the hydrophilic group of surfactant molecules. Moreover, other ions with a relatively high concentration are contained in sea water solution, and they are competitively adsorbed on the surface of the reservoir rock. Therefore, the adsorption of surfactants does not occur or is very weak in sea water.

The study of the adsorption of surfactant solution compositions on the surface of the basement reservoir rocks in terms of the concentration of surfactant compositions and the temperature of thermal exposure gave similar results. After two days of exposure at 50, 70 and 91 °C, the pH of surfactant solutions decreased slightly – in the range of 7-8, at that time, the concentration of surfactant sulfate increased significantly compared to the initial value.

Thus, the obtained result shows no adsorption, or in other words – negative absorption.

The experiments demonstrate that the adsorption of the surfactant used during injection does not occur on the surface. The calculated amount of surfactant after injection is higher than at the beginning due to the fact that oil from the Dragon field contains high amount of resinous and asphaltene substances that can act as natural surfactants.

In order to study adhesion-wettability of rocks the contact angle is measured at the oil-water-rock boundary of the Dragon field basement using OSA-20 on 2x1x0.2 cm tiles from basement rocks.

The results show that the reservoir rock has a very strong wettability to oil – the contact angle is about 148 °C (after exposure in basement oil), and after processing with the surfactant solutions system it becomes transient by wettability with a contact angle of about 85 °C. The change in surface wettability of the reservoir rock was caused by the interaction of the surfactant with it under the influence of polarization, which increases the solubility of polarization constituents from oil to surfactant solutions. It is this process of solubility that leads to a change in the wettability of the reservoir rock surface.

Thus, the selected surfactant composition fully meets the technical requirements for oil displacement from the Dragon field basement.

To determine the effect of surfactant compositions on capillary autopermeability, rock samples are saturated with sea water and then oil. The samples are then held in a glass of oil at 91 °C in the oven for 12 days in order to restore the wettability of the rock. After that, the samples will be placed in bottles with surfactant solutions and sea water. Over time, oil is displaced from the rock by capillary autopermeability of water in the rock and collected at the top of the bottle, and the amount of displacement oil is recorded.

The results of the studies show insignificant amount of oil was extracted from sea water, and on the third day it almost stopped, and in the case of surfactants, the amount of extracted oil during thermal exposure and at all the tested concentrations (350, 500 and 1000 ppm) increased sharply. The permeability increases over time under the influence of capillary autopermeability.

Thus, under the influence of a surfactant, the wettability of reservoir rocks changes from oil wetting to transition or water wetting (change in contact angle), which leads to an increase in autopermeability, therefore, an improvement in oil recovery.

To obtain high efficiency of oil extraction, the contact time of surfactant solutions with oil-bearing core plays a decisive role. This can be done by adjusting the injection rate or core exposure time to maximize oil displacement to the surface.

The process of oil displacement by water and the rim of surfactant compositions on the reservoir model of the basement deposit was studied.

The results of the thermal stability tests of the ethoxylated and propoxylated sulfate, α-olefin sulfonate and alkylphenol ethoxylate solution compositions with a concentration of 500 ppm in sea water before and after exposure at 91 °C show that the surface tension at the oil-surfactant composition boundary before exposure to temperature is 0.19 mN/m, and after exposure at 91 °C after 31 days – 0.21 mN/m.

On the basis of the above results to determine the oil displacement coefficient and data on reservoir models, the following can be noted (Table 2):

- recovery of sample system permeability in models 2 and 3 is very good compared to model 1, where permeability decreased by 40%;
- at the same surfactant content (1000 ppm) the model with higher share of macroscopic voids (0.16 versus 0.15) has lower oil displacement coefficient (6.2% versus 8.3%);
- when a surfactant content is doubled (from 500 ppm to 1000 ppm), the oil displacement coefficient increases by 2.3 times.

Table 2. Results of oil displacement ratio

Parameter	Model 1	Model 2	Model 3
Initial core permeability – K1, 10-3 μm2	41.3	291.0	411.0
Permeability for water before surfactant injection – K2, 10-3 μm2	8.4	31.3	115.0
Permeability for water after surfactant injection – K3, 10-3 μm2	4.9	49.4	228.0
Permeability recovery factor for water before surfactant injection, K	0.6	1.58	1.98
Increase of oil displacement ratio	3.5	8.3	6.2

In the system of samples in models 2 and 3, the permeability did not only recover, but also increased by quite a large amount compared to their water permeability before the injection of surfactant solutions. This indicates that the surfactant solution displaces most of the oil from the cracks (especially microcracks), thereby expanding their dimensions, which leads to an increase in the bond between the cracks, and as a result, fluids more easily flow in the sample system, the water permeability increased compared to the initial value.

Besides, the value of the ratio of macroscopic to microscopic voids significantly affects the increase in the oil displacement coefficient. In the system where the micro-voids are larger, the residual oil in the reservoirs after sea water injection will be large, therefore, when surfactant solutions are injected, the displaced part of oil is larger than in another system.

On the other hand, the change in the concentration of surfactant in the injection solutions also plays a major role in the growth of the oil displacement coefficient. This phenomenon may be explained based on the surface tension change. The results of determining the surface tension of the reservoir oil-surfactant composition in sea water at different concentrations show that the critical micelle concentration (CMC) of the AMS-2 surfactant composition is achieved at 1000 ppm, since when the concentration is increased above this value, the surface tension does not change.

It can be noted that the oil displacement coefficient increases significantly at a surfactant concentration close to CMC (1000 ppm), while at a surfactant concentration of 500 ppm (much lower than CMC), this ratio increases slightly. In comparison with water injection, when the AMS-2 surfactant solutions are injected the oil displacement coefficient increases by:

- 3.5% at a concentration of 500 ppm;
- 6.2% and 8.3% (average 7.2%) at a concentration of 1000 ppm.

The results of an experimental study on the injection of surfactant compositions consisting of a mixture of ethoxylated and propoxylated sulfate, α-olefin sulfonate and alkylphenol ethoxylate at the Dragon field basement model show that at a relatively low concentration (1000 ppm) and a small volume injection mode of the unit amounting to 0.15 of the pore volume without the use of alkaline or polymer solutions, the oil recovery factor increase up to 7.2%. This indicates a good possibility of applying the method of the surfactant system injection. Therefore, it is advisable to continue the study of injection conditions, the influence of the core feature, the surfactant concentration, the thermal exposure time, and the modeling of the oil displacement process in wells in order to choose the optimal mode for practical application thus achieving the most efficient oil displacement.

In order to study the effect of reservoir waters on thermal and chemical resistance of experimental surfactant compositions, sea water was mixed with reservoir water from the Dragon field at a ratio

IOP Conf. Series: Earth and Environmental Science 1021 (2022) 012069 doi:10.1088/1755-1315/1021/1/012069

from 0 to 100%, and the issue of the change in the main components (basic ions, salinity, pH, etc.) was analyzed.

After obtaining the results of the analysis of water samples, 1000 ppm of the AMS-2 surfactant was added to them and kept for 31 days at 91 °C. The monitoring results of pH and surface tension for each week of exposure are given in Figure 1.

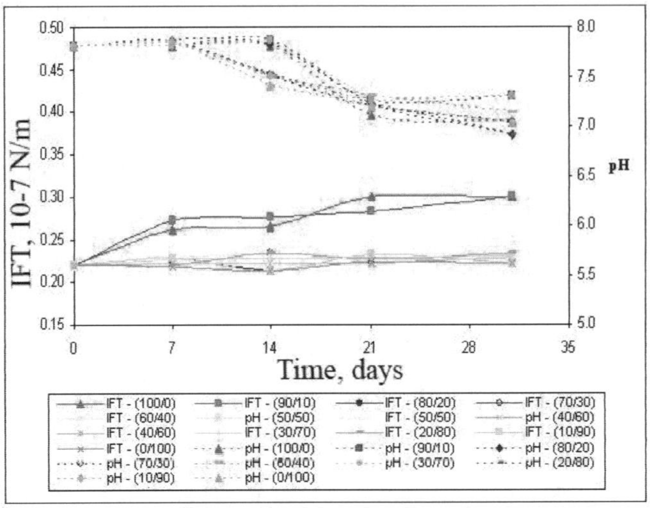

Figure 1. Effect of reservoir water on pH, surface tension (IFT) of oil-surfactant solution composition during exposure at 91 °C.

As can be seen, in the presence of reservoir waters with their different contents, no particular decreases in the degree of surfactant compositions were observed throughout the entire exposure period compared to exposure in sea water. This can be explained by the fact that the reservoir water of the Dragon field has a high Ca2+ content (1595 ppm), low Mg2+ (33 ppm), while sea water contains low Ca2+ (351 ppm) and high Mg2+ (1244 ppm) content, and with mixing at different component ratios, the total content of divalent ions does not exceed their total content in sea water and does not lead to salting of surfactant compositions. The HCO3- content in the reservoir waters is quite high, but it will be diluted when the surfactant solution is injected, which will not lead to the sedimentation of the surfactant.

At high reservoir temperatures (> 90 °C) the thermal stability of the polymer used is quite critical. In this case, it is advisable to use sulfonated polyacrylamide, which is a copolymer of acrylamide and acrylamide propyl sulfonic acid. In addition to thermal stability, such polymers are more stable to divalent ions such as Ca2+ and Mg2+ present strong brines and sea water. They are also less adsorbed than hydrolyzed polyacrylamides. Although the molecular weight of these polymers is not as large as that of conventional polyacrylamides, it can also reach 14 mln Da.

The study on the interaction and heat resistance of some heat-resistant polymer additives at 128 °C made it possible to select a sulfonated polyacrylamide polymer AN-125 (dissolution time in fresh water – 120 minutes, effective viscosity at 170 s-1 – 19 mPa·s, gelling time at 90 °C – 0.8 hours, gel stability at 70-90 °C – more than 30 days, sulfonation degree – 25 mole%, approximate molecular weight – 2 mln Da) as an additive for improving viscosity of the AMS-2 surfactant composition with test samples. Their advantage is high thermal and salt resistance.

The sulfonate groups present in such PAA macrochains shield amide groups of acrylamide links thus preventing their hydrolysis at high temperatures, thereby increasing the resistance of the dissolved polymer to thermal-oxidative breakdown.

The results of the study of the change in viscosity and surface tension at the oil-surfactant solution boundary during the exposure at 91 °C in the reservoirs of the Dragon field show that the sample No.

3 containing a surfactant composition of a mixture of ethoxylated and propoxylated sulfate, a-olefin of sulfonate and alkylphenol ethoxylate with the concentration of 800 ppm and polymer sulfonated polyacrylamide AN-125 with the concentration of 200 ppm are the most stable in terms of both the reduction of surface tension and viscosity throughout the exposure process (Figure 2 and 3), and thus it was selected for further study on injection on the core model of the Dragon field basement rock.

A series of direct determinations of oil displacement coefficients during water and surfactant injection was carried out on 5 prefabricated models of the reservoir (taking into account the results obtained above, compiled from core samples of rocks of the Dragon field basement and covering a wide range of reservoir properties: void ratio – 3.44...9.82% (average – 5.84%) and oil permeability – 11.1...411·10-3 μm2 (average – 221·10-3 μm2).

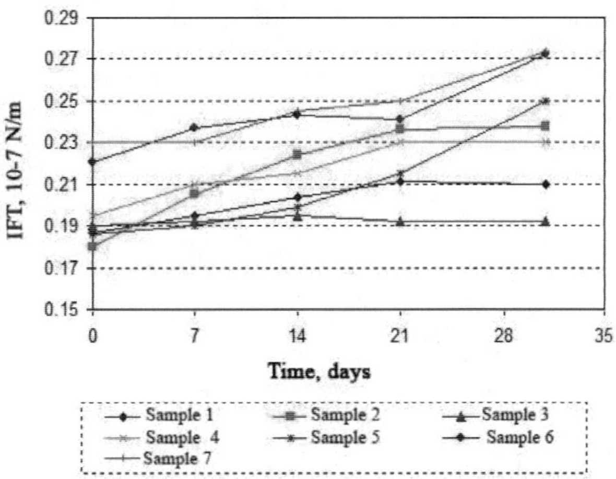

Figure 2. Change of surface tension at oil-surfactant solution boundary during exposure at 91 °C.

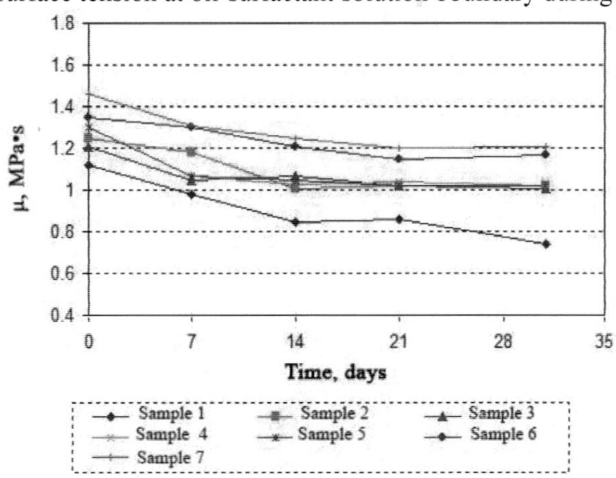

Figure 3. Change of viscosity of surfactant solutions during exposure at 91 °C.

The studies were carried out at a steady state and in vertical direction of filtration from the bottom upwards and in horizontal direction using WATERFLOODING, FDESS-100 and COREFLOODING equipment, reservoir pressure – 10.0 MPa and temperature – 91 °C. Oil models were used (75% of oil from well No. of 201+25%, oil viscosity – 1.95 MPa·s at a reservoir temperature), sea water (viscosity – 0.328 MPa·s at a reservoir temperature, a surface tension on oil-water boundary at 30 °C – 31.54 mN/m) and solution composition from the mixture of ethoxylated and propoksilirovanny sulfate, sulphonate α-olefin, alkylphenol ethoxylate and sulfonated polyacrylamide polymer AN-125 (surface tension on oil-solution boundary at a concentration of 1000 ppm and temperature of 30 °C: 0.19mN/m).

The injection of the surfactant rim begins after the oil is displaced by water in the amount of more than 6 pore volumes, i.e. when the reservoir model is completely watered. Once the surfactant was injected, the reservoir model was held at specified conditions for 36 hours in order to obtain the surfactant effect on the capillary displacement of oil and subsequent injection of water. The reservoir model is composed of several core samples of the Dragon field basement rocks.

The results of the tests to determine the oil displacement coefficients during water and surfactant injection are given in Table 3.

The average arithmetic value of the oil displacement coefficient increase was 0.069 (0.0604...0.0829) and the permeability recovery coefficient for water during the injection of the surfactant solution was 1.25 (0.71...1.98).

The obtained results to determine the increase of oil displacement coefficient and permeability recovery coefficient upon the injection of AMS-2P solution (0.15 pore volume) showed that the increase of the oil displacement coefficient with the AMS-2P solution at a concentration of 1000 ppm on the model of the Dragon field basement rocks under reservoir conditions was equal to 7.78% and the permeability recovery coefficient for water after surfactant injection was 0.91 (Figure 4).

Table 3. Experimental results of surfactant composition injection on core models

Model No.	Oil permeability, 10-3 μm2	Porosity F, unit fractions	Residual water saturation, Srws, unit fractions	Residual oil saturation after displacement by water S1ros, unit fractions	η1, water injection, unit fractions	Residual oil saturation after surfactant injection S2ros, unit fractions	η2, surfactant injection, unit fractions	Δη, unit fractions	Permeability recovery factor, K, unit fractions
1	291	0.0344	0.3956	0.3721	0.3843	0.3220	0.4672	0.0829	1.58
2	411	0.0622	0.4091	0.3725	0.3696	0.3359	0.4316	0.0620	1.98
3	11.1	0.0495	0.3618	0.3924	0.3851	0.3537	0.4458	0.0607	1.09
4	27.4	0.0477	0.3575	0.3681	0.4271	0.3175	0.5059	0.0778	0.91
5	365	0.0982	0.2836	0.3502	0.5112	0.3069	0.5716	0.0604	0.71
Average	221.1	0.0584	0.3615	0.3711	0.4155	0.3272	0.4844	0.0690	1.25

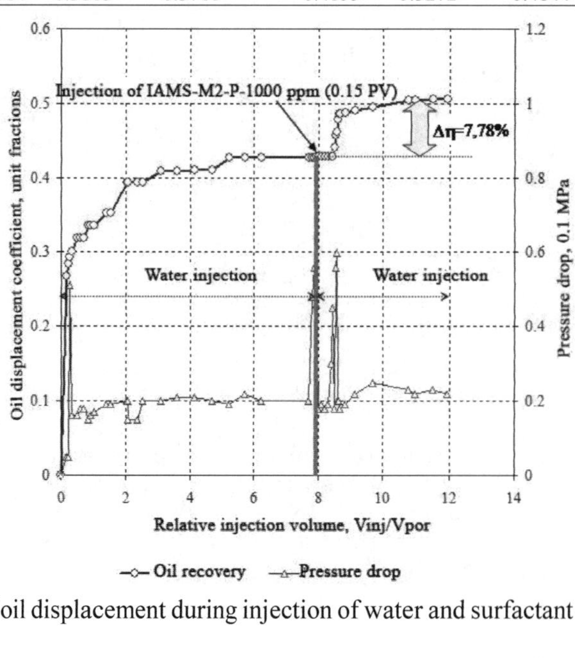

Figure 4. Dynamics of oil displacement during injection of water and surfactant compositions to model No. 4

4. Conclusion

1. Based on the results of the study on the interaction and heat resistance of some heat-resistant polymer additives at 128 °C, a sulfonated polyacrylamide polymer AN-125 (dissolution time in fresh water – 120 minutes, effective viscosity at 170c-1 – 19 MPa·s, gelling time at 90 °C – 0.8 hours, gel stability at 70-90 °C – more than 30 days) is proposed as an additive.

2. The results of experimental studies of a new surfactant composition prepared on the basis of a mixture of ethoxylated and propoxylated sulfate, α-olefin of sulfonate and alkylphenol ethoxylate with the concentration of 500 ppm and additives in order to improve viscosity in the reservoir model from basement samples with the length of >200 mm and diameter of 50 mm in sea water before and after exposure at 91 °C show that the surface tension at the oil-solution boundary prior to thermal exposure is 0.19 mN/m, and after exposure at 91 °C for 31 days – 0.21 mN/m.

3. The experimental studies of water and surfactant compositions injection on 5 models of basement rock core samples covering a wide range of reservoir properties (void ratio – 3.44...9.82% and oil permeability – 11.1...411·10-3 μm2) showed that when injecting compositions based on a mixture of ethoxylated and propoxylated sulfate, α-olefin sulfonate and alkylphenol ethoxylate with a concentration of 800 ppm and an additive of polymerasulfonated polyacrylamide with a concentration of 200 ppm in a volume of 0.15, the pore volume gives an increase in the oil displacement coefficient of 7.78%, and the water permeability recovery coefficient of the reservoir model after the injection of surfactant compositions is 0.91.

References

[1] Armstrong L J, Edmunds J, and Clare J B 2008 Environmental Considerations for Transferring Offshore Facilities to National Oil Company Operators *SPE International Conference on Health, Safety, and Environment in Oil and Gas Exploration and Production* (Nice, France, 15-17 April 2008) DOI: 10.2118/111704-MS.

[2] Akhmetov R T, Mukhametshin V V, Kuleshova L S, Grezina O A, and Malyshev P M 2020 The generalized correlating function of capillary curves and the relationship of the filtration-capacitive parameters of reservoirs in Western Siberia with the size distribution of pore channels *Journal of Physics: Conference Series* **1661** 012016 DOI: 10.1088/1742-6596/1661/1/012016.

[3] Mukhametshin V Sh, Andreev V E, and Yaskin S A 2020 Designing measures to increase oil recovery based on the identification and grouping of deposits *IOP: Earth and Environmental Science* **579** 012013 DOI: 10.1088/1755-1315/579/1/012013

[4] Alvarado V and Manrik E 2011 *Methods of increasing oil recovery. Planning and application strategies* (Moscow: Premium Engineering).

[5] Zeigman Yu V, Mukhametshin V V, and Kuleshova L S 2020 Management of flooding of low-production deposits according to geological and field data *IOP: Earth and Environmental Science* **579** 012019 DOI: 10.1088/1755-1315/579/1/012019

[6] Malyarenko A M, Kotenev Yu A, Mukhametshin V Sh, Bogdan V A, Kotenev A Yu, Blinov S A, and Umetbaev V G 2020 The influence of the rocks material composition on the reservoir properties of the reservoir *Journal of Physics: Conference Series* **1582** 012061 DOI: 10.1088/1742-6596/1582/1/012061.

[7] Alvarado V, Thyne G, and Murrell G R 2008 Screening Strategy for Chemical Enhanced Oil Recovery in Wyoming Basin *SPE Annual Technical Conference and Exhibition* (Denver, Colorado, USA, 21-24 September), DOI: 10.2118/115940-MS.

[8] Kadyrov R R, Mukhametshin V V, Galiullina I F, Kuleshova L S, and Safiullina A R 2020 Prospects of applying reservoir water and heavy brines derived therefrom in oil production and national economy *IOP Conference Series: Materials Science and Engineering* **905** 012081 DOI: 10.1088/1757-899X/905/1/012081

[9] Khormali A, Sharifov A R, Torba D I 2018 Investigation of barium sulfate precipitation and prevention using different scale inhibitors under reservoir conditions *International Journal of Engineering, Transactions A: Basics* **31(10),** pp. 1796-1802 DOI: 10.5829/ije.2018.31.10a.24.

[10] Kuleshova L S and Mukhametshin V V 2020 Justification for the selection of distances between wells to increase the efficiency of flooding by deposits in terrigenous reservoirs of Western Siberia *Journal of Physics: Conference Series* **1582** 012057 DOI: 10.1088/1742-6596/1582/1/012057

[11] Veliev M M, Ivanov A N, Bondarenko V A, Grishenko E N, Bahtizin R N, and Veliev E M, 2020 Experience of asphaltene sediments control during oil production at White Tiger field *Oil Industry* **6** 84-89

[12] Nur Wijaya, Sheng J J 2020 Comparative study of well soaking timing (pre vs. post flowback) for water blockage removal from matrix-fracture boundary *Petroleum,* **6(3)** 286-292 DOI: 10.1016/j.petlm.2019.11.001

IOP Conf. Series: Earth and Environmental Science 1021 (2022) 012070 doi:10.1088/1755-1315/1021/1/012070

Experience of applying the shock wave impact method for the bottomhole zone

R R Khusnutdinova, L V Petrova, L F Yusupova, R A Gilyazetdinov

Ufa State Petroleum Technological University, Branch of the University in the City of Oktyabrsky, Oktyabrsky, Republic of Bashkortostan, Russian Federation

E-mail: xusnutdinova.1986@mail.ru, info@of.ugntu.ru

Abstract. Improving the processes of exploration and development of fields, as well as increasing the productivity of oil reserves using physicochemical methods are one of the main tasks in stabilizing oil recovery and increasing the oil recovery factor. The most rational method used for hard-to-recover reserves is the shock-wave impact technology. This method can stimulate the release of energy from anomalously stressed zones and intensify the influx of hydrocarbons from deep layers of the Earth along the tectonic faults and sub-vertical inversion ring structures. The article presents the results of a study of the shock-wave impact on production wells of the Sabanchinskoye field and discusses factors that reduce the profitability of this technology.

1. Introduction

The stabilization of oil reserve development in Tatarstan is dependent on the process of exploration and productivity of fields that are at a late stage of their exploitation.

Oil producing enterprises have accumulated experience in operating wells in difficult conditions (with ARPD deposits, high-viscosity oil deposits, etc.).

An analysis of the operating conditions of mechanized wells at TATNEFT LLC shows that they can be attributed to the three main groups: wells exploiting D0–DI (Devonian) horizons, Lower Carboniferous (bobrik, tourney) horizons, Middle Carboniferous (Bashkir, Verei) horizons. High-viscosity oil wells, low-production wells, high-water-cut wells can be referred to separate subgroups. This classification makes it possible to simplify and specify recommendations for choosing geological and technical measures. At the same time, this approach does not ignore individual characteristics of wells, conditions for applying the shock-wave impact to the Sabanchinskoye field [1].

The technology of treatment of the bottomhole zone using a shock-depression device is a progressive technology for influencing productive formations in order to ensure high efficiency of field development and oil extraction.

Reservoir productivity is influenced by the geodynamic factors that determine the reservoir elasticity, saturating it with fluid, accumulation of natural energy in the reservoir, which is spent on displacing the fluid.

In order to solve the above problems, the shock-wave impact on the near-wellbore zone of the productive formation in the perforation interval is used to create an impact on the formation that simulates the geodynamic impact. The method was tested at the Sabanchinskoye field. The practice of using this method has shown that its effect lasts for more than a year [2].

2. Methods and materials

Geological exploration, research and experience in the development of oil fields show that the bottomhole zone (BHZ) is always characterized by low filtration properties due to the formation of a colmatation zone as a result of physicochemical processes occurring in the pore space of productive formations during drilling and subsequent operations [3]. To restore the well productivity, various methods of BHZ treatment are used: chemical, hydromechanical, impulse and combined ones. However, their effectiveness is relatively low. In 2003, on the basis of pilot works performed at six sites and more than 200 demonstration wells of Bavlyneft and RF patents 2117145, 2117146, 2151863, the instruction on the shock wave impact technology was developed, in which four options were described: three options - for producing wells, including one for the Carboniferous terrigenous reservoirs, the second - for the Devonian terrigenous reservoirs, and the third – for the carbonate reservoirs. The fourth option was given for injection wells.

The method involves the regular depression impulses of pressure created in the well at the level of the productive formation as a result of which the rocks are under the alternating dynamic loading. Paraffin solvent produced by oil treatment units of Bavlyneft and marketable oil in the volume of the well to push the solvent into the reservoir are used. With a negative impulse (a decrease in the bottomhole pressure), the rocks experience a compressive stress, and with a positive impulse (an increase in the bottomhole pressure), they experience a tensile stress. This change in the stress state of rocks causes the formation of cracks, and the wave motion of the fluid helps cleanse the pore channels. The shock wave technology was simple to implement and did not require any changes in the development system applied at the Bavlyneft fields.

The process of shock-wave impact is often called implosion. Implosion is the process of creating an instantaneous drawdown on the productive formation with a special device, which helps increase the rate of movement of formation fluid from the formation to the wellbore, which cleans the bottomhole zone from contamination. Implosion is accompanied by a water impact, which causes the formation of microcracks [4].

A depression (implosion) device is placed in the well in the interval of the treated formation; the treatment composition is pumped into the bottomhole zone, and the bottomhole zone is treated by creating multiple shock-depression effects with simultaneous cleaning of reaction products. Before pumping the treatment composition into the bottomhole zone, the intervals of the drained and compacted zones are determined. The drained zones are filled with the viscoelastic material inert to the treatment composition. The depression device is installed against the compacted formation zone and subjected to shock wave impacts. The treatment composition is injected in the process of exposure. In new wells, the method allows one to get rid of bridging agents that have penetrated into the formation during well construction; and in old wells, the method helps get rid of stripping agents that have penetrated into the formation during operation and repair works. The productivity of wells increases due to the intensification of oil production during shock-wave impacts on the bottomhole zone, taking into account the reservoir properties of the formation. The consumption of the treating composition is reduced.

There is a BHZ treatment method that involves the impact of multiple depressions and repressions using a jet pump lowered to the bottom on the tubing string. The method is used to clean the bottomhole formation zone from contamination. The disadvantage of this method is the limited use, because most of the productive formations are heterogeneous in permeability, and the depression-repression effect does not give positive results.

There is a BHZ treatment method that involves isolating the well above the treated interlayer with a solution for treating the bottomhole formation zone, technological holding, flushing the zone of the treated interlayer, creating a vacuum by pumping fluid into the zone above the place of interaction and along the annular space to the wellhead wells, recovering pressure in the zone of the treated interlayer by fluid overflow from the zone above the point of separation into the zone of the treated interlayer, repeating operations until the oil inflow into the well and oil production from the zone of the treated interlayer through the zone above the place of separation along the annular space of the well.

The method allows you to combine the depression-repressive action with the formation treatment. The disadvantage of this method is the low BHZ efficiency, because solution is absorbed by highly permeable sections of the formation, and low-permeability sections remain untreated [5].

The method for treating the bottomhole based on the vacuum-impulse (implosion) effect involves the simultaneous cyclic injection of solution, while each portion of the solution is pushed with oil into the formation, and each subsequent portion is increased by 1.15 - 1.25 in comparison with the volume of the previous portion. The method is implemented in the repression mode, i.e. when the bottomhole pressure (Pbh) is greater than formation pressure (Pf); therefore, the treatment composition comes from the well into the formation. The efficiency of the method is short-term due to the re-contamination of the porous medium of the bottomhole zone after the well is put into operation. The bridging agents pushed into the formation will return to the bottomhole zone with oil coming from the formation and reduce its inflow into the wellbore.

However, the result of formation treatment is reduced due to the fact that the bottomhole zone is cleaned from paraffin, resins, salts, clay particles, mechanical impurities, silt deposits and other contaminants only in a purely mechanical way due to the removal of formation fluid from the formation into the well. Most of the contaminants remain on the walls of the pore channels, mainly adsorbed particles, near-wall hydrophilic layers that cannot be carried away by the formation fluid even with multiple depression and hydraulic effects caused by the implosion effect. As a result, the potential treatment efficiency is not achieved either by water impacts or by depression. In addition, cleaning the repression bottomhole formation zone by the implosion method is ineffective, because Under these conditions, the borehole fluid from the annular space enters the imlosion chamber, and the partial cleaning of the bottomhole formation zone is caused by the water and wave impacts arising from this water impact [6].

The technical task of this method is to increase the productivity of wells due to the intensification of oil production with the shock-wave (implosion-shock) impact on the bottomhole formation zone, taking into account the reservoir properties of the formation [7].

To implement measures for shock-wave exposure using a shock-depression device, the following technical means and materials are required:

- a shock-depression device;
- a lifting structure used for well workover (A-50, AzINMASH-37A, AR-32, UPI-60A, etc.);
- pumping unit TsA-320 (TU 26-02-30-75);
- tank truck AC-10 (TU 26-13-32-77);
- tubing (GOST 633-80), steel D, K, E, L, M, R, with a diameter of 73 or 89 mm;
- pump rods (GOST 1054-74M, GOST 4543-71 and GOST 633-80), steel grade - 40, 20N2M6, 15NZMA, 30XMA and 152GNMF;
- HH2C sucker rod pumps (OST 26-16-06-86), with a diameter of 38, 44 or 56 mm;
- a metal container with a volume of 15 - 20 m^3;
- universal detergent ML-80 (TU 2481-007-48482528-99);
- distillate (a wide fraction of light hydrocarbons produced by TATNEFT;
- marketable oil in the volume of the well.

The device that creates depression impulses consists of a converted deep-well pump NSN-56; 43; 38; 32.

The pump cylinder was converted into a vacuum pump, i.e. the suction valve was eliminated, the lower end of the cylinder was plugged with a plug, and two through holes with a diameter of 18 mm to 25 mm were drilled in the middle or in the upper part of the cylinder body, depending on the diameter of the pump cylinder. This design makes it possible to simultaneously produce oil and treat the bottomhole zone with dynamic loads, develop wells coming out of drilling, clean the bottom from water-emulsion mixture and sludge [8, 9].

The is installed in the well at a maximum depth (in the perforation zone). The liner, attached to the body, is assembled by screwing pipes long enough to ensure that the end of the liner is at the bottom of the well. When the oil level is below the level of perforations, the annular space is filled with oil and the device is put into operation. When oil is withdrawn from the bottomhole, the annulus is periodically replenished with fluid. The operation of adding fluid continues up to a dynamic level above the inlet openings, after which the pouring is stopped and switched to the normal oil production mode, but with the simultaneous bottomhole zone treatment.

The method was tested at well No. 130A, deposit 1 of the Elkhovskaya area. In flushing the bottom, the degree of injectivity was determined - 185 m^3/day at 4.0 MPa. The downhole flow meter was lowered into the well and the reservoir injectivity profile was taken.

It was found that out of 5.5 m of the productive stratum, the upper zone of the reservoir with a thickness of 1.2 m has injectivity, the rest of the productive stratum does not receive the injected fluid. The body (cylinder) of the implosion percussion device was lowered into the well on the tubing string, installed in the lower zone of the productive formation, pumped into the tubing and pushed into the10 m^3 formation (into the drained zone) of high-viscosity (360 cP) invert emulsion; the treatment composition (hydrocarbon solvent-nefras 120/200) was injected into the annular. The plunger was lowered into the well and introduced into the cylinder (body) of the implosion-percussion device. The plunger was set to the extreme, lower position. Then the plunger was slowly moved to the extreme upper position, at which the water impact occurs. During the water impact, the plunger was stopped lifting and a mark was made on the rod (a landmark of the plunger lifting height). After that, the implosion-impact action (reciprocating movement of the plunger) was started and lasted for 6 hours. During this time, 720 double strokes of the plunger, i.e. 720 shock-wave impacts on the bottomhole zone were performed. During the implosion-shock impact along the annulus, nefras was injected; after exposure to the contaminated zone of the formation, it was pumped out by the implosion-shock device into the tubing string. The solvent consumption was 3 m^3.

After that, the implosion-percussion device was removed from the well, the pumping equipment was lowered and the well was put into operation. 7 days after, hydrodynamic studies were carried out and the inflow profile was taken. The data analysis showed that after treatment, the static level rose by 15 m (25 m versus 10 m), the dynamic level decreased by 490 m (370 m versus 860 m), the fluid flow rate increased almost twice (4.9 m^3 / day versus 2.5 m^3 / day), water cut decreased from 17% to 8%, productivity coefficient decreased from 0.025 t / day MPa to 0.35 t / day MPa.

The change in the inflow profile indicates the inclusion of the low-permeability zone of the formation; the operating thickness was 5.5 m (versus 1.2 m) [10].

All these indicators characterize the high productivity of this method.

The method makes it possible to reduce the treatment solution consumption due to its slow feeding.

3. Results

The shock-wave impact (SWI) on the bottomhole zone has the longest effect and is simple. The indicators of the technology are presented in Figure 1.

Table 1 shows the data on the shock-wave impact. It can be seen that withdrawals of fluid and oil significantly. Thanks to the shock-wave impact method, it was possible to increase the volume of oil production. The cumulative production of wells No. 123, 150, 151, 170, 175, 130, 139 and 120 amounted to 11,450 tons. The average daily increment per well is two tons per day, the average duration of the effect is two years.

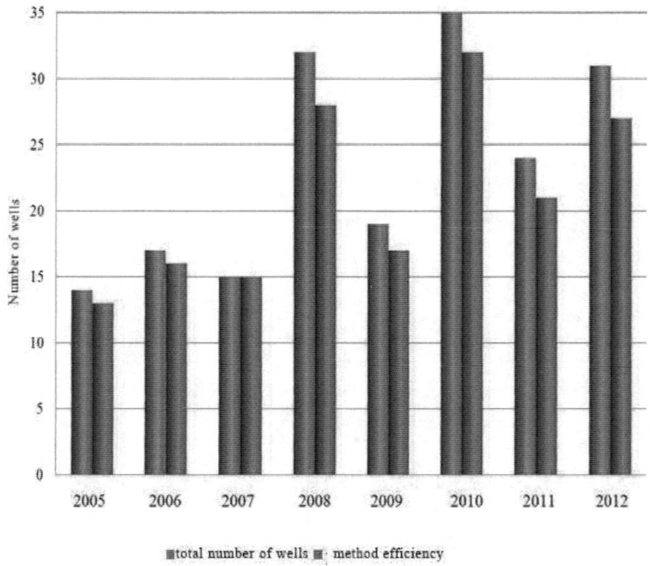

Figure 1. Results of application of the shock-wave impact on the bottomhole zone at the Sabanchinskoye field.

Table 1. Results of SWI method application at the Sabanchinskoe field

No of well	**Before application**			**After application**			**Additional volume, tons**
	Qz, t/days	*Qo, t/days*	*W, %*	*Qz, t/days*	*Qo, t/days*	*W, %*	
100	4.7	3.7	21.3	7.2	5.4	25	274
101	7.4	3	59.5	10.1	4.7	53.5	75
120	10.8	2.4	77.8	15.6	5.2	66.7	350
123	7.4	5.2	29.7	12.3	7.2	41.5	1571
130	17.1	6.3	63.2	22.2	8.4	62.1	752
135	5.1	1.6	68.6	5.9	1.8	69.5	228
139	9.6	3.5	63.5	13.2	6.1	53.8	813
145	0.9	0.5	44.4	1.9	0.8	57.9	540
150	2.6	2	23.1	6.7	4.2	37.3	6221
151	7	2.9	58.6	11.8	4.5	61.9	858
164	9.1	3.4	62.6	12.1	4.8	60.3	51
170	12.4	4.1	66.9	13.1	4.4	66.4	158
175	6.4	2.3	64.1	7.6	3.7	51.3	727

The advantage of the method is that the BHZ is treated along the way with the current well workover, if necessary, associated with the replacement of underground equipment [11].

Thus, a combination of an oil production device and a bottomhole zone treatment device equipped with a liner and a casing increases the productivity of deep wells and treatment intensity.

Based on these data, the dependence of changes in oil production rates and water cut was built (Figures 2, 3). The figures show a linear relationship.

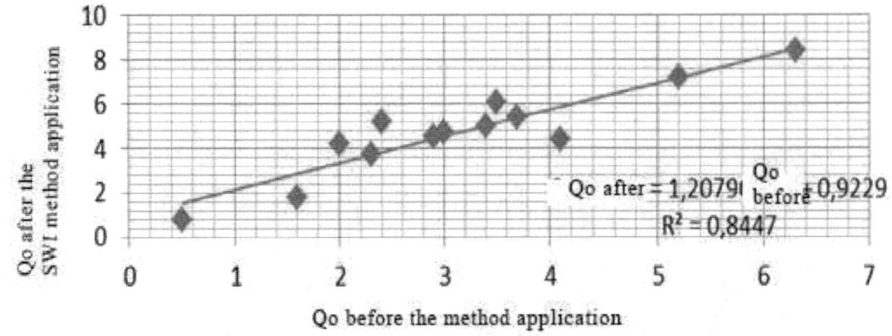

Figure 2. Dependence of the average daily increase in oil production after the SWI method application

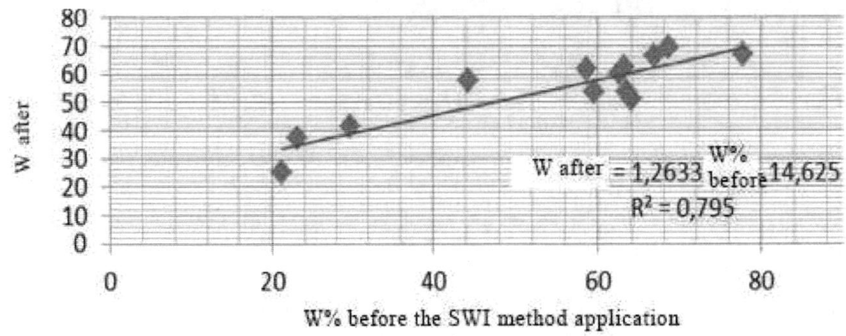

Figure 3. Dependence of the change in water cut after the SWI method application

To clean the bottomhole zone by the SWI method, it is necessary to calculate the parameters and select the layout of the sucker rod pumping unit. First of all, it is necessary to determine basic technological characteristics of the well.

Calculations of productivity of a downhole sucker rod pumping unit are of an estimate nature based on approximate formulas [12-14]. It is difficult to calculate the effect of a complex spatial profile of real wells on the amount of deformation of rods and pipes, which makes the expediency of using exact formulas questionable.

It is calculated from the formula

$$Q = 1440 \cdot F_N \cdot S_0 \cdot n \cdot K_2 \cdot \frac{s_{eff}}{s_0}, \quad (1)$$

In (1), S_{eff} and S_0 are lengths of the segments on the dynamometer chart corresponding to the movement of the plunger with an open discharge valve and the movement of the wellhead stem. Dynamogram performance calculation gives satisfactory results in the absence of leaks in the pump and tubing.

Performance is calculated on a dynamometer chart by the TATNEFT method.

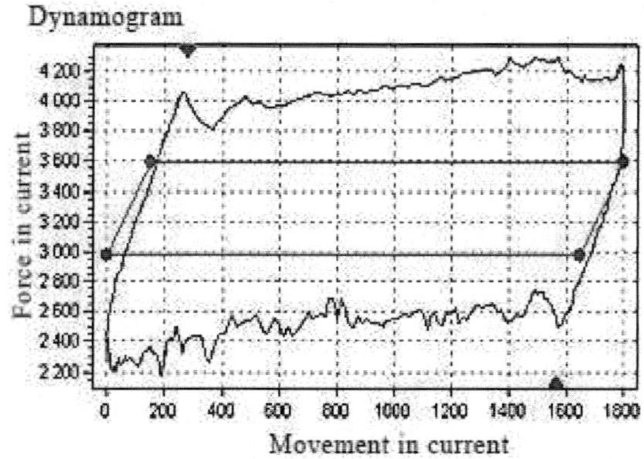

Figure 4. Dynamogram of changes in the force on the wellhead stock of well No. 100 of the Sabanchinsky field before the SWI method application

The productivity and the actual flow rate of pumps for a TATNEFT oil is calculated from the task "Analysis / Operational analysis of wells" in ARMITS. At each underground repair and significant changes in operating conditions, it is necessary to carry out a check calculation for selecting the downhole pumping equipment, taking into account the TATNEFT rod movement rules.

4. Conclusion

Experience in the implementation of shock-wave technologies for stimulating oil production has been accumulated.

The effectiveness of dilatation-wave impacts is characterized by an increase in the volume of oil production, a decrease in cost and an increase in profits. The method is easy to use, safe, environmentally friendly and is used at fields with a high water cut level and viscous oil reserves.

The shock-wave technology has been used at the Sabanchinskoye oil field since 1996. During this period, 149 treatments were carried out using the method, of which 138 were successful, the success rate of treatments was 94.4%. As a result of this treatment, additional 87642 tons were produced.

References

[1] Petrova L V,.Yarulli D R 2019 Evaluation of the effect of asphalt resin paraffin deposits on oil well performance *IOP Conference Series: Materials Science and Engineering* **560** 012084 DOI: 10.1088/1757-899X/560/1/012084

[2] Almukhametova E M, Zakirov A I, Fattakhov D I, Stepanova R R 2019 «Simulating Geological Features of Torrential Deposits», Advances in Engineering Research **182,** 9-13 DOI: 10.2991/ciggg-18.2019.2

[3] Mukhametshin V V, Kuleshova L S 2019 Justification of Low-Productive Oil Deposits Flooding Systems in the Conditions of Limited Information Amount *SOCAR Proceedings* **2** 16–22 DOI: 10.5510/OGP20190200384.

[4] Gimaltdinov I K, Gizzatullina A A, Gimaltdinova A A 2020 The dynamics of sound waves in a tube with elastic walls filled with gas-liquid mixture *IOP Conference Series: Materials Science and Engineering* **919** 062062 DOI 10.1088/1757-899X/919/6/062062.

[5] Mingulov Sh G and Mukhametshin V Sh 2020 Restoring the intake capacity of injection wells by means of shock-wave exposure *IOP Conference Series: Materials Science and Engineering* **952** 012060 DOI: 10.1088/1757-899X/952/1/012060

[6] Yusupova L F, Khalikova K M, Khusnutdinova R R 2021 Technological feature of water shutoff operations *IOP Conference Series: Earth and Environmental Science,* **868,** 012086 DOI:10.1088/1755-1315/868/1/012086

[7] Nie Z, Liu Z and Sun X 2018 A Multirelay Cooperation Method for Wireless Transmission of MWD and LWD Signals *IEEE Transactions on Geoscience and Remote Sensing* **56** 1229-37.
[8] Sun W and Mun-Hong H 2017 Forecasting and uncertainty quantification for naturally fractured reservoirs using a new data-space inversion procedure *European Assoc Geoscientists & Engineers Computational geosciences 15th Conference on the Mathematics of Oil Recovery (ECMOR)* (Amsterdam, Netherlands) 21(5-6) 1443-1458.
[9] Iurinak I I 1991 Oilfield applications of colloidal silica gel *SPE Production Engineering* **6(4)** 406-412
[10] Edmunds W M, Shand P, 2008 *Natural groundwater quality* (Moscow: Nedra)
[11] Akhmetov R T, Mukhametshin V V, and Kuleshova L S 2019 Simulation of the absolute permeability based on the capillary pressure curves using the dumbbell model *Journal of Physics: Conference Series* **1333** 032001 DOI: 10.1088/1742-6596/1333/3/032001.
[12] Akhmetov R T, Mukhametshin V V, Kuleshova L S, Mingulov Sh G, and Malyshev P M 2020 The choice of the correlating function of capillary pressure curves under conditions of reservoirs in Western Siberia *IOP Conference Series: Materials Science and Engineering* **905** 012095 DOI: 10.1088/1757-899X/905/1/012095.
[13] Almukhametova E M, Yusupova L F 2021 Theoretical study of the change in the concentration of demulsifier due to the phase ratio in the product *AIP Conference Proceedings* **2410** 020002 https://doi.org/10.1063/5.0067638
[14] Yusupova L F, Almukhametova E M, Faizullin I A 2021 Sources of oily waste affecting the Earth's crust interior *AIP Conference Proceedings* **2410** 020034 https://doi.org/10.1063/5.0067642

Possibilities of using fiber-optic sensors in the study of oil and gas wells

E O Shiryaev [1], V F Rybka [1], I Yu Kolychev [2]

[1] LLC "PITC Geofizika", Electronic passage, 3, Dobryanka, Polazna district, 618703, Russia
[2] Perm National Research Polytechnic University, 29, Komsomolsky prospect, Perm, 614990, Russia

E-mail: doc_galkin@mail.ru

Abstract. With an increase in the accuracy of fiber-optic sensors, the possibility of their use in various fields is being formed. In this article, the authors present their own experience in the implementation of fiber-optic technologies in geophysical surveys of wells, indicate current problems and prospects for the development of this direction in the region of activity. Since 2014, LLC "PITC Geofizika" has been actively developing the direction of fiber-optic temperature measurement systems. To date, these are measurements in more than 40 wells (15 wells are being measured at present time). As a result, we can say that today there is already a certain niche in the basket of the geophysical sector, confidently occupied with fiber-optic sensors technologies, but the prospects for the development of this technology are significant and are fully revealed only in the process of work. However, it is worth approaching the implementation of this method only with a complete understanding of the processes occurring in the well.

1. Introduction

Fiber-optic temperature measurement systems (OVSt) confidently solve the problem of determining the fluid level in the wellbore using a mechanized production method [2] (Fig. 1-2). It was tested in dozens of wells with both ESP and SRP [3-5]. In this case, the level is determined during the operation of the pump and its stops. It is the liquid level that is determined, not the foam level, which is a significant advantage over echo sounders.

If the subsoil user wishes and there is a possibility of mass introduction of the technology of this logging during production, it is possible to automate this process online without any special difficulties. In addition, the pump temperature can be monitored. When the level drops and the pump heats up (Fig. 1), the pump should be automatically transferred to a lower capacity, not "blindly" as it happens now, but with an assessment of the result of changing the operating mode.

Separately, it is worth touching on the topic of determining the flow of gas. Since the throttling effect during gas flow has a negative value, the identification of gas-fueled interlayers is a confidently solvable problem. During the current studies of an operating well (perforation under the pump), an exemplary example of the formation of a gas hydrate plug due to a poor-quality coupling was observed [1].

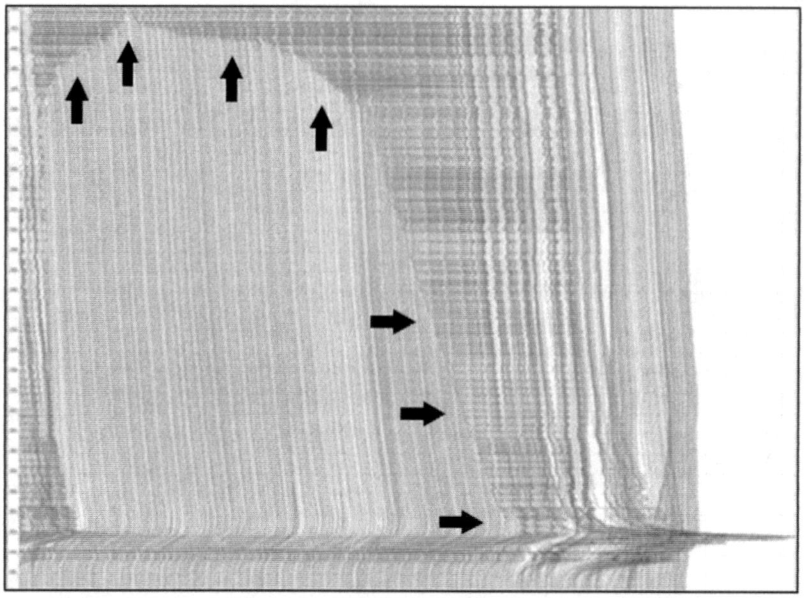

Fig. 1. The dynamic level when the ESP operates at 45 Hz; a decrease in the level when switching to 50 Hz

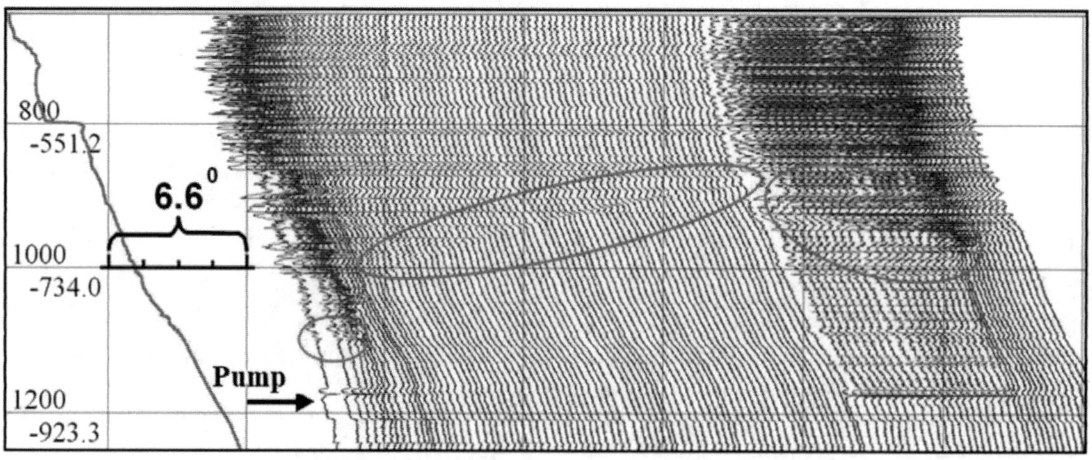

Fig. 2. Dynamic level during operation and shutdown of the SRP

2. Experience in using the fiber-optic temperature measurement systems

When the ESP was started, in the process of monitoring by OVSt, opposite the upper perforated formations, a cooling anomaly was observed, caused by gas throttling. Heating occurred in the area of the engine (SEM) of the ESP. The pump has stopped automatically due to low pressure. The gas, rising upward, began to displace the liquid in the annulus, where the pressure began to rise. At the same time, the OVSt revealed a leakage in the tubing sleeve, through which liquid and then gas began to flow into the tubing, throttling due to leakage and creating first a positive (liquid flow) and then negative (gas flow) anomaly.

After calling the inflow, the lower perforation interval was shot. Running the rock drill downhole was difficult due to the large angles in the well. The pressure in the annulus was reduced from 80 to 20 atm., but it was not possible to reduce it any more.

Further, when trying to start the ESP, it was automatically stopped due to lack of pressure. Perforated strata at that time worked with gas. It continued to flow into the annulus and throttled through a leaky coupling. The amplitude of the negative anomaly increased and reached 7 ° C.

The SEM was warmed up, and above the pump, apparently at the check valve, a negative gas-throttling anomaly was created and gas hydrates began to form (Figure 3).

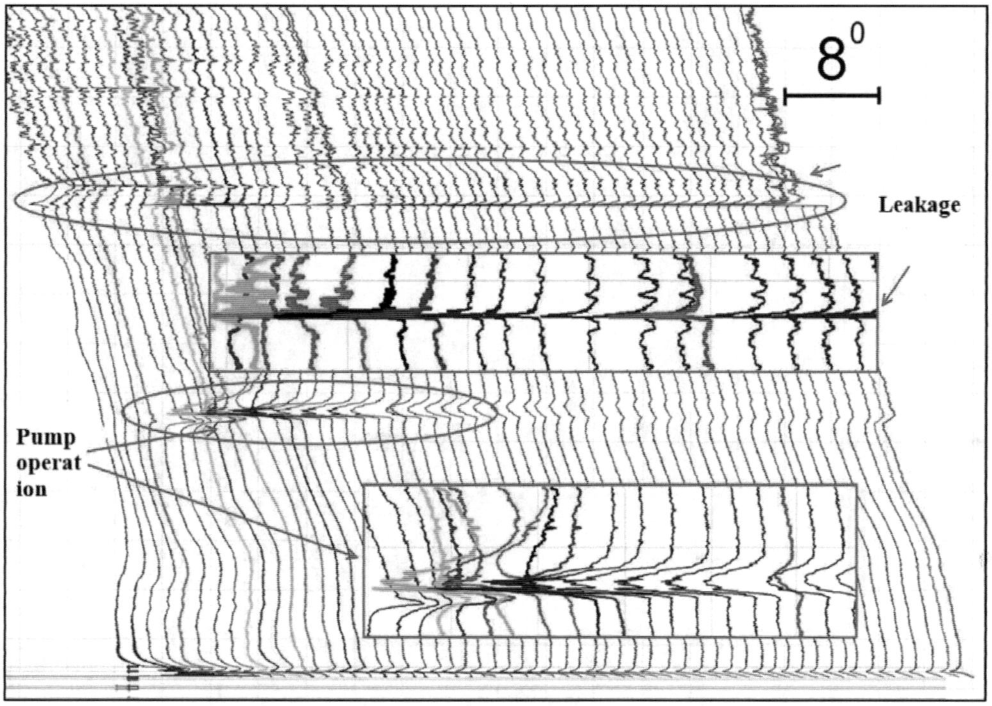

Fig. 3. Attempt to start the ESP. Formation of gas hydrates on it.

Further attempts to start the ESP and appearance of the bleed gas from the annulus only complicated the situation. The temperature in the area of tubing leaks has reached minus 12 ° C. The amplitude of the anomaly above the ESP reached 5 ° C. At the same time, the heating from the submersible motor has ceased to spread upward, which, with the poor thermal conductivity of gas hydrates, indicates the presence of gas hydrate in the ESP area (Fig. 4).

A decrease in temperature above the ESP and with a leaking coupling, as well as an increase in pressure in the annulus, created conditions for the occurrence of gas hydrate.

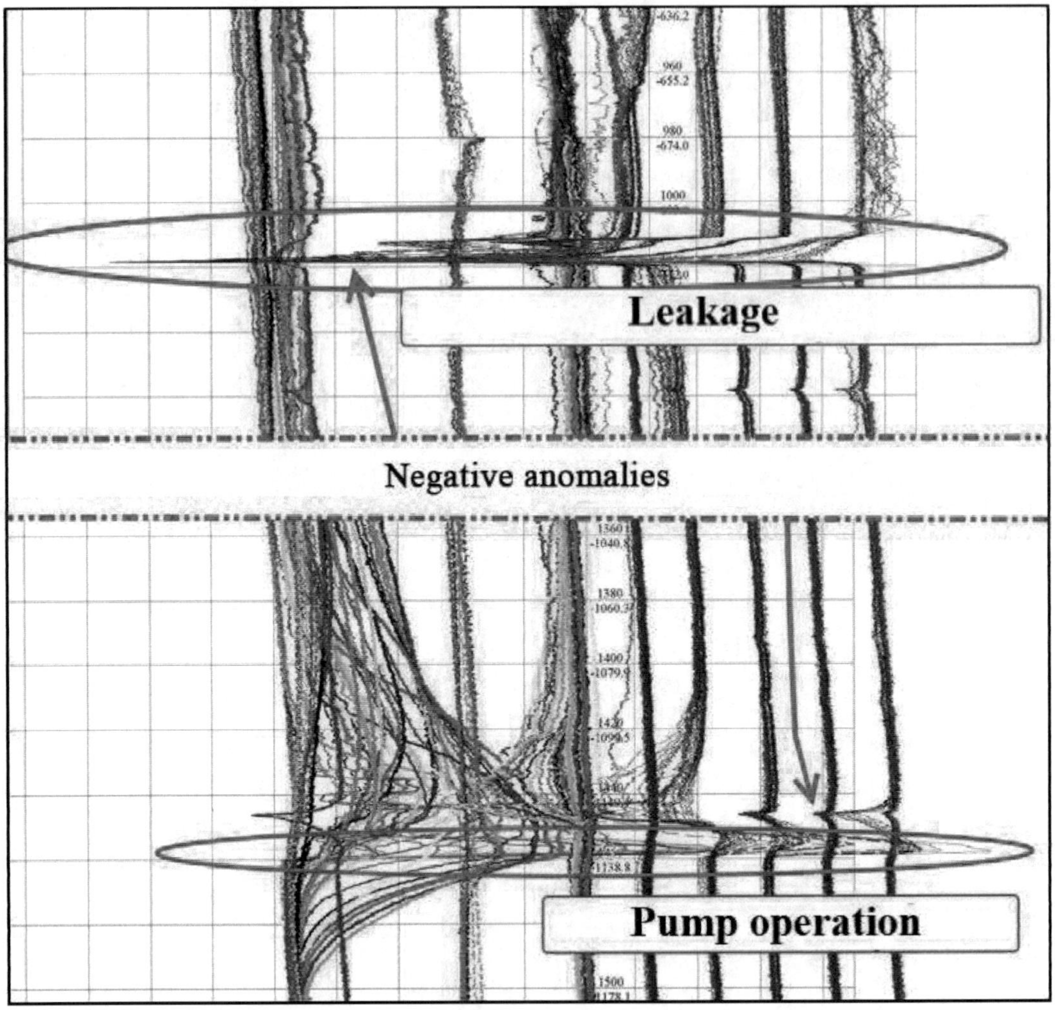

Fig. 4. Attempts to start the ESP.

Then attempts were made to inject oil into the annulus, the pressure rose to 80 atm., 7 m^3 were injected per day. Most of the oil was pushed through the tubing sleeves above the leak interval. The annular space, starting from the place of leakage of the tubing to the ESP, was filled with gas hydrate.

When creating geological and technological models of oil fields, one of the important tasks is to determine the working oil-saturated reservoirs [6-8]. With the help of OVSt, under certain conditions associated with the limitation of thermometry, it is possible to determine the production rate of the reservoirs. More precisely, according to the thermometry data, it is possible to determine the percentage of the flow rates of various interlayers or to determine the maximum and minimum working intervals, provided that the incoming fluid is homogeneous.

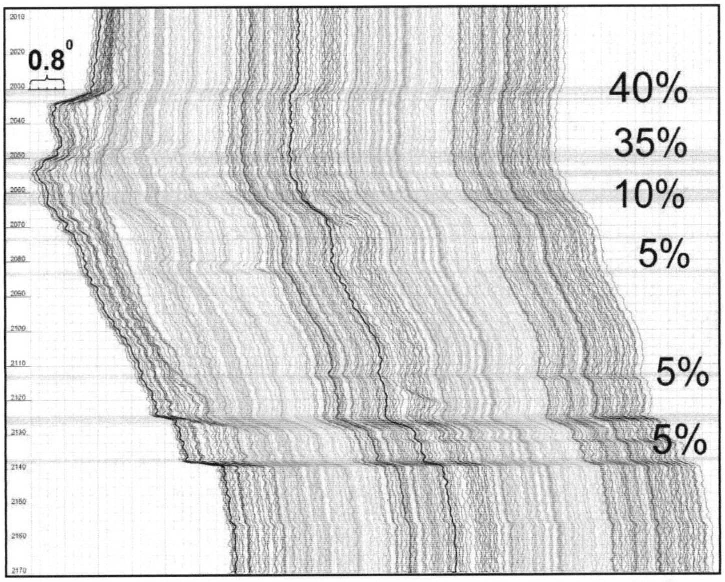

Fig. 5. Separation of flow rates according to thermometry data

However, the accuracy and reliability of such determinations are low due to the complex dependence of the throttling effect on the inflow volume and the superposition of other thermodynamic effects. To increase the accuracy of flow rate measurements, it is necessary to heat the cable by 2 - 3 ° C by passing an electric current through its current-carrying conductors. This will allow obtaining the so-called STD mode and measuring the degree of cable cooling, which depends on the inflow volume.

Solution of technological problems is as follows. In this block, it is possible to combine the capabilities of the OVST method when solving one-time non-standard problems - during the production of work, experience has accumulated in solving the following problems:

- Determination of a casing leakage location using a cable packer. The work is carried out according to the following scheme:

1) Planting the packer above the perforation intervals - injection with registration.

2) Rise of the packer above the place of detected leakage - injection with registration.

3) In case of revealing a new leak - repeat paragraph 2, in the absence of a leak - completion of work.

The efficiency of these works in relation to the standard method of performing these works by the workover crew is obvious - setting the packer on the tubing-injection-lifting by 50/100 m - pumping and so on to the wellhead.

Here we can state that the time spent on solving this problem has been reduced several times.

- Monitoring the operation of leaks during well development. The work is carried out according to the following scheme:

1) Joint running with the tubing of the OV-cable located in the annulus.

2) Production of well development by swabbing with recording OVSt data.

3) Issuance of an opinion with the provision of data on the processes occurring in the annulus.

Carrying out these works makes it possible to judge about the possibility of contaminants entering the well product not from the perforation interval.

- Hot wash control. In the process of well workover, there is a need to clean up the wellbore, for which the hot flushing technique is used, the efficiency of which does not always correspond to expectations. The task was set to evaluate the effect of this technological operation using the OVST (Fig. 6).

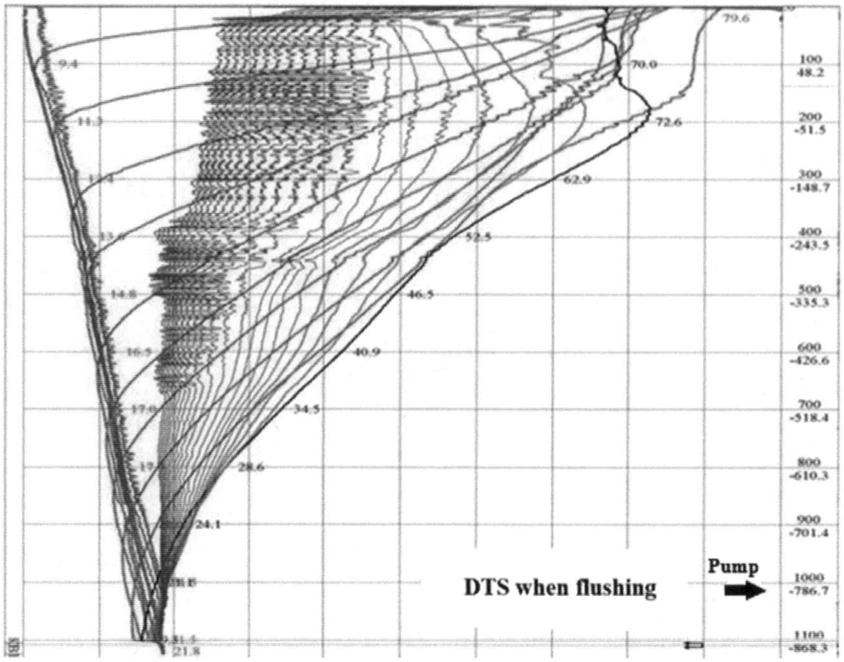

Fig. 6. Registration of OVSt in the process of hot flushing of the well

As a result of the control, it was revealed that the required heating to 50 C° was carried out only to a depth of 450 m. At the same time, there is no heating at the depth of the sucker-rod pumping unit, insignificant cooling is observed due to the fluid shift down the wellbore. To implement the elimination of paraffins by this method, it is necessary to increase the volume of liquid at least 3 times and the rate of injection – 2 times.

Summarizing all of the above, it can be argued that the available technology already now allows solving a certain range of tasks, which makes it possible to have this technology in the arsenal of GIS methods. Let us compare OVSt with the available technologies.

3. Comparison of fiber-optic temperature measurement systems with the available technologies. There is a need to make a reservation. At present, the technologies of the inflow profile (departure) are developed, which are carried out on a geophysical cable. Moreover, their area of application is well repair. Those technologies are applicable when pumping equipment is removed from the wells. The well is plugged and the purpose of the work is to search for a problem that has arisen in the well. Sometimes a preventive search for the problem is carried out before lowering the pumping equipment. There are only 2 exceptions - flowing wells and wells equipped with a Y-tool. Moreover, when a problem arises, flowing wells, as a rule, stop producing and must be stopped and "treated".

OVSt will not be able to compete with a complex cable device in the near future. The total cost of the equipment is significantly higher, and even the additional equipment of the OVSt with distributed pressure and liquid composition will not allow solving the problem at the same technical level as a standard complex device. Consequently, where a complex device "walks", OVSt are less effective.

The scope of OVSt is monitoring the well performance. That is the area of "prevention", not "treatment". Let us highlight the main advantages:

- no moving parts - monitoring is easy to implement without the participation of a geophysical crew, using only data transmission systems;
- it is easier to organize automatic processing and decision-making;

- there are no connecting elements and sensors - it is possible to work in an aggressive environment for the entire overhaul period and longer.

However, OVSt and other logging methods in the development process are mostly applicable not for finding problems with the well. Their purpose is to manage the operation of the field. It is practically pointless to control the operation of one well or pad.

Having decided on the scope of the OVSt, let us move on to the problems of technology implementation.

4. Solutions and development prospects

The systems of smart wells and smart fields will eventually take a leading position in oil and gas production. Production optimization is impossible without these systems.

Already now, we can state a confident solution of the following tasks within the framework of an intelligent field:

1. Underbalanced perforation followed by well performance monitoring. In this case, the perforation system and the pressure gauge are lowered under the ESP or SRP on a fiber-optic geophysical cable together with a submersible pump. Fastening to tubing ensures cable run. Further, the perforation system is positioned against the formation, and a depression is created by the pump. The perforator is undermined and the pressure is recorded. Next, the perforator is lowered to the bottom of the well, the pump is started and the reservoir performance is monitored using the OVSt and a downhole pressure gauge for the entire development period. Working intervals, water intake intervals, hydrodynamic parameters are determined. The optimal operating mode of the pump is selected. After the well reaches the mode, the geophysical cable remains in the well for subsequent monitoring of the operation, which is recommended to be performed on a quarterly basis or when the productivity (production) of the well changes. Advantages - formation cleaning during drilling-in, the operation of the reservoirs in the operational mode is fully controlled without spending time on stopping and killing the well and well logging.

2. Monitoring the operation of flowing wells. The pressure gauge is lowered on a fiber-optic geophysical cable with a rigid attachment behind the tubing. The well is controlled in terms of the regime, the operation of the reservoirs, the hydrodynamic parameters. The optimal well operation mode is selected. The cable and pressure gauge remain in the well for the entire overhaul period. The only unresolved issue is a cable bore packer.

Further development of the OVSt technology implies the organic growth of the technology, which consists in expanding the capabilities of the OVSt method.

At present date, a positive experience has been obtained when testing the cable by heating - in fact, a STD sensor distributed along the wellbore is obtained.

Prototypes of composition sensors (resistivity meter, moisture meter) were manufactured and tested in the form of separate modules, fixed on an optical fiber cable, assembled according to a specific well.

Also, in addition to organic growth, work is underway to further develop technology in the field of control of technological operations - today a solution is ready to control the cementing process of a well with further monitoring of its work. Here it should be borne in mind that cementing today is an absolutely unexplored process and it will take significant time only to understand the processes occurring in the WCC process.

Today there is already a certain niche in the basket of the geophysical sector, confidently occupied with OVS technologies, but the prospects for the development of this technology are significant and are fully revealed only in the process of work. However, it is worth approaching the implementation of this method only with a complete understanding of the processes occurring in the well. The lack of competencies forms a negative attitude towards the technology as a whole, from an innovation, in the eyes of a subsoil user. OVSt has turned into a non-working incomprehensible old one; our task is to correct this opinion by solving specific tasks.

Acknowledgments
The study was carried out within the framework of the state task of the Ministry of Science and Higher Education of the Russian Federation, FSNM-2020-0027 for the implementation of fundamental scientific research.

References

[1] Rybka V F, Vasyutinskaya S I 2018 Fiber optic well thermometry. Monitoring the formation of a gas hydrate plug *Scientific journal of the Russian gas society,* **1** 43-46 DOI 536.51.083/ 2412-6497
[2] Kremenetskiy M I, Ipatov A I, Gulyaev D N 2012 Information support and technologies of hydrodynamic modeling of oil and gas deposits, (Moscow — Izhevsk: RKhD Publ.)
[3] Kaeshkov I S, Kremenetskiy M I, Buyanov A V 2014 Horizontal well monitoring based on temperature profile measuring with Distributed Temperature Sensors (DST) (In Russ.), *SPE* 171236
[4] Ipatov A I, Kremenetskiy M I, Kaeshkov I S et al. 2014 Undiscovered DTS potential of horizontal well inflow profile monitoring (In Russ.), *Neftyanoe khozyaystvo = Oil Industry* **5** 96–100
[5] Yarullin R K, Valiullin R A, Sadretdinov A. A. et al. 2014 Fiber optic technologies for monitoring of operating horizontal wells (In Russ.), *Karotazhnik* **9** 38–46.
[6] Cherepanov S S, Ponomareva I N, Erofeev A A, Galkin S V 2014 Determination of fractured rock parameters based on a comprehensive analysis of the data core studies, hydrodynamic and geophysical well tests *Neftyanoe Khozyaistvo – Oil Industry* **2** 94-96.
[7] Kozyrev N D, Vishnyakov A Yu, Putilov I S 2020 Assessment of the Uncertainty Parameters Influence on the Development Indicators Forecasting *Perm Journal of Petroleum and Mining Engineering* **20(4)** 356-368. DOI: 10.15593/2712-8008/2020.4.5
[8] Repina V A, Galkin V G, Galkin S V 2018 The use of integrated accounting of petrophysical characteristics in the adaptation of geological and hydrodynamic models (using the example of the Visean reservoir of the Gondyrevskoye oil field) *Journal of Mining Institute* **231** 268-274 DOI: 10.25515 / PMI.2018.3.268.

IOP Conf. Series: Earth and Environmental Science 1021 (2022) 012072 doi:10.1088/1755-1315/1021/1/012072

Analysis and forecasting of the success of hydraulic fracturing treatment in the selection of candidate wells

V V Anisimov, V Sh Mukhametshin and A I Valeev

Ufa State Petroleum Technological University, Branch of the University in the City of Oktyabrsky, 54a, Devonskaya St., Oktyabrsky, Republic of Bashkortostan, 452607, Russian Federation

E-mail: vsh@of.ugntu.ru

Abstract. At present, a huge part of oil and gas reserves are hard-to-recover, that is, they are located in deposits being classified as low-permeability reservoirs. A detailed analysis of the success of previously conducted hydraulic fracturing treatment of reservoirs in the wells of the explored production facility, as well as the methodology approbation for the assessment and selection of design candidate wells for further HFT in this field are considered in the paper.

1. Introduction

Improving the efficiency of managing the oil deposits development is based on the study and formalization of various processes aimed at extracting liquid hydrocarbons [1–10]. One of the given processes is the intensification of oil production with the use of impact on the bottomhole zone using hydraulic fracturing treatment.

Hydraulic fracturing treatment (HFT) is a method of mechanical impact on the bottomhole formation zone (BFZ), the essence of which is to break the rock of the bottomhole zone along the regions of least resistance when exposed to high pressure created when a fluid is pumped into the well that carries energy into bottomhole zone, while its consumption should exceed the formation absorption value. The technology is among the most efficient methods for increasing the flow in wells, but also HFT is one of the most costly methods. In this regard, the task of analyzing and predicting the success of HFT is extremely urgent.

2. Materials and methods

According to published data, the success rate of HFT in many fields is 40–90%. Moreover, their effectiveness varies within fairly wide limits. The increase in HFT success is associated with both deeper laboratory studies and the experience generalization of applying diverse technologies for such treatments in various field conditions. Identification of the parameters that have a prevailing effect on the success, and the process subsequent formalization based on statistical modeling allow carrying out a scientifically grounded choice of wells and select the optimal treatment technology in order to increase HFT efficiency [11–16].

One of the methods for predicting the success of hydraulic fracturing treatment is the Kullback method. The approach makes it possible to assess the influence of various parameters on the success of a hydraulic fracturing operation by calculating and comparing their information content. The given

IOP Conf. Series: Earth and Environmental Science 1021 (2022) 012072 doi:10.1088/1755-1315/1021/1/012072

technique, with all its inherent algorithms, serves as one of the additional tools of the neural network program for the selection of candidate wells for HFT.

Production wells of the operating stock were selected in the considered sample of wells in the studied field, namely, the HFT location, since this object has more complete information about the performed operations.

For further calculations, 155 wells were selected for "training" and 17 for "forecasting" (assessing the method validity). Besides, the criteria for assessing the HFT success of the operations carried out on wells were adopted on the basis of the history analysis of HFT operations in the considered field, and already drilled wells. The analysis of previous periods at the "operating functioning" fund has revealed that the average oil gain after HFT in 3 months is 7 tons/day for all reservoirs, and the average water-cut in 3 months after HFT is no more than 40%. After analyzing the historical data, the given indicators are "good." Consequently, wells with an increase in production by 7 tons/day and more, and a water-cut value of up to 40% are included in the computation for the wells separation "with successful HFT" – "A", and "unsuccessful HFT" – "B", in "A" category. The rest of wells fall into category "B."

The task is to select the most informative features based on an array of data on wells, and then obtain a forecast of the hydraulic fracturing efficiency for the next wells (Table 1).

We select the following criteria for assessing the impact:

– reservoir geological parameters: the reservoir efficient thickness, m; permeability, 10-3 μm^2; oil saturation, un. fract.; sand content, un. fract.; stratification factor, un;

– technological impact indicators: mass of injected proppant, t; maximum proppant concentration, kg/m^3; oil production rate 3 months before HFT, tons/day; water-cut 3 months before HFT, %.

Table 1. Data on wells and attributes influencing HFT success

Well number	Efficient thickness, m	Oil saturation, un. fract.	Permeability, 10^{-3} μm^2	Sand content, un. fract.	Stratification factor, un.	Injected proppant weight, t	Maximum proppant concentration, kg/m^3	Oil production rate 3 months before HFT, tons/day	Water-cut 3 months before HFT, %	Category
437	14.2	0.47	2.6	0.50	5	80.0	1300	2.9	19.0	B
447	5.4	0.57	0.4	0.16	1	10.4	800	5.0	2.7	A
447	5.4	0.57	0.4	0.16	1	11.0	800	7.9	2.3	B
626	4.3	0.46	0.6	0.13	1	11.0	960	4.0	3.4	A
626	5.1	0.46	0.6	0.15	2	11.1	554	2.1	8.7	B
653	6.6	0.51	0.5	0.19	1	20.6	710	2.0	2.9	A
703	12.4	0.53	0.9	0.38	3	13.0	880	5.0	5.7	A
1006	7.6	0.41	3.4	0.28	5	50.0	900	0.9	5.9	B
1031	8.2	0.57	8.5	0.27	3	13.0	950	2.3	4.3	A
1031	8.2	0.57	8.5	0.27	3	40.4	1190	3.4	8.3	A
1036	8.6	0.44	1.0	0.26	5	40.0	1100	1.7	50.6	B
1067	7.3	0.53	0.5	0.20	5	35.0	1100	1.8	7.0	A
1093	3.0	0.50	1.2	0.10	1	20.6	710	6.0	11.2	B
1093	3.0	0.50	1.2	0.10	1	30.0	1100	2.9	7.6	B
1100	15.7	0.71	2.6	0.54	7	40.0	1170	8.4	15.3	A
1129	11.0	0.65	2.7	0.36	2	60.3	1032	5.9	10.2	B
1129	14.6	0.63	2.2	0.48	3	60.0	1000	2.7	58.1	B
1130	9.7	0.61	1.7	0.29	6	20.5	700	7.3	7.9	A
1130	9.7	0.61	1.7	0.29	6	24.0	1100	3.2	5.5	A
1791	3.5	0.44	0.7	0.14	2	40.0	1200	4.2	35.4	B

Table 1 shows 1/9 of the analyzed wells.

56 wells are included in group "A", and 99 - in group "B" as a result of assessment and distribution of wells.

Here is the calculation of the informative value for the “efficient thickness” feature (Table 2), which varies in the range from 3 to 24.1 m. We divide this range into equal intervals with a step of 2.11 m; 10 intervals are obtained. Further, we determine the frequency of wells hitting one of the groups “A” or “B.”

Table 2. Informative value determination of the “efficient thickness”

Interval	Range of change	Number of wells		Relative frequency, %		Smoothed frequency		$\bar{y}_A/\bar{y}_B$	DC	J_t
		A	B	y_A	y_B	$\bar{y}_A$	$\bar{y}_B$			
-1		0	0	0	0	0.36	0.91	-	-	-
0		0	0	0	0	2.32	3.43	-	-	-
1	2.99–5.11	2	9	3.57	9.09	7.50	9.09	0.825	-1	0.007954545
2	5.11–7.22	9	16	16.07	16.16	15.18	14.14	1.073341837	0	0
3	7.22–9.33	16	22	28.57	22.22	21.61	18.59	1.16255823	1	0.015106421
4	9.33–11.44	13	14	23.21	14.14	21.25	18.18	1.16875	1	0.015340909
5	11.44–13.55	11	27	19.64	27.27	16.96	17.98	0.943519262	0	0
6	13.55–15.66	4	10	7.14	10.10	9.46	10.91	0.867559524	-1	0.007224026
7	15.66–17.77	1	0	1.79	0.00	4.11	4.75	0.865121581	-1	0.003201659
8	17.77–19.88	0	0	0.00	0.00	1.07	1.11	0.964285714	0	0
9	19.88–21.99	0	0	0.00	0.00	0.18	0.20	0.883928571	-1	0.000117244
10	21.99–24.10	0	1	0.00	1.01	0.00	0.40	0	0	0
Total		56	99	100.00	100.00				-2.00	0.05

Calculation of the weighted smoothed frequency is required in order to align the impact indicators on the intervals of distribution processes; the frequency of a specific feature for this purpose is taken into consideration in 2 previous intervals and 2 further.

The weighted smoothed frequency is computed by the formula:

$$\bar{y}_n=(y_1 + 2y_2 + 4y_3 + 2y_4 + y_5)/10, \quad (1)$$

where $y_1 - y_5$ – frequency in intervals.

Fictitious intervalsshould be included for calculating the specified indicator in the 1st interval: 0 and –1. The frequencies are equal to 0 in the given intervals, since not a single observation is recorded. In this case, the smoothed frequency value is determined as follows in the 1st and 2nd intervals for “A” group:

$$\bar{y}_{1A}= (0 + 0 + 4y_3 + 2y_4 + y_5)/10, \quad (2)$$

$$\bar{y}_{2A} = (0 + 2\, y_2 + 4y_3 + 2y_4 + y_5)/10. \quad (3)$$

Smoothed frequency ratio $\bar{y}_A/\bar{y}_B$ is recorded in the next column of the Table. Further, the diagnostic coefficients (DC) are calculated. DC is the logarithm of the smoothed frequency ratios multiplied by 10 and rounded to 1.

$$DC=10\log_{10}\left(\frac{\bar{y}_A}{\bar{y}_B}\right). \quad (4)$$

According to the Kullback formula, the value of J informative value of the *i-th* range of the *j-th* feature is:

$$J(x_j^i) = DC(x_j^i)\frac{1}{2}\left[P\left(\frac{x_j^i}{A}\right)-P\left(\frac{x_j^i}{B}\right)\right], \quad (5)$$

where $DC(x_j^i)$ – DC diagnostic coefficient of the *i-th* range of the *j-th* feature; $P\left(\frac{x_j^i}{A}\right)$ – the probability (smoothed frequency) of falling into group A of the *i-th* range of the *j-th* feature, the one that is indicated $\bar{y}_{Ai}$:

$$P\left(\frac{x_j^i}{A}\right) = \bar{y}_{Bi} . \quad (6)$$

It is required to calculate the information value level of the characteristic for all ranges in order to form a diagnostic table. Then, the information value is computed for the entire x_j feature, which is equal to the sum of the nformation values of its ranges:

$$J(x_j^i) = \sum_i J(x_j^i) . \quad (7)$$

Next, we carry out the calculation and analysis according to other criteria specified earlier, in the same way.

According to the calculation results, the most informative features according to the sum J_t are: water-cut 3 months before HFT, Oil saturation, the amount of proppant injected, and the maximum concentration of the proppant injected. The least informative are the following parameters: stratification factor, permeability, efficient thickness, sand content, and oil production rate 3 months before HFT, which can be neglected in the subsequent analysis in order to exclude any influence on the estimation reliability.

Then, we draw up a general table 3 for wells with parameters and diagnostic coefficients.

Further, we plot the distribution of the sums of diagnostic indicators for HFT in wells that did not have a result, and in wells that brought this result, shown in the Figure.

Table 3. Forecasting HFT decision based on the sums of diagnostic coefficients

Well number	Oil saturation, interval	Oil saturation, DC	Proppant injected, interval	Proppant injected, DC	Maximum proppant concentration	Maximum proppant concentration, DC	Water-cut, interval	Water-cut DC	Total DC	Decision fact
437	4	-1	8	-6	9	-2	2	1	-8	-
447	7	2	1	7	3	5	1	3	17	+
447	7	2	1	7	3	5	1	3	17	-
626	4	-1	1	7	5	3	1	3	12	+
626	4	-1	1	7	1	3	1	3	12	-
653	5	1	2	5	2	5	1	3	14	+
703	6	1	1	7	4	5	1	3	16	+
1006	2	-3	5	0	5	3	1	3	3	-
1031	7	2	1	7	5	3	1	3	15	+
1031	7	2	4	1	8	-2	1	3	4	+
1036	3	-2	4	1	7	0	6	-7	-8	-
1067	6	1	3	3	7	0	1	3	7	+
1093	5	1	2	5	2	5	1	3	14	-
1093	5	1	3	3	7	0	1	3	7	-
1100	10	5	4	1	8	-2	2	1	5	+
1129	9	3	6	-2	6	2	1	3	6	-
1129	8	3	6	-2	6	2	6	-7	-4	-
1130	8	3	2	5	2	5	1	3	16	+
1130	8	3	2	5	7	0	1	3	11	+
1791	3	-2	4	1	8	-2	4	-2	-5	-

From the data presented in the Figure, it can be concluded that the DC value equal to 1 stands between the zones in which the indicators of groups A and B are concentrated, as the optimal boundary. The range limits are taken as thresholds:

- threshold B = (1), in case DC sum is less than 1 (do not conduct HFT);
- threshold A = (2), in case DC is more than 2 (conduct hydraulic fracturing);
- if $1 \leq DC \leq 2$ (no single solution).

Thus, there is "no solution" for 23 out of 155 wells where "training" has been carried out, which is only 14% of the total amount, and is a low indicator of the error in this method.

Further, we select candidate wells for hydraulic fracturing operations based on the obtained outcomes.

Initial data for 17 wells are presented in Table 4.

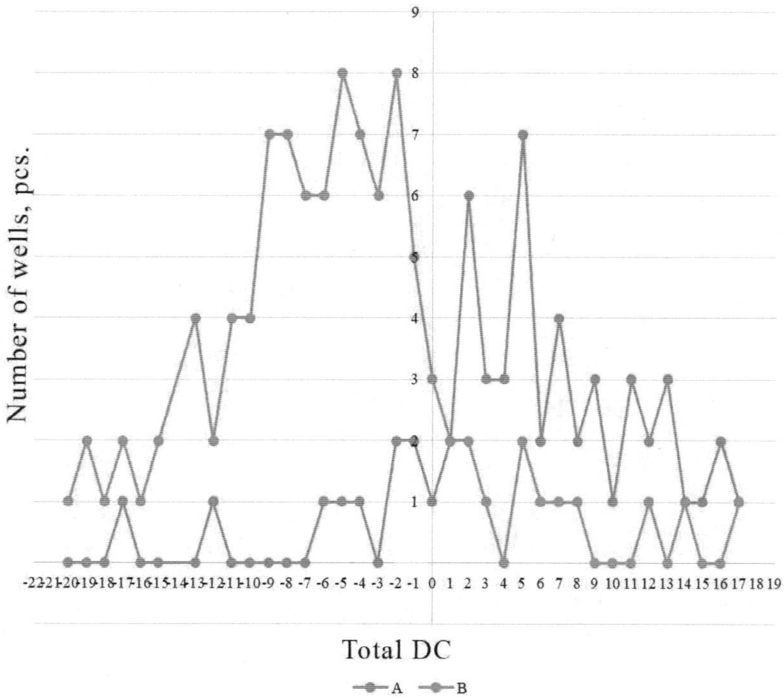

Figure 1. Distribution of wells by the total of diagnostic coefficients

Table 4. Data on candidate wells and features influencing HFT

Well number	Oil saturation, un.fract.	Proppant injected, t	Maximum proppant concentration, kg/m^3	Water-cut, %
7093	0,45	60,0	1300	28,2
7101	0,50	30,0	1000	20,9
7104	0,53	40,0	1400	31,0
7113	0,41	49,5	1000	30,6
7136	0,51	80,0	1400	44,5
7142	0,58	70,0	1300	12,8
7170	0,57	50,0	1400	27,7
7339	0,51	50,0	1200	43,5
7354	0,40	70,0	1300	59,4
7396	0,43	40,0	1200	48,8
7764	0,42	50,0	985	17,2
7818	0,49	50,0	1200	20,6
7832	0,56	50,0	1100	12,1
7853	0,47	40,5	1000	9,2
7871	0,54	60,0	1300	15,8
1770n	0,34	60,0	900	63,6
228r	0,43	40,3	1025	40,2

IOP Conf. Series: Earth and Environmental Science 1021 (2022) 012072 doi:10.1088/1755-1315/1021/1/012072

The selection of diagnostic coefficients was carried out based on the calculation according to the data of the "training" wells group.

Table 5 shows the diagnostic coefficients and their sum for the forecasting.

Table 5. Diagnostic coefficients by characteristics and their sum for the forecasting

Well number	Oil saturation, interval	Oil saturation, DC	Proppant injected, interval	Proppant injected, DC	Maximum proppant concentration, interval	Maximum proppant concentration, DC	Water-cut, interval	Water-cut, DC	Total DC	Decision fact	forecasting	Outcomes
7093	3	-2	6	-2	9	-2	3	0	-6	-	-	+
7101	5	1	3	3	6	2	2	1	7	+	+	+
7104	6	1	4	1	10	-2	3	0	0	-	-	+
7113	2	-3	5	0	6	2	3	0	-1	-	-	+
7136	5	1	8	-6	10	-2	5	-5	-12	-	-	+
7142	7	2	7	-4	9	-2	2	1	-3	-	-	+
7170	7	2	5	0	10	-2	3	0	0	-	-	+
7339	5	1	5	0	8	-2	5	-5	-6	-	-	+
7354	2	-3	7	-4	9	-2	6	-7	-16	-	-	+
7396	3	-2	4	1	8	-2	5	-5	-8	+	-	-
7764	3	-2	5	0	6	2	2	1	1	-	No decision	?
7818	5	1	5	0	8	-2	2	1	0	+	-	-
7832	7	2	5	0	7	0	2	1	3	+	+	+
7853	4	-1	4	1	6	2	1	3	5	+	+	+
7871	6	1	6	-2	9	-2	2	1	-2	+	-	-
1770n	1	-4	6	-2	5	3	7	-10	-13	-	-	+
228r	3	-2	4	1	6	2	4	-2	-1	-	-	+

3. Results and Discussion

The results of the candidate wells assessment using the Kullback method have revealed that the validity of the given study is quite high, since only 4 wells out of 17 candidate wells do not have convergence of the HFT data. Thus, the forecast accuracy for this dataset is 76%, which is a high indicator. It means that the technique can be applied as a supplementary assessment tool for the selection of candidate wells for HFT.

Therefore, the use of additional tools for the analysis of historical data sets on the performed HFT operations is an integral part of the selection of candidate wells. The analysis results should be taken into consideration in further GTM planning.

4. Conclusion

The paper presents the selection of design wells for HFT and the forecast of the event success using the Kullback method. The technique can be applied as a supplementary tool for a neural network program for the selection of HFT wells candidates.

References

[1] Gabzalilova A Kh, Batalov D A, Mukhametshin V Sh, and Andreev V E 2021 Geological and technological justification of the parameters of acid-clay treatment of wells. *IOP Conference Series: Materials Science and Engineering* **1064** 012058 DOI: 10.1088/1757-899X/1064/1/012058

[2] Rodrigues V F, and Medeiros A C R 2011 Limits of Fracture Conductivity Correlation Improvement Through Acid Fracture Surface Characterization *OTC Brasil, Rio de Janeiro, Brazil, October 2011.* DOI: 10.4043/22337-MS

[3] Mukhametshin V V, and Kuleshova L S 2021 Using the method of canonical discriminant functions for a qualitative assessment of the response degree of producing wells to water injection during the development of carbonate deposits. *IOP Conference Series: Materials Science and Engineering* **1064** 012069 DOI: 10.1088/1757-899X/1064/1/012069

[4] Economides M J, and Martin A N 2010 How to decide between horizontal transverse, horizontal longitudinal and vertical fractured completion *SPE Annual Technical Conference and Exhibition, held in Florence, Italy.* DOI: 10.2118/134424-MS.

[5] Khuzin R R, Mukhametshin V Sh, Salikhov D A, Andreev V E, Pepelyaev D V, Stefanovich Yu N 2021 Improving the efficiency of horizontal wells at multilayer facilities. *IOP Conference Series: Materials Science and Engineering* **1064** 012066 DOI: 10.1088/1757-899X/1064/1/012066

[6] Litvin V T, Strizhnev K V, Shevchuk T N, and Roshchin P V 2018 The acid treatment of the well on Bazhenov oil formation after hydraulic fracturing operation *Oil Industry* **4** 70-73 DOI: 10.24887/0028-2448-2018-4-70-73.

[7] Batalov D A, Andreev V E, Mukhametshin V V, and Kuleshova L S 2021 Development regulation of oil and gas reservoirs based on effective geological and geophysical information *IOP Conference Series: Materials Science and Engineering* **1064** 012055 1-9 DOI: 10.1088/1757-899X/1064/1/012055

[8] Dikshit A, Woiceshyn G, Agnihotri V, Chochua G, and Md. Noor M N B 2021 Quantifying Erosion of Downhole Solids Control Equipment during Openhole, Multistage Fracturing *SPE Drill & Compl.* **36 (02)** 413–426 DOI: 10.2118/203096-PA

[9] Malyarenko A M, Bogdan V A, Blinov S A, Kotenev Yu A, Mukhametshin V Sh, and Kotenev M Yu 2021 Improving the reliability of determining physical properties of heterogeneous clay reservoir rocks using a set of techniques. *Journal of Physics: Conference Series* **1753** 012074 DOI: 10.1088/1742-6596/1753/1/012074

[10] Ghaleb Al Habsi, Soheil Ghanbarzadeh, Siyavash Motealleh, and Badar Al Busafi 2019 Hydraulic Fracturing: Best Remedy for Condensate Banking Effects in Tight Gas-Condensate Reservoirs. *Abu Dhabi International Petroleum Exhibition & Conference, Abu Dhabi, UAE, November 2019.* DOI: 10.2118/197422-MS

[11] Mukhametshin V G, Dubinskiy G S, Andreev V E, Mukhametshin V V, and Kuleshova L S 2021 Geological, technological and technical justification for choosing a design solution for drilling wells under different geological conditions. *IOP Conference Series: Materials Science and Engineering* **1064** 012061 DOI: 10.1088/1757-899X/1064/1/012061

[12] Kaiyi Zhang, Bahareh Nojabaei, Kaveh Ahmadi, and Russell T. Johns. 2020. Effect of Gas/Oil Capillary Pressure on Minimum Miscibility Pressure for Tight Reservoirs. *SPE J.* **25 (02)** 820–831. DOI: 10.2118/199354-PA

[13] Zeigman Yu V, Mukhametshin V Sh, Khafizov A R, Kharina S B, Abutalipova E M, Avrenyuk A N 2017 Peculiarities of selecting well-killng fluids composition for difficult conditions. *Oil Industry* **1** 66–69.

[14] Piyush Pankaj, Alhad Phatak, and Sandeep Verma. 2018 Evaluating Natural Gas-Based Foamed

Fracturing Fluid Application in Unconventional Reservoirs. *SPE Asia Pacific Oil and Gas Conference and Exhibition, Brisbane, Australia, October 2018.* DOI: 10.2118/192042-MS
[15] Yaskin S A, Mukhametshin V V, Andreev V E, and Dubinskiy G S 2018 Forecasting the parameters of non-stationary water flooding of oil deposits. *IOP Conference Series: Earth and Environmental Science* **194** 062037 DOI: 10.1088/1755-1315/194/6/062037
[16] Jin F, Shunyuan Z, Bingshan L, Chen C, and Kedi M 2017 Green Fracturing Technology of Shale Gas: LPG Waterless Fracturing Technology and its Feasibility in China *SPE Latin America and Caribbean Petroleum Engineering Conference, Buenos Aires, Argentina, May 2017.* DOI: 10.2118/185500-MS

Assessment of applicability of preformed particle gels for Perm region oil fields

Yu A Rozhkova[1], V Sh Gurbanov[2], G M Efendiyev[2], S V Galkin[1]

[1] Perm National Research Polytechnic University, 29, Komsomolsky prospect, Perm, 614990, Russia
[2] Oil and Gas Institute of Azerbaijan National Academy of Sciences, 9, F. Amirova St., Baku, AZ1000, Azerbaijan

E-mail: doc_galkin@mail.ru

Abstract. The application of preformed particle gels (PPG) in Russian oil fields is very limited. In this research the authors tried to find tools for primary screening of oil wells candidates for PPG application. For a fractured reservoir, the Warren-Rutt method was considered for estimating fracture sizes in injected oil wells. Taking into account the mechanical properties of the developed PPG, a powder fraction is recommended for starting treatment of oil wells. An express method for assessing the pressure index (PI) was considered for the primary selection of well candidates. PI provides information on reservoir heterogeneity. Using this method, the most promising candidates for the implementation of PPG technology were selected.

1. Introduction

The annual increase in the number of mature oil fields is the reason for the search for new technologies for researvoir conformance control. A proven technology is the injection of preformed particle gels (PPG). The effects of using this technology are increasing the production of residual oil and involving for development of previously not drained reservoir zones [1, 2]. However, the success of treating the reservoir by PPG depends on the correct well candidate selection. The authors of the publication have considered methods for the primary screening of candidate wells.

2. Materials and methods

Significant experience of the PPG successful application has been accumulated in the oil fields of China. PPG of both types (macrogels and microgels) are used [3]. Usually the candidates for the PPG treatment are the mature oil fields with hydrodynamic connection between the injection and productive wells, high horizontal and vertical anizotropy [4]. The recommended permeability of the reservoir for PPG ingection is more than 500 md. The analysis of PPG injection practice cases led to the following conclusions: PPG suspension is pumped alternately with water; the concentration of the suspension varies on average from 2000 - 3500 mg/l, the volume of the suspension for one treatment varies from 2500-4300 m^3; the mass of PPG powder is 5-15 tons. PPG treatment is controlled by three main parameters: particle size, injection rate, and concentration. The particle size is selected based on the characteristics of the reservoir, as well as the characteristics of the gel (strength, absorption capacity). The general treatment algorithm includes initial injection of PPG suspension with the small particles. If the pressure does not change, the PPG concentration increases. If there is still no effect, then a larger PPG powder is used. There is also another treatment strategy that is used if there are superchannels in the formation. In this case, in order to increase the efficiency of treatment, the PPG of a larger fraction is injected first, and then start to inject the PPG with a smaller size. If there is sharp increase of injection pressure at the initial stage of treatment, then the injection rate and PPG concentration are reduced. The treatment parameters are selected and adjusted during the treatment to maintain the injection pressure no more than 80% of the fracturing pressure. In general, the duration of PPG treatment can range from 1.5 to 6 months. A breaker is used to clean the bottomhole formation zone. Application of PPG technology gives the following effects: the predicted decrease in water cut in the production of adjacent production wells reaches 8-15%; additional oil production is 120-250 tonnes of oil per tonne of PPG powder injected [3].

When the PPG particle enters the oil reservoir, the following mechanisms of particle movement are possible:

• Transport of a particle through the pore space with water flow (PPG suspension dispersion medium), if the particle has a smaller diameter.

• Adsorption of a particle on the pore surface if its size is too small.

• Stuck particles in the throat of the pore and its blockage.

• Particle deformation and passage through the pore space with the flow of the dispersion medium.

• Compression of particles and passage through the pore space with the flow of the dispersion medium.

• Breaking large particles into small ones and pumping them through the pore space with the flow of the dispersion medium [5]. In fig.1 there is a picture of PPG particle movement through the fracture model.

Traditionally PPG synthesized based on the polyacrylamide with the addition of N, N'-methylene(bis) acrylamide as crosslinker. Authors tried to receive the PPG based on acrylamide and acrylic acid, with 3D structure formed during imidization process.

Earlier studies of the obtained PPG showed high elasticity of the particles [6]. Due to its mechanical characteristics, particles are able to compress and penetrate into holes 20 times smaller than the diameter of the swollen particle. We used this fact in order to predict at which operational facilities the developed PPG could be applied to.

The minimum size of the PPG gel fraction obtained by mechanical grinding of a dry polymer block after synthesis is about 100 μm. After contact with water with a salinity of more than 1.5%, the particle diameter increases by an average of 5 times. It means that the dry particle with a diameter of 100 μm after swelling has the diameter of about 500 μm. This particle can pass into a hole (or crack) with a diameter (or opening) of about 25 microns.

Thus, taking into account the minimal size of the fracture, we tried to find the production object with expressed fracturing in the near-wellbore zone of the injection well and with a heterogeneity of the reservoir matrix.

Fractured reservoirs are located on whole territory of the Perm region: both on the platform part and on the territory of the Pre-Ural Foredeep. They are mainly represented by Tournaisian and Famennian deposits. First of all, we considered reservoirs with a high coefficient of residual oil saturation and a high water cut of the well production. To assess the bottomhole zone fracture size of the injection wells, hydrodynamic studies were analyzed using the Warrenn-Root method [7-9].

An example of the fractured reservoir for PPG treatment is the Tournaisian oil reservoir of the Opalikhinskoe field. This production facility is lockated on the territory with the current development of reserves 46.8%, more than 60% of the wells operate with a water cut of more than 50%, which indicates the feasibility of using tertiary oil recovery methods. The density of the formation water is more than 1.1 g/cm^3. Three injection wells were considered as candidate wells for PPG treatment. The fractured reservoir type, according to the Warren-Root model, is confirmed by the characteristic inflection of the pressure drop curve of the well. The fracture openig in reservoir perforation interval varies from 30 to 50 microns. At this operational facility, it is possible to use a PPG powder with a particle size distribution of 100 to 200 microns.

The Warrenn-Rutt method allows one to compare and PPG particle size with the reservoir fracture and predict the minimum PPG powder fracture for treatment start. However, for porous reservoirs it is necessary to consider other algorithms that would allow preliminary identification of potential candidate wells.

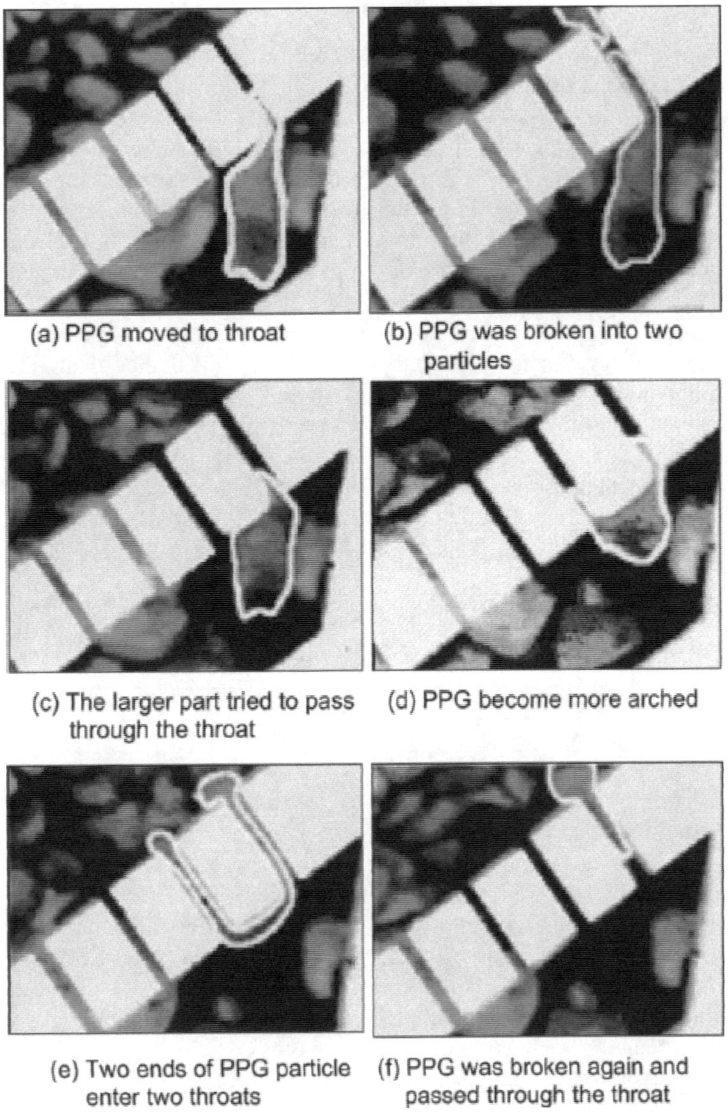

Fig. 1. PPG particle movement in the fractured space [5]

In the foreign practice of oil production, a short-term express method is used for oil well hydrodynamic testing, which allow one to calculate the pressure index (PI). The technology for determining the PI index consists of several stages. First, water is pumped into the injection well during the day at a constant rate. Then the well is shut off for 90 minutes, recording the pressure from the moment of shut-in and then every 5-10 minutes. Pressure measurement intervals are adjusted depending on the dynamics of its fall. If the pressure drops quickly, then the time intervals between measurements are shortened. The pressure drop curves are plotted in the interval of 90 minutes. The PI index, which is calculated by the formula (1), is equal to the area of the figure, limited by the pressure drop curve during 90 min and the coordinate axes.

$$PI = \frac{\int_0^T p(t)}{T}, \tag{1}$$

where PI is the pressure index, MPa; p(t) is a pressure as a function of time during shut-in of the injection well; T is the period of shut-in time [4].

Analysis of the obtained PI values provides information on the heterogeneity of the reservoir over the area of the reservoir. The greater the difference between the maximum and minimum PI values of the wells, the more laterally heterogeneous the reservoir is. If for a group of wells, the difference between the minimum and maximum PI is more than 5 MPa, this is a sign for conformance control technology application. The lower the PI index, the higher the predicted efficiency of the planned treatment.

Based on the proposed approach, the authors considered an example of calculating the PI index for the Visean production facility of the Shagirtsko-Gozhanskoye field. The main characteristics of this deposit are the following: oil viscosity - 38 mPa · s, average formation permeability - 1.2 D, residual oil reserves are 60.9%, formation water density is more than 1.1 g/cm^3. The water cut of wells in the area is more than 85%; the density of residual oil reserves is (0.2-0.6 t/m^2). This oil field has a potential for application of conformance control technology.

The results of PI index analysis based on well test data for 90 minutes for a group of injection wells are presented in Table 1.

Table 1. Characteristics of injection wells in the Shagirtsko-Gozhanskoye field

Injection well	PI, units
302	110.6
370	83.5
1113	104.8
1128	113.8
1133	78.9

The values of PI index of production facility injection wells were compared. The difference between the maximum and minimum PI values (Table 1) is about 35 units, which indicates the pronounced heterogeneity of the reservoir and confirms advisability of selective isolation of high-permeability intervals. Injection wells No. 370 and No. 1133 have the lowest PI index, which makes them the most promising candidate for the implementation of PPG technology.

It should be noted that the application of the PI index in RF fields is currently very limited. Taking into account the effective world practice, this approach requires detailed study and adaptation to various geological and technological conditions of domestic oil fields.

In the Perm region the PPG technology has a great potential. The authors suggest that this technology in combination with surfactants can especially help in the development of deposits with an oil viscosity of more than 5 mPa·s, because in this oil fields the advance flooding of high permeable reservoirs zones is typical [10]. During planning of PPG technology, the assessment of reservoir heterogeneity is important. In the future, the authors plan to conduct the geological and hydrodynamic modeling using the example of algorithms from the article [11].

Acknowledgments
The study was carried out according to the state task of the Ministry of Education and Science of the Russian Federation FSNM-2020-0027 for performing academic studies for 2020 and the planning period of 2021 and 2022.

References

[1] Bai B, Li L, Liu Y, Liu H, Wang Z, You C 2007 Preformed Particle Gel for Conformance Control: Factors affecting its properties and application *SPE reservoir evaluation & engineering* **10 (4)** 1-10
[2] Bai B, Wei M, Liu Y Injecting Large 2012 Volumes of preformed particle gel for water conformance control *Oil & gas science and technology* **67 (6)** 941-952
[3] Yao C, Lei G, Li L, Gao X 2012 Selectivity of pore-scale elastic microspheres as a novel profile control and oil displacement agent *Energy Fuels* **26** 5092-5101
[4] Liu Y, Bai B, Wang Y 2010 Applied technologies and prospects of conformance control treatment in China *Oil & gas science and technology* **65 (6)** 859-878.
[5] Bai B, Liu Y, Coste J, Li L 2007 Preformed particle gel for conformance control: transport mechanism through porous media *SPE reservoir evaluation & engineering* **10 (2)** 1-9
[6] Ketova Yu, Galkin S, Kolychev I 2021 Evaluation and X-Ray tomography analysis of super-absorbent polymer for water management in high salinity mature reservoirs *Journal of Petroleum Science and Engineering* **196** 107998
[7] Martyushev D A 2020 Improving the geological and hydrodynamic model of a carbonate oil object by taking into account the permeability anisotropy parameter *Journal of Mining Institute* **243 (3)** 313-318
[8] Martyushev D A, Galkin S V, Shelepov V V 2019 Influence of the stressed state of rocks on matrix and fracture permeability under the conditions of various lithological-facies zones of the Tournai-Famennian oil deposits of the Upper Kama region *Bulletin of Moscow University. Series 4: Geology* **5** 44-52
[9] Mengaliev A G, Martyushev D A 2020 Accounting the parameter of anisotropy of permeability in geological and hydrodynamic models of carbonate objects (by the example of the Gagarin deposit). *Bulletin of Tomsk Polytechnic University. Geo assets engineering* **5** 7-17
[10] Poplygln V V, Galkin S V 2011 Forecast quick evaluation of the indices of the development of the oil deposits *Oil Industry* **3** 112–115
[11] Repina, V A., Galkin, V I, Galkin S V 2018 Complex petrophysical correction in the adaptation of geological hydrodynamic models (by the example of visean pool of Gondyrev oil field) *Journal of Mining Institute* **231** 268–274

Telemetry system based on fiber-optic gyroscopes

A A Melekhin, E P Riabokon, S N Krivoshchekov

Perm National Research Polytechnic University, 29 Komsomolsky ave., Perm, 614990, Russia
E-mail: melehin.sasha@mail.ru

Abstract. The article presents a telemetry system for monitoring the bit trajectory during well drilling. Fiber-optic closed-loop gyroscopes are the main sensitive element in the telemetry system. The gyroscopes allow determining the azimuth and zenith angles of the well with a high accuracy exceeding most magnetometers. There is no need to use the expensive non-magnetic steels in the body of the telemetry system and non-magnetic drill pipes in the drill string. Fiber-optic closed-loop gyroscopes allow using the telemetry system when drilling near the casing of adjacent wells and magnetized rocks. An inertial navigation system is installed in the developed telemetry system as a measuring unit containing three angular velocity sensors and three accelerometers mounted rigidly. The sequence and results of simulation of a hydraulic link for data transmission from the bottom of the well to the top are described. Pilot tests of the telemetry system were carried out at well No. 450 of the Tatsuksinskoye field. The obtained values of the spatial intensity of the borehole curvature in depth are close in values to the directional survey data for the borehole. The trajectory obtained during the lift is close to the trajectory gathered from directional survey data.

1. Introduction

All modern wells for hydrocarbon production are built along a complex multi-interval profile. To control the trajectory of the rock cutting tool during well drilling telemetry systems (TMS) are used. A telemetry system with a diameter of 180 mm based on fiber-optic gyroscopes was developed in Perm National Research Polytechnic University [1].

2. Problem statement

In the majority of telemetry systems, magnetometers are used as inclinometric sensors. Magnetometers have a number of disadvantages such as sensitivity to magnetic interference from drilling equipment [2] (which leads to the need to use a non-magnetic tool and thereby move the measuring devices away from the bit) and a low accuracy when working in latitudes close to the poles (caused by bending of the earth's magnetic field). The developed TMS uses fiber-optic closed-loop gyroscopes [3]. The gyroscopes have several advantages. Firstly, borehole azimuth and zenith angles can be determined with high accuracy, exceeding accuracy of the most magnetometers. Secondly, there is no need to use the expensive non-magnetic steels in the TMS casing and non-magnetic drill pipes in the drill string. Thirdly, fiber-optic closed-loop gyroscopes allow using TMS when drilling near casing pipes of adjacent wells and magnetized rocks [4, 5].

A number of telemetry systems is battery powered. It means that when the batteries are discharged, it is necessary to lift the tool up to the surface for its replacement.

During rotary drilling and rotation of the entire tool in the borehole, the sensitive elements of telemetry systems can increase the measurement error or completely stop making measurements due to loss of orientation.

3. Methods of solution

An inertial navigation system is installed in the developed TMS as a measuring unit containing three angular velocity sensors and three accelerometers mounted rigidly. The sensitivity axes of accelerometers are oriented along the axes *Ox*, *Oy*, *Oz* of the rectangular coordinate system *Oxyz*, connected to the body of the inertial navigation system. These devices react to the angular velocities of rotation (*ωx*, *ωy*, *ωz*) and accelerations (*ax*, *ay*, *az*) of the base, which are the projections of the vector *ωa* of the absolute angular velocity of rotation and the vector *Ak* of the apparent acceleration of the object on the axes *Ox*, *Oy*, *Oz*, and then generate signals that are primary for solving the orientation problem. Three accelerometers, as part of a measuring unit, form an orthogonal system for measuring linear accelerations of an object along two horizontal and vertical channels (*X*, *Y*, *Z*). Three angular velocity sensors, built according to the Sagnac ring interferometer scheme based on a fiber-optic loop with tracking feedback, form an orthogonal system for measuring the angular velocities of an object. The method of imparting the compass properties to the inertial navigation system is based on the use of the readings of the angular velocity and linear acceleration sensors with their subsequent processing. The method to determine the North direction is shown in Figure 1.

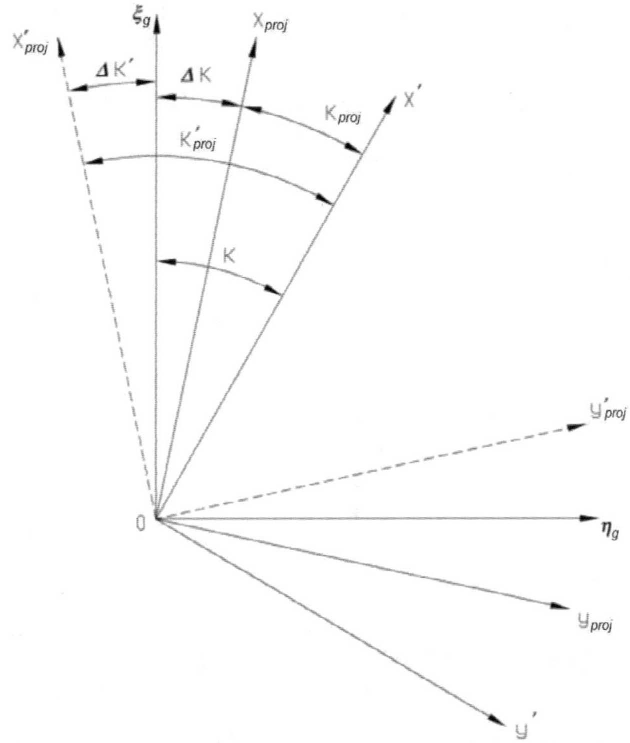

Figure 1. Method for determination of the North direction.
where $0\xi_g$ and $0\eta_g$ – geographic axes; $0x'$ and $0y'$ – projections of the object's axes onto the plane; $0x'_{proj}$ and $0y'_{proj}$ – instrument axes at $\Delta K'<0$; $0x_{proj}$ and $0y_{proj}$ – instrument axes at $\Delta K>0$.

True heading K differs from instrumental K_{proj} by ΔK_i, called the correction in the course and is calculated by the formula (1):

$$K = K_{proj} + \Delta K_i , \tag{1}$$

where K_{proj} is the value of the heading determined by the device, degree;
ΔK_i is correction in the heading, radian.

Components of the vectors of the angular velocity of the Earth's rotation $U\xi_g$ and $U\eta_g$ at zero velocities of the object is determined by the formula (2):

$$U\xi_g = U\cdot\cos\varphi;\ U\ U\eta_g = 0, \tag{2}$$

where U – vector of the angular velocity of the Earth's rotation, radian/s;
φ – latitude of the object location, radians.
Then the projection of the vector U onto the axis Oy_{proj} will be equal to:

$$Uy_{proj} = -\ U\cdot\cos\varphi\cdot\sin\Delta K_i. \tag{3}$$

It follows that Uy_{proj} at a known latitude φ contains information about the heading correction ΔK_i.
Uy_{proj} is a projection of the angular velocity of the Earth's rotation onto the instrument axis Oy_{proj}, measured by the yaw rate sensor.
From the last ratio, the value of the correction in the course will be equal to:

$$\Delta Ki = -\ \arcsin\frac{Uy_{proj}}{U\cdot\cos\varphi}. \tag{4}$$

When substituting the found value ΔK_i into formula (1), the calculation algorithm forms the value of the "instrumental" heading, "converging" to the value of the true heading.
Possible change in the sign of the correction in the course ΔK_i+1 is fixed by the algorithm and does not affect the "approach" of the axes $0x_{proj}$ and $0\xi_g$ in the calculation algorithm, since the sign of the right-hand side in formula (3) will change (the position of the axes $0x'_{proj}$ and $0y'_{proj}$ shown in Figure 1).
To correct the system readings for the heading, roll and trim angles while the object is moving, the on-board computer can receive latitude signals φ, longitude λ, velocity V, height h from the satellite navigation system and the distance traveled S from the wheel sensor installed on the object.
The system provides the user with the values of the orientation angles ψ, γ, θ in the analog form and in the digital form according to an agreed protocol.
The developed TMS includes a borehole turbogenerator, which ensures continuous operation at the bottomhole, as well as a set of rechargeable batteries charged from the turbogenerator for backup power supply.
The developed TMS uses a unique system for ensuring the geostationarity of sensitive elements, which allows it to be used in rotary drilling with an angular speed of up to 250 rpm.
The developed telemetry system is attached to the drill pipe string and is located above the downhole motor. A flushing fluid pressure sensor is installed in the discharge line of the mud pump. The ground equipment of the system contains a receiver connected to a flushing fluid pressure sensor.

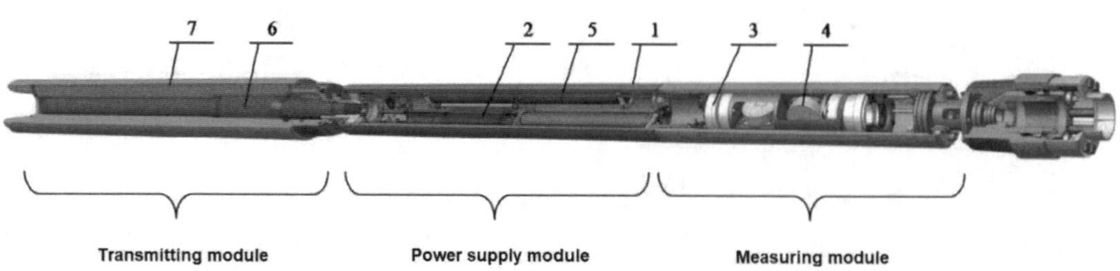

Figure 2. Telemetry system composition.

The bottom-hole part of the telemetry system (Figure 2) contains a measuring module, a power supply module and a transmitting module. All mentioned modules for protection are installed in the

IOP Conf. Series: Earth and Environmental Science 1021 (2022) 012074 doi:10.1088/1755-1315/1021/1/012074

inner case 1, centered in the outer case of the system. An annular gap is made between the bodies for the passage of the drilling fluid. The measuring module contains a control board 2, a geostationary system 3 and various sensors, for example, directional sensors 4.

The power supply module contains batteries 5 and a pulsator-electric generator 6 installed in a separate housing 7. The generator is connected to the control board.

In the generator body 7 there is a pulsator-electric generator 6 for a hydraulic communication channel.

While drilling, the control board reads information about the borehole angles and the characteristics of the drilling mode, analyzes the data, determines the average values over a certain period of time and encodes the data packet for transmission to the surface.

During well drilling, a mud pump is pumped with flushing fluid inside the drill string. On the way of the drilling fluid there is a pulsator-electric generator 6, which, when transmitting information, depending on the signals received from the control board 2, periodically blocks the flow of drilling fluid. That creates positive pulses of hydraulic pressure in the drill string, recorded by a pressure sensor located in the pump discharge line.

Simulation of a hydraulic channel is a complex task that requires complex debugging of software and hardware, testing algorithms for data exchange between the wellhead and drilling equipment in the well.

Simulation of a hydraulic communication channel is as follows: the data transmission medium through the hydraulic channel is emulated by generating rectangular pulses of standard amplitude of a certain frequency. The parameters of the signals then change in such a way as the real transmission medium could distort them: a well with drilling fluid (work of the bit, pumps, rotation of the string etc.); before they are recorded by the pressure sensor of the receiving device. To do this, the noise signal from the noise generator is mixed into the original signal entering the input of the buffer amplifier after passing through the attenuator. The received signal passes through the voltage-current converter to the output and through the 4-20 mA current loop interface is fed to the central processing unit of the physical layer, where it is restored to its original state using noise-immune coding algorithms.

The installation for simulating a hydraulic data link is portable and can be used both in laboratory conditions and directly during the process of TMS manufacturing for its testing before shipment to the customer (Figure 3).

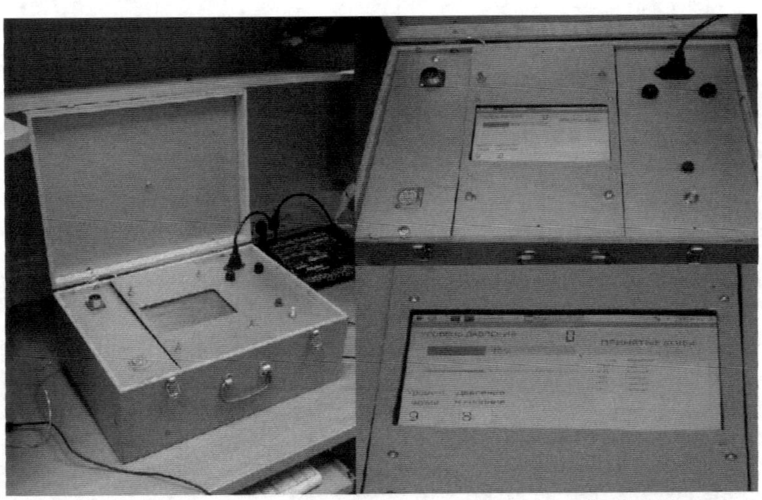

Figure 3. Exterior view of the installation for simulating a hydraulic data link.

The testing of the process of simulating the hydraulic data link was carried out in the following sequence. The signal of the upper level was formed (in real conditions it is created by the pulsator) and the oscillogram was monitored at the lower level (pressure sensor), while noises characteristics of the process of drilling wells were introduced into the signal passing through the stand (pumps, vibrations of the string, bits, etc.).

4. Discussion of the results

Table 1. Parameters of the developed TMS.

Parameter	Value
Maximum pressure of a drilling fluid column, MPa	40
Drilling fluid consumption, l/s	
– minimum	15
– maximum	35
Maximum rotation frequency, rpm	200
Maximum axial load, kN	180
Maximum torque, kN·m	10
Operating temperature, °C	
– minimum	80
– maximum	-20
Determination of the zenith angle, degrees	
– range	0-180
– accuracy	±0.1
Determination of the azimuth angle, degrees	
– range	0-360
– accuracy	±0.5
Supply voltage, V	24
Power consumption, W	300
Dimensions, mm	
– length	5995
– diameter	180
Mass, kg	743
Connecting thread to drill pipe and DDM in accordance with GOST R 50864	Z-147
Make-up torque, kN·m	40^{+4}
Reliable operation under vibrations and shock loads, g	up to 10
Maximum permissible level of vibrations and shock loads, g	30

Pilot tests of the TMS were carried out at well No. 450 of the Tatsuksinskoye field, which was under the operation of PJSC TATNEFT, NGDU Aznakaevskneft. The work started on December 5, 2019 at 00:30 and finished at 06:30. The values of the spatial intensity of the borehole deviation in depth were close in values to the directional survey data for the borehole, although they have small jumps. The trajectory obtained during the lift is close to the trajectory according to the directional survey data.

5. Acknowledgments

The investigation was financially supported by the grant from the President of the Russian Federation (Agreement No. 075-15-2020-150 dated 17 March 2020).

References

[1] Krivoshchekov S N, Melekhin A A, Turbakov M S, Shcherbakov A A, Krysin N I 2017 Development of a telemetric system for monitoring downhole parameters in the course of wells construction *Neftyanoe Khozyaystvo - Oil Industry* **9** 86-88

[2] Ivanova T, Korshunov A 2021 Telemetry Tracking During Directional Drilling *Acta montanistica slovaca* **26** 127-138

[3] Weis R S, Bachim B L, Beadle B M, Gearhart M 2000 Optical fiber telemetry systems for measurement-while-drilling applications *Optical engineering* **39** 1591-1596

[4] Kychkin A V, Volodin V D, Sharonov A A, Belonogov A V, Krivoshchekov S N, Turbakov M S, Shcherbakov A A 2016 The synthesis of the hardware and software system structure for remote monitoring and control of the wellbore trajectory while drilling by rotary steerable system *Neftyanoe Khozyaystvo - Oil Industry* **11** 128-132

[5] Krysin N I, Dombrovsky I V, Krivoshchekov S N, Melekhin A A, Riabokon E P, Shcherbakov A A, Nestertov I I 2016 Study of fiber optic gyroscopes for telemetry systems of well trajectory monitoring *Neftyanoe Khozyaystvo - Oil Industry* **12** 102-105

Influence of oil structures on the component composition of free gases of potash reservoirs at the BPPU-4 mine field of PJSC "Uralkali"

A S Papulov[1], S S Andreyko[1], O V Ivanov[2]

[1] Perm National Research Polytechnic University, 29, Komsomolsky prospect, Perm, 614990, Russia
[2] Mining Institute of the Ural Branch of the Russian Academy of Sciences, 78a, Sibirskaya St, Perm, 614007, Russia

E-mail: aspapulov@gmail.com

Abstract. The influence of oil structures on the content of heavy hydrocarbon gases in the component composition of free gases in the rocks of the overlying silvinite formations is considered. There is an assumption about the influence of reef formations on the overlying sedimentary deposits and the development of sub-vertical cracks, through which heavy hydrocarbon gases can penetrate into the overlying potash layers. The aim of the study is to study the relationship between the nature of the distribution of heavy hydrocarbon gases and geological disturbances and the subsequent prediction of the gas-dynamic situation during the development of potash reservoirs. The method consists of drilling holes into the roof of the mine, sealing the hole with a sealer, and then taking gas from the sealed part to the sampler and analyzing it on a gas chromatograph. The highest values of heavy hydrocarbon gases were determined for the B_c, АБ and КрII formations at the BPPU-4 mine field of PJSC Uralkali.

1. Introduction

Improving the safety of mining operations is an important part of potash mining. The component composition of the free gas contained in potash reservoirs is represented by the following gases: methane, hydrogen, heavy hydrocarbon gases, nitrogen, carbon dioxide, etc. At that, their presence and share in the total volume can vary widely. At present, there is no single theory about the source of generation of heavy hydrocarbon gases and the nature of their distribution in potash reservoirs. The authors of this article proceed from the assumption that the source of heavy hydrocarbon gases in potash formations can be oil and gas-bearing structures.

Determining the relationship between geological disturbances and the distribution of heavy hydrocarbon gases in potash formations will allow us to refer the latter to "markers" indicating various geological disturbances. This article presents the results of studies of the distribution of heavy hydrocarbon gases in the conditions of geographically combined oil and potash deposits.

2. Materials and methods

Gas content is the amount of gas contained in a unit of mass or volume of a rock, as well as its component composition. The study of gas content by free gases consists in mine instrumental observations [1,2]. To determine the quantitative component of the gas content in the mine conditions, drilling of research holes or wells is carried out and their sealing is performed at a distance of 0.5 m from the bottom of the hole using a sealer. To determine the qualitative component of the gas content of rocks from the boreholes, samples of gases released from the walls and bottom of the sealed part of the borehole are taken, and the subsequent study of the component composition of the free gas is performed on a gas chromatograph. The objects of the study are heavy hydrocarbon gases (ethane (C_2H_6), propane (C_3H_8), isobutane (i-C_4H_{10}), n-butane (n-C_4H_{10}), isopentane (i-C_5H_{12}), n-pentane (n-C_5H_{12})), which are part of the component composition of free gases in the rocks of the silvinite formations of the $В_с$, АБ and КрII.

3. Research results

Verkhnekamskoye field is being developed at 6 mines at a depth of 300-400 m. Within the boundaries of these mines, there are more than 15 oil structures with a depth of more than 1500 m. The location of the oil structures relative to the mine fields is shown in Figure 1. At present, there are ideas that reef formations create prerequisites for the formation of naturally weakened zones in the upper part of sedimentary deposits and the development of sub-vertical cracks [3]. Based on this, we can assume an upward migration of heavy hydrocarbon gases from oil-bearing reef structures to the overlying rock strata containing potash layers. Taking into account the significant depth of oil structures (about 2000 m), the distribution of heavy hydrocarbon gases in potash formations lying at a depth of about 350-400 m can vary widely in area, which will largely be determined by the position in the geological section and the area of the mine field of permeable zones. It also does not exclude the lateral migration of heavy hydrocarbon gases over the area of the mine field. The results of experimental studies showed an increased content of heavy hydrocarbon gases near the axes of syncline structures [4]. Studies on organic matter inclusions in natural salts have shown that the presence of methane and increased concentration of heavy hydrocarbon gases in the composition of bound gases in salt rocks can serve as one of the signs of possible oil and gas activity of subsalt deposits. Variegated and striped silvinite are characterized by increased hydrocarbon contents [5]. The results of studies of the content of heavy hydrocarbon gases in potash formations in the mine field of the BPPU-4 mine are shown in Figure 2. Different colors indicate the places of research in different layers of the BPPU-4 mine field.

As can be seen from Figure 2, the maximum content of heavy hydrocarbon gases in the АБ formation is 0.120 m^3/m^3. This value was recorded during the study of the 1st northwestern panel of the 5th eastern block and near the geological disturbance. The average heavy hydrocarbon content for the АБ formation was 0.053 m^3/m^3. The maximum content of heavy hydrocarbon gases in the КрII formation is 0.150 m^3/m^3, with an average value of 0.061 m^3/m^3. This value is recorded on the 2nd north-west panel of the 3rd eastern block. The maximum value of the content of heavy hydrocarbon gases in the $В_с$ formation is noted on the 4th - 5th north-western panel of the 2nd western block and is 0.182 m^3/m^3, which is also the maximum of the observed values. However, there is not enough data for a full analysis of the data on the $В_с$ formation, since the studies of this formation are troubled. In addition, all studies were conducted near geological disturbances.

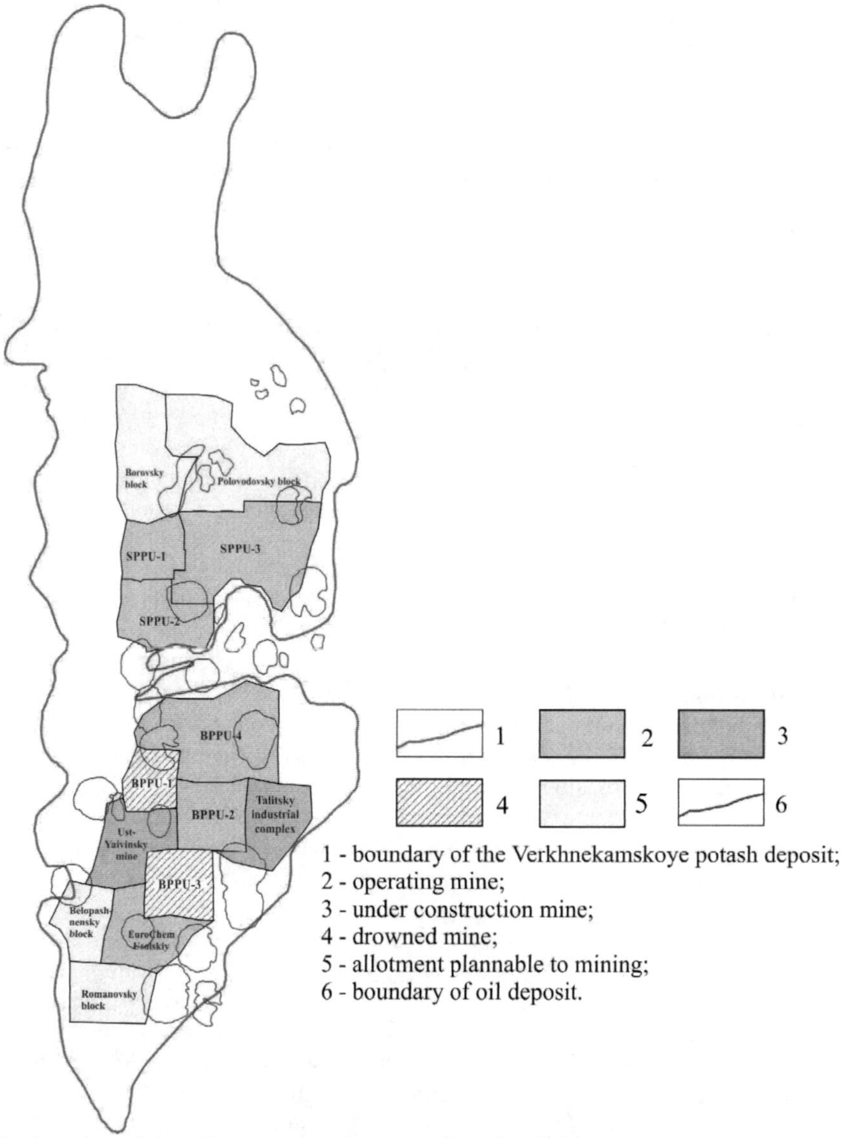

Figure 1. The location of the oil structures relative to the mine fields.

4. Conclusion

The results of experimental studies showed an extremely uneven distribution of the content of heavy hydrocarbon gases in potash formations in the mine field of the BPPU-4 mine, which varies from 0.001 m^3/m^3 to 0.182 m^3/m^3. The gas was selected and studied at 29 sites of the mine field of the BPPU-4 mine. The uneven distribution of heavy hydrocarbon gases in the rocks of potash beds may be due to the discrete distribution of permeable zones through which upward migration from oil-bearing reef structures (Lehchim and Chashkinskoe) occurred, as well as lateral migration within the salt column along cracks, tubular channels, layers and interlayers of salt-bearing clays, microcracks and intergranular boundaries.

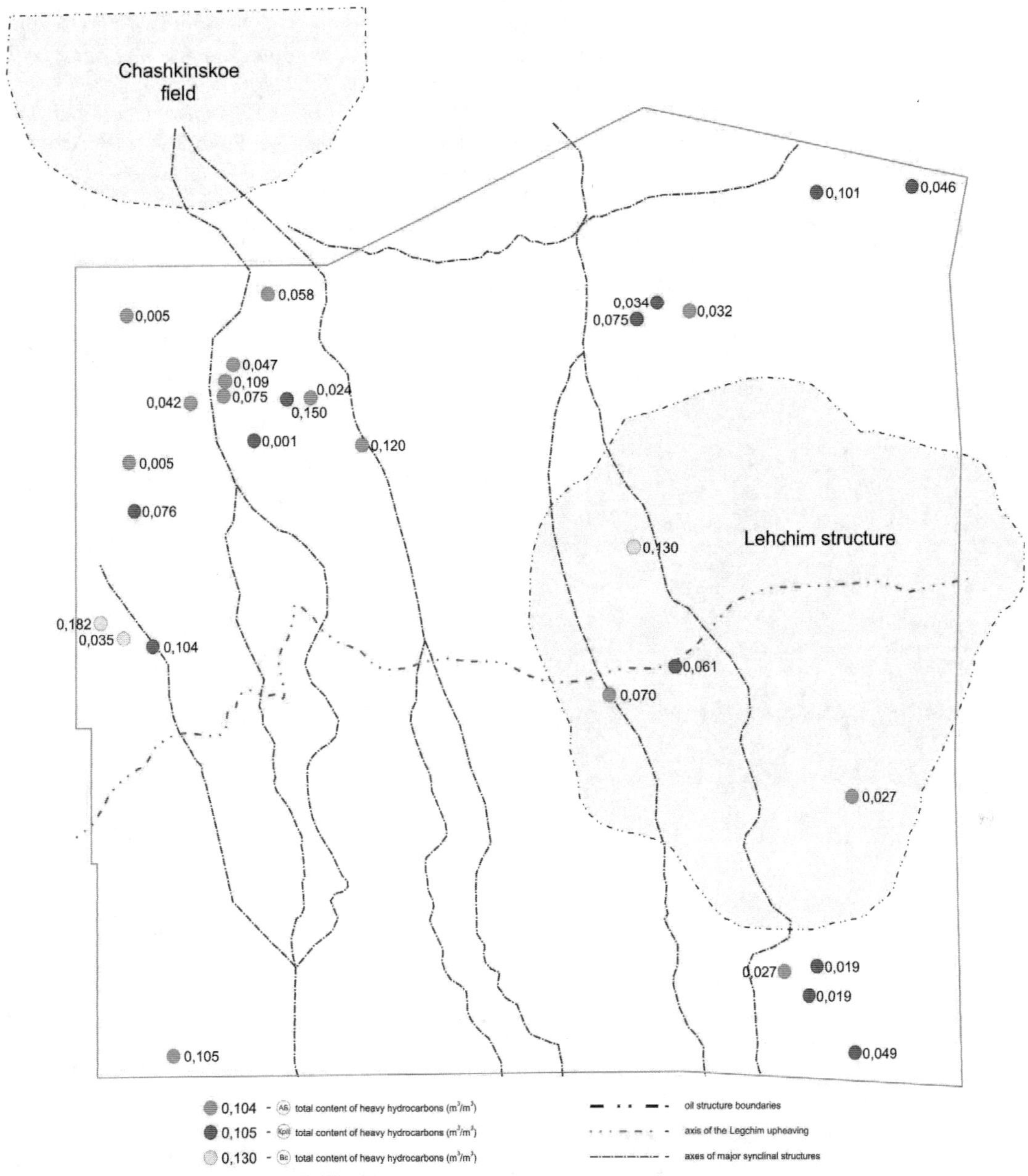

Figure 2. Distribution of the content of heavy hydrocarbon gases in potash formations at the BPPU-4 mine field of PJSC Uralkali.

5. Acknowledgments

The research was supported by the RFBR grant No. 20-45-596017.

References

[1] Nesterov E A, Makarenko M V 2018 Study of the gas content of salt rocks of productive layers in the conditions of the SPMPU-3 mine of PJSC Uralkali *Actual problems of improving the efficiency and safety of operation of mining and oilfield equipment* **1** 183-189

[2] Andreiko S S, Litvinovskaya N A 2013 Local forecast of zones dangerous for gas-dynamic

phenomena from the soil of the АБ formation mining workings in the southern part of the BPPU-4 mine field of the Verkhnekamskoye potash deposit *Mining information and analytical bulletin (scientific and technical journal)* **4** 205

[3] Fedoseev A K 2011 *Geomechanical analysis of the formation of geological anomalies in the sedimentary cover*, abstract of the dis. Candidate of Technical Sciences, Perm, MI UB RAS,

[4] Andreiko S S, Lukyanets E V 2017 Analysis of gas-bearing factors and the tectonic structure of silvinite layers in thrust and shear zones of the mine fields of the BPPU-2 and BPPU-4 mines of PJSC Uralkali *Bulletin of the Kuzbass State Technical University* **6 (124)**, 31-41

[5] Shanina S N 2000 *Inclusions of organic matter in natural salts*, abstract of the dis. Candidate of Technical Sciences, Syktyvkar, Institute of Geology Komi SC UB RAS,

Impact of hydraulic fracturing on the creation of highly permeable channels within the Earth's interior

K M Khalikova, E M Almukhametova, A Y Gutorov

Ufa State Petroleum Technological University, Branch of the University in the City of Oktyabrsky, Oktyabrsky, Republic of Bashkortostan, Russian Federation

E –mail: 6karina_xalikova@mail.ru

Abstract: in modern times hydrocarbon production in fields with hard-to-recover reserves is actively being implemented. This trend is caused by the depletion of conventional reserves, which increases every year of operation. In Russia, most of the hard-to-recover reserves belong to deposits with low-permeability channels and deteriorated reservoir properties. The cost-effective utilization of such subsoil resources is only possible using special technologies that can increase well productivity. One of the most commonly used technologies is hydraulic fracturing (HFT). This study analyzed the impact of HFT on the creation of highly permeable channels as a stimulation method. The methodology for the selection of well candidates for HFT was identified, its effectiveness and practical significance were evaluated.

1. Introduction

The history of the practical application of hydraulic fracturing (HFT) in oil and gas industry dates back to more than 70 years. During this period, HFT has become one of the main technologies for the penetration of low-permeability formations, the development of which by "traditional" is unprofitable due to small well production rates.

As a rule, reservoirs with low permeability are mainly characterized by a pronounced macroinhomogeneity of reservoir properties both in height and in formation direction. In this case, the purpose of hydraulic fracturing is to increase the productivity of wells by impacting on the bottomhole zone of a well – a change in the properties of the porous medium and fluid (the properties of the porous medium change during hydraulic fracturing due to the formation of a crack system) [1, 3, 7].

Indeed, as the field experience shows, the productivity of wells after hydraulic fracturing sometimes increases several dozens of times. This indicates that the formed cracks are connected to the previously existing ones, and the fluid influx comes from highly productive zones that are remote and isolated from a well prior to hydraulic fracturing. The opening of natural or formation of artificial cracks is analyzed by the schedules of flow rate Q and pressure P during the process [4]. The formation of artificial cracks on a graph is characterized by a pressure drop at a flow rate of injection, and when revealing natural cracks, the flow rate of the fracturing fluid increases disproportionately to the pressure increase.

The main principle of hydraulic fracturing is that high pressures are created at the bottomhole by pumping viscous fluid, which 1.5-2 times exceed the formation pressure, as a result of which the formation is stratified thus creating cracks [2, 3]. HFT crack allows combining a set of thin layers into a single productive formation and improving the quality of penetration in the conditions of deep

IOP Conf. Series: Earth and Environmental Science 1021 (2022) 012076 doi:10.1088/1755-1315/1021/1/012076

penetration zones, increasing the drainage area compared to drilling without HFT. The degree of hydraulic fracturing efficiency is determined by the diameter and extent of the created cracks and, therefore, by the increased permeability. The larger the diameter and length of cracks, the higher the treatment efficiency [5].

The complexity of analyzed objects (heterogeneous formation structure, low reservoir properties, complex methods of well completion) determined the need to improve the existing approaches to the selection of HFT candidate wells of these objects and, accordingly, the methodology of complex interpretation of historical data at previous operations [6, 8].

The development and implementation of methods for evaluating and interpreting historical data on the performed HFT, as additional analysis tools, is an urgent task of modern field development control in evaluating and selecting candidate wells.

2. Methods and materials

First, let us study the geological setting of the considered field and analyze the lithological-stratigraphic characteristic of the section [5].

The deposits of the Jurassic system transgressively occur on folded basement rocks and are represented by all three sections: lower, middle and upper. The lower and middle sections consist of burnt rocks and rocks of the Tyumen suite.

Due to the complex relief of the pre-Jurassic surface the burnt suite is not pervasive. Basal conglomerates and breccia-conglomerates lay at the base of the suite, light, greenish and dark gray, fine-medium-grained, large and coarse-grained polymyctic, micaceous, tight and hard sandstones are located upper in the section. Reflective seismic horizon “A” is confined to the bottom of the suite. The thickness of the suite varies from 80 m to 120 m [3-6].

The Tyumen suite is lithologically represented by the rhythmic layering of sandstones, aleurolites and argillites according to a regressive-transgressive scheme. A transgressive sandstone formation (JV_9 formation) is at the base of the suite. JV_2 productive formation is at the top of the suite. The reflecting horizon “T” is confined to the top of the Tyumen suite. According to seismic exploration, the thickness of suite sediments at the field varies from 580 to 700 m.

Marine deposits of the upper Jurassic system transgressively occur and lay accordingly on Middle Jurassic deposits. According to the characteristic lithological composition, three suites are distinguished from bottom in the Upper Jurassic sediments: Vasyugan, Georgiev and Bazhenov.

According to lithological composition, the Vasyugan suite is represented by marine deposits and is divided into two subsuites: lower, mainly clay, and upper siltstone-sandstone. The JV_1^1 productive formation represented by interlayering of sandstones and siltstones is distinguished in the upper subsuite. According to drilling data, the thickness of the Vasyugan suite deposits varies from 50 to 80 m.

The JV_1 formation is developed in reservoirs over most of the area of the considered field.

Physicochemical properties of formation fluids and gas of the considered field were also studied.

According to the study of surface samples of oil and gas oil, the oil density at 20 °C – 840.8 kg/m^3, average kinematic viscosity at 20 °C – 6.27 mm^2/sec, at 50 °C – 2.84 mm^2/sec, molar weight – 187.6 g/mol. The yield of light fractions at up to 300 °C makes up to 52.89%, average sulfur content – 0.62%, resins – 6.13%, asphaltenes – 0.72%, paraffins – 2.04%, and oil solidification temperature – (-) 16.1 °C. Thus, oil deposits belong to heavy, resinous, paraffinic.

Next, let us consider the current state of development of field N represented by 9 productive formations with by 31 oil deposits: four deposits are identified within BV_8, 10 deposits – in formations of the Achimov sequence, 15 deposits – in JV_1 formation and each in JV_0 and JV_2 formations.

Since the beginning of the field development 207,354 thousand tons of oil have been produced. The extraction from the initial recoverable reserves is 74.5% with the current water production of 79.3%, the achieved oil recovery factor is 0.273. Since the beginning of development 431416 thousand tons of fluid were collected, accumulated water-oil factor – 1.1 [7-10].

Accumulated compensation for fluid extraction by injection in the field as a whole is 113.0% with water injection in the amount of 603652 thousand m^3. 2.9 m^3 of water injected into the formation accounts for 1 ton of produced oil, 1.4 m^3 – per ton of fluid.

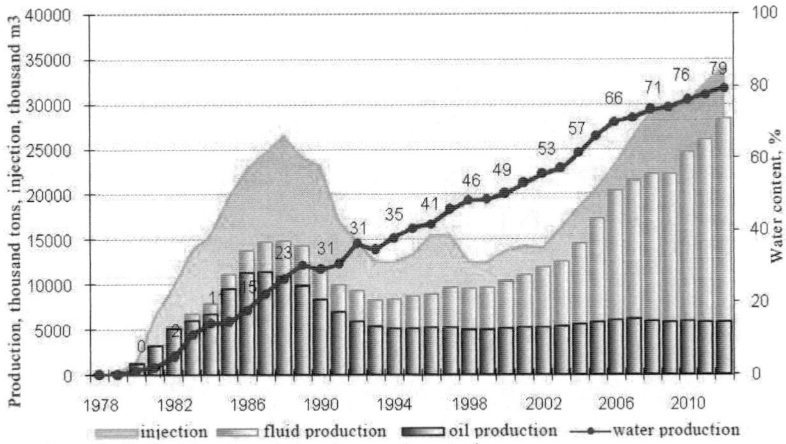

Figure 1. Dynamics of oil, fluid, water injection and watering

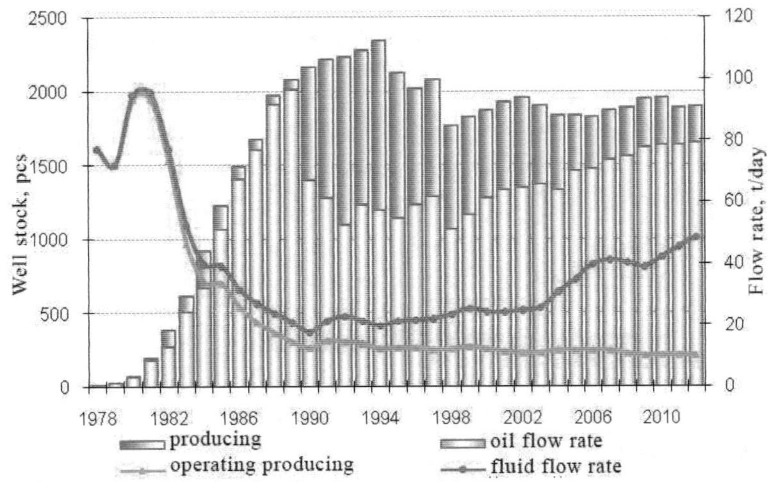

Figure 2. Dynamics of oil production by development objects

Oil production is carried out from three production objects: BV_8, Ach and JV_1. The main object determining oil production at the field is BV_8 (96% of the total, 77% of the current oil production). It is the most developed (at the moment, the production of recoverable reserves makes 78.8%, the share of remaining recoverable reserves in the total volume has decreased from 90.7% to 75.6%). Differences in the degree of their production are primarily determined by different commissioning periods [1, 5, 10].

IOP Conf. Series: Earth and Environmental Science 1021 (2022) 012076 doi:10.1088/1755-1315/1021/1/012076

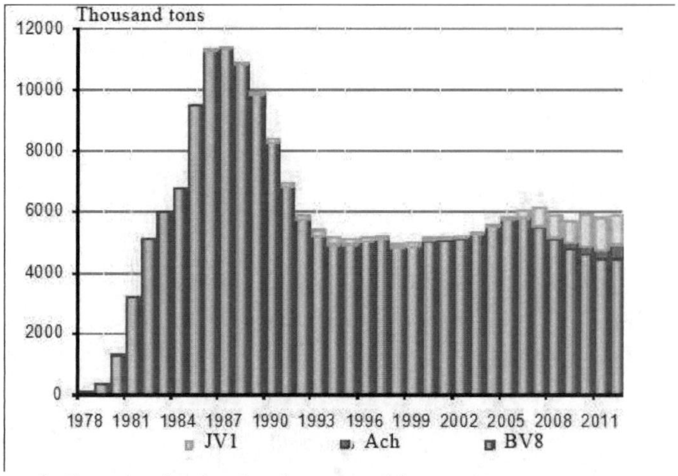

Figure 3. Dynamics of oil production by development objects

To date, all highly productive reserves have already been put into development, drilling of new wells is carried out in lower productivity zones, in peripheral areas.

Let us consider JV_1 and analyze the HFT efficiency on the studied object.

The application of HFT at the studied field began in 1989 (well No. 487, put into operation – 30.01.1990). As of 01.01.2018, 3,205 HFT were performed [3].

JV_1 – 284 HFT (incl. 184 HFT – at the operating well stock, 55 HFT – at the wells during commissioning from drilling (of which 10 HFT – during sidetracking and 1 HFT – during sidetracking with horizontal end), 8 HFT – when transferring to the HFT, 4 HFT – from suspension after drilling and 33 HFT – at the injection well stock) [2-4].

The current incremental oil production due to 251 HFT at the production well stock of the JV amounted to 2122 thousand tons or 8.5 thousand tons/well, incl. 1204 thousand tons or 6.5 thousand tons/well at the operating well stock (of which 7.7 or 5.6 thousand tons/well, respectively), at wells from drilling – 745 thousand tons or 13.5 thousand tons/well (of which 537 thousand tons or 12.2 thousand tons/well, during sidetracking – 201 thousand tons or 20.1 thousand tons/well and during sidetracking with horizontal end – 6.9 thousand tons). The share of incremental oil production in the total accumulated extraction amounted to 31% [6].

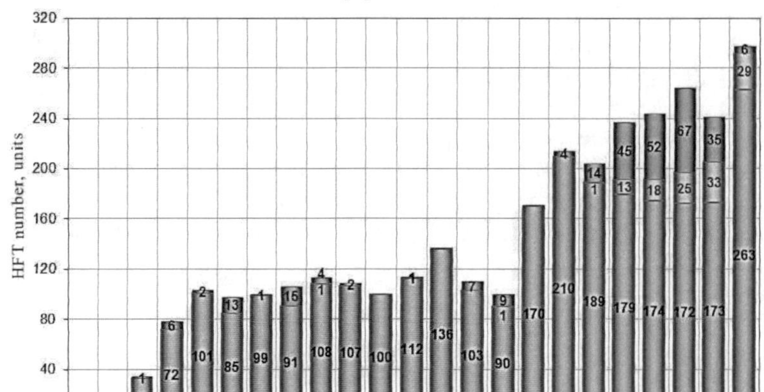

Figure 4. Distribution of HFT impact on the creation of highly permeable channels by years, by well categories and by impact objects

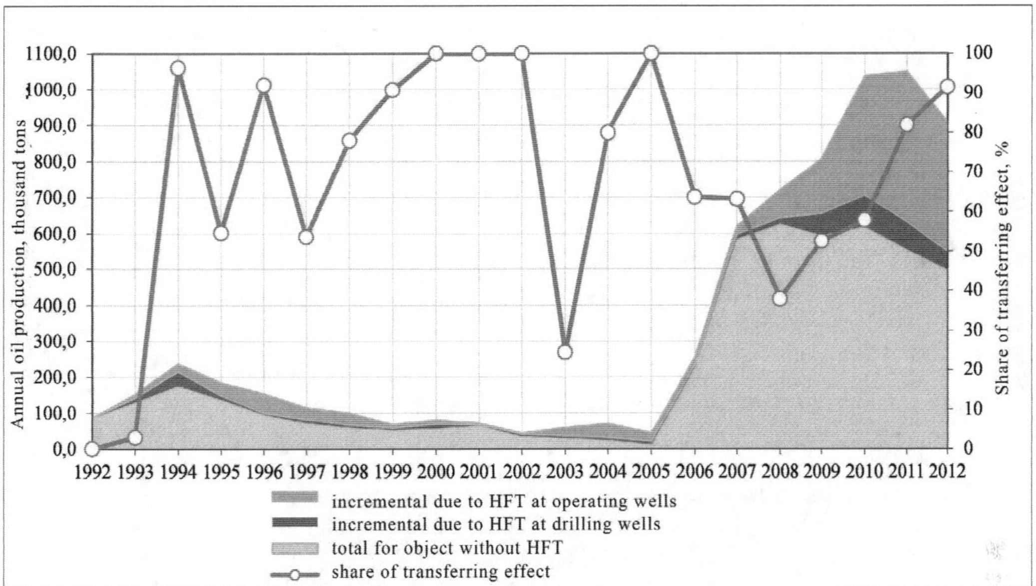

Figure 5. Incremental oil production due to the creation of highly permeable channels during HFT, annual oil production without HFT by year (JV)

3. Results

It is well known that the main indicator of the HFT efficiency is the subsequent increase in oil production from the design well. It is worth noting that the HFT efficiency to a large extent depends on the correct selection of a well for the HFT.

Based on the obtained data, the HFT carried out at the JV object was analyzed and the impact of various geological and technological parameters on the success of future operations was estimated.

Collected data on the already drilled 43 wells where the HFT was carried out. At the same time, 32 wells for "training" and 11 for "forecasting" (assessment of the methodology validity) will be selected for further calculation. Also, based on the analysis of the HFT history at the considered field, and the already drilled wells of the JV group, criteria were adopted for evaluating the efficiency of the performed HFT at the wells [4, 5]. Based on the analysis carried out for previous periods for JV at the "out of drilling" stock, as well as horizontal sidetracking, the average oil production after the HFT for all deposits after 3 months was 25 tons/day, the average water content 3 months after the HFT was not more than 30%. After analyzing the historical data, we can conclude that these indicators are "good", therefore, in the future, in the calculation for the separation of wells "with effective HFT" – "A", and "inefficient HFT" – "B", wells with production increase over 25 tons/day and a watering value up to 30% will fall into A category. The remaining wells belong to B category.

According to the results of evaluation of candidate wells via the Kulbak method, the validity of this study is quite high [11], since out of 11 candidate wells, only one remains unclear concerning the HFT. Thus, the accuracy of the forecast for this data set is 90%, which is a high indicator, which means that this methodology can be used as an additional assessment tool for the selection of candidate wells for the HFT. Besides, if we analyze a larger number of wells, for example, in operation, then due to a larger data set, the error can still be reduced by several percent [10-12].

The HFT effect was estimated for 43 wells and the decision for 11 candidate wells for hydraulic fracturing was made on the basis of the forecast [7].

IOP Conf. Series: Earth and Environmental Science 1021 (2022) 012076 doi:10.1088/1755-1315/1021/1/012076

Table 1. The HFT effect

Well number	Net thickness, m	Oil content, unit fraction	Porosity, unit fraction	Permeability, mD	Sandiness, unit fraction	Weight of injected proppant, t	Maximum proppant concentration, kg/m^3	Category
2724	3.1	0.57	0.18	34.5	0.12	7.5	450	B
2736	1.8	0.64	0.19	61.7	0.06	6.0	500	B
2762	5.4	0.46	0.17	51.3	0.15	10.0	700	B
2928	3.0	0.61	0.19	127.3	0.31	42.3	1200	B
2941	10.4	0.58	0.19	153.5	0.84	29.6	1200	B
2942	4.8	0.72	0.19	93.1	0.29	29.0	1100	A
2952	6.6	0.57	0.17	30.6	0.43	30.3	1020	B
2953	6.0	0.65	0.17	39.3	0.32	30.0	1100	A
2954	5.4	0.78	0.20	243.8	0.27	15.1	850	A
2955	6.8	0.72	0.16	6.0	0.35	25.0	1100	A
2963	5.2	0.62	0.18	34.1	0.29	30.0	1300	B
2964	6.4	0.64	0.2	76.4	0.48	30.1	1200	B
2980	17.2	0.50	0.14	41.5	0.82	30.1	1200	B
7057	2.2	0.48	0.20	149.8	0.19	6.0	790	B
7084	6.0	0.65	0.19	103.0	0.18	3.4	600	A
7089	8.0	0.61	0.17	20.8	0.52	10.2	800	A
7122	8.6	0.60	0.20	218.0	0.21	12.0	1000	A
7135	8.9	0.54	0.17	18.4	0.57	12.0	1000	A
7149	8.9	0.54	0.18	31.5	0.54	20.0	1 200	A
7152	8.5	0.56	0.19	115.5	0.23	15.0	900	A
7173	8.9	0.58	0.18	23.7	0.54	12.3	1300	B
7184	5.0	0.64	0.18	52.6	0.35	7.0	600	A
7185	8.0	0.69	0.19	82.0	0.33	12.1	750	A
7186	7.8	0.62	0.18	52.6	0.30	12.1	750	B
7193	13.2	0.71	0.18	54.8	0.55	4.2	700	B
7206	6.4	0.66	0.19	126.1	0.55	8.0	520	B
7211	6.6	0.65	0.19	91.3	0.59	8.0	950	A
7238	7.4	0.57	0.18	76.4	0.77	6.0	650	B
7251	3.6	0.72	0.20	187.3	0.17	3.4	550	A
7256	2.8	0.69	0.20	228.9	0.17	4.4	600	A
7276	2.4	0.73	0.20	267.2	0.32	2.8	500	B
7302	5.1	0.72	0.16	10.0	0.73	12.0	1000	B

The next step in assessing the applicability of the methodology is the analysis of the performed HFT at the operating well stock.

Wells of the JV object, among which there are vertical, directional wells, were also selected for analysis. Data were collected for 95 wells where HFT was carried out, and for which all geological and technological information is available.

According to the results of evaluation of candidate wells via the Kulbak method, the validity of this study is quite high, since out of 15 candidate wells, only 4 remain unclear concerning the HFT. Thus, the accuracy of the forecast for this data set is 74%, which is a high indicator, which means that this methodology can be used as an additional assessment tool for the selection of candidate wells for the HFT.

Next, the results of this procedure were tested, which was used to predict the success of the HFT operation at well No. 7084.

As a result of the evaluation of the candidate well according to the methodology of this study, it was concluded that the HFT at the well should not have been carried out. However, despite this, the HFT was still performed on September 30, 2020.

As a result of the HFT operation, the design values of oil production after the HFT amounting to 16 tons/day were not achieved, and the current production for October and November was not more than 6 tons/day, which is only 1 ton/day more than in the last 3 months before the HFT. Water content slightly increased – by half a percent – but in general remained at the same level. In general, the HFT did not fulfill the task of oil production increase by 2 times.

Thus, when estimating the efficiency of the performed hydraulic fracturing at the design well No. 7084 of the field under consideration, a negative result of the performed measure was obtained by using the studied evaluation method, thus confirming the fact that the use of additional tools for analyzing historical data sets for the performed HFT operations is an integral part of the selection of candidate wells and will further reduce the risk of performing the "unsuccessful" operation, thereby saving company's material and technical resources.

4. Conclusion

Based on the study and analysis of the obtained data, the following conclusions can be made.

1. The JV wells selected for the analysis are important from the point of view of the immediate plans for field development, since at the moment and in the near future the JV group will be actively drilled using horizontal technologies and HFT (MHFT).
2. The geological-field information is studied on the basis of mathematical analysis for subsequent assessment of the efficiency of candidate wells for the HFT.
3. The methodology confirmed that its informative capabilities allow it to be used as an additional analysis tool to increase the accuracy of determining specific candidate wells, thereby reducing the risks of an erroneous decision and misuse of material resources for a geological and technical operation.
4. This methodology can be used as additional tools on a number of other algorithms for selecting HFT candidate wells in a complex of software product based on neural networks and machine learning, which will cover an even larger volume of geological and field data.

5. References

[1] Almukhametova E M, Soloviev N N 2019 Changes in formation pressure and technological efficiency of unsteady flooding in double permeability reservoirs having different productivity rates" *IOP Conference Series: Earth and Environmental Science* **378(1)** 012036 DOI: 10.1088/1755-1315/378/1/012036.

[2] Almukhametova E M 2018 Expanding the experience of using non-stationary waterflooding technology with changing direction of the filtration flow in the example of the Northern Buzachi field *Georesursy* **20** 115-121.

[3] Almukhametova E M, Gizetdinov I A 2018 Optimization of FPM system in Barsukovskoye deposit with hydrodynamic modeling and analysis of inter-well interaction *Journal of Physics: Conference Series* **1015,** 1-4

[4] Zainagalina L.Z, Petrova L V, Petrov V A 2019 Analysis of the efficiency of telemetric systems for drilling wells *IOP Conference Series: Materials Science and Engineering* **560** 012096 DOI: 10.1088/1757-899X/560/1/012096.

[5] Mukhametshin V V, Andreev V E 2018 Increasing the efficiency of assessing the performance of techniques aimed at expanding the use of resource potential of oilfields with hard-to-recover reserves *Bulletin of the Tomsk Polytechnic University. Geo Assets Engineering* **329(8)** 30–36.

[6] Abdyukova R Ya 2019 Studies on operation and types of drilling pump valves IOP Conference Series: Materials Science and Engineering **560** 012050 DOI: 10.1088/1757-899X/560/1/012050.

[7] Mukhametshin V V 2017 The need for creation of a unified comprehensive method of geological and field analysis and integration of data on effective influence on the bottom-hole formation zone *Oil Industry* **4** 80–84 DOI: 10.24887/0028-2448-2017-4-80-84.

[8] Tyncherov K T, Selivanova M V 2017 Modular wavelet approach for reinforced plastic pipelines operation quality evaluation, International Conference on Industrial Engineering, Applications and Manufacturing (ICIEAM) Saint Petersburg, 16-19 May DOI: 10.1109/ICIEAM.2017.8076360

IOP Conf. Series: Earth and Environmental Science 1021 (2022) 012076 doi:10.1088/1755-1315/1021/1/012076

[9] Brannon H D, Malone M R, Rickards A R, Wood W D, Edgeman J R, Bryant J L 2004 *SPE Annual Technical Conference and Exhibition,* pp 1-23
[10] Abdyukova R Ya 2018 Improving the design of poppet valve in piston mud pump *Advances in Engineering Research* **157** 1-4 DOI:10.2991/aime-18.2018.1
[11] Reisberg J, Djsher T M 2012 *Interfacial Phenomena in Crude-Oil-Water Systems* **21** 43
[12] Kimbler O K, Reed R L, Silberberg I H 2011 Physical Characteristics of Natural Films Formed at Crude Oil-Water-Interfaces *Society of Petroleum Engineers Journal* **6** 153-165

Improvement of operational efficiency of high water-cut oil wells

I R Raupov, J Milic

St Petersburg Mining University, 2, 21st Line, St Petersburg 199106, Russia

E-mail: inzirrr@yandex.ru

Abstract. This paper describes a new algorithm for selecting candidate wells for water shut-off works. A new comprehensive methodology for the selection of candidate wells for water shut-off technology is proposed, consisting of 4 stages, where wells with excessive water cut are identified. The reasons for the water cut are determined, and the effectiveness of water shut-off technology on a potential candidate well is substantiated and predicted. The proposed method has been tested by the example of two wells in terrigenous reservoirs of a gas-oil field.

1. Introduction

Currently, more than half of the stock of production wells in Russia needs the use of water shut-off technology (WSOT) [2,4,10,11,17,21,23]. The main prerequisites for conducting WSOT are the transfer of fields to the final stage of development, and, accordingly, an increase in the water cut of the produced products [18].

Work on limiting water inflows during isolation of individual watered intervals of productive formations has a success rate of 50-67%, while eliminating behind-the-casing water flows - 60-75% [9]. The share of work, the costs of which were paid off by the sale of additionally produced oil and a decrease in the cost of extracting a reduced volume of associated water production, is no more than 30-40% [9]. In this regard, the issue of improving the approach to the selection of candidate wells for WSOT through the introduction of new techniques in order to increase the efficiency of oil wells operation in conditions of high water cut of the produced products is especially urgent.

2. Methods

Understanding the reasons for watering and solving them through a high-quality selection of candidate wells for water shut-off works becomes a key point in the operation of wells in complicated conditions.

Analysis of the available both domestic and foreign scientific and technical literature [1,3,5,13,14,25-27] does not allow us to identify a universal approach to the selection of candidate wells for WSOT.

Currently, there are many methods for the selection of candidate wells [1,3,5,12-15,25-27]: express analysis of the state of the well stock, diagnostics of water inflows by a graphical method, methods of statistical and probabilistic modeling, borehole diagnostics of water inflows. They also include construction waiting functions "productivity-water cut" and "reserves-water cut", selection based on modeling, selection of candidate wells due to water cut, an integrated and step-by-step approach to solving the problem of selection of candidate wells for WSOT.

Modern methodological aspects of improving the efficiency of WSOT design involve an integrated approach to the selection of candidate wells. This selection consists of 4 stages, where wells with excess water cut are identified, the reasons for the water cut are determined, and the effectiveness of WSOT performance at a potential candidate well is substantiated and predicted [3, 14, 27].

Analysis of all known techniques allows us to conclude that they are based on three main principles, namely: comparison of field development indicators, application of boundary conditions for the use of WSOT technologies and identification of the causes of watering [3]. Each of the above methods has certain advantages and disadvantages, and the choice of one of them depends on a number of factors such as the quantity and quality of initial data and resources, time constraints for the selection of candidate wells.

The lack of a universal algorithm, large uncertainties and risks when choosing one of the methods indicate the need to create simultaneously both a unified and a detailed approach to the selection of candidate wells for WSOT testing.

This paper proposes a new comprehensive methodology for selecting candidate wells for WSOT, consisting of 4 stages, in which the first stage, in addition to identifying candidate wells by technical parameters, includes the selection of potential candidate wells through a preliminary economic assessment. Analysis of economic indicators includes an operational calculation of the profitability of a potential candidate well in the current period, critical flow rates and losses due to well shut-in.

At the second stage, the main reasons for watering wells are analyzed by applying both geophysical, hydrodynamic, chemical and graphical-analytical methods [1,27], and probabilistic-statistical analysis of field data. The use of statistical methods [6,12,15,20,22,28] allows using the previously derived regression equations, discriminant analysis and scatter diagrams of values to determine the sources of water cut in the well.

The third stage involves the selection and justification of the WSOT technology. Currently, many different approaches have been proposed, based on decision making based on the boundary conditions for the application of a particular technology [25]. The correct choice and justification of technology depending on specific geological, physical and technological conditions is a capacious complex problem and deserves special attention, respectively, is not the subject of this study.

At the last stage of the proposed algorithm for assessing the technological efficiency of WSOT at a potential candidate well, the equation of regression analysis is determined for the considered area of the field. This method will allow one, on the basis of previously conducted WSOTs, to determine the expected technological effect of water shut-off works on new wells and is an addition to standard methods for predicting technological efficiency, such as "direct" counting, displacement characteristics and the method of production decline curves. The use of probabilistic and statistical methods at the fourth stage will make it possible to more quickly determine the effectiveness of the WSOT with high accuracy.

3. Results and Discussion

To substantiate the possibility of using the algorithm proposed in this work in order to improve the efficiency of oil wells operation in conditions of high water cut in the produced product, the selection of candidate wells was carried out on the basis of the proposed methodological aspects.

At the first stage of selection, based on a preliminary economic assessment, field and data on relative phase permeabilities (RPP), wells No. 3 and No. 4 were selected as candidate wells, which penetrated the terrigenous reservoirs of the gas-oil field.

Since wells No. 3 and No. 4 are close to each other, the average values of oil-saturated thickness and permeability are equal, respectively, well No. 4 with a higher value of the current water cut is given an advantage in ranking.

Ranked wells are further considered based on analysis of production data, which includes changes in water cut and fluid production dynamics in both the candidate well and offset wells.

To determine the causes of water cut using the methods of mathematical statistics, an analysis was carried out on the WSOT field. From WSOT at the field are carried out:

- elimination (E) of behind-the-casing fluid flows to the perforated interval from the underlying or overlying formations (45% of jobs);

• isolation (I) of water inflows in the perforated interval of the productive formation or below it using polymer (32%);

• isolation below or above previously perforated intervals (23%).

Based on the presented analysis, the first two groups of water cut causes were selected for discriminant analysis. For this analysis, the following parameters are highlighted: interval thickness, m; permeability, μm^2; porosity,%; perforation interval, m; current reservoir pressure, MPa; current density of water, g/cm^3; oil viscosity in reservoir conditions, mPa·s. Based on the calculated functions, canonical values for wells of groups 1 and 2 were determined and their scatter diagram was constructed (Figure 1).

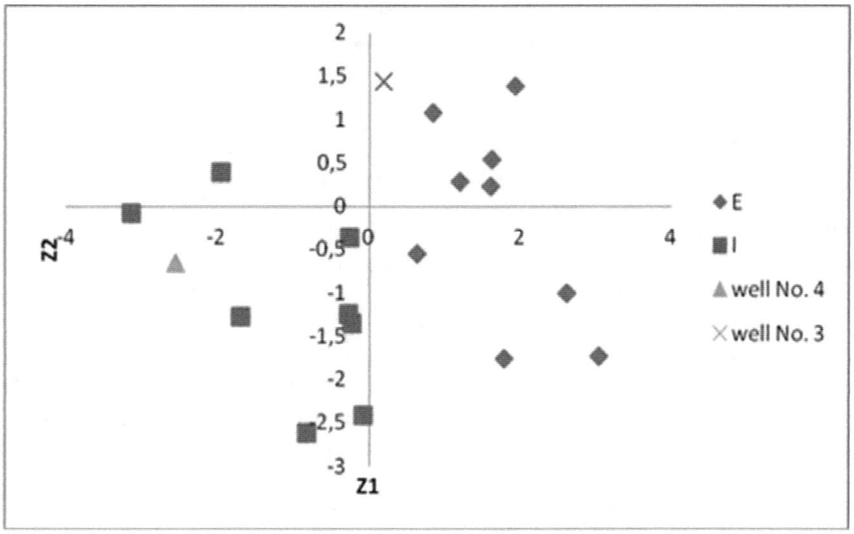

Figure 1. Scatter diagram for wells groups 1 and 2

Based on Figure 2, it can be concluded that the reasons for the watering of the wells under consideration are different. Using the methods of mathematical statistics, it was possible to prove that the increased water cut in well No. 4 is a consequence of the overlying water breakthrough due to coning. The data obtained can be confirmed by one of the standard methods for determining the causes of flooding [1,27] - graphical analytical.

Well No. 4 will be used as an example for a quantitative assessment of both technological and economic effect, since during the previous stages it was selected as a potential candidate well using the algorithm proposed in the work. For further forecasting, a similar WSOT technology will be used, as in other wells, where water shut-off work was carried out in the area under consideration.

Only 7 factors were used in the regression analysis. However, each of them has a different effect on the analyzed value, i.e. has different significance and degree of influence. For clarity, a histogram is built (Fig. 2), which shows the degree of influence of one factor or another and the direction of this influence (minus - negative, plus - positive).

The relative importance of the variable was estimated by the value of the contribution of a particular factor to the total proportion of variance explained by the model. The study of the most significant of them (i.e. those included in the mathematical model) confirmed the theoretical justification of the qualitative assessment.

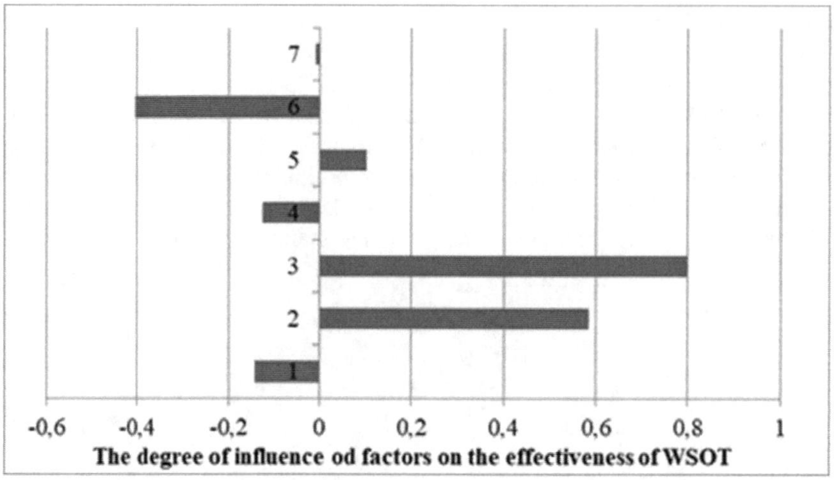

Figure 2. Histogram for assessing the influence of factors on the effectiveness of WSOT
1 - reservoir depth H_r, m; 2 - reservoir pressure P_r, MPa; 3 - Oil flow rate before WSOT Q_{before}, t/day; 4 - Water cut before carrying out WSOT B_o, %; 5 - perforation interval h_{perf}, m; 6 – injectivity Q_{inj}, m^3/ day; 7 – volume of the waterproofing composition V_{comp},m^3

Geological factors, such as reservoir pressure, have the greatest impact on the effectiveness of WSOT. The technological factor, such as the volume of the waterproofing composition, has a secondary impact, since almost everywhere the same amount of the waterproofing composition was used during the WSOT.

The obtained multiple regression equation for determining the well flow rate after testing (Formula 1) made it possible to compare the predicted and actual [24] oil flow rates after testing (Figure 3).

$$Q_{pred} = -0.141 \cdot H_r + 0.584 \cdot P_r + 0.799 \cdot Q_{before} - 0.123 \cdot B_o + 0.103 \cdot h_{perf} - 0.405 \cdot Q_{inj} -0.009 \cdot V_{comp} \quad (1)$$

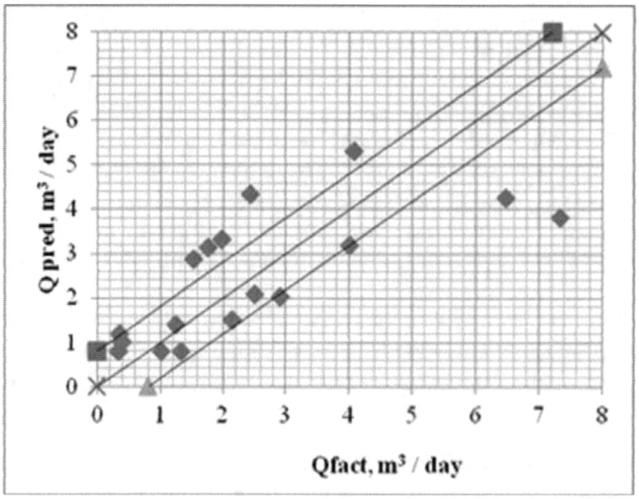

Figure 3. Comparison of actual and calculated production rates after WSOT

The calculation of economic efficiency was carried out for the period of the average duration of the effect of water shut-off works, determined by the method of analogy with other wells. Using the application for predictive modeling, forecasting and optimization Oracle Crystal Ball by the Monte Carlo method [16,19], the economic effect was simulated taking into account the risks. Based on the values obtained by the Monte Carlo simulation, it can be concluded that conducting WSOT in the case under consideration with a positive effect will pay off in any case.

4. Conclusion

Currently, only 1/3 of all waterproofing works receive a technologically and economically positive effect. One of the ways to increase the efficiency of the work may be to improve the algorithms for the selection of candidate wells for water shut-off works in order to increase the efficiency of oil wells operation in conditions of high water cut.

Modern methodological aspects of improving the efficiency of WSOT design involve an integrated approach to the selection of candidate wells. In this paper, a new approach to the selection of candidate wells is substantiated by combining methods at separate stages of selection.

The proposed selection technique within the framework of the work was tested by the example of one section of a gas and oil field at a late stage of development. At the first stage of selection, based on a preliminary economic assessment, production and RP data, two wells were selected as candidate wells. Based on discrete analysis, it is shown that the cause of watering in one of the wells is water breakthrough due to coning. This analysis could not show with high reliability what is the reason for the watering of the second well, and, accordingly, the need for integration at the 2nd stage of the selection of methods of mathematical statistics with standard methods for determining the causes of watering is shown. The results of the regression-correlation analysis made it possible to determine the expected technological effect on the first well after the WSOT. To determine the full technological effect under dynamic conditions, the statistical method was integrated with the methods of the production decline curve and displacement characteristics.

References

[1] Bailey B, Crabtree M, Tyree D et al. 2001 Diagnostics and limitation of water inflows *Oil and Gas Review* **1** 44-67

[2] Dvoynikov M V, Kuchin V N, Mintzaev M S 2021 Development of viscoelastic systems and technologies for isolating water-bearing horizons with abnormal formation pressures during oil and gas wells drilling *Journal of Mining Institute* **247** 57-965 https://doi.org/10.31897/PMI.2021.1.7

[3] Gabdulov R R, Nikishov V I and Cream P I 2009 Generalization of the experience of selecting potential candidate wells and technologies for carrying out repair and isolation works *Scientific and technical bulletin of OJSC "NK" Rosneft* **4** 22-7

[4] Gabdrakhmanov T N 2015 Application of reagents of LLC NPF "InTechService" in technologies for limiting water inflow into wells (water isolation) *Engineering practice* **8** 80

[5] Gazizov A Sh and Gazizov A A 1999 *Improving the efficiency of field development based on limiting the movement of water in layers* (Moscow: Nedra-Business Center)

[6] Galkin V I and Koltyrin A N 2020 Investigation of probabilistic models for forecasting the efficiency of proppant hydraulic fracturing technology *Journal of Mining Institute* **246** 650-9 DOI: 10.31897 / PMI.2020.6.7

[7] Galkin V I, Ponomareva I N, Koltyrin A N 2018 Development of probabilistic and statistical models for assessing the effectiveness of the use of proppant hydraulic fracturing (by the example of the Tl-Bb object of the Batyrbayskoye field) *Perm Journal of Petroleum and Mining Engineering* **18/1** 27-33 DOI: 10.15593 / 2224-9923 / 2018.1.4

[8] Derendyaev R A and Derendyaev K A 2019 Evaluation of the effectiveness of water isolation works at the Visean facility of the Perm Territory field *Master's Journal* **2** 41-50

[9] Zemtsov Yu V 2020 *Scientific and methodological foundations for the use of polyfunctional organosilicon and polymer compounds for effective limitation of water inflows in oil wells.* Doctoral

dissertation: Ufa State Petroleum Technological University.
[10] Ketova Yu A, Bai B, Khizhnyak G P, Gladkikh Ye A, Galkin S V 2020 Testing of preformed particles polymer gel technology on core filtration models to limit water inflows *Journal of Mining Institute* **241** 91-6 DOI: 10.31897/pmi.2020.1.91
[11] Korolev M, Rogachev M, Tananykhin D 2020 Regulation of filtration characteristics of highly watered terrigenous formations using complex chemical compositions based on surfactants *Journal of Applied Engineering Science* **18(1)** 147-56 doi:10.5937/jaes18-24542
[12] Kudryashova D A 2018 The use of probabilistic and statistical methods to determine the sources of watering of candidate wells for water shut-off works (by the example of the Visean object of the Perm Territory field) *Perm Journal of Petroleum and Mining Engineering* **17/1** 26–36 DOI: 10.15593 / 2224-9923 / 2018.1.3
[13] Krivoshchekov S N, Galkin V I and Kozlova I A 2012 Determination of promising areas of geological exploration for oil by probabilistic and statistical methods by the example of the Perm Territory *Perm Journal of Petroleum and Mining Engineering* **11/4** 7–15
[14] Lapytov A R, Kulikov A N et al 2009 Improvement of the methodology for selecting wells for carrying out waterproofing works *Oil and Gas Business* **7/2** 46-50
[15] Legkokonets V A, Mardashov D V and Morozov A V 2018 Forecast of the effectiveness of water isolation works by the example of the Yamburgskoye oil and gas condensate field *PRONEFT* **3/9** 51-5
[16] Mikhalevich I M and Primina S P 2006 *Application of mathematical methods in the analysis of geological information (using computer information)* (Irkutsk: Irkutsk Publishing House)
[17] Fan Guangli, Li Meng, Chen Xinxiao, et al 2021 Polymer-Nanosilica-assisted to evaluate oil recovery performances in sandstone reservoirs *Energy Reports* **7** 2588-93 DOI: 10.1016/j.egyr.2021.04.047
[18] Filimonova I V, Nemov V Yu et al 2019 *Oil and Gas Complex of Russia - 2019: Part 1. Oil Industry* (Novosibirsk: INGG SO RAN)
[19] Pangina N N and Pangin A A 2002 *Static modeling: Monte Carlo method* Computer tools in education **5** 30-43
[20] Prischepa O M, Nefedov Y V and Ibatullin A K 2020 Raw material source of hydrocarbons of the arctic zone of Russia *Periodico Tche Quimica* **17(36)** 506-26
[21] Raupov I R, Shagiakhmetov A M 2019 The results of the complex rheological studies of the cross-linked polymer composition and the grounding of its injection volume *International Journal of Civil Engineering and Technology* **2(10)** 493-509
[22] Sabukevich V S, Podoprigora D G, Shagiakhmetov A M 2020 Rationale for selection of an oil field optimal development system in the eastern part of the pechora sea and its calculation *Periodico Tche Quimica* **17(34)** 634-55
[23] Shagiakhmetov A M, Podoprigora D G, Terleev A V 2020 The study of the dependence of the rheological properties of gelforming compositions on the crack opening when modeling their flow on a rotational viscometer *Periodico Tche Quimica* **17(34)** 933-9
[24] Silaev K O and Silaeva A N 2016 Methods for the Analysis of Production Decline Curves Used in the Fuel and Energy Sector *Economics and Society* **9 (28)** 543-6
[25] Strizhnev K V 2010 *Repair and insulation works in wells: Theory and practice* (Saint Petersburg: Nedra)
[26] Fattakhov I G 2011 Methods for identifying ways to flood oil wells *Oil and Gas Business* **3** 154-160
[27] Yakubov R N et al 2020 Methodological aspects of repair and insulation operations imrpoving efficiency *Problems of gathering, treatment and transportation of oil and gas products* **4(126)** 50-9 DOI: 10.17122/ntj-oil-2020-4-50-59
[28] Zhenzhen Wei, Shanyu Zhu, Xiaodong Dai, Xuewu Wang, Lis M. Yapanto, Raupov I R 2021 Multi-criteria decision making approaches to select appropriate enhanced oil recovery techniques in petroleum industries *ENERGY REPORTS* **7** 2751-58

Diagnostics of traction motor of 1DT.003.11 type of electric train EP3D by modeling fuzzy linguistic parameters in matlab and fuzzy tech environment

I A Menshikov[1], A.V. Shorina[1], V A Kargin[2], A V Volgin[3], A P Moiseev[3], E V Volgina[4]

[1] Saratov State Technical University named after Gagarin Y.A., 77 Politechnicheskaya street, Saratov, 410054, Russia
[2] Moscow State University of Food Production, Moscow, Volokolamskoe shosse, 11, 125080, Russia
[3] Saratov State Agricultural University named after N.I. Vavilov, 1, Theatralnaya Square, Saratov, 410012, Russia
[4] Saratov State Law Academy, 1, Volskaya street, Saratov, 410056, Russia

E-mail: saratov-79@list.ru

Abstract. The article presents the results of research in the field of diagnostics of a traction electric motor of the 1DT.003.11 type of the alternating current electric train of the EP3D series by simulating emergency modes of functioning of the fuzzy logic of a modular dynamical neural network. It is proposed to solve the problems of assessing the reliability of traction electric motors in railway transport with an increased service life using simulation modeling of the ANFIS modular neural network in the MATLAB package. The result of modeling the statistical diagnostic parameters of the traction motor is an assessment of the change in the linguistic coefficients of the residual life of the traction motor of the EP3D electric train and recommendations for the maintenance of traction electric motors, depending on their technical condition at the current time of operation.

1. Introduction

The main purpose of the traction electric motor (TEM) diagnostics is to determine its technical condition, as well as to search for malfunctions and failures with a given accuracy and selectivity using diagnostic parameters. The efficiency of the on-board diagnostic system depends on the number of diagnostic parameters that determine the stability of the TEM operation over a long period of service.

The task of diagnosing TEM, with a monotonic change in diagnostic parameters, is to develop some expert assessment that would be implemented as a fuzzy inference system and would allow determining the operable state of TEM and the frequency of maintenance based on subjective estimates of the output diagnostic parameters. The solution to this problem can be carried out using the dynamic modular neural network model (DMNN).

The basic idea behind the DMNN model is to use an existing sample of data to determine the parameters of membership functions that best fit some fuzzy inference system. At the same time, well-known neural network training procedures are used to find the parameters of membership functions [1, 3, 4].

In the Fuzzy Logic Toolbox package of the MATLAB system, artificial neural networks are implemented in the form of the so-called adaptive neural fuzzy inference system - ANFIS fuzzy inference. ANFIS is an adaptive neural network based on a fuzzy inference system (Adaptive Network-based Fuzzy Inference System).

The aim of the work is to study the process of diagnosing a traction motor using a fuzzy model of a dynamic modular neural network.

The task of the diagnostic study is to develop some expert assessment that would be implemented in the form of a fuzzy inference system and would allow determining the performance of the traction motor and the power control circuit during the movement of an electric train based on subjective estimates of the input diagnostic parameters [1, 2, 4, 8, 10, 11, 12].

2. The object and method of research

The object of the research is the traction motors of the 1DT.003.11 type of the EP3D series alternating current electric trains operating in severe service conditions.

The subject of research is the theory, methods, algorithms and tools for constructing an adaptive on-board diagnostic system of a traction motor, Figure 1.

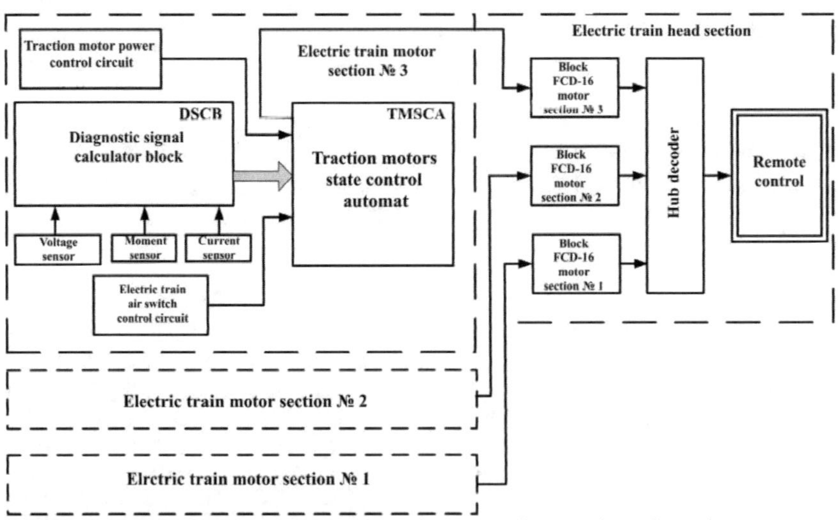

Figure 1. Block diagram of the adaptive on-board system diagnostics of traction motors and power contactors of the EP3D electric train control circuit

As input parameters of the fuzzy inference system, we will consider 3 fuzzy linguistic variables (passport values of a traction motor of 1DT.003.11 type):

- the voltage on the armature windings, the fluctuation of which is permissible within 10% of the nominal value – 825 V;
- change in the static moment on the armature shaft of the traction motor within 10% of the nominal value -1068 Nm.

The function of the fuzzy measure of the term-set of the first linguistic variable T_1 is defined using the graph (Figure 2).

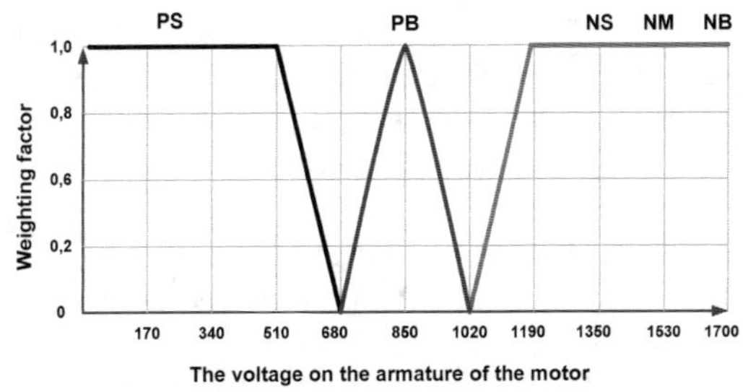

Figure 2. Graphs of the membership function for the input term «Armature voltage of the TEM»

The function of the fuzzy measure of the term-set of the second linguistic variable T_2 is defined using the graph (Figure 3).

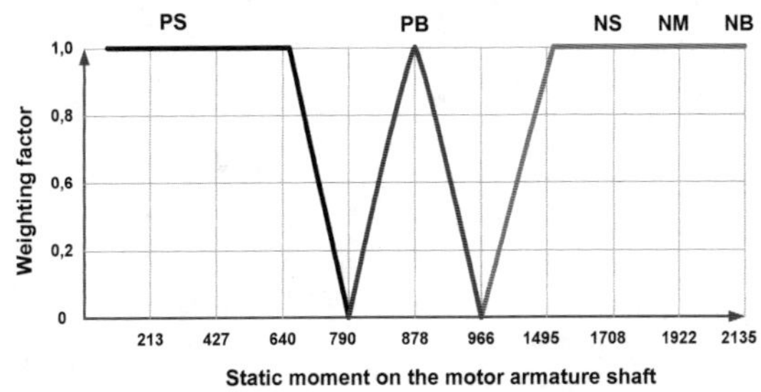

Figure 3. Graphs of the membership function for the input term "Static moment on the armature shaft of the TEM"

As a term-set of the first linguistic variable «Armature winding voltage of the TEM» we will use the set T_1 = {«Minimum armature voltage U_{min}», «Nominal armature voltage U_{nom}», «maximum armature voltage U_{max}»}. As a term-set of the second linguistic variable «Static moment on the armature shaft of the TEM» we will use the set T_2 = {«Minimum static load M_{min}», «Nominal static load M_{nom}», «Maximum static load M_{max}»}.

The degree of this influence in a diagnostic system based on the use of fuzzy logic is assessed using linguistic truth values, presented in table 1.

Table 1. Linguistic values of the output diagnostic parameters of the traction motor.

Name of diagnostic parameter	**PB** Positive big (all parameters at the standard level)	**PS** Positive small (deviation of diagnostic signs at the acceptable level)
The value of the current in the armature circuit of the traction motor I_a, (A)	<270> 265; 277	<265
The value of the electromagnetic moment of the traction motor M_{el}, (Nm)	<878> 790; 966	<790
The value of the angular frequency of rotation of the traction motor armature ω, (rad/s)	<158> 153; 187	<153
The value of the active power of the traction motor P, (kW)	<170> 153; 187	<153
The value of the total power of the traction motor S, (kVA)	<200> 180; 220	<180
The value of the normalized power loss factor K_m	<0,85> 0,76; 0,93	<0,76

As the term-set of the output linguistic variable «Stability of the traction motor operation» we will use the term-set T_3 = {«The operable state of the traction motor», «Emergency operation of the traction motor (turn-to-turn closure of the armature winding sections of the TEM)»}. In this case, each of the terms of the first and second variables will be estimated using weight coefficients from 0 to 1, on a 10-point ordinal scale, in which the number 0 corresponds to the worst estimate, and the number 1.0 corresponds to the best estimate of the input parameter. As for the terms of the output variable T_3, we will assume that: if the traction motor operates stably, then the estimate is equal to 1.0; if the traction motor operates in emergency mode, the score is 0.

With the help of the proposed theory of fuzzy logic, the problems of detecting dangerous failures in the armature winding in the presence of an implicitly expressed relationship between the defect and the output measured diagnostic parameters are solved in the traction motor technical diagnostics systems, since complex defects simultaneously affect several diagnostic signs [2, 3, 4, 5, 6].

Table 2. Linguistic values of the output diagnostic parameters of the traction motor

Name of diagnostic parameter	**NS** Negative small (presence of diagnostic signs of an accident at the initial stage of development)	**NM** Negative mean (presence of stable diagnostic signs of a defect)	**NB** Negative big (presence of stable diagnostic signs of significant development of the defect)
The value of the current in the armature circuit of the traction motor I_a, (A)	>297	>304	>311
The value of the electromagnetic moment of the traction motor M_{el}, (Nm)	>1708	>1922	>2135
The value of the angular frequency of rotation of the traction motor armature ω, (rad/s)	>126	>110	>94.8
The value of the active power of the traction motor P, (kW)	>204	>221	>280
The value of the total power of the traction motor S, (kVA)	>240	>260	>280
The value of the normalized power loss factor K_m	>0.68	>0.59	>0.51

3. Results and discussion

Using a fuzzy model of a modular dynamic neural network (Figure 4) of neuro-fuzzy inference of diagnostic parameters, we will consider the process of diagnosing a traction motor of 1DT.003.11 type [2, 3, 4, 5, 6, 13, 14].

Knowing the dynamics of the change in the linguistic coefficients $\mathbf{K}_l$ of the traction motor (PB, PS, NS, NM, NB), which characterizes the change in diagnostic parameters at a point in time $(t-t_0)$, it is possible to predict the values of the linguistic coefficients $\mathbf{K}_l$ at a certain point in the future.

The value of linguistic coefficients can be defined as:

$$K_1 = \frac{PB_i}{NS_i}, \tag{1}$$

where: PB_i is a positive linguistic value of diagnostic parameters, while most of the diagnostic parameters correspond to the normative values. This means that the traction motor is operational; NS_i is a negative small linguistic value of diagnostic parameters, meaning that the traction motor has signs of insulation aging at the initial stage. NM is a negative average linguistic value of diagnostic parameters, means that the traction motor has signs of persistent defects in the armature winding insulation, for example, intensive moistening of the armature winding or a sharp decrease in the resistance of the insulation armature winding relative to the magnetic core of the armature of the traction motor. NB is a negative large linguistic value of diagnostic parameters, means that the traction motor has signs of persistent significant defects in the insulation of the armature winding or a short interturn circuit of the armature winding of the traction motor.

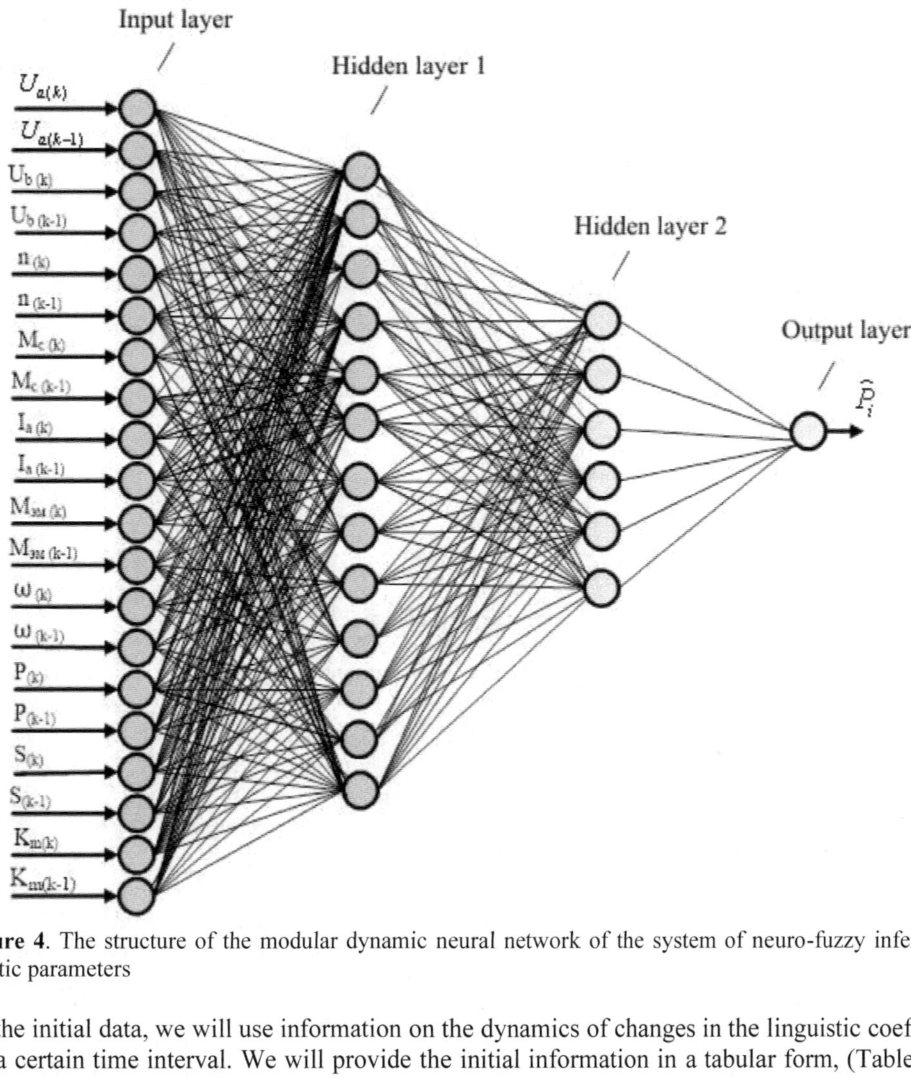

Figure 4. The structure of the modular dynamic neural network of the system of neuro-fuzzy inference of diagnostic parameters

As the initial data, we will use information on the dynamics of changes in the linguistic coefficients **K₁** for a certain time interval. We will provide the initial information in a tabular form, (Table 2) and (Table 3) we will take the time T = 0 as the starting point.

Table 3. Initial information of the dynamics of changes in the parameters of the control of an efficient traction electric motor of the 1DT.003.11 type

No.	T, month	K_I	Linguistic value	No.	T, month	K_I
1	0	1	PB	9	48	0.985
2	6	1	PB	10	54	0.985
3	12	0.995	PB	11	60	0.983
4	18	0.994	PB	12	66	0.978
5	24	0.994	PB	13	72	0.978
6	30	0.991	PB	14	78	0.964
7	36	0.991	PB	15	84	0.964
8	42	0.991	PB	16	90	0.952

Table 4. Initial information of the dynamics of changes in parameters of control of the emergency mode of operation of the traction motor of the 1DT.003.11 type

No.	T, month	Linguistic value	No.	T, month	K_I	Linguistic value
1	0	PB	17	96	0.939	PB
2	6	PB	18	102	0.927	PB
3	12	PB	19	108	0.915	PB
4	18	PB	20	114	0.910	PB
5	24	PB	21	120	0.870	PS
6	30	PB	22	126	0.834	NS
7	36	PB	23	132	0.769	NM
8	42	PB	24	138	0.765	NB

To create a training sample of the values of the linguistic coefficients K_I we assume that the fuzzy DMNN model will have 4 input variables. In this case, the first input variable will correspond to K_I at the current time moment, or the 132nd month of TEM operation, the second variable – K_I for the previous time (i -1), that is, the 126th month of TEM operation. The third input variable will correspond to K_I for the time (i -2), that is, the 120th month of TEM operation. The fourth variable K_I will correspond to (i -3), i.e. the 114th month of TEM operation.

To solve the problem of diagnosing a traction electric motor, various models of technical analysis can be applied, based on the use of various monotonically changing parameters (indicators) of the technical state of the insulation of the TEM winding.

However, the presence of implicit potential in the dynamics of changes in the linguistic coefficient K_I, allows us to use the model of adaptive dynamic neural networks.

4. Conclusion

The fuzzy modeling process involves the analysis of the results of predicting the diagnostic fuzzy inference coefficient K_{I38} after 138 months of TEM operation in order to establish the adequacy of the developed fuzzy model (in this case, the fuzzy inference system predicted the coefficient $K_{I38} = 0.764$).

Comparing the obtained value of the fuzzy inference coefficient with the corresponding table value from (Table 2) and (Table 3), it can be stated that with an increase in the service life of traction motors, the insulation of the armature winding ages and the value of the diagnostic parameter decreases from the stable operation mode $K_{42} = 0.952$ to the emergency operation mode $K_{138} = 0.764$.

Thus, the proposed fuzzy model of a modular dynamic neural network shows a sufficiently high degree of adequacy to real initial data, which allows us to conclude that it is possible to use it practically for diagnostics and organization of maintenance of a traction motor for a given resource of elements and to prevent emergency failures. The simulation results indicate that the maintenance of the traction electric motor is necessary according to the actual state after 99360 operating hours, which is 11.5 years.

According to the instructions of the Traction Directorate of the branch of JSC Russian Railways «Rules for the repair of electrical machines of electric rolling stock», on average, the following mileage rates are established on the Russian railway network between types of repair of electrical machines in each cycle interval: for electric machines of electric locomotives - 350 thousand km; for electric cars of electric trains - 300 thousand km; but not more than 8 years of operation [7, 8, 9, 15, 16].

To increase the efficiency of maintenance of electric vehicles of electric rolling stock, reduce operating costs for repairs, it is possible to use the considered model of fuzzy forecasting of monotonic changes in diagnostic parameters of TEM, which is a promising direction for organizing maintenance of electrical machines according to their actual state.

References

[1] Parkhomenko P P 1981 *Fundamentals of technical diagnostics* (Moscow: Energy)
[2] Volkov A K 1988 *Increasing the operational reliability of traction engines* (Moscow: Transport)
[3] Klepikov V B, Sergeev S A, Makhotilo K V, Obruch I V 1999 Application of methods of neural networks and genetic algorithms in control systems for electric drives *Elektrotekhnika* **5** 2-6.
[4] Chernykh I V 2001 *Simulink: environment for creating engineering applications* (Moscow: Peace)
[5] Sergienko A B 2006 *Digital signal processing: A textbook for universities* (Saint-Petersburg: Peter)
[6] Diakov V P 2015 *Simulink: Tutorial* (Moscow: DMK Press)
[7] Richard Lyons 2009 *Digital Signal Processingy* (Moscow: LLC «Binom-Press»)
[8] Volkov A K 1988 *Improving the operational reliability of traction motors* (Moscow: Transport)
[9] Dmitriev A K 1981 *Recognition of failures in electrical systems* (Moscow: Energoatomizdat)
[10] Neyman V Yu, Rogova O V 2016 New construction types of a linear electromagnetic motor with the active teeth-slot zone *Proceedings of IFOST-2016 11th International Forum on Strategic Technology* **2** 28-31
[11] Neyman L A, Neyman V Yu, Shabanov A S 2016 Simulation of processes in an electromagnetic converter with energy loss in the massive magnetic core. *17th International Conference of Young Specialists on Micro/Nanotechnologies and Electron Devices (EDM -2016) Conference Proceedings* **2** 522-525
[12] Shabanov A S, Neyman V Yu 2016 The effect of the structure of the magnetic circuit on the traction characteristics of the electromagnetic press *Science. technology. innovation is a collection of scientific papers in 9 parts ed. by E. G. Gurova* 94-95
[13] Moshkin V I, Ugarov G G 2015 The concentration of magnetic energy in the working clearances pulse linear electromagnetic engine at the stage of its electromagnetic conversion *Journal of Electrotechnics* **4(9)** 20-26
[14] Moshkin V I, Ugarov G G 2014 Main dimensions and their ratios for the magnetic system of the pulsed linear electromagnetic engine *Journal of Electrotechnics* **1 (2)** 71-78
[15] Moshkin V I, Ugarov G G 2016 Electromechanical characteristics of pulse linear electromagnetic engines of the longitudinal magnetic field as movement *2016 International Conference on Actual Problems of Electron Devices Engineering (APEDE)* **2** 1-8
[16] Usanov K M, Ugarov G G, Moshkin V I. 2006 *Linear pulse electromagnetic drive of machines*

with autonomous power: monograph. (Kurgan: Publishing house of Kurgan state University press)

IOP Conf. Series: Earth and Environmental Science 1021 (2022) 012079 doi:10.1088/1755-1315/1021/1/012079

Temperature-baric behavior of the effective thermal conductivity of sandstones

S N Emirov [1,2], **A A Aliverdiev** [1,3,*], **R M Aliev** [1,2], **V D Beybalaev** [1,3], **E N Ramazanova** [2], **and AA Amirova** [4]

[1] Institute for Geothermal Researches and Renewable Energy - a branch of the Joint Institute for High Temperatures of Russian Academy of Sciences, 30A Pr. Shamilya, Makhachkala, 367030, Russia
[2] Dagestan State Technical University, 70A Pr. Shamilya, Makhachkala, 367030, Russia
[3] Dagestan State University, 43A Gadzhieva Str., Makhachkala, 367025, Russia
[4] Institute of Physics DFRC of Russian Academy of Sciences, 94 Yaragskogo Str., Makhachkala, 367003, Russia

E-mail: aliverdi@mail.ru

Abstract. The paper presents the results of measurements of the effective thermal conductivity of three sandstone samples exhibiting different structural ordering in the temperature range 273-523 K and pressures from atmospheric to 400 MPa, as well as (2) a low-parameter model, described as a temperature dependence at a fixed pressure and the pressure dependence at a fixed temperature. It also describes the influence of pressure on the nature of the temperature dependence. The samples from the following deposits were studied: (1) Kochubei, the Republic of Dagestan, (2) Buinaksk, the Republic of Dagestan, and (3) the Tyumen superdeep well. It was shown that the cumulative increase in temperature and pressure for sandstones with different structural ordering (even when the absolute values of the effective thermal conductivity at atmospheric pressure and room temperature are close) can have a qualitative difference and with a predominantly amorphous ordering, it must be taken into account when compiling temperature models in the terrestrial bark.

1. Introduction

Monitoring and forecasting the temperature field in the reservoir and the bottomhole zone are of great importance at all stages of preparation, drilling and operation of any well and in many tasks related to the study and development of subsoil [1,2]. All this aspects impose additional requirements on models describing temperature fields, including a deep understanding of the thermal properties of rocks in general and their effective thermal conductivity in particular.

As it is known, the process of heat transfer in solids with an ordered crystal structure has a wave nature, which leads to an inversely proportional dependence of thermal conductivity ( ) on temperature, while the process of heat transfer in solids with an amorphous structure has an activation nature [3,4] and $\lambda \propto T^{0.5}$ [5]. In disordered crystalline solids, the masses of atoms and their force constants randomly change from site to site. It causes a weakening of phonon scattering of heat waves, which, together with the influence of block boundaries and crystal lattice defects, has a rather complex nature of the temperature dependence of effective thermal conductivity [6]. Nevertheless, the analysis of many experimental and theoretical works ([7, 8] and references therein) shows that the temperature dependence of the effective thermal conductivity of a large class of compounds and, first of all, rocks is well described by the power law:

$$\lambda(T) = const \cdot T^n \quad (1)$$

Here, the exponent n in ranges from is 0.5 to +0.5.

On the other hand, hydrostatic pressure significantly affects the effective thermal conductivity of rocks. When constructing the pressure dependences of the effective thermal conductivity of rocks (for example, [9, 10]), a linear approximation is usually used after a certain pressure (~ 100 MPa) without taking into account the effect of pressure on the temperature dependence. Despite satisfactory agreement at sufficiently high pressures, this may be insufficient when considering thermal processes occurring in the range from atmospheric to 200 MPa.

This work is aimed to study the thermal conductivity of sandstones in a fairly wide range of temperatures and pressures characteristic of natural conditions in the temperature range 273-523 K and pressures from atmospheric to 400 MPa.

2. Experiment

Effective thermal conductivity measurements were carried out in the temperature range of 273-523 K in the range of hydrostatic pressures from atmospheric to 400 MPa using the absolute stationary compensation method of flat plates on a setup consisting of a Unipress (Poland) NMR-10 high gas pressure chamber, a GCA-10 Unipress high-pressure gas compressor, an air thermostat with electronic temperature control up to 0.1 K and a measuring cell. Argon gas with low thermal conductivity (=0.018 W/m·K) was used as a pressure-transmitting medium [7.8].

The samples from the following deposits were studied: Kochubei, the Republic of Dagestan (sample 1, ρ=2.27·g/cm^3, k=7%), Buinaksk, the Republic of Dagestan (sample 2, ρ=2.17·10^3 g/cm^3, k=16.2%) and the Tyumen superdeep well (sample 3, ρ=2.29·g/cm^3, k=11%).

3. Results and discussion

For an effective comparative analysis, in addition to sample (1) showing a more typical ("polycrystalline") temperature dependence of the effective thermal conductivity, two samples were selected (samples 2 and 3). The temperature dependence of the effective thermal conductivity corresponds to amorphous structure. The absolute values of the effective thermal conductivity of samples (1) and (3) at atmospheric pressure and temperature of 300 K were close (Table 1) and differed from the corresponding value for the second sample by more than 3.5 times.

The experimental values of the effective thermal conductivity for all samples, together with the calculated curves according to equation (1) with the coefficients obtained by the least squares method, are shown in Figure 1. As it can be seen from the figure, the power-law temperature dependence of the effective thermal conductivity is performed with very good accuracy for each sample at any fixed pressure.

Table1. Summary table of the samples thermal conductivity data

Sample	λ_0,(T_0=300K) W/(m·K)	n_0	n_{250MPa}	T^*_0 K	α	P^*_0 MPa
1	2.693	–0.208	–0.184	3000	1.197	100
2	0.631	0.319	0.204	450	1.442	60
3	2.330	0.429	0.056	155	(ν) 1.92 (δ) 1.98	(ν) 10 (δ) 20

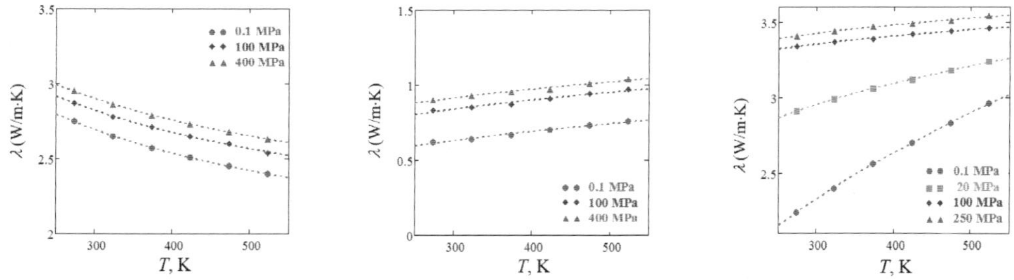

(a) (b) (c)

Figure 1 (color on-line). The experimental values of the effective thermal conductivity of sandstone samples together with the calculated curves according to equation (1): a) sample (1); b) sample (2); c) - sample (3).

Applying equation (1) to the current (T) and fixed (T_0) temperatures, we can use the following:

$$\lambda(T,P)=\lambda(T_0,0)\cdot\left(1+\delta_{T_0}(P)\right)\cdot\left(\frac{T}{T_0}\right)^{n_0\cdot(1-\nu(P))} \qquad (2)$$

This representation makes it possible to single out only two reference quantities $n_0=n(0)$ and $\lambda(T_0,0)$, and two dimensionless pressure functions $\nu(P)$ and $\delta_{T_0}(P)$:

$$\delta_{T_0}(P)=\frac{\lambda(T_0,P)}{\lambda(T_0,0)}-1, \qquad (3)$$

$$\nu(P)=1-\frac{n(P)}{n_0}. \qquad (4)$$

For highly ordered (monocrystalline) materials and, conversely, for completely amorphous materials, the baric dependence $\delta_{T_0}(P)$ is often close to linear, but for complex multicomponent structures, which include sedimentary rocks, this dependence is highly nonlinear especially in the initial section. At the same time, in the range of pressure from atmospheric to several hundred MPa (the most interesting in geological applications), it can be fairly well estimated by empirical and semi-empirical two-component dependences ([7, 8] and references therein), in particular:

$$\delta_{T_0}(P)=\frac{1+\alpha P/P_0^*}{1+P/P_0^*}-1 \qquad (5)$$

In this case, the coefficients included in it have a rather clear physical meaning. Indeed, substituting (5) into (2), it is obvious that at a fixed temperature the coefficient α (α>1) has a physical meaning of the ratio of the maximum effective thermal conductivity (at $P\rightarrow\infty$) to the thermal conductivity at zero pressure, and P_0^* is the pressure at which the ratio of effective thermal conductivity to the thermal conductivity at zero pressure reaches the value $(1+\alpha)/2$.

IOP Conf. Series: Earth and Environmental Science 1021 (2022) 012079 doi:10.1088/1755-1315/1021/1/012079

The experimental dependences $\nu(P)$ and $\delta_{T_0}(P)$ (for $T_0 = 300K$) are shown in Figures 3 and 4, respectively. As it can be seen from the figures, for all studied samples the dependences are quite strongly correlated, and, therefore, we can admit an empirical representation similar to equation (5):

$$\nu(P) = \frac{1 + \alpha_\nu P/P^*_{\nu 0}}{1 + P/P^*_{\nu 0}} - 1 \quad (6)$$

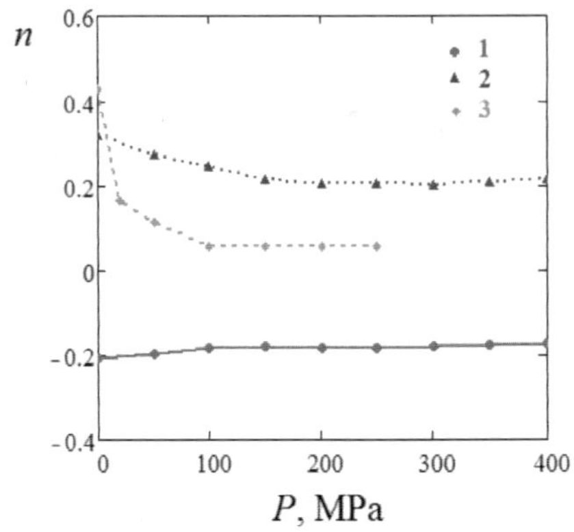

Figure2 (color on-line). Baric dependences of the coefficient *n* for three samples.

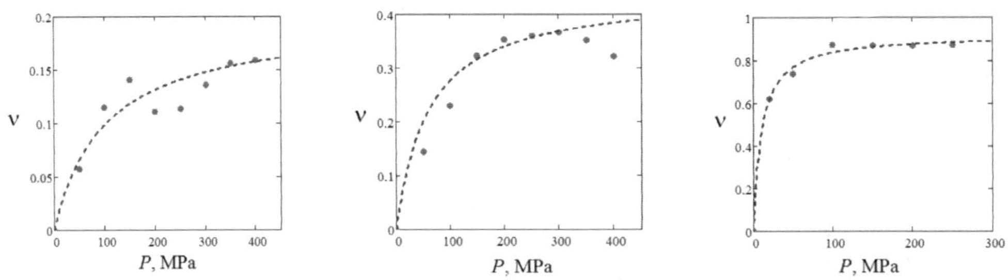

(a) (b) (c)

Figure 3(color on-line). Calculated values of ν, obtained for each experimentally investigated pressure, according to equation (4) (red symbols) and approximation (blue dotted line) made according to equation (6): a) sample (1); b) sample (2); c) - sample (3).

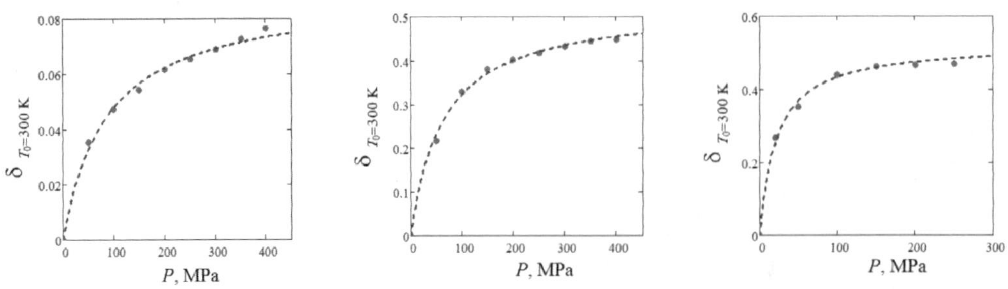

(a) (b) (c)

Figure 4 (color on-line). The calculated values $\delta_{300K}(P)$ performed for each experimentally investigated pressure, according to the equation (3) (red symbols) and approximation (blue dotted line), made according to the equation (5): a) sample (1); b) sample (2); c) sample (3)

Dependence $\nu(P)$ (as well as value n_0) is an objective parameter that depends only on pressure, while $\delta_{T_0}(P)$ also depends on parametric value T_0. The relationship between $\delta_{T_0}(P)$ for different T_0 (T_0 and T_0^* correspondently) is obviously expressed as:

$$\delta_{T_0}(P) = \left(1 + \delta_{T_0^*}(P)\right) \cdot \left(\frac{T_0}{T_0^*}\right)^{-n_0 \cdot \nu(P)} - 1 \tag{7}$$

Thus, for some temperature T_0^*, we can find the maximum correlation. At the same time, for a significant part of the rock samples studied by us, it turned out to be possible to represent both dependences $\nu(P)$ and $\delta_{T_0}(P)$ by the one function. For samples (1) and (2), the parameters α and P_0^* for such a function, together with the values T_0^*, are presented in Table 1.

For the third sample, the correlation turned out to be insufficient for using one function to set both functions $\nu(P)$ and $\delta_{T_0}(P)$ even for the temperature with the maximum similarity, therefore, for each quantity, its own coefficients are indicated.

Thus, the data presented in Table 1 are sufficient to calculate the effective thermal conductivity for the presented samples in a wide temperature-baric range. Indeed, substituting α and P_0^* in equations (5) and (6) we can calculate $\nu(P)$ and $\delta_{T_0^*}(P)$, then, according to equation (7), we calculate $\delta_{300K}(P)$, and then substituting the obtained functions $\nu(P)$ and, as well as λ (300K, P), $\delta_{300K}(P)$ and n_0 (value of n at zero pressure) into equation (2) we can calculate the effective thermal conductivity over the entire temperature-baric range. Atmospheric pressure can usually be neglected in these calculations. The calculation results are shown in Figure 5.

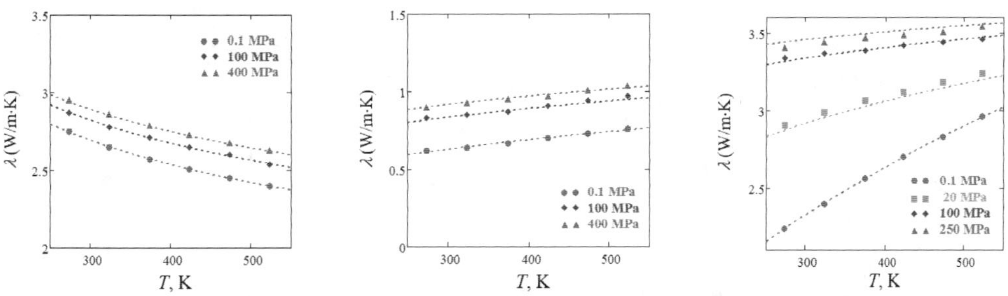

(a) (b) (c)

Figure 5 (color on-line). The values of the effective thermal conductivity of sandstone samples (lines), made according to equations (2), (5) - (7) with the coefficients presented in table (1) together with the experimental points (symbols): a) sample (1); b) sample (2); c) - sample (3).

It should be noted that, in contrast to the data for granites given in [8], for sandstones, we can not determine a single value T_0^* for different samples. Moreover the obtained value can be far beyond the reasonable extrapolation of the power-law dependence (it is the reason why we do not present calculated value $\lambda(T_0^*,0)$ in table 1). Nevertheless, the obtained value T_0^* can be used as a calculated parameter for $\nu(P)$ and $\delta_{T_0^*}(P)$.

4. Conclusion

Thus, we presented the experimental temperature-pressure dependences of the effective thermal conductivity of sandstones and given a low-parameter expression that brilliantly described the experimental data.

We studied three samples from different deposits, one of which showed a predominantly polycrystalline structure, the other two - close to amorphous. It should be especially noted that, that despite the closeness of the absolute value of the effective thermal conductivity at 300 K and atmospheric pressure, the difference in the ordering leads to a radical change in the effect of the cumulative increase in temperature and pressure. If with a more widespread polycrystalline nature (at which $n<0$), this cumulative effect can often really be neglected due to mutual compensation, which was noted by a number of authors [1, 11], then with an amorphous nature of the effective thermal conductivity temperature dependence ($n> 0$), this can not be done

5. Acknowledgments

The research was performed with a partial support of the RFBR Grants 18-08-00059a and 20-08-00319a.

6. References

[1] Norden B, Förster A, Förste, H-J, Fuchs S 2020 Temperature and pressure corrections applied to rock thermal conductivity: impact on subsurface temperature prognosis and heat-flow determination in geothermal exploration *Geothermal Energy* **8(1)** 1-19. doi: 10.1186/s40517-020-0157-0

[2] Sun Q, Zhang W, Zhu Y, Huang Z 2019 Effect of high temperatures on the thermal properties of granite. *Rock Mech. Rock Eng.* **52** 2691–2699 doi: 10.1007/s00603-019-1733-0

[3] Eucken A 1911 About the temperature dependence of the thermal conductivity of solid non-metals (in German language) *Ann d Phys* **34** 185 DOI: 10.1002/andp.19113390202

[4] Debye P 1912 The Theory of Specific Heat (in German language) *Ann Phys.* **39** 789. doi: 10.1002/andp.19123441404

[5] Ziman J M 1962 *Electrons and Phonons. The Theory of Transport Phenomena in Solids* (Clarendon Press: Oxford)

[6] Klemens P G 1967 Decay of High Frequency Longitudinal Phonons. *J. Appl. Phys.* **38** 4573 doi: 10.1063/1.1709187

[7] Emirov S N, Aliverdiev A A, et al 2020 Some features of the temperature-pressure dependence of the effective thermal conductivity of rocks. *J. Phys.: Conf. Ser.* **1683** 032044. doi: 10.1088/1742-6596/1683/3/032044

[8] Emirov S, Aliverdiev A, Beybalaev, V, Amirova A 2020 On the temperature and pressure dependences of the effective thermal conductivity of granites. *Thermal Science* **25(4),** Part A, 2493-2501 doi: 10.2298/TSCI200408176E

[9] Kukkonen I T, Jokinen J, Seipold U 1999 Temperature and Pressure Dependencies of Thermal Transport Properties of Rocks: Implications for Uncertainties in Thermal Lithosphere Models and new Laboratory Measurements of High-Grade Rocks in the Central Fennoscandian Shield *Surveys in Geophysics* **20** 33-59. doi: 10.1023/A:1006655023894

[10] Fu H, Zhang B, Ge J, et al. 2019 Thermal diffusivity and thermal conductivity of granitoids at 283–988 K and 0.3–1.5 GPa *American Mineralogist* **104 (11)** 1533–1545. doi: 10.2138/am-2019-7099

[11] Merriman J D, Whittington A G, Hofmeister A M, et al. 2013 Thermal transport properties of major Archean rock types to high temperature and implications for cratonic geotherms. *Precambrian Research* **233** 358-372 doi: 10.1016/j.precamres.2013.05.009

AUTHOR INDEX